Themes in Contemporary Archaeology

The Themes in Contemporary Archaeology series provides cutting edge perspectives on key areas of debate in current archaeological enquiry, with a particular emphasis on European archaeology. The series has a broad coverage, encompassing all archaeological periods and all approaches. Examples of topics welcome in the series include, but are not limited to:

- from theoretical debate to archaeological practice
- landscape studies
- bioarchaeology
- issues of cultural heritage

The volumes are based on research presented at the Annual Meetings of the European Association of Archaeologists. They include proceedings of individual sessions, which can be enlarged if necessary to provide coherence for publication. Each volume undergoes a strict peer-review process, ensuring volumes of high quality that capture current debates in the field.

SERIES EDITORS

The Series editors are Peter Attema, Agathe Reingruber and Robin Skeates (themes@e-a-a.org).

DISCOUNT FOR MEMBERS

All EAA members can buy print and e-copies of all THEMES volumes at 30% discount - please email administrator@e-a-a.org who will assist you.

Please refer the following link for the EAA guidelines

https://www.e-a-a.org/EAA/Navigation_Publications/THEMES.aspx?WebsiteKey=59c4f6fb-bee0-4278-9c2f-9529288deb09&THEMES=4#THEMES

Laura Coltofean • Bettina Arnold
László Bartosiewicz
Editors

Connecting People and Ideas

Networks and Networking in the History of Archaeology

Springer

European Association of Archaeologists

Editors
Laura Coltofean
Research Cluster 5: History of Archaeology
German Archaeological Institute
Berlin, Germany

Bettina Arnold
Department of Anthropology
University of Wisconsin–Milwaukee
Milwaukee, WI, USA

László Bartosiewicz
Department of Archaeology and Classical Studies
Stockholm University
Stockholm, Sweden

ISSN 2730-7441 ISSN 2730-745X (electronic)
Themes in Contemporary Archaeology
ISBN 978-3-031-81005-3 ISBN 978-3-031-81006-0 (eBook)
https://doi.org/10.1007/978-3-031-81006-0

This Springer imprint is published by the registered company Springer Nature Switzerland AG
The registered company address is: Gewerbestrasse 11, 6330 Cham, Switzerland

Preface

"Networks are present everywhere. All we need is an eye for them."—writes physicist Albert-László Barabási in his acclaimed book *Linked: The New Science of Networks* (2002: 7).[1] He is right to say so. As our book demonstrates, networks can also be found everywhere in archaeology, making it a highly interconnected research field.

In its 16 chapters, this volume explores social networks (e.g., scientific collaboration, academic, intellectual, professional, affiliation, acquaintance networks) and the various networking modes that formed during the history of archaeology in distinct geographical settings in Europe, North America, and South Asia in the nineteenth and twentieth centuries. Using a wide array of examples from diverse geopolitical, cultural, and social contexts, the book reveals how essential social networks and networking have been to the development of archaeology; to the production, transfer, exchange, and dissemination of archaeological and cross-disciplinary knowledge; and to the formation, upward mobility, barrier transcendence, research, and association of archaeological practitioners.

The volume grew out of a one-day session[2] on "Networks and networking in the history of archaeology" at the 26th (Virtual) Annual Meeting of the European Association of Archaeologists (EAA), which took place on August 24–30, 2020, in the first year of the COVID-19 pandemic. All three session organizers have dealt with networks on various occasions in their research and the Annual Meeting's theme on #Networking was the perfect opportunity to gain insight into the emergence, structure, dynamics, and contribution of knowledge transfer systems to archaeology's evolution at a larger scale and from multiple perspectives.

Just before the EAA Annual Meeting, Christi Lue from Springer had contacted and invited us to consider publishing the papers of our session in an edited collection. A few days later, the success of the session, the relevance of its topic, and the quality of the papers had convinced us and the speakers that it would indeed be a good idea to produce such a volume. We invited more authors to join the project to ensure that the contributors come from a wide range of geographical settings, academic traditions, career stages, and professional paths. Step by step, the book connected people and ideas, forming a network of editors, authors, reviewers, and publishers around it. We have devoted the past three and a half years to assembling the volume in what has been an enriching experience of knowledge sharing.

We would like to thank the authors for their contributions and for wholeheartedly committing their time in the middle of a pandemic to this volume, as well as to Nathan Schlanger (École nationale des chartes, Paris) and the numerous anonymous peer reviewers who were willing to provide constructive feedback on the chapters and the volume itself, often against tight deadlines. We are also grateful to the EAA "Themes in Contemporary Archaeology" Series Editors Agathe Reingruber, Robin Skeates, and Peter Attema, as well as to Publishing Editor Christi Lue and Production Editors Arunsanthosh Kannan and Manonmani Pradeepkumar from Springer for the opportunity to publish this volume and for assisting us during its

[1] Barabási, A.-L. (2002). *Linked: The new science of networks*. Cambridge, MA: Perseus Publishing.

[2] Session 265: "Connecting people and ideas: Networks and networking in the history of archaeology", organized by Bettina Arnold, Laura Coltofean-Arizancu, and László Bartosiewicz on August 27, 2020. For more details, see https://www.e-a-a.org/EAA2020virtual.

preparation and publication. Finally, we thank our families, friends, and colleagues for their support, patience, and understanding while putting together the book.

Braşov, Romania
Milwaukee, WI, USA
Stockholm, Sweden
July 2024

Laura Coltofean
Bettina Arnold
László Bartosiewicz

Contents

Contributors

Bettina Arnold Department of Anthropology, University of Wisconsin-Milwaukee, Milwaukee, WI, USA

Aleksandar D. Bandović National Museum of Serbia, Belgrade, Serbia

László Bartosiewicz Department of Archaeology and Classical Studies, Stockholm University, Stockholm, Sweden

Cristina Bodó Museum of Dacian and Roman Civilization, Deva, Romania

Ionuț-Cosmin Codrea Museum of Dacian and Roman Civilization, Deva, Romania

Laura Coltofean Research Cluster 5: History of Archaeology, German Archaeological Institute, Berlin, Germany

Antonino Crisà Department of Humanities and Social Sciences, Prince Mohammad Bin Fahd University, Al Khobar, Saudi Arabia

Martijn Eickhoff NIOD Institute for War, Holocaust and Genocide Studies, Amsterdam, The Netherlands

Anna Charlotta Gustavsson Department of Historical Studies, University of Gothenburg, Gothenburg, Sweden

Črtomir Lorber Department of Archaeology, University of Ljubljana, Ljubljana, Slovenia

Ana Cristina Martins IHC NOVA FCSH – University of Évora Pole and IN2PAST, Évora and Lisbon, Portugal

Heidi J. Miller Division of Liberal Arts, Middlesex Community College, Bedford, MA, USA

Fedra Alessandra Pizzato Department of Cultures and Civilization, University of Verona, Verona, Italy

Judith A. Rasson Department of Medieval Studies, Retired, Central European University, Budapest, Hungary

Kathleen L. Sheppard Department of History and Political Science, Missouri University of Science and Technology, Rolla, MO, USA

Mathijs Smith Independent Researcher, Leiden, The Netherlands

James E. Snead Department of Anthropology, California State University, Northridge, CA, USA

Andrew Whitefield School of Geography, Archaeology and Irish Studies, University of Galway, Galway, Ireland

About the Editors

Laura Coltofean is an archaeologist and historian whose research focuses on the development of interdisciplinary practices, knowledge production and transfer, social networks, and political ideologies in the history of European archaeology, as well as on gender issues, oppressive behaviors, safety, and well-being in the current archaeological profession. She is a member of Research Cluster 5: History of Archaeology of the German Archaeological Institute. Previously, she was a research assistant and postdoctoral research fellow at the Römisch-Germanische Kommission of the German Archaeological Institute, Germany (2021–2022), a researcher at the University of Barcelona, Spain (2018–2020), and a museum curator at the Brukenthal National Museum in Sibiu, Romania (2012–2018). She is an active member of international professional and scientific networks, such as the European Association of Archaeologists, the Archaeology and Gender in Europe Community, the European Society for the History of Science, and the International Union for Prehistoric and Protohistoric Sciences, some of which she has served as co-chair and secretary. Her recent publications include the edited collections *Interdisciplinarity and Archaeology: Scientific Interactions in Nineteenth- and Twentieth-Century Archaeology* (Oxbow Books, 2021; with M. Díaz-Andreu), *Gender Stereotypes in Archaeology: A Short Reflection in Image and Text* (Sidestone Press, 2021; with B. Gaydarska and U. Matić), *The Oxford Handbook of the History of Archaeology* (Oxford University Press, 2024; with M. Díaz-Andreu), *Gender Trouble and Current Archaeological Debates* (Springer, 2024; with U. Matić, B. Gaydarska and M. Díaz-Guardamino), and *Archaeology and the Global Cold War: Scientific Practices and Political Ideology* (Routledge, forthcoming 2025; with S. Grunwald, F. Link and K. Rösler).

Bettina Arnold obtained her BA in Archaeology from Yale University and her MA and PhD degrees in Anthropology from Harvard University. She is a Full Professor of Anthropology at the University of Wisconsin-Milwaukee and Adjunct Curator of European Archaeology at the Milwaukee Public Museum. Her research interests include pre-Roman Iron Age Europe, the archaeological interpretation and analysis of complex societies as reflected in mortuary contexts and the production and consumption of alcoholic beverages; the archaeological interpretation of gender; and the socio-political history of archaeology and museums in nineteenth- and twentieth-century nationalist and ethnic movements in Europe and the United States. She published a ground-breaking article on the use and abuse of archaeology for political purposes in Nazi Germany in *Antiquity* in 1990 that has been reprinted repeatedly since first appearing. As a bilingual dual citizen who conducts research in southwest Germany but was raised and educated in the United States, she can relate to the experience, both positive and negative, of serving as a conduit for trans-Atlantic information transfer. The final report of the excavations she co-directed in two early Iron Age burial mounds near the Heuneburg hillfort on the Danube River in Swabia will appear in two languages next year.

László Bartosiewicz is an archaeozoologist working at the Osteoarchaeological Research Laboratory at Stockholm University, Sweden. He studied Animal Science at the University of Gödöllő, Hungary. He was granted a DSc degree by the Hungarian Academy of Sciences for his work on the archaeology of cattle exploitation. Between 1978 and 1995 he worked as full-time researcher at the Archaeological Institute of the Hungarian Academy of Sciences. His research interests include human-animal relationships throughout history, as well as the variability of using animals as media in human communication. His specialization in animal palaeopathology helps fine-tuning the osteological evidence of such attitudes toward animals. Bartosiewicz taught archaeozoology at the University of Budapest, Hungary, from 1996 to 2015 and Edinburgh University, United Kingdom, from 2004 to 2015. In 2015 he received the chair in osteoarcheology at Stockholm University. Bartosiewicz has been analyzing animal remains from an archaeological perspective in Hungary, Switzerland, Slovenia, Serbia, Bulgaria, Turkey, Israel, France, and Bolivia for over 40 years. He was elected for two terms president of the International Council for Archaeozoology (2006–2014).

Social Networks and Networking in the History of Archaeology: An Introduction

1

Laura Coltofean, Bettina Arnold, and László Bartosiewicz

Abstract

This chapter presents the major themes addressed by the volume while concurrently providing an overview of social networks and networking in the history of archaeology from the nineteenth century to the end of the twentieth. The chapter first introduces the rationale of the volume and positions it within the history of archaeology as a research field. Informed by network science, geographies of science, and actor-network theory, the chapter then explains the three major concepts around which the book is built—network, social network, and networking. This is followed by an overview of the types of social networks that have formed throughout the history of archaeology, along with why and how these networks were created, maintained, and used in different personal, professional, institutional, and geopolitical settings. The introduction also reflects on the role played by social networks and networking in the development of archaeology and in archaeological actors' personal and professional trajectories and practices.

Keywords

Social network · Networking · History of archaeology · Network science · Geographies of science · Actor-network theory

L. Coltofean (✉)
Research Cluster 5: History of Archaeology, German Archaeological Institute, Berlin, Germany

B. Arnold
Department of Anthropology, University of Wisconsin-Milwaukee, Milwaukee, WI, USA
e-mail: barnold@uwm.edu

L. Bartosiewicz
Department of Archaeology and Classical Studies, Stockholm University, Stockholm, Sweden
e-mail: laszlo.bartosiewicz@ofl.su.se

1.1 Introduction

Network science is a science of the twenty-first century. The study of networks, however, is much older and its incipient forms date back to the eighteenth century (Barabási 2002: 10–12). Nevertheless, in the past two decades research on networks has generated increasing interest, well beyond the core disciplines which have mainly dealt with it so far, like mathematics, computer science, physics, and sociology. The interest in networks has also surged in archaeology, but it is only in recent years that its impact has been felt in specific archaeological branches and research topics (Peeples 2019), culminating in the development of network analysis methods that are especially suitable for archaeology (e.g., Knappett 2011, 2013; Brughmans 2013; Collar et al. 2015; Brughmans et al. 2016; Brughmans and Brandes 2017).

Compared to other sub-fields in archaeology, the history of archaeology so far has been less influenced by network thinking and methods *per se*. This is quite curious, because—as universalist as this statement may appear—nearly every topic in the history of archaeology can be fruitfully and explicitly studied through the lens of networks. These topics include the biographies of archaeological actors, the study of the emergence, institutionalization, professionalization, and internationalization of archaeology, the interaction of archaeology with other disciplines, archaeological knowledge production, transfer, and exchange, as well as the impact of political ideologies on archaeological discourse and practice. The history of archaeology is, in fact, a rich field for (social) network studies. Eight years ago, Brughmans et al. noted that there was a general reluctance to utilize network concepts and methodologies in archaeology and history, partly because these are considered unsuitable for such research, and partly because of the challenges that their application poses (2016: 3–4). The reduced presence of network perspectives in research on our disciplinary past is perhaps explained by the same reluctance but also by the lack of awareness of the investigative possibilities that such a perspective can pro-

L. Coltofean et al. (eds.), *Connecting People and Ideas*, Themes in Contemporary Archaeology,
https://doi.org/10.1007/978-3-031-81006-0_1

vide. The challenges posed by such an approach are still emerging and the chapters in this volume highlight both the limitations and potential of network studies in archaeology.

Several studies in the history of archaeology have recognized the importance of reconstructing networks (e.g., Bentz 2008: 101–2; Tomášková 2011: 222–23) and even more have touched upon or dealt with networks (see the references throughout this chapter), although often not explicitly and/or without direct reference to network thinking, theories, methods, or models. Exceptions are the edited volume *Communities and Knowledge Production in Archaeology* (Roberts et al. 2020), which applies scientific network theories to various case studies from the history of archaeology; historian of archaeology Amara Thornton's work (e.g., 2011, 2015) on British archaeological networks in the Eastern Mediterranean and the Middle East; archaeologist Sébastien Plutniak's exploration (Plutniak 2014, 2017) of the interactions between engineering and archaeology in France in the second half of the twentieth century; as well as archaeologist Alice Stevenson's analysis (Stevenson 2019) of the distribution of Egyptian archaeological finds resulting from British excavations across the globe, from the late nineteenth century onwards. The current volume builds upon this earlier research and contributes to the use of networks as an investigative construct in the history of archaeology with an expanded thematic, geopolitical, social, and cultural focus.

The chapters in this book are case studies that reveal how various types of social networks in the history of archaeology emerge, develop, and behave, and how the interactions which take place between network components and within network clusters impact the development of archaeology on a local, regional, national, and international level. What this volume demonstrates especially clearly is that networks are an integral and ubiquitous part of archaeology and that analyzing the history of the field through the interpretive lens of networks helps us gain a more nuanced understanding of the discipline's evolution and of the human and non-human actors involved in it.

This chapter lays out the theoretical foundations of the book. The three major concepts around which the volume is built—network, social network, and networking—are discussed first, followed by the theoretical aspects of the formation, structure, and behavior of social networks. The history of archaeology is presented from the perspective of network science, geographies of science, and sociology. In the field of network science, this volume was especially inspired by the work of scholars such as Albert-László Barabási (2002, 2016) and Mark E. J. Newman (2001, 2018), while in the fields of geographies of science and sociology several chapters draw upon the theories of David Livingstone (2003) and Bruno Latour (2005). This introduction applies these theoretical approaches to the chapters in this volume and other case studies from the existing literature on the history of archaeology in order to provide an overview of why and how social networks in archaeology were created, maintained, and used in different personal, professional, institutional, and geopolitical settings. Finally, the chapter reflects on the roles played by social networks and networking in the development of archaeology and the various processes that took place in it during the nineteenth and twentieth centuries.

1.2 Social Networks and Networking—Theoretical Foundations

The contributors to this volume were trained in a range of academic and research traditions and this is reflected in the diversity of the approaches they have chosen when addressing the networking theme. In writing their chapters, the authors had the freedom to explore networks and networking in their historical case studies as they wished, without a theoretical or methodological framework being imposed on them. Some of the authors have therefore opted for grounding their work theoretically, while others have chosen a more narrative or biographical framework for their case studies. Tim Murray and Christopher Evans have pointed out that biographies of individuals and histories of institutions are useful but limited as a means of understanding larger social and intellectual contexts, partly because they tend to focus on major figures in the field while neglecting the minor players (Murray and Evans 2008: 9–10). However, other historians of archaeology argue that biographies provide insight into subjects' various types of networks, among other aspects of their personal and professional lives, as well as "into the macroscopical level of archaeological practice and of society at a certain point in time" (Coltofean 2024: 112), especially when approached as microhistories (see Kaeser 2008, 2013; Díaz-Andreu and Portillo 2021; Coltofean 2024). The chapters in this volume exemplify more recent efforts to move beyond the 'Great Man' approach to archaeological biography by thinking of individuals like Oscar Montelius (1843–1921) (Gustavsson, Chap. 2) as analogous to strange attractors in mathematics: what is pulled into their orbit is as interesting as the strange attractors themselves. Also worth mentioning is the extent to which early archaeological networks overlapped with one another: Oscar Montelius is mentioned in three of the chapters in this volume (Gustavsson, Chap. 2; Pizzato, Chap. 8; and Whitefield, Chap. 9), Gabriel de Mortillet (1821–1898) in two (Pizzato, Chap. 8; Bandović, Chap. 14) and Édouard Desor (1811–1882) in two (Arnold, Chap. 6; Pizzato, Chap. 8), for example. Given the diversity of the contributions to the volume, this introduction will provide an overarching theoretical reflection on (social) networks and networking, along with insights into the theories and methods that can be used for researching them with reference to the individual chapters and how they relate to one another.

Physicist Mark Newman defines networks as "a collection of points joined together in pairs by lines." The points are known as "nodes" or "vertices," while edges are referred to as "lines" (Newman 2018: 1). There are various types of networks and Newman divides them into four main categories: technological (e.g., the Internet, networks of roads, distribution networks like electrical grids); information (e.g., the World Wide Web, citation networks); social; and biological (e.g., neural and metabolic networks, food webs) networks (Newman 2018, especially pp. 1–7 for categorization).

This volume deals with various types of social networks, specifically those that developed throughout the formative decades of the discipline of archaeology in the nineteenth and twentieth centuries. Social networks are defined by Newman as "a collection of people, each of whom is acquainted with some subset of the others" (Newman 2001: 404). More technically speaking, a social network is "any network in which the nodes represent people, and the edges represent some form of connection between them" (Newman 2018: 47; also see Newman 2001: 404). Nodes are not only individual people but can also be groups of people (Newman 2018: 4–5). Edges, for instance, can take the form of friendships, "professional relationships, exchange of goods or money, communication patterns, romantic or sexual relationships, or many other types of connection" (Newman 2018: 48–49). In sociology, the terminology is slightly different: the nodes are called "actors" and the edges "ties" (Newman 2018: 47).

Further expanding the definition of social networks, network scientist Albert-László Barabási highlights the fact that a social network "is the fabric of the society and determines the spread of knowledge, behavior and resources" (Barabási 2016: 6). With this definition in mind, the social networks that are presented in this volume could be described as personal, friendship, and acquaintance networks, in addition to professional, intellectual, academic, and scientific collaboration networks (Newman 2001). The nodes or actors within these networks are generally individuals (e.g., professional or amateur archaeology practitioners as well as archaeology enthusiasts outside the discipline) but also institutions, associations, and even ideas. This is in line with sociologist and philosopher Bruno Latour's idea that an actor is not necessarily a person; it can be anything from an excavation to an idea as long as it "do[es] things" and "make[s] a difference" by producing a transformation (Latour 2005: 128) within a network. For example, William Frankfurth's (1829–1891) late nineteenth-century investigations of pre-Roman sites in the South Tyrol illustrate how the disruptive presence of a non-local individual may influence the direction and nature of future scholarly activity in a region (Arnold, Chap. 6, this volume).

In addition to 'network' and 'social network', another term which is foundational for this book is 'networking'. This is a term that originates in the field of computer science, where it refers to connecting computers and other electronic devices (e.g., printers) so that they can communicate with one another in order to share information and resources (e.g., printers, applications) (Ciccarelli and Faulkner 2004: 1–3). The meaning of this term has since been expanded to include human social life, where it refers to meeting and talking to people, most often with a specific purpose (e.g., sharing and exchanging information, ideas, further contacts) and especially in a professional context (Cambridge dictionary n.d.). It is this latter meaning that this volume employs.

Some of the chapters in this book are studies of "scientific collaboration networks" Newman (2001). However, in these cases, the concept of "collaboration" is not limited to joint authorship of publications but goes beyond such interactions to include other forms of scientific cooperation, such as participating in the same excavations or research projects. Several chapters are close-up studies of subsets of larger social networks due to their focus on archaeology practitioners' so-called personal or ego-centered networks. These are networks "surrounding one particular individual" and consist of the individual, i.e., the ego, and the "immediate contacts", i.e., the alters (Newman 2018: 55). Other chapters (e.g., Pizzato, Chap. 8; Bartosiewicz, Chap. 16) also deal with or touch upon affiliation networks, in which "actors are connected via their membership in groups of some kind" (Newman 2018: 60). Networks can also include "clusters" or "communities" which are "groups of people who are closer to each other within the network" (e.g., close friends within acquaintance networks). Such clusters can provide information about "levels and concepts of organization" in networks (Newman 2018: 10–11). A good example of such a cluster is represented by the German American scholars who established what would become the Milwaukee Public Museum (see Arnold, Chap. 6, this volume). Their community identity was based on language and recent immigrant status as well as shared progressive ideas about the value of science education, civic engagement, and exposure to other cultures, past and present.

"Centrality" is an important aspect to consider in the study of networks, as this makes it possible to quantify the relative importance of nodes within a network (Newman 2018: 9). One of the ways of measuring centrality is by degree, which is determined based on the number of edges that a node has within a network. The higher the number of edges, the more central the node is to the network (Newman 2018: 9). For instance, within a social network of professionals, the degree of an archaeologist would be determined by the number of connections they have within the network. Nodes which have an exceptionally high degree are called "hubs" (Newman 2018: 9). In the case of archaeology, these are very well-connected archaeological practitioners, like Gerhard Bersu (1889–1964) in Germany, who had an astonishing number of contacts in Europe and beyond (see Krämer 2002; von Schnurbein 2002; Grunwald et al. 2022; Whitefield, Chap. 9, this volume), or Alfred V. Kidder (1885–1963; e.g., Woodbury 1973; Givens 1992), Mortimer Wheeler (1890–1976; e.g., Hawkes 1982), and Grahame Clark (1907–1995; e.g., Fagan 2018), among many others. Examples of hubs in this volume are archaeologists Marija Gimbutas (1921–1994) and Colin Renfrew (see Rasson, Chap. 15), archaeozoologists Anneke T. Clason (1932–2008) and Sándor Bökönyi (1926–1994) (see Bartosiewicz, Chap. 16), and Egyptologist Caroline Ransom Williams (1872–1952) (see Sheppard, Chap. 3). Importantly, the chapters in our volume reveal that it was not only 'Great Men' who could be hubs, as mentioned previously, but also women. Therefore, looking at the history of archaeology through a network perspective is a useful method to discover previously unknown and/or marginalized archaeology practitioners, including women, and to reveal their contributions to the discipline.

Hubs can also be non-human actors such as institutions and associations; the Italian Alpine Club discussed by Fedra Pizzato (Chap. 8, this volume) is a good example. Furthermore, hubs play a significant role in the ways in which networks and the systems they represent function and perform; in fact, they can have "a disproportionate effect" on them (Newman 2018: 9), as the examples of Bersu and the Italian Alpine Club illustrate.

Hubs can have a strong impact on a network, but the patterns and intensity of interactions can as well. In Newman's words,

> (…) the pattern of interactions can have a big effect on the behavior of a system. (…) The connections in a friendship network affect how people learn, form opinions, and gather news, as well as other less obvious phenomena, such as the spread of disease. Unless we know something about the structure of these networks, we cannot hope to understand fully how the corresponding systems work. (Newman 2018: 7)

This also applies to the social networks that have existed throughout the history of archaeology. It is therefore important to study archaeological actors' patterns of interactions within various social networks and their structure in order to understand how, for instance, the institutionalization and internationalization of archaeology took place, or how archaeological knowledge was produced and shared over time. All this makes it possible to reveal how archaeology emerged, developed, and advanced as a discipline.

"Interconnectivity induces a remarkable non-locality" writes Albert-László Barabási. Thus, whatever happens in a network, whether it is good or bad, can spread and impact the entire network (Barabási 2016: 4). This means that networks can be both beneficial and detrimental to the individuals and non-human actors who participate in them and are potentially vulnerable as a result (Barabási 2016: 4). The chapters in this book are generally examples of the benefits that networks can confer on their constituent elements and the systems they represent. However, Črtomir Lorber's analysis (Chap. 13, this volume) of the vulnerability of networks highlights how the unfavorable conditions (e.g., lack of state support, political and economic instability) affecting the archaeological system in the Kingdom of Yugoslavia between 1918 and 1941 also influenced the behavior and performance of learned societies' internal and external networks, leading to periods of stagnation and failed attempts to merge diverging archaeological traditions into a unitary whole at a national level.

Several chapters in the volume (e.g., Arnold, Chap. 6; Gustavsson, Chap. 2; Lorber, Chap. 13; Snead, Chap. 5) exemplify geographer and historian David Livingstone's theory (2003) that place and space influence how scientific knowledge is generated and how scientific enterprises are carried out. It is not only the practice of science that is determined by geographies, but also social interaction, which is especially dependent on the ever-changing and imbricated spaces within which quotidian transactions are carried out. These places and spaces both enable and constrain habitual social relations (Livingstone 2003: 6–7). Livingstone's theory can also be applied to the types of interactions between nodes and the degree assigned to each node, as these are also dependent on geographies, as the chapters in this book show. In addition to topology, geographies influence the formation, dynamics, and behavior of networks. It is therefore essential to examine social networks and networking in the historiography of archaeology from the perspective of geographies of science.

There are several methods of collecting data for social network analysis. These include the analysis of questionnaires and interviews, direct observation, archival records, and other documentary sources, both primary and secondary. Other techniques include experimental methods and techniques such as the "small world experiment," snowball sampling, contact tracing, random-walk sampling, and respondent-driven sampling (Newman 2018: 51–69). Archival records in social network analysis not only comprise historical archival materials in the traditional sense, but include email and call logs, along with online data records of social media platforms, websites and blogs, among other more recent sources (Newman 2018: 58–60). Historical archives are especially useful for reconstructing networks which no longer exist (Newman 2018: 58), as in the case of most of the social networks analyzed in this book. Exceptions are the extant International Council for Archaeozoology (ICAZ), discussed by archaeozoologist László Bartosiewicz (Chap. 16, this volume) and, partially, the network of American-Yugoslav archaeologists formed on joint excavations in the Socialist Federal Republic of Yugoslavia during the Cold War, examined by archaeologist Judith Rasson (Chap. 15, this volume).

Historical archival records—particularly correspondence—are the main source of data for the qualitative studies of social networks included in this volume. Nevertheless, the other types of archival sources mentioned above will be useful in a few decades—in some cases already—for future researchers studying the history of archaeology from the 1980s onwards. Historical social networks are often time-resolved, in the sense that their characteristics are fixed and cannot change anymore, in contrast to time-varying networks. Archival materials often also allow us to reconstruct the time and duration of contacts within social networks (Newman 2018: 59–60), and epistolary exchanges are specifically useful in this regard. Other sources used by the studies in this book include print media (e.g., newspapers), institutional reports, comparative analyses of museum collections, and scientific publications from the nineteenth and twentieth centuries.

When carrying out research in archives, one must not forget that these are collections of documents which have not been objectively assembled and organized; there is a human component behind them which selects and classifies files based on specific aims and criteria (see e.g., Schlanger and Nordbladh 2008) derived from personal or institutional to third-party associations. In some ways, research into the history of archaeology is comparable to mortuary archaeology, where the adage "the dead don't bury themselves" complicates interpretation. What is preserved and what makes its way into repositories, which are themselves subject to the vicissitudes of time and chance, are processes subject to serendipity and the arbitrary decisions of descendants and custodians of the material, who may not recognize its value. For example, Agda Montelius (1850–1920) destroyed part of her and Oscar Montelius' correspondence (perhaps for privacy reasons) and, following her death, her niece also selectively curated the couple's archive (Gustavsson, Chap. 2, this volume). The type of medium carrying information is also prone to selective loss: financial records stand a better chance of surviving than family postcards as they tend to be stored and curated more systematically. In the digital age, the differential preservation of electronic documents has become an underappreciated problem as rapid changes in both soft- and hardware as well as the elimination of publishing platforms due to institutional changes may result in collateral damage in the form of the random erasure of information. Thus, documentary information can be fully or partially lost, and this includes erasing or obscuring individuals, institutions, actions, stories, and events. This selective elision influences whom and what researchers discover during their archival pursuits. The chapters by Anna Gustavsson (Chap. 2) and Kathleen Sheppard (Chap. 3) discuss this selectivity in their pursuit to give voice to female archaeological actors Agda Montelius and Caroline Ransom Williams.

In addition to network science and geographies of science, sociology has also made important contributions to understanding networks. One such contribution is the so-called actor-network theory (ANT) which was developed by a group of thinkers in the 1980s, including Michael Callon, John Law (Latour 2005: 10–11), and Bruno Latour. The theory has subsequently generated discussions and varied applications in multiple fields. Latour's approach of ANT—especially as presented in his book *Reassembling the Social: An Introduction to Actor-Network-Theory* (Latour 2005)—is perhaps the best known of these approaches and his work informs several of the chapters in this volume, either explicitly or implicitly.

Contrary to the previously discussed perspectives on networks grounded in network theory and science, in Latour's view a network is not a "a thing out there that would have roughly the shape of interconnected points, much like a telephone, a freeway, or a sewage 'network'" (Latour 2005: 129). Instead, it is "a string of actions where each participant is treated as a full-blown mediator" who produces transformation; participants are therefore not intermediaries who merely transport without also inducing transformation (Latour 2005: 39). In Latour's view, a network is "a tool to help describe something, not what is being described"; it should not be explained, nor analyzed but traced (Latour 2005: 128, 131). This approach, therefore, is useful to understanding the structure, dynamics, and behavior of networks, especially because accounts employing ANT should focus on "the work, and the movement, and the flow, and the changes" of the actors (Latour 2005: 143). Another important point that Latour makes—and this contradicts general practices in historiographic research, including in archaeology—is that in ANT-inspired works, actors should be listened to and allowed to unfold themselves without the researcher imposing predetermined contexts, agendas, frameworks, or categories upon them. Therefore, actors' actions should not be explained in pre-established social, societal, or cultural contexts by the researcher, as these are, in fact, the constructions and results of the actors' actions (Latour 2005: 11–23). In the case of unexpected events (wars, natural disasters, etc.) or the sudden death of the actor, they may be the consequences of interrupted or unfinished actions.

1.3 Social Networks and Networking Throughout the History of Archaeology—An Overview

Drawing upon the chapters in this volume and the existing literature, this section provides an overview of how social networks were created, developed, maintained, and used in different personal, professional, institutional, and socio-political settings throughout the history of archaeology in the nineteenth and twentieth centuries. It also reflects on the structure, functioning, and dynamics of these networks, and the effects they had on the evolution of archaeology as a discipline and on its practitioners.

Throughout the history of archaeology, networking—and other forms of social interactions—took place in various settings, including conferences, congresses, and society meetings; in the field on excavations, study trips, excursions, and site visits; in universities, summer courses, museums, and exhibitions; during private meals, in salons, pubs, and cafés (various chapters, this volume; for more examples on interactions occurring in such venues, see Sweet 2004; Díaz-Andreu 2007a, 2012; Coltofean 2015; Coltofean-Arizancu 2022, among others). The chapters of Aleksandar Bandović (Chap. 14) and Judith Rasson (Chap. 15) in this book bring to the surface the complex interactions that occurred on American and German archaeological expeditions and American-Yugoslav joint excavations in Yugoslavia in two especially interesting times of the twentieth century: the interwar period in the 1930s and the Cold War in the late 1960s. Their thorough analyses reveal the domestic and foreign research interests involved in the archaeology of Yugoslav territories, as well as the intricacies of the personal, scientific, economic, political, and ideological aspects behind these projects. These two chapters are complemented by Črtomir Lorber's (Chap. 13) insights into networking within and between archaeological learned societies in Yugoslavia from 1918 to 1941, and how this influenced the development of archaeology in various parts of the country (also see Novaković 2021).

The places mentioned above are spaces where archaeological practitioners met one another as well as people outside the discipline, interacting and making contacts that added to their social networks during the nineteenth and twentieth centuries, as well as today. These encounters consisted of formal and informal in-person interactions which took place at local, regional, national, and international levels. Antonino Crisà's study (Chap. 12) of social networks in Sicily during the Second World War explicitly identifies these four scales of spatial analysis; if combined with a system of temporal categorization, his approach could be fruitfully applied in other contexts as well.

In addition to face-to-face interactions, contacts were also initiated, maintained, and reinforced throughout the history of archaeology via written communication in the form of correspondence and telegrams. Depending on place and space, social networks at all levels were maintained through correspondence until the 1980s–1990s. This again highlights the importance of archival sources in studying networks. However, this analog communication method was gradually replaced after the Second World War by much faster and less tangible phone calls, emails, and other forms of electronic and virtual communication. The last of these is the most ephemeral form of communication and the least likely to be accessible to researchers, partly because the looming problem of curating and providing access to the vast quantities of email generated in academia as well as between scholars and individuals outside the Ivory Tower in the past decades has so far not been adequately addressed. Essentially, however, all these forms of interaction enabled the formation and expansion of various social networks (e.g., intellectual, professional, scientific collaboration, friendship) within archaeology and in so doing directly influenced the course of the discipline.

Correspondence compensated for the lack of in-person meetings, especially when it came to large distances from one side of the European continent to the other, including transatlantic networks such as those discussed by Bettina Arnold (Chap. 6, this volume). In person meetings were important in maintaining networks (see Bartosiewicz, Chap. 16, this volume) but these were more frequent when the actors were geographically close to each other. For most international actors until recently, in person meetings were limited to encounters at scientific events like congresses, research stays, study trips, summer courses and personal visits, which often depended on funding through grants, institutional support, and personal wealth, as is the case today. Although traveling possibilities, especially transatlantic ones, were limited and occasional, they could be transformational for the archaeology practitioners in question (see Arnold, Chap. 6; Rasson, Chap. 15, this volume). These bidirectional and often multi-lingual migrants also served as catalysts for information transfer on various levels.

Social networks can be built from scratch or inherited and further expanded. The case of Portuguese archaeologist João Manuel Bairrão Oleiro (1923–2000) is a good example (see Martins, Chap. 10, this volume). Through the summer schools and courses that he attended in Spain and the research stays that he carried out there at key institutions, he actively sought during his formative years to associate himself with well-known scholars. He did so in order to build an academic network that would provide him with opportunities to access knowledge, research infrastructure and resources, along with expanding his knowledge and skills, and advancing in his career. This is just one of the many examples which illustrates that going abroad for international experiences and engaging in networking during such travels have been essential for the personal as well as professional development of archaeological practitioners. The international contacts established abroad become an integral part of their professional, academic, and intellectual networks that serve as a resource on various occasions for the archaeologists whose careers are the subject of several chapters in this volume.

Bairrão Oleiro, along with several other archaeological actors (see Rasson, Chap. 15; Bartosiewicz, Chap. 16, this volume), was in contact with scholars who were hubs within their own networks. Actors like Bairrão Oleiro were initially nodes in the European international network of archaeological practitioners but their interactions with hubs—among other factors—significantly contributed to gradually transforming their own networks, especially in their native national contexts. This granted them prestige and status, empowered them to access positions and resources, and impacted the trajectories of other archaeological practitioners and the development of archaeology in their own countries. This illustrates how hubs and nodes interact with and influence each other and how, as an effect, nodes can gradually become hubs themselves in the same network and/or in smaller networks at a national, regional or local level. In addition, these case studies also show how the limitations of places can be overcome through networking, especially internationally.

In her analysis of the life and career of Caroline Ransom Williams, Kathleen Sheppard (Chap. 3, this volume) exemplifies how high degree nodes within networks can preserve their position and influence even when the settings in which they operate change. More specifically, through the network Ransom Williams created while working at the Metropolitan Museum of Art, she was able to maintain her central position in the network of Egyptologists and in the field of Egyptology after leaving her full-time position at the museum and marrying in the first half of the twentieth century. The durability of social networks is also highlighted by James Snead (Chap. 5, this volume) in his discussion of anthropologist Lewis Henry Morgan's (1818–1881) field networks and the challenges of gathering first-hand, high quality ethnological and archaeological data in the United States during a period of expanding frontier zones and unrest in a rapidly changing environment.

Building networks, especially in unknown and unfamiliar places, was and still is especially challenging. The analysis of Agda and Oscar Montelius' travels to Italy in the second half of the nineteenth century (see Gustavsson, Chap. 2, this volume) shows how difficult this was. In order to carry out research on the archaeological material held by institutions and private collectors, relying on local networks was fundamental. Knowing the right people and contacting them in advance was not always possible. Therefore, these contacts had to be built gradually on site, and often by word of mouth. This was the Montelius couple's networking practice, and its overall success largely depended on their physical presence in the region they were researching, in addition to their perseverance in the face of logistical and other difficulties. Lewis Henry Morgan's attempts to obtain information about indigenous communities in the southwestern United States in the 1860s and 1870s is also interesting in this context (Snead, Chap. 5, this volume). When Morgan tried to collect information by sending out a circular inviting anyone interested to voluntarily contribute to his research, his attempts were largely unsuccessful. However, when he visited some of the southwestern territories and approached local contacts in person, he was able to acquire at least some of the information he was interested in. Thus, the examples of Agda and Oscar Montelius and Lewis Henry Morgan show how important physically accessing places and people can be for building networks and conducting research in remote or less well-known areas.

The examples described above also indicate that collecting artifacts, images, and information requires a network of suppliers and buyers, patrons and clients, performers and audiences who engage in mutually reinforcing behaviors that mark them as members of a community. Collector networks in the nineteenth century were more than a market in art and antiquities, they were systems of information sharing, exchange of knowledge, and source bases facilitated by publication. Collectors were often members of various professional associations which "attested to the existence of the network and increased its viability and effectiveness" (Crane 2000: 64).

Social networks were used for multiple purposes throughout the history of archaeology, whether those purposes were personal or professional, academic or popular. Whether through correspondence or face-to-face encounters, social networks were most often conduits for discussing ideas, practicalities, and publications; disseminating research and findings; and producing, transferring, and exchanging knowledge (various chapters, this volume; for case studies, see e.g., Díaz-Andreu 2007a; Schlanger and Nordbladh 2008; Díaz-Andreu 2012; Delley et al. 2014; Roberts et al. 2020; Coltofean-Arizancu and Díaz-Andreu 2021). Social networks also fostered pluri-, multi- and interdisciplinary collaborations, projects, and practices in archaeology (see e.g., Plutniak 2014, 2017; Delley 2015; de Beaune et al. 2021; Coltofean-Arizancu and Díaz-Andreu 2021). In addition to ideas and knowledge, artifacts such as photographs, drafts of publications, journals, and books, and especially archaeological objects and replicas were also circulated. The distribution of Egyptian antiquities across the globe and their entrance into private and public collections is perhaps the example of this (see Stevenson 2019). The chapters of Kathleen Sheppard (Chap. 3) and Bettina Arnold (Chap. 6) in this book touch upon the transatlantic circulation of Egyptian material and of finds from European lake-dwelling sites, respectively; they also highlight how this circulation impacted the archaeological actors who excavated and/or researched them, the museum collections they became part of (i.e., the Metropolitan Museum of Art and the Milwaukee Public Museum), and the museum audiences and communities for which they were exhibited and/or procured.

Social networks have been essential to acknowledging archaeology as a discipline in various countries, as well as in its institutionalization, internationalization, and professionalization (see e.g., Kaeser 2002; Lorber and Novaković 2020; Coltofean-Arizancu 2022). In fact, the Congrès international d'anthropologie et d'archéologie préhistoriques (CIAAP: International Congress of Prehistoric Anthropology and Archaeology), conceived in 1865—which was the "first institutional body in the field of prehistoric research" (Kaeser 2009: 1) and played an important role in the recognition of prehistoric archaeology as a subfield—was also the result of networking and the joint efforts of like-minded nineteenth-century scholars with a shared interest in the past (Kaeser 2009; also see Kaeser 2002). In addition, the support of social networks can facilitate—and is often fundamental to—the recognition and spread of novel ideas, methods, and research within archaeology, as well as "newly emerging archaeologies" and "alternative archaeologies;" they also enable the maintenance and continuation of already existing archaeological traditions (Tomášková 2011: 222).

Social networks connecting archaeologists, publishers, editors, curators, and journalists were significant in popularizing archaeology, disseminating archaeological knowledge and discoveries among domestic and international non-specialist audiences. This process of increased exposure for the field often involved enhancing personal prestige, gaining public support for archaeology as a discipline, obtaining funding for excavations, and gaining the recognition of archaeological communities (see e.g., Tomášková 2011; Carr 2012; Snead 2018; Thornton 2018; Coltofean-Arizancu 2022; Pizzato, Chap. 8, this volume).

As emphasized over several decades of scholarship (e.g., Díaz-Andreu and Champion 1996/2015; Härke 2000; Díaz-Andreu 2007b; Lozny 2016; Effros and Lai 2018; Eickhoff et al. 2023), archaeological practitioners and enthusiasts frequently employed archaeology to serve political ideologies and agendas, often with the help of their networks. Andrew Whitefield's chapter (Chap. 9, this volume) illustrates this particularly well. It traces the ideas and transnational networks of European antiquarians, archaeologists, folklorists, and philologists who developed ideas related to Celticism and the *Volk* from the eighteenth century until the interwar period. In the process, these networks and ideas transformed archaeology in Ireland, generating a legacy that persists within the discipline up until today. While the use and abuse of archaeology for political purposes was often overt, it is to some extent a subtext in most of the case studies presented in this book, particularly those in countries attempting to craft an identity after the collapse of the Austro-Hungarian Empire (various chapters, this volume).

At a more individual level, social networks in archaeology were used by actors as a means of achieving upward mobility by transcending socio-economic, class, professional, and gender barriers on their way to achieving respectability and prestige, or for building legitimacy and gaining recognition within the discipline. The latter was often the case for practitioners at an early stage in their careers (see Martins, Chap. 10, this volume; Plutniak and Tarantini 2014) or female archaeologists like the Hungarian Zsófia Torma (1832–1899), whose presence on the international and domestic archaeological scene in the nineteenth century was enabled and legitimized by the support of her international scholarly network—the core of which she established at the 1876 CIAAP in Budapest (Coltofean 2017).

Social networks can be accompanied by advantages like protection and disadvantages such as exclusion, marginalization, seclusion, and elision if one does not fit into the societal or organizational norms of a certain place and time. In this regard, Martijn Eickhoff and Mathijs Smith's discussion (Chap. 4, this volume) of German Egyptologist and collector Friedrich Wilhelm Freiherr von Bissing (1873–1956) illustrates how his network facilitated his admission into Dutch society and academia in the 1920s and offered him protection by covering problematic aspects of his life, including his nationalist political views. However, the same network no longer supported him when his anti-Semitism and homosexuality became public. His network quietly sanctioned him, leading to his return to Germany; nevertheless, it also salvaged and to some extent white-washed his image as an intellectual and collector. Von Bissing's example also highlights the conflictual component of social networks. Networking and interactions in general are not always sources of pleasant or positive experiences. They are also accompanied by various forms of friction, often informed by nationalist sentiments, like those between the French and German participants at various CIAAP editions (Müller-Scheessel 2011; for another example, see Coltofean 2015, 2017; Coltofean-Arizancu 2022). The conflicts that arose between the wealthy German-American antiquarian William Frankfurth and the antiquarian establishment in the South Tyrol in the late nineteenth century are another good example. The initial enthusiastic support he received quickly soured after he publicly criticized the lack of interest on the part of the local intelligentsia in the prehistoric past of the region (Arnold 2019; Arnold, Chap. 6, this volume).

As Nathan Schlanger noted in his peer review of this volume, nineteenth- and early twentieth-century antiquarian and scientific networks were not just collaborative and cooperative; they were very much concerned with claiming primacy in the discovery of sites as well as capacities of collection and exchange. Networks then and now can be combative, competitive and exclusionary spaces and individual scholars may form strategic or tactical alliances to navigate them or refuse to participate in them entirely (see e.g., Arnold, Chap. 6; Miller, Chap. 11, this volume). Linguist and prehistorian Gustaf Kossinna's (1858–1931) domination of German prehistoric archaeology and the rivalries he developed with those who questioned his Germanocentric views of the past and methodology illustrate this well (see Brather 2008; Klejn 2008). Another more recent example is the introduction of costly scientific methods in archaeology, such as human and non-human archaeogenetics, which contributes to an increasingly competitive atmosphere as resources (both funding and valuable samples) are in relatively short supply (for a discussion on this, see Gaydarska et al. 2024). Thus, a major intellectual development may have an impact on the re-alignment of existing and formation of new professional networks.

Social networks are also important in establishing hierarchies and power relationships between archaeological actors, civilians, and different authority levels within various archaeology-related settings in times of conflict and war. Ultimately, these interactions directly impact the preservation of archaeological heritage and the practice of archaeology, as shown in Antonino Crisa's chapter shows (Chap. 12, this volume) on the protection of antiquities by diverse intersecting networks of archaeologists, locals, and authorities in Sicily during the Second World War. Power relationships and hierarchies are further explored by Heidi Miller (Chap. 11, this volume). Her chapter reveals how British archaeologists Mortimer Wheeler and Stuart Piggott (1910–1996) supported each other in creating disciplinary authority and the impression that they were introducing true scientific practices in South Asian archaeology; this was enabled by the fact that they operated within the unequal power structures of colonial India.

Social networks shaped different actors' interest in researching, protecting, and preserving the remains of the past, in pursuing archaeology (see e.g., Schlanger and Nordbladh 2008; Coltofean 2015; Szabó et al. 2017; Szabó 2024), as well as in employing archaeology as a form of civic engagement and a pedagogical tool, especially in the context of museums (see Arnold, Chap. 6, this volume). The chapter on the work of Hungarian illustrator and photographer Gábor Szinte (1855–1919) (Codrea and Bodo, Chap. 7, this volume) illustrates how his membership in the National Monuments Commission in Budapest and the History and Archaeology Society of Hunedoara County enabled his participation in archaeological excavations and documentary field trips aimed at researching, documenting, and protecting medieval and Roman buildings in Transylvania, Hungary. Several of these monuments have since been destroyed or the images they contained are no longer preserved; thus, Szinte's professional drawings and photographs literally serve as the sole surviving link between these periods in Transylvanian history and the present day.

1.4 Conclusions

This chapter—like the volume that it introduces—highlights the rich variety of ways in which social networks and networking are intrinsic to the history of archaeology but have not been explicitly recognized in some of the areas of the world represented by the selected case studies. Networks have contributed to the emergence and shaping of archaeology as a discipline, to the various processes that took place within the field (e.g., professionalization, institutionalization, internationalization, knowledge production and exchange), as well as influencing archaeological actors' personal and professional trajectories and practices. Networks are essentially everywhere—not only in archaeology but also in other realms of science and intellectual life in general (Barabási 2002: 7). The chapters in this book support this view by providing insights into various facets of social networks and networking in distinct geopolitical, cultural, and chronological settings throughout the history of archaeology. They also reveal that by examining archaeology's disciplinary history from a networks and networking perspective, one can gain a deeper and more nuanced understanding of the field and its place in the societies that both enable and are created by it.

The chapters in this book reveal how the exploration and combination of different types of archival records as well as print media and scientific publications can further our understanding of the social networks and networking modes that formed during the nineteenth- and twentieth-century emergence of archaeology in Europe and beyond as an intellectual endeavor, a scientific discipline, and a profession. The contributions are by no means exhaustive with regards to the topics and the geographical areas that they cover or the theoretical frameworks and analytical methods that they apply. However, the cross-section of case studies is rich and varied enough to challenge the reader to think about the history of archaeology in a different way—as an interconnected community of like-minded, curious people who have continued to exchange ideas and insights in volumes such as this one. It also underscores the fact that this community is interconnected spatially as well as temporally, and by mapping past connections we form new networks in the present.

In order to remain relevant and produce novel research in the future, the history of archaeology will need to foster more effective and stronger pluri-, multi- and interdisciplinary interactions with other fields. It is hoped that this volume will inspire additional studies to explore networks and networking in the history of archaeology and to employ a wide range of theories, methods, and techniques from other disciplines (e.g., network science, history of science, history of knowledge) in this pursuit.

References

Arnold, B. (2019). "Der stete Antiquar": William Frankfurth (1829–1891) aus Milwaukee, WI, USA. In G. Kaufmann, & A. Putzer (Eds.), *Lost and found: Die Anfänge der Archäologie im Süden Tirols: Forschungsgeschichte bis 1919* (Schriften des Südtiroler Archäologiemuseums, 6) (pp. 289–300). Bozen: Athesia.

Barabási, A.-L. (2002). *Linked: The new science of networks*. Cambridge, MA: Perseus Publishing.

Barabási, A.-L. (2016). *Network science*. Cambridge: Cambridge University Press.

Bentz, M. (2008). 'More than a village'. On the medieval countryside as an archaeological field of study. In N. Schlanger, & J. Nordbladh (Eds.), *Archives, ancestors, practices: Archaeology in the light of its history* (pp. 97–107). New York and Oxford: Berghahn Books.

Brather, S. (2008). Virchow and Kossinna: From science-based anthropology of humankind to the culture-historical archaeology of peoples. In N. Schlanger, & J. Nordbladh (Eds.), *Archives, ancestors, practices: Archaeology in the light of its history* (pp. 317–334). New York and Oxford: Berghahn Books.

Brughmans, T. (2013). Thinking through networks: A review of formal network methods in archaeology. *Journal of Archaeological Method and Theory*, *20*(4), 623–662. https://doi.org/10.1007/s10816-012-9133-8

Brughmans, T., & Brandes, U. (2017). Visibility network patterns and methods for studying visual relational phenomena in archeology. *Frontiers in Digital Humanities*, *4*, article 17. https://doi.org/10.3389/fdigh.2017.00017

Brughmans, T., Collar, A., & Coward, F. (Eds.). (2016). *The connected past: Challenges to network studies in archaeology and history*. Oxford and New York: Oxford University Press.

Cambridge dictionary. (n.d.). Networking. In *Cambridge dictionary*. Retrieved July 24, 2022, from https://dictionary.cambridge.org/dictionary/english/networking

Carr, L. C. (2012). *Tessa Verney Wheeler: Women and archaeology before World War Two*. Oxford: Oxford University Press.

Ciccarelli, P. & Faulkner, C. (2004). *Networking foundations*. San Francisco and London: Sybex.

Collar, A., Coward, F., Brughmans, T., & Mills, B. J. (2015). Networks in archaeology: Phenomena, abstraction, representation. *Journal of Archaeological Method and Theory*, *22*(1), 1–32. https://doi.org/10.1007/s10816-014-9235-6

Coltofean, L. (2015). Importanța ediției a opta a *Congrès international d'anthropologie et d'archéologie préhistoriques* în dezvoltarea arheologiei preistorice în Transilvania secolului al XIX-lea. In S. Forțiu, & A. Stavilă (Eds.), *ArheoVest III: In Memoriam Florin Medeleț. Interdisciplinaritate în arheologie și istorie, Timișoara, 28 noiembrie 2015* (Vol. 2: *Metode interdisciplinare în arheologie și istorie* (pp. 1035–1055). Szeged: JATEPress Kiadó.

Coltofean, L. (2017). Zsófia Torma: A pioneer of prehistoric archaeology in nineteenth-century Transylvania. In Cs. Szabó, V. Rusu-Bolindeț, G. T. Rustoiu, & Mircea Gligor (Eds.), *Adalbert Cserni and his contemporaries. The pioneers of archaeology in Alba Iulia and beyond* (pp. 327–354). Cluj-Napoca: Editura MEGA.

Coltofean-Arizancu, L. (2022). Displaying archaeology and circulating knowledge: The exhibition at the 1876 International Congress of Prehistoric Anthropology and Archaeology. *Nuncius*, *37*(3), 529–563. https://doi.org/10.1163/18253911-bja10039

Coltofean, L. (2024). Biographical research in the history of archaeology. In M. Díaz-Andreu, & L. Coltofean, The *Oxford handbook of the history of archaeology* (pp. 98–120). New York: Oxford University Press.

Coltofean-Arizancu, L., & Díaz-Andreu, M. (Eds.). (2021). *Interdisciplinarity and archaeology: Scientific interactions in nineteenth- and twentieth-century archaeology*. Oxford and Philadelphia: Oxbow Books.

Crane, S. A. (2000). *Collecting and historical consciousness in early nineteenth-century Germany*. Ithaca and London: Cornell University Press.

de Beaune, S. A., Guidi, A., Moro Abadía, O., & Tarantini, M. (Eds.). (2021). *New advances in the history of archaeology. Proceedings of the XVIII UISPP World Congress (4-9 June 2018, Paris, France)* (Vol. 16: *Sessions organised by the History of Archaeology Scientific Commission at the XVIII World UISPP [4-9 June 2018, Paris, France]*). Oxford: Archaeopress.

Delley, G. (2015). Au-delà des chronologies: Des origines du radiocarbone et de la dendrochronologie à leur intégration dans les recherches lacustres suisses. Neuchâtel: Office du patrimoine et de l'archéologie.

Delley, G., Díaz-Andreu, M., Djindjian, F., Fernández, V. M., Guidi, A., & Kaeser, M.-A. (Eds.). (2014). *History of archaeology: International perspectives. Proceedings of the XVII UISPP World Congress (1–7 September 2014, Burgos, Spain)* (Vol. 11: *Sessions A8b, A4a and A8a organised by the History of Archaeology Scientific Commission*). Oxford: Archaeopress.

Díaz-Andreu, M. (2007a). Christopher Hawkes and the International Summer Courses of Ampurias. *Bulletin of the History of Archaeology*, *17*(1), 19–34. https://doi.org/10.5334/bha.17104

Díaz-Andreu, M. (2007b). *A world history of nineteenth-century archaeology: Nationalism, colonialism, and the past*. Oxford: Oxford University Press.

Díaz-Andreu, M. (2012). *Archaeological encounters: Building networks of Spanish and British archaeologists in the 20th century*. Newcastle upon Tyne: Cambridge Scholars Publishing.

Díaz-Andreu, M., & Champion, T. (Eds.). (2015). *Nationalism and archaeology in Europe*. London and New York: Routledge. (Original work published 1996)

Díaz-Andreu, M., & Portillo, M. (Eds.). (2021). *Arqueología e interdisciplinariedad: La microhistoria de una revolución en la arqueología española (1970–2020)*. Barcelona: Edicions Universitat de Barcelona.

Eickhoff, M., Modl, D., Meheux, K., & Nuijten, E. (Eds.). (2023). *National-socialist archaeology in Europe and its legacies*. Cham: Springer. https://doi.org/10.1007/978-3-031-28024-5

Effros, B., & Lai, G. (Eds.). (2018). *Unmasking ideology in imperial and colonial archaeology: Vocabulary, symbols, and legacy*. Los Angeles: Cotsen Institute of Archaeology Press, University of California.

Fagan, B. M. (2018). *Grahame Clark: An intellectual life of an archaeologist*. New York and London: Routledge.

Gaydarska, B., Matić, U., Coltofean, L., & Díaz-Guardamino, M. (2024). Let's talk about gender—The place of gender in current archaeological debates. In U. Matić, B. Gaydarska, L. Coltofean, & M. Díaz-Guardamino (Eds.), Gender trouble and current archaeological debates (Themes in Contemporary Archaeology). Cham: Springer.

Givens, D. R. (1992). *Alfred Vincent Kidder and the development of Americanist archaeology*. Albuquerque: University of New Mexico Press.

Grunwald, S., Bánffy, E., Gramsch, A., Hofmann, K. P., & Rasbach, G. (Eds.). (2022). Digging Bersu. Ein europäischer Archäologe [Special issue]. *Bericht der Römisch-Germanischen Kommission*, *100*/2019–2022.

Hawkes, J. (1982). *Mortimer Wheeler: Adventurer in archaeology*. London: Weidenfeld and Nicolson.

Härke, H. (Ed.). (2000). *Archaeology, ideology and society: The German experience*. Frankfurt am Main, Berlin, Bern, Bruxelles, New York, Oxford and Wien: Peter Lang.

Kaeser, M.-A. (2002). On the international roots of prehistory. In N. Schlanger (Ed.), *Ancestral archives: Explorations in the history of archaeology* [Special section]. *Antiquity*, *76*(291), 170–177. https://doi.org/10.1017/S0003598X0008995X

Kaeser, M.-A. (2008). Biography as microhistory: The relevance of private archives for writing the history of archaeology. In

N. Schlanger, & J. Nordbladh (Eds.), *Archives, ancestors, practices: Archaeology in the light of its history* (pp. 9–20). New York and Oxford: Berghahn Books.

Kaeser, M.-A. (2009). Establishing prehistory: The foundation of the international congress (1865/1866). In M. Babeş, & M.-A. Kaeser (Eds.), *Archaeologists without boundaries: Towards a history of international archaeological congresses (1866–2006)/Archéologues sans frontières: Pour une histoire des Congrès archéologiques internationaux (1866–2006)* (BAR International Series, 2046) (pp. 1–2). Oxford: Archaeopress.

Kaeser, M.-A. (2013). Biography, science studies and the historiography of archaeological research: Managing personal archives. *Complutum*, 24(2), 101–108.

Klejn, L. (2008). Gustaf Kossinna (1858–1931). In T. Murray, & C. Evans (Eds.), *Histories of archaeology: A reader in the history of archaeology* (pp. 312–327). Oxford and New York: Oxford University Press.

Knappett, C. (2011). *An archaeology of interaction: Network perspectives on material culture and society*. Oxford: Oxford University Press.

Knappett, C. (Ed.). (2013). *Network analysis in archaeology: New approaches to regional interaction*. Oxford: Oxford University Press.

Krämer, W. (2002). Gerhard Bersu—ein deutscher Prähistoriker, 1889–1964. In 100 Jahre Römisch-Germanische Kommission [Special issue]. Bericht der Römisch-Germanischen Kommission, 82/2001, 5–101.

Latour, B. (2005). *Reassembling the social: An introduction to actor-network-theory*. Oxford and New York: Oxford University Press.

Livingstone, D. N. (2003). *Putting science in its place: Geographies of scientific knowledge*. Chicago and London: The University of Chicago Press.

Lorber, Č., & Novaković, P. (2020). Internationalisation as a long-term strategic project of the post-war renewal of the Yugoslav archaeology (1950–1971). *Etnoantropološki Problemi/Issues in Ethnology and Anthropology*, *15*(3), 689–715. https://doi.org/10.21301/eap.v15i3.3

Lozny, L. R. (Ed.). (2016). *Archaeology of the communist era: A political history of archaeology of the 20th century*. Cham: Springer. https://doi.org/10.1007/978-3-319-45108-4

Müller-Scheessel, N. (2011). '... dem Romanismus entgegentreten': National animosities among the participants of the *Congrès International d'Anthropologie et d'Archéologie Préhistoriques*. In A. Gramsch, & U. Sommer (Eds.), *A history of Central European archaeology. Theory, methods, and politics* (pp. 57–87). Budapest: Archaeolingua.

Newman, M. E. J. (2001). The structure of scientific collaboration networks. *Proceedings of the National Academy of Sciences*, *98*(2), 404–409. https://doi.org/10.1073/pnas.98.2.404

Newman, M. (2018). *Networks* (2nd ed.). Oxford and New York: Oxford University Press.

Murray, T., & Evans, C. (Eds.). (2008). *Histories of archaeology: A reader in the history of archaeology*. Oxford and New York: Oxford University Press.

Novaković, P. (2021). *The history of archaeology in the Western Balkans* (Pontes academici). Ljubljana: Ljubljana University Press, University of Belgrade, University of Rijeka, and University of Split.

Peeples, M. A. (2019). Finding a place for networks in archaeology. *Journal of Archaeological Research*, *27*(4), 451–499. https://doi.org/10.1007/s10814-019-09127-8

Plutniak, S. (2014). Interests on the margins of the disciplines: Computing, engineers and archaeologists in France (1950–2000). In In G. Delley, M. Díaz-Andreu, F. Djindjian, V. M. Fernández, A. Guidi, & M.-A. Kaeser (Eds.), *History of archaeology: International perspectives. Proceedings of the XVII UISPP World Congress (1–7 September 2014, Burgos, Spain)* (Vol. 11: *Sessions A8b, A4a and A8a organised by the History of Archaeology Scientific Commission*) (pp. 221–232). Oxford: Archaeopress.

Plutniak, S. (2017). The archeological operation. A sociohistorical perspective on a discipline faced with developments in automatics and mathematics. France, Spain, Italy, in the second half of the 20th century (L'opération archéologique. Sociologie historique d'une discipline aux prises avec l'automatique et les mathématiques. France, Espagne, Italie, 2e moitié du XXe siècle) [Unpublished doctoral dissertation]. École des hautes études en sciences sociales, France.

Plutniak, S., & Tarantini, M. (2014). An influential outsider. Georges Laplace between French institutions and Italian prehistory. In G. Delley, M. Díaz-Andreu, F. Djindjian, V. M. Fernández, A. Guidi, & M.-A. Kaeser (Eds.), *History of archaeology: International perspectives. Proceedings of the XVII UISPP World Congress (1–7 September 2014, Burgos, Spain)* (Vol. 11: *Sessions A8b, A4a and A8a organised by the History of Archaeology Scientific Commission*) (pp. 79–89). Oxford: Archaeopress.

Roberts, J., Sheppard, K., Hansson, U. R., & J. R. Trigg (Eds.). (2020). *Communities and knowledge production in archaeology*. Manchester: Manchester University Press.

Schlanger, N., & Nordbladh, J. (Eds.). (2008). *Archives, ancestors, practices: Archaeology in the light of its history*. New York and Oxford: Berghahn Books.

von Schnurbein, S. (2002). Abriß der Entwicklung der Römisch-Germanischen Kommission unter den einzelnen Direktoren von 1911 bis 2002. In *100 Jahre Römisch-Germanische Kommission* [Special issue]. *Bericht der Römisch-Germanischen Kommission, 82/2001*, 137–289.

Snead, J. E. (2018). *Relic hunters: Archaeology and the public in nineteenth-century America*. Oxford and New York: Oxford University Press.

Stevenson, A. (2019). *Scattered finds: Archaeology, Egyptology and museums*. London: University College London Press.

Sweet, R. (2004). *Antiquaries: The discovery of the past in eighteenth-century Britain*. London and New York: Hambledon and London.

Szabó, Cs. (2024). Alfred von Domaszewski and Béla Cserni: Authority and academic networks in the late 19th century Transylvanian research. In S. Nemeti, & C. Timoc (Eds.), *Alfred von Domaszewski: Latin epigraphy in the Roman Empire. Acts of the colloquium held in Timişoara in December 14th–17th 2022* (pp. 93–110). Cluj-Napoca: Editura MEGA.

Szabó, Cs., Rusu-Bolindeţ, V., Rustoiu, G. T., & Gligor, M. (Eds.). (2017). *Adalbert Cserni and his contemporaries. The pioneers of archaeology in Alba Iulia and beyond*. Cluj-Napoca: Editura MEGA.

Tomášková, S. (2011). Archaeology in a middle country. In L. R. Lozny (Ed.), *Comparative archaeologies: A sociological view of the science of the past* (pp. 221–242). New York: Springer. https://doi.org/10.1007/978-1-4419-8225-4

Thornton, A. (2011). *British archaeologists, social networks and the making of a profession: The social history of British archaeology in the Eastern Mediterranean and Middle East, 1870–1939* [Unpublished doctoral dissertation]. University College London, United Kingdom.

Thornton, A. (2015). Social networks in the history of archaeology. Placing archaeology in its context. In G. Eberhardt, & F. Link (Eds.), *Historiographical approaches to past archaeological research* (Berlin Studies of the Ancient World, 32) (pp. 69–94). Berlin: Edition Topoi.

Thornton, A. (2018). *Archaeologists in print: Publishing for the people*. London: University College London Press.

Woodbury, R. B. (1973). Alfred V. Kidder. New York: Columbia University Press.

In the Archive of Agda Montelius, or How to Find the People and Practices Behind a Great Man's Work

2

Anna Charlotta Gustavsson

Abstract

In June 1876, Agda Montelius (1850–1920) embarked on a steamship and left Stockholm for a six-month long journey in Europe together with her husband, the famous Swedish archaeologist Oscar Montelius (1843–1921). This was the beginning of their research activities, leading to the publishing of Oscar's *magnum opus La civilisation primitive en Italie depuis l'introduction des métaux*, two decades later. This chapter highlights the presence of Agda in her husband's international (net)work and illustrates how even less visible actors can be identified and the formation of networks reconstructed by drawing on a combination of sources. Furthermore, the text addresses archival practices that have contributed to the (in)visibility of certain actors in intellectual networks and knowledge transfer systems. The focus is on the Montelius spouses' travels and work in Italy in the late 1870s and early 1880s, drawing on Agda Montelius' diaries. Today, Agda is known for her engagement in human rights, but she was also an experienced traveler, amateur archaeologist, and illustrator. By delving into the work of a 'Great Man' like Oscar Montelius and his relationship with a supportive spouse, it is possible to come to a more nuanced understanding of how the creation of networks and archaeological knowledge transfer took place, and to highlight the collective effort behind all such projects.

Keywords

Agda Montelius · Oscar Montelius · Archaeological practices · Archives · Italy · Knowledge production · Networks

A. C. Gustavsson (✉)
Department of Historical Studies, University of Gothenburg, Gothenburg, Sweden
e-mail: anna.gustavsson@gu.se

2.1 Introduction

Women as independent actors in scientific networks and as practitioners of research and knowledge production in the long nineteenth century have been studied for several decades, but historians of science (in a wider sense) have often ignored the women who went on study trips and/or conducted different kinds of research in fields such as botany, geology and archaeology. The historian Bonnie Smith (1998: 1) has argued concerning the writing of history itself that the development of modern scientific methodology, epistemology, professional practice and writing is closely tied to evolving definitions of masculinity and femininity. It is surprising, as the archaeologist Laura Nicotra writes (2004: 19–20), that women's activities in the archaeological field have been considered so marginal that none of the standard works on the histories of modern archeology include more than a handful of female actors. This androcentric perspective of the history of archeology has focused on the great male archaeologists without exploring or even recognizing the value of the participation of the women who dedicated their time to archaeological work. The contributions of female archaeology practitioners are therefore less documented and visible but important, nevertheless.

It is true that women were most often excluded from formal and professional institutions of science and knowledge until the late nineteenth century (Livingstone 2003: 44). However, it has been suggested for some time now that women played an informal, yet important, role in the mid-nineteenth century when archaeology was developing into a modern scientific discipline (Sørensen 1998: 46). There is still much to investigate when it comes to the few women performing archaeological research and publishing in their own names in the late nineteenth century and the first years of the twentieth century. Examples include Johanna Mestorf (1828–1909), Zsófia Torma (1832–1899), Ersilia Caetani Lovatelli (1840–1925) and Harriet Boyd Hawes (1871–1945), as well as the wives, sisters and daugthers of male scholars.

L. Coltofean et al. (eds.), *Connecting People and Ideas*, Themes in Contemporary Archaeology,
https://doi.org/10.1007/978-3-031-81006-0_2

Fig. 2.1 Agda Montelius (born Reuterskiöld) c. 1869. (Photo: Gustaf Joop & Co.; reproduced with the permission of the Kungliga biblioteket [Royal Library], Stockholm, Sweden; KB, KoB Alb. 95:110)

In the second category we find Ida Nilsson (1840–1920), the daughter of the famous Swedish naturalist and prehistorian Sven Nilsson (1787–1883), Alice Evans (1856–1887), the daugther of John Evans (1823–1908) and sister of Arthur Evans (1851–1941), and Hilda Petrie (1871–1957), the wife of Flinders Petrie (1843–1952)—although the latter eventually directed excavations and published her work (see Evans 1943; Drower 1985; Allsebrook 2002; Johnsen and Welinder 2002; Nicotra 2004; Coltofean 2014, 2017; Nicklasson 2021 for studies on the aforementionend individuals; see also Claassen 1994; Díaz-Andreu and Sørensen 1998). Archaeologist Marie Louise Stig Sørensen has argued that in writing historiographies of women it becomes possible to identify the types of women who were involved in archaeology (Sørensen 1998: 47). So, how do we find these women in the history of archaeology? Because they often are invisible in offical documents and publications, it might be fruitful to turn to personal records in the archives such as letters and journals. What can the documents in their private archives—if these exist and can be accessed—add to the understanding of their lives, as well as to the formation of professional networks and archaeological practice?

In this chapter, the reader will be introduced to the Swedish Agda Montelius (1850–1920; Fig. 2.1). She was one of the many nineteenth-century women who were involved in the archaeological work and research activities of a husband, brother, or father. The great man in question is the internationally known archaeologist Oscar Montelius (1843–1921). The aim of this contribution is to highlight Agda Montelius' presence in her husband's professional network and scholarly output. This case study demonstrates how less visible actors can be identified and their activities revealed by reconstructing practices connected to the formation of networks based on a combination of archival sources. Furthermore, the text addresses practices, such as the organization of archives, which might contribute to the (in)visibility of certain individuals in research on the creation of archaeological knowledge. Agda Montelius' personal archive provides insights into her work and travel activities while illuminating the participation of women in archaeological and antiquarian practices in the nineteenth century in general.

2.2 Agda and Oscar Montelius

Agda Montelius, born Reuterskiöld, is primarily known today for her strong engagement in humanitarian and social issues, and her charity work. She grew up in an aristocratic environment. Her mother came from a wealthy family, while her father was in the military and served as a minister in the Swedish government in the 1860s. Agda was well-educated and studied French, German and English, as well as Swedish, history, church-history, literature, physics, natural sciences, and drawing in school. The family moved frequently during Agda's childhood, due to her father's work (Bokholm 2000: 47).

Agda and her husband Oscar Montelius met for the first time in 1866 when Agda was only 16 years old, or possibly even earlier. They became a couple officially in 1868 and married in 1871. At the time Agda was 21, and Oscar 28 years old. The spouses moved into the Montelius family home on the Södermalm island in Stockholm. Agda's interests in welfare issues as well as in archaeology seem to have started early and developed rapidly after she married into and lived with the Montelius family. Oscar's mother had been engaged for many years in charity work organized by a local women's society. In the decades that followed, Agda took on various responsibilities and posts in several associations. In addition to her charity work, she participated in the ongoing emancipation debates. The spouses were convinced liberals; like his wife, Montelius supported women's rights and female suffrage (Arwill-Nordbladh 1989; Bokholm 2000; Baudou 2012a). Like many women of her time, Agda can be described as a difference feminist. In short, this means she believed that the differences between men and women should be viewed in a gender-complementary manner, with the proviso that women were best qualified for caring for the household, the family and children. She was convinced that making the best use of the best qualities of both genders would improve society, i.e., what benefited women would benefit everyone (see Bokholm 2000; Baudou 2012a; Nordström 2014;

Arwill-Nordbladh 2018 for biographical information on Agda and Oscar Montelius).

The Swedish scholar Oscar Montelius was one of the most central and best-known Scandinavian archaeologists in the early days of modern archaeology. He was a pioneer in prehistoric archaeology in the late 1800s and is mainly known for his work on typology and chronology; his *Om tidsbestämning inom Bronsåldern med särskildt afseende på Skandinavien* (1885) is still frequently cited.[1] Montelius held positions at the Nationalmuseum (National Museum) of Stockholm, and eventually served as *Riksantikvarie*, the director of the Riksantikvarieämbetet (Swedish National Heritage Board), between 1907–1913. He was involved in numerous excavations in Sweden. He also traveled all over Europe—perhaps more than many other scholars at the time—to compile and study archaeological finds and is still one of the few scholars to have documented and published such vast numbers of archaeological artifacts (e.g., Montelius 1880–1882, 1895–1910, 1912, 1924–1928). Montelius became affiliated with Nationalmuseum as a young scholar in the mid-1860s, together with his friend and colleague Hans Hildebrand (1842–1913). They worked together with the then *Riksantikvarie*, Bror Emil Hildebrand (1806–1984), who also happened to be Hans Hildebrand's father and a long-time friend of the Montelius family.

One of the earliest assignments for the young scholars was the preparation of a new display of the archaeological collections at the museum, which was completed in 1866.[2] Montelius studied history, Nordic languages, and natural sciences before turning to archaeology. He was fluent in spoken and written English, German, French and Italian. He was awarded a doctorate in history since archaeology was not yet an established academic discipline. His dissertation, entitled *Från Jernåldern*, published in 1869, was an overview of current research on how Iron Age culture spread from Egypt, via Greece, Rome, and the Hallstatt zone to Scandinavia (Montelius 1869a). It was translated into English (not German or French, *nota bene*!) in the same year and was published with the title *Remains from the Iron Age of Scandinavia* (Montelius 1869b). One of the defining characteristics of his research method from the beginning was to collect the most comprehensive record of objects possible for each object type (Baudou 2012b: 181).

Archaeological sites and finds in Italy were of fundamental importance not only for the general development of prehistoric archaeology as a discipline, but also for Montelius in his efforts to use the prehistory of Europe, including Scandinavia, to develop and construct chronologies for the Metal Ages. He was the first scholar to produce a comprehensive synthesis of the Italian Bronze and Iron Ages, *La civilisation primitive en Italie depuis l'introduction de métaux*, which was published in five volumes (1895–1910). Even though Oscar Montelius is a well-known scholar, little is known today about his international reputation and massive work on European and especially Italian prehistory, as well as the dynamics behind those efforts.

Agda became involved in her husband's work and travels from the start and documented their whereabouts in her journals. She was referred to as Sweden's best-traveled woman, and in 1879, at the age of 29, she was elected to the Svenska Sällskapet för Antropologi och Geografi (Swedish Society for Anthropology and Geography). This membership provided her with access to a scientific and almost exclusively male arena (Bokholm 2000: 51). Over time, she served as a stand-in for her husband at the museum office, where she conducted a range of administrative tasks and handled financial transactions. She took over his duties on journal boards, serving as editor and proof-reader. This enabled Oscar to travel and conduct excavations in Sweden, and to go on study trips abroad. Letters could be addressed to the Professor's alter ego, which indicates that Agda was trusted. Beginning in the 1890s Agda stayed in Stockholm when Oscar traveled. At the same time, she was very busy with her work on welfare and social issues in Stockholm (Bokholm 2000: 51–52). Agda and Oscar Montelius are buried at Norra begravningsplatsen in Solna. Their funerary monument, which was erected roughly a decade after their deaths, is a pastiche of a Stone Age chamber tomb. An inscription acknowledges Oscar's archaeological achievements, while Agda's efforts in life are not recognized in the same way.

The only monograph on Agda Montelius published to date is the biography *En kvinnoröst i manssamhället* by the historian Sif Bokholm (2000). Bokholm's analysis focuses on Agda's charity work and her active participation in women's rights and social welfare initiatives. Her life with Oscar and her involvement in his work and travels is also discussed but not in-depth. By contrast, scholars who have studied Oscar have commented on Agda in relation to her husband's professional activities (Bokholm 2000; Baudou 2012a; Nordström 2014) but do not represent her as a whole person with a fully realized life of her own. Early on in my ongoing research, it was clear that Agda had played an important role in her husband's research activites, as had been mentionend

[1] The book was first published in 1885 in Swedish with a French summary. A century later it was translated into English and republished with the title *Dating in the Bronze Age with special reference to Scandinavia* (O. Montelius 1986).

[2] 1866 has been proposed as the unofficial founding year of Historiska Museet, today the Staters historiska museum (Swedish History Museum). In the late 1930s the archeological collections were moved to a new building hosting the Historiska museet, which was properly inaugurated in 1943. For simplicity and as the collections were exhibited in Nationalmuseum and belonged to the same administrational body in Montelius' days, the author has chosen to use Nationalmuseum in this text.

by Bokholm and others, but to what extent and exactly how, remained to be investigated.[3]

2.3 In the Montelius Archives

The growing number of studies on different aspects of the history of archaeology has led to an increasing interest in and research on archival source material. Archives in general, including those dedicated to archaeological documentation, have been the subject of critical debates for the past two decades. The "archival turn" is characterized by a shift from using the archives as (a neutral) source, to viewing the archives as a (not neutral) subject (Stoler 2009: 44). Discussions have focused on the constitution of archives and how archives can and should be consulted (e.g., Delsalle 1998, 2018; Derrida 1996; Schlanger and Nordbladh 2008; Stoler 2009). Acknowledging that an archive has been created, curated, organized, and classified by several individuals over the years can contribute to the effort of extending the scholarly gaze beyond organisational structures and categorization.

The extensive material related to Oscar and Agda Montelius is mainly preserved in two large archives, the Oscar Montelius arkiv (Oscar Montelius Archive) and the Montelius-Reuterskiöld samling (Montelius-Reuterskiöld Collection), at the Riksantikvarieämbetet (RAÄ: Swedish National Heritage Board) and the Antikvarisk-topografiska arkivet (ATA: Antiquarian Topographical Archive) in Stockholm. The archives are not digitalized but include a synoptic index of the archive categories and folders. Agda's private archive, as well as many folders with photos from their journeys, are filed under the Montelius-Reuterskiöld samling. We know that Agda destroyed an unknown number of letters (probably mainly letters between her and Oscar from certain time periods), and that her niece sorted and organized the records after Agda's death (Bokholm 2000: 19). Nevertheless, 38 small notebooks/travel diaries from the years 1876 to 1918 have survived, as have four archive boxes with material related to her social work, three boxes with mixed material such as newspaper-clips, and ten boxes with letters written to Agda. There is only one box with correspondence from foreign letter writers. Some of the letters add a lot of information about her work and international contacts.

Oscar Montelius' archive is at least ten times larger than his wife's. It is sorted into a great number of indexed categories that list letters, sketches, photos, notes, drafts and manuscript pages of books, articles, and lectures (and a range of other documents e.g., invitations, dinner menus, theatre programs) that are more (or less) sorted. Many of the documents fit under several of the indexed categories, and it is not clear if the archivist wanted to keep the 'order' of the documents as left by Oscar Montelius or tried to sort the documents into potentially useful categories. Hence, letters are found under "Correspondence", as expected, but are also filed with other types of documents (e.g., sketches, notes, photos) under "Work material". These letters cannot be located without skimming each box and folder. The foreign correspondence of Oscar Montelius includes thousands of letters.

While surveying and scanning the archive documents related to Oscar Montelius' research on Italian prehistory, I discovered that the "Work material" folders in Oscar Montelius' archive contained many drawings of artifacts made by his wife Agda. Eager to find out more, I turned to her archive and the preserved travel journals. The archive material produced by Agda (Fig. 2.2) is an important source of information about the couple's travels and archaeological work, including Oscar's professional network, as it complements the other categories (e.g., letters, notes) of source material related to him. While Oscar had been the focus of my research initially, it soon became clear that Agda, and her private archive, were indispensable in understanding his scientific life, which had to be viewed from the perspective of their lives as a couple. For instance, Agda's notes are often the only source of information on how and when her husband met colleagues and provide essential insights into her participation in the networking activities of scholars at the time in which wives played a significant role. Related archival issues focusing on methodology and the potential of personal archives have been discussed recently by archival scholars Jennifer Douglas and Allison Mills. They reject the idea that there is an opposition between organizational and personal archives, and state that the imagined gap between the two categories can be bridged depending on how an archive is activated (Douglas and Mills 2018: 258). The study of the preserved records of Agda Montelius is an example of how consulting materials in a personal archive can contribute to the state of knowledge of a research field on several levels, from the individual to the general.

The methodological inspiration for this investigation draws on a geographical conceptualization of the formation of networks and archaeological knowledge production (see especially Livingstone 2003; Livingstone and Withers 2011; Roberts and Sheppard 2020). It is widely accepted among scholars today that the places and spaces where research activities take place matter. A geographical approach facilitates the understanding of how scientific networks emerge

[3] The author has previously touched upon on the role of Agda Montelius in her husband's work (Gustavsson 2018, 2020). It should be noted that some of the research results and source materials presented in this chapter are part of an ongoing PhD dissertation project that analyzes dimensions of knowledge production and the development of archaeology during the late nineteenth century. The objects of study are Oscar and Agda Montelius, their work and travels in Italy, and Montelius' massive publication on prehistoric Italy.

Fig. 2.2 Agda Montelius diaries from 1876 to 1879. (Photo: Anna Ch. Gustavsson; RAÄ, ATA, Montelius-Reuterskiöld Collection, folder F2 B, vol. 1A)

and function, depending on where they were formed and how they produced knowledge. To investigate such matters, it is essential to identify the actors and their interactions at a variety of locations, such as universities, archaeological sites, museums, private homes, hotels, restaurants, or formal scientific meetings (Roberts and Sheppard 2020: 5). The travel journals of Agda Montelius turned out to be a useful source of information when considering such issues in relation to the research and network activities she and her husband engaged in during their journeys to Italy.

2.4 Traveling and Networking in Italy

From the 1860s onwards, many archaeological discoveries were made in Italy that were important in various ongoing international debates among scholars and for the development of prehistoric archeology as a science. A large part of Oscar's research on the chronology of the Bronze and Iron Ages drew on material from Italy and he published the first extensive work on the country's prehistoric periods at that time (Montelius 1895–1910). Therefore, the spouses traveled there many times over the years and the lion's share of their studies and collection of information were carried out 'on site' in Italy. The events described below are derived mainly from Agda's travel journals from the years 1876, 1879 and 1881. Agda began accompanying Oscar on research trips immediately after their marriage in September 1871. In fact, they spent part of their honeymoon at the Congrès international d'anthropologie et d'archéologie pre-historiques (CIAAP: International Congress of Prehistoric Anthropology and Archaeology) in Bologna in October 1871. Unfortunately, none of Agda's diaries or other information about this sojourn in Italy have survived. However, based on other available sources, such as the travel notes of Swedish delegate Magnus Lagerberg (1844–1920) (published in the book *Brev från Söderns länder*, Lagerberg 1872), it is possible to make some qualified suggestions about the scholars they likely met in person at the conference. Among the scholars present at CIAAP in 1871 who became part of Montelius' Italian network were Edoardo Brizio (1846–1907), Giovanni Capellini (1833–1922), Giovanni Gozzadini (1810–1887), Luigi Pigorini (1842–1925), and Antonio Zannoni (1833–1910).

The next trips to Italy took place in 1876 and 1879. Oscar had received a rather unique grant, the Letterstedtska resestipendiet (Letterstedtska travel grant), which paid for a twelve-month study trip throughout Europe for the lucky candidate. Oscar and Agda completed the Letterstedt journey in several stages during the years 1876 and 1878–1879 (Baudou 2012a: 182–87). In 1881 they both took off again, and in 1895 they made their last joint journey to Italy.[4] Agda's diaries from these years are a rich source of information about the ancient monuments and museum collections they visited as a couple, providing details about how they gained knowledge about and got access to archaeological sites and material. Because Oscar's notes are much sparser, one gets the impression that it was Agda's 'official' task to keep a more detailed travel description, and her notes contain many elements of an ethnographic nature. Agda not only wrote about her own and their shared experiences, but she also mentioned what Oscar did and who he met on the occasions when she chose to stay at their hotel or engaged in her own activities. What at first reading seems purely entertaining and anecdotal can often provide information about the conditions of the time and tells us about how the spouses experienced the communities

[4]Oscar Montelius made a few journeys to Italy without Agda in the late 1890s and early 1900s. Those research visits will not be addressed here but are included in the analysis in the forthcoming doctoral thesis.

and people they met. In addition, the reader can vicariously travel with them through the descriptions of landscapes, weather conditions, modes of transport and observations regarding the local population. The fact that the couple could travel together for long periods of time is usually explained by the fact that they had no children, and thus Agda was not tied to the home, as most women of that time would have been. While their family situation naturally affected the practical possibilities of traveling, it is obvious that Agda was interested, knowledgeable and involved in Oscar's work from the start. In the section below, the reader will be provided with a few glimpses of the information that can be extracted from Agda's travel diaries from the years 1876, 1879 and 1881. Her travel notes help to shed light on practical and logistical dimensions of knowledge creation, as well as providing insight into Agda's often-detailed journaling routine. References to her own and her husband's personal interactions with scholars during their study trips are an important source of information on the formation and constitution of their networks. Records of such contacts between scholars are often difficult to trace (Roberts and Sheppard 2020: 6), one of the reasons this particular archival source is both unusual and important.

2.5 In the Diaries: Places, Practices, and People

In the fall of 1876, Agda and Oscar crossed the border of Italy on September 30 and stayed in the country until December 14. They visited Trieste and Venice along the way, but concentrated on museums and collections in Bologna, Florence, Perugia, Rome, and Naples (Montelius 1876: August 25–December 14). They left "la Bella" Venezia and arrived in "la Grassa" Bologna on October 5, where they spent eight intensive days studying and drawing archaeological artifacts from the pre-Roman eras of Bologna. The finds were kept in storage at the San Francesco church, the Archiginnasio library and the Palazzo Gozzadini—the family home of the scholar Count Giovanni Gozzadini. According to Agda the studied material mainly belonged to the Villanovan period (900–700 BC) (Montelius 1876: October 9–15). The next stop was Florence, where Oscar attempted to meet with the anthropologist Paolo Mantegazza (1831–1910), without success. A few days later, Agda noted that Oscar was out on his "vanliga fruktlösa letande efter arkeologer" (usual fruitless search for archaeologists) (Montelius 1876: October 19). Trying to locate scholars and owners of archaeological collections would turn out to be a recurrent activity for Oscar during their journeys. It is not possible from the archival material to verify which Florentine scholars Oscar encountered, but within the next few days he was granted permission to document the finds at the "Etruskiska Museet" (Etruscan Museum) Agda and Oscar both visited the museum and made drawings on the days that followed. On October 29 they left Florence and moved on to Perugia. The next day Oscar was picked up at the hotel by the collector and scholar Giuseppe Bellucci (1844–1921) who wanted to show him a collection. In the entry from the following day, Agda wrote that Oscar went back to Bellucci in the morning. In the afternoon he and Agda went to the university to draw finds. On November 1 Oscar went to study Guardabassi's collection (Montelius 1876: October 30–31, November 1).[5]

On November 5, Agda and Oscar reached Rome, where they stayed at the Hotel Minerva, a stone's throw from the Pantheon. They spent a whole month in the city, enabling them to visit famous sites and churches as well as studying archaeological finds and meeting with scholars (Montelius 1876: November 5–December 2). From the diary we learn that Oscar met frequently with Augusto Castellani (1829–1914), Wolfgang Helbig (1839–1915), and Luigi Pigorini. This is not surprising, as they all were key figures in Rome at the time.[6] However, here an important observation regarding the value of combining different categories of archival material is worth making. If one consulted only the correspondence between Oscar Montelius and Castellani, and not the diaries of Agda Montelius, it would not be possible to confirm the onsite interactions between these scholars during these research stays. The networking activities involved meetings with Castellani and Pigorini to discuss artifacts at "deras" (their) museums, but also included private dinners and cultural *salons* at the Helbig's, who lived in the beautiful Villa Lante located on the Gianicolo Hill in Rome. Rome was a destination and center for many European travelers, artists, and intellectuals, so Agda and Oscar also frequented famous cafés such as the Café degli Artisti, and the Circolo Scandinavio (Scandinavian Association), which served as a hub for Nordic intellectuals as well as writers and artists (Montelius 1876: November 18 and 30, December 1). We must assume that the couple received tips on which Roman and Rome-based scholars to contact from their fellow Scandinavians, even those who were not archaeologists themselves. During all their research trips, Agda made notes of the many meetings, excursions and dinners that took place in the company of other European, Scandinavian, especially Swedish, individuals. These persons were sometimes travelers like the couple themselves or expats who lived temporarily or permanently in Italy. They all belonged to a privileged

[5] This passage refers to the private collection of Mariano Guardabassi (1823–1880), an artist from Perugia with a great interest in archaeology.

[6] Castellani was an art dealer/collector and one of the directors of the Capitoline museums. Helbig, an art dealer/collector and archaeologist, held important positions at the Istituto di Corrispondenza Archeologica, later the German Archaeological Institute. Pigorini had transferred to Rome and was the director of the prehistoric national museum there.

stratum of the population, with access to significant financial, cultural, and/or scientific resources. Even though they were not archaeologists or researchers, they were part of the general intellectual scene and often belonged to more general cultural circles, such as those centered on the Gozzadini family in Bologna and the Helbig family in Rome.

In early December 1876, the couple was in Naples for nine days. In the diary passages describing these dates, Agda recounts how they spent their first days in the city at the archaeological museum, getting an overview of the collections, after which selected sections were studied more carefully. On the following days Agda worked at the museum, producing sketches of a vast number of artifacts. In the meantime, Oscar was out scouting for original artifacts, mainly Bronze and Iron Age *fibulae* (brooches), for purchase at the local antique markets. He also bought souvenir photographs of finds and monuments to add to his growing reference collection of illustrations. Oscar managed to locate one Neapolitan scholar, Guglielmo Guiscardi (1821–1885), whom he met on at least two occasions while Agda was working at the museum (Montelius 1876: November 10–11). The couple seem to have been extremely efficient, as they also found time to go on excursions to Pompeii and Sorrento (Montelius 1876: December 4–12). The journal passages from Naples provide great examples of some of the core practices performed by the spouses, highlighting their joint productivity as well as Agda's autonomy during her solo activities at the museum. The drawings she produced would later serve as drafts for the illustrations in her husband's publications. They left Naples on December 12, and after a few short stops on the trip north, they left Italy two days later.

The next portion of the Letterstedtska journey took place in 1878 and 1879. In early March 1879, the couple arrived in Naples again, to visit or re-visit the national museum, and tour Pompeii, Herculaneum, Paestum, and other famous locations. The diary covering the Italian part of the 1879 research trip starts on March 1.[7] This time they spent their time on the South Italian mainland in the company with some friends and including for a time the young Crown Prince of Sweden, Oscar Gustaf Adolf (1858–1950). These arrangements meant that the stay involved mostly social or tourist-related sightseeing and related events, with less focused research visits to archaeological sites (Montelius 1879: March 1–9). In mid-March, they left their Swedish travel companions behind, and traveled by boat across the Strait of Messina to Sicily, where they visited most of the known archeological sites. The couple arrived in Taormina on March 14:

> Vägen till Taormina (ung. 1½ timme) var utmärkt vacker, följande hafvet så nära som möjligt under det djerfva klippformationer på andra sidan erbjöda ett ständigt vexlande pittoreskt panorama. Vid Taormina d.v.s. Giardini = stationen lemnade vi en del saker och åkte derpå en mycket vacker väg upp till Taormina som ligger högt upp på klipporna. Ännu högre ligger den lilla staden Mola, båda vigtiga orter på sin tid, men nu temligen öde och förfallna. Vi tog in på Hotell Bellevue och fingo rum med präktig utsigt öfver hafvet, strax derpå gingo vi ut till theatern, med grekisk underbyggnad, men vad som nu synes romerskt.
> (The road to Taormina (approx. 1½ hour) was excellently beautiful, closely following the sea under bold rock formations, on the other side offering an ever-changing picturesque panorama. At Taormina, i.e., Giardini = the station, we left some things and then walked the very beautiful road up to Taormina which is situated high up on the cliffs. Even higher up lies the small town of Mola, both important places in their time, but now rather deserted. We entered Hotel Bellevue and found a room with a magnificent view of the sea, shortly afterwards we went out to the theater, which has Greek substructure, but now seems Roman.) (Montelius 1879: March 14)

On most occasions the couple wasted no time, but immediately started to explore sites and museums after arriving at a new location. Agda wrote about visits to a few museums and collections in Sicily, but there is no clear evidence of which scholars Oscar and Agda might have interacted with. On March 28 Agda mentioned their visit to the museum in Catania, where they spent the whole morning, but only made drawings of a small number of artifacts (Montelius 1879: March 28). In most places they visited, Oscar went out to purchase photographs. In addition to pictures of buildings, art objects and artifacts, the couple also bought individual landscape and portrait images. They noted and reflected on the customs and living conditions of the locals. This is well illustrated by Agda's notes from their time on the neighboring island of Sardinia, where they arrived at Cagliari in the southern part of the island on the evening of March 30. During the first two days Agda was busy at their hotel mainly mending their clothes, while Oscar met with local researchers. The travel journal reveals that Oscar had planned to meet with the museum director Giovanni Spano (1803–1878). Unfortunately, Oscar learnt that he had died the previous year, and no new museum director had been appointed. Instead, Oscar contacted the university principal Paolo Lioy (1834–1911) (for more details about Lioy and his network, see Pizzato, Chap. 8, this volume) and Vincenzo Crespi, "en för dessa saker intresserad" (a person interested in such matters—i.e., archaeology). Oscar also met Professor Filippo Vivanet (1836–1914), who, according to Agda had an interesting collection of bronze figurines (Montelius 1879: March 30–April 3). Over the course of the next few days Oscar and Agda set out to discover the city. They were very surprised by the traditional costumes, which Agda wrote about in detail in her journal. Agda also described the lively crowds at the city markets, and how bare footed water carriers went to

[7] The previous diary ends on January 27, when the couple were still in Spain, and it has not been possible to find any notes from the last days of January or the month of February.

the town wells to fetch water. They both studied the collections at the museum in Cagliari, where Agda made some sketches of finds, and admired a cork model of a *nuragh*[8] (Montelius 1879: April 2). Oscar also met Efisio Timon (1833–1914), a man who had many artifacts of importance and is described in the diaries as a friendly and welcoming Sardinian (Montelius 1879: April 3).

During their stay in Sardinia, the couple went on an excursion inland. The excerpts below are included to illustrate the richness of detail contained in Agda's diaries, but also to provide examples of the travel conditions and of what the Montelius spouses experienced, in addition to searching for and meeting scholars and studying museum collection. From Cagliari, they made a four-hour train journey to the town of Oristano, located on the northeastern part of the island, which was as far as the railway stretched at the time. From there they arranged for a guide and a horse-drawn carriage to take them on to the village of Macomer in the interior of the island.

The aim of this rather tiresome journey inland was to visit archeological remains such as *nuraghes* and prehistoric graves that they must have learned about during their conversations with scholars and while visiting the museum in Cagliari. After eight hours of terrible weather conditions, they reached the village of Macomer on the evening of April 4. The next day they set out on a real adventure with horseback riding and hiking in extremely difficult terrain but managed to reach and study the archaeological sites they had wanted to see. On the way back to Macomer from the wilderness they had to ride in the dark. On several occasions, they rode past groups of people sitting around fires who had set up camps for the night. At one point they were followed by a lone rider who looked spooky in the moonlight, wearing a long black coat with the hood pulled down over his face. When they finally managed to get some distance from the mysterious figure, Oscar exclaimed "Så, nu äro vi ur skotthåll!" (So, now we are out of shooting distance!). Maybe their humble appearance and means of transport had kept potential attackers at bay. When they left Macomer the next morning, they met a foreign gentleman in a carriage who was escorted by two *carabinieri* (police officers) for safety and protection, and this must have made the couple a bit wary of their arrangements and the conditions on the island. The couple learnt that the roads were never safe, and certainly not at night (Montelius 1879: April 5–6). Agda's diary entries describe how the spouses were told many terrible stories and reports of robberies, murders, and ravages by bandits, but they doubted the information was reliable, dismissing it as exaggerated. Especially Oscar comes across as a bit gullible or naive in the diary passages recounted above. One might think that the couple should have taken far more precautions during their trips in the countryside. What is most striking, however, is their curiosity, perseverance and what they were prepared to endure to get to remote areas, to visit archaeological sites that were unknown to the international research community at the time. Oscar later used text passages written by Agda in an article published in the magazine *Ymer* in 1883, in which he described life in Sardinia (Montelius 1883b; Gustavsson 2020: 121). Southern Italy was still an unknown part of Europe for many northerners around 1880, including the scholarly community (Baudou 2012a: 210). After the excursion they stayed a few more days in Cagliari, during which Oscar and Agda revisited the museum and met the above-mentioned scholars on several occasions (Montelius 1879: April 6–9). On April 10, it was time to go back to the mainland.

The couple arrived in Rome three days later, where they stayed for two weeks. They then continued north, visiting Florence and a few other places along the way. When returning to Bologna around lunch time on April 30, Oscar immediately went to the museum in search of his colleagues Antonio Zannoni and Giovanni Gozzadini. He managed to find them later that day accompanied by the French scholars Ernest Chantre (1843–1924), a geologist and archaeologist, and Elysée Pelagaud (1848–1919), a geologist, who had been invited by Oscar by letter. Based on Agda's notes, it is plausible that Oscar made a greater effort in contacting the colleagues he wanted to meet in advance this time than in 1876 (Montelius 1879: April 30). The next day both Agda and Oscar went to the museum to count and draw (Fig. 2.3) artifacts from the San Francesco hoard. Among the 14,000 bronze items, were 3000 *fibulae*. After a few hours they were joined by Gozzadini, Chantre and Pelagaud, all of whom engaged in discussions about the finds. Unfortunately, Agda rarely documented explicit arguments or details from the scientific discussions she and Oscar were involved in. After working well into the evening, the couple had dinner with Chantre and Pelagaud. While Agda and Oscar were out, the Capellini spouses had been searching for them at their hotel, leaving them a note with an invitation to visit the Capellini home the next day (Montelius 1879: May 1). A few days later they moved on to the city of Reggio Emilia, where they both met up and worked together with the priest and scholar Gaetano Chierici (1819–1886), who had a large collection of archaeological objects (Montelius 1879: May 3; see Pizzato, Chap. 8, this volume, for more detailed biographical information on Chierici). Some days later, Agda and Oscar left Italy and started their journey back home with several stops along the way. The couple arrived in Stockholm on May 29.

[8]A *nuragh* is a circular building constructed of stone with a dome-shaped roof. These buildings have been dated to the period 900–700 BC, and they are exclusive to Sardinia. Some consist of only a tower or building, others of a whole cluster of joined circular 'rooms.' To date a total of about 7000 *nurgahi* have been documented on the island. Discussions about their exact function are still ongoing. They have been interpreted as fortifications, religious buildings, gathering places, settlements, or a combination.

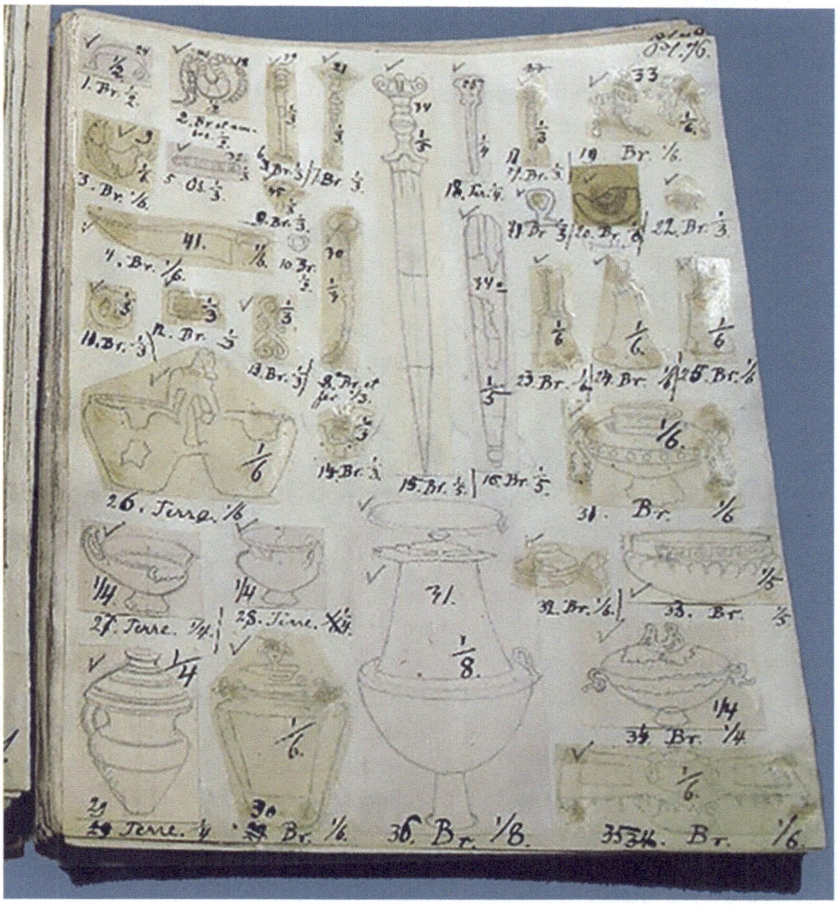

Fig. 2.3 Drawings made by Agda in Bologna in 1879 or 1881, which were to be incorporated in *La civilisation primitive en Italie depuis l'introduction des métaux*. (Photo: Anna Ch. Gustavsson; RAÄ, ATA, Oscar Montelius Archive, Folder F1a, vol. 108; see Montelius c. 1879/1881)

The third research journey to Italy took place in late spring and summer of 1881. The couple left Stockholm on May 5 and arrived in the north of Italy on May 17. They stayed in the country until mid-July (Montelius 1881). They worked their way southwards, stopping in Bologna, but then moved on rather rapidly into the Puglia region, where they stayed for two weeks. On June 11, they took the train to Rome and spent the rest of the month of June visiting locations in Latium and Tuscany. They worked at the museums in Rome and Oscar met with Pigorini and Helbig. On June 15, Agda and Oscar went to Orvieto to visit the ongoing excavations of Etruscan tombs on Helbig's recommendation (Montelius 1881: June 13–15). Oscar wanted to be as up to date as possible on recent finds to add as much new information as he could to his publication on prehistoric Italy. On the return to the north, there was (again) a focus on revisiting the collections in Bologna and the surrounding area, but this time the couple also spent time engaging in social encounters. In Agda's diary there are several references to dinners and meetings with the archaeologist Edoardo Brizio, both with and without his family, and with the Gozzadini family. Oscar also met up with Antonio Zannoni again (Montelius 1881: June 19–22). Even though they most likely met several times before 1881, this is the first time Edoardo Brizio and his family are mentioned more frequently in the diaries. Brizio had been appointed director of the archaeological collections at the museum in Bologna in 1878, which is probably why there are more descriptions of Oscar and Agda's interactions with him in 1881 than during the previous journeys. Judging from the preserved drafts in the Oscar Montelius Archive, it was after the 1881 research trip that his work on Italy and plans for the publication became more systematic. The first preserved draft pages of *La civilisation primitive en Italie depuis l'introduction des métaux* date to 1883 (Montelius 1883a).

2.6 Corresponding Careers

An interesting example of the results of the couple's joint networks and careers unrelated to their activities in Italy can be found in the correspondence between Agda and the Evans family.[9] Sir John Evans (1823–1908) was a geologist, prehistorian, and numismatist. He was married three times and became a widower twice. His son from his first marriage was Sir Arthur Evans (1851–1941), who would become world famous for his excavations at Knossos in Crete and Mycenae on the Greek mainland. Oscar corresponded with them both

[9] The contact between the families has previously been highlighted in Gustavsson 2018.

and was especially inspired by John Evans' ideas about the Bronze Age.

The Evans and Montelius families certainly met on various occasions from the 1870s onwards, initially at international congresses and other scientific meetings. In Agda's archive, we find letters from John Evans, his second wife Frances (1826–1890), and his daughter-in-law Margaret (1848–1893), who was married to Arthur Evans. A close reading of the letters reveals interesting information on a range of topics. In the first letter from Frances to Agda, Frances writes about the upcoming wedding of her daughter Alice, and how dearly missed she would be in the family home (Evans 1882). She would be missed especially by John, as he would lose her assistance in the organization and curation of his archaeological collections. It is therefore clear that Alice had been involved in research on her father's collections to such an extent that her moving out of the parents' home would significantly affect her father's work situation (see also Evans 1943: 143, 149, 161). The prospect of continuing to work with her father does not seem to have been a possibility. Frances did not primarily write to tell Agda about the marriage, but to inform the Montelius spouses that the Evans couple would arrive in Stockholm in a few weeks and that they were hoping to meet. Through this correspondence, the reader understands that Frances, like Agda, traveled together with her husband and that she probably had a good insight into his scholarly activities.

Both Frances and Margaret wrote letters to Agda in which they informed her of different charity and women's associations (e.g., Evans and Evans c. 1882–1890). They advised her to come to London and Oxford to participate in meetings held by these organizations to receive first-hand information and gain insight into their work.[10] Although this activity is not directly related to Oscar and Agda's archaeological research, this example is included to emphasize the wide spectrum of information and analytic possibilities represented by personal archives. What might look like secondary information in a letter can reveal, as shown in the Evans example, how women participated in their male family members' work activities while also providing insights into the interaction between individual family members' networks and engagement in different areas in the community.

2.7 Conclusions

In this chapter, I have examined Agda Montelius' role in her husband's research and networks to demonstrate how such a study can be methodologically conducted by drawing on archival sources that include personal documents. The method presented here requires a thorough and impartial survey of a variety of document categories and a close reading of selected material. The Montelius archives include an online index that lacks detail and is not digitalized. Visiting the archive and consulting the material hands-on is therefore necessary, not only to determine what has been preserved, but also to understand the constitution of the archives. In the case of Agda Montelius, it has been possible to go beyond the narrative of the so-called 'Great Man' interpretation of Oscar Montelius' impact on the field of archaeology by revealing her role as an amateur antiquarian/archaeologist in the 1870s and 1880s. This was accomplished by carefully considering the organization of the archives, by realizing the value of combining different types of source material, and by actively searching for secondary information in letters and diaries. It would certainly have been interesting if Agda had provided more details of the scholarly discussions she mentioned in her journal. A possible way forward to increase our understanding of the direct relationship between the couple's research activities and the impact of their networking efforts on Oscar's scholarly output might be to compare the publications and lectures Oscar produced during and after their travels. Brief references to certain people and places in the diaries are often the only clue to when and how Oscar acquired valuable information about current archeological discoveries and finds.

Perceptions of women's participation in public life had begun to change by the late nineteenth century, and women gradually became professionals in several fields (Sørensen 1998: 47). While we can only guess how Agda viewed herself, she never officially gained professional recognition in the archaeological field and her work was mostly unremunerated. However, she must have been respected enough to be allowed to fill in for her husband, as described here, engaging in activities that involved access to collections and being entrusted with various archaeological tasks. It might be worth attempting to search for references to Agda in the archives of the Italian scholars who were known to have interacted with the couple during their travels in Italy to get a sense of how she was viewed by Oscar's colleagues.

Agda was both unique and representative of her time. Many of the activities she participated in were common practice for her female peers. Nevertheless, her extensive traveling and her work alone or alongside Oscar, documenting finds in European museums, was still uncommon among her Scandinavian female peers, and was perceived as unorthodox at the time, at least in some countries (Baudou 2012a: 192). On many occasions, she was the only woman present when she and Oscar worked in different museums. There are few examples of Agda acting as an independent/professional scholar within the Montelius international/Italian network, but she did work autonomously when documenting artifacts, for example while working in the archaeological museum of Naples. Although she was never credited as an author/co-

[10] As mentioned previously, Agda wrote about people's living conditions among other topics in their travels in Italy, but apart from the descriptions of inspirational visits to England that followed, I have not found any evidence so far of her engagement in charity work, or specific comments on welfare issues in other countries in the travel journals.

producer in any of her husband's publications, she contributed to them in many ways. The drawing of artifacts is probably the best example of an indisputable professional archaeological practice directly associated with her.

Based on Bokholm's study and my research in the archives so far, it seems that Agda did not work with find drawings and archaeological documentation from sites and museums in Sweden except occasionally; she mainly engaged in this activity when the couple were abroad on their joint journeys. Agda had professional knowledge and was more of a co-producer than she was an assistant to her husband. Like Hilda Petrie (Drower 1985: 243, 248), Agda assisted her husband with rewriting and proof-reading throughout his career, but unlike her, Agda did not publish any archaeological texts or illustrations in her own name.[11] Another important aspect of Agda's participation in these research journeys is that she seems to have taken on the main responsibility for documenting their activities. This means that even if Agda did not participate in all of Oscar's meetings or networking occasions herself, her diaries reveal how her husband established and expanded his scholarly network. Even if we cannot assume that every personal meeting is mentioned in the travel journals, Agda's notes include a significant number of examples of how these contacts and interactions functioned, and they reveal the importance of meeting in person. With few exceptions, Agda did not correspond with male scholars when the couple were not traveling and when she did write it was not about archaeology. Unfortunately, in-depth comments and details from the discussions she and Oscar engaged in are usually missing from her notes, but they reveal that she was very active and involved in meetings and discussions, including socializing with scholars during their journeys.

It might be difficult to state to exactly what extent, and how, traveling in the company of his wife might have benefitted Oscar in establishing his professional relationships. Agda played an important role in social networking, meeting the wives and families of foreign scholars, especially in Bologna. A close reading of the archival material demonstrates how the involvement of wives and daughters in the activities of their more famous husbands and fathers can be revealed by such research. It is also obvious from the numerous passages in Agda's diaries that describe the many hours she worked alone in museums that her contribution to the documentation of finds gave Oscar extra time to seek out and engage in conversations with fellow scholars. That said, and after taking the social conventions of the elite and bourgeois groups of the time into consideration, it is safe to say that her presence was strongly beneficial to his career and reputation. As many of the wives and daughters of the male scholars and collectors at the time were involved in archaeological research and/or the curating of collections, and as cultural circles were a la mode, it is not far-fetched to assume that the women also were engaged in discussions about archaeology and other scientific and cultural topics. Although the women in the history of science and archeology are often invisible, this is not necessarily because of a lack of source material; rather it is due to the ways in which the material has been handled, including ignoring it entirely. Delving into the work of a 'Great Man' like Oscar Montelius, has made it possible to come to a deeper understanding of how the creation of networks and archaeological knowledge took place at a particular place and time, thereby highlighting the collective effort represented by all such projects. This means that it might be fruitful to embrace a slightly different, more inclusive approach when studying scholarly networks and the exchange and transmission of archaeological knowledge. Finally, historians of archaeology should recognize the potential of personal archives and the methodological possibilities of combining different and unexpected categories of sources. Time is a scarce commodity for researchers today, but the value of conducting basic research that includes going down the rabbit hole of digging in the archives cannot be underestimated.

[11] It should be noted that Agda published reports and articles related to her charity work and the associations she was engaged in (Bokholm 2000; Arwill-Nordbladh 2018).

References

Primary Sources

Archival Material

Evans, F. (1882, October 5). [Letter from Frances Evans to Agda Montelius]. The Montelius-Reuterskiöld Collection (Folder F2 C, vol. 9, Evans correspondence, letters to Agda Montelius), Antikvarisk-topografiska arkivet (ATA), Riksantikvarieämbetet (RAÄ), Stockholm, Sweden.

Evans, F., & Evans, M. (c. 1882–1890). [Letters from Frances Evans and Margaret Evans to Agda Montelius]. The Montelius-Reuterskiöld Collection (Folder F2 C, vol. 9, Evans correspondence, letters to Agda Montelius), Antikvarisk-topografiska arkivet (ATA), Riksantikvarieämbetet (RAÄ), Stockholm, Sweden.

Montelius, A. (1876). [Agda Montelius' diary from 1876]. The Montelius-Reuterskiöld Collection (Folder F2 B, vol. 1A, [1864–1879]), Antikvarisk-topografiska arkivet (ATA), Riksantikvarieämbetet (RAÄ), Stockholm, Sweden.

Montelius, A. (1879). [Agda Montelius' diary from 1879]. The Montelius-Reuterskiöld Collection (Folder F2 B, vol. 1A, [1864–1879]), Antikvarisk-topografiska arkivet (ATA), Riksantikvarieämbetet (RAÄ), Stockholm, Sweden.

Montelius, A. (c. 1879/1881). [Drawings made by Agda Montelius]. The Oscar Montelius archive (Folder F1a, vol. 108), Antikvarisk-topografiska arkivet (ATA), Riksantikvarieämbetet (RAÄ), Stockholm, Sweden.

Montelius, A. (1881). [Agda Montelius' diaries from 1881]. The Montelius-Reuterskiöld Collection (Folder F2 B, vol. 1B, [1880–1918]), Antikvarisk-topografiska arkivet (ATA), Riksantikvarieämbetet (RAÄ), Stockholm, Sweden.

Montelius, O. (1883a). [Manuscript/draft of *La civilisation primitive en Italie depuis l'introduction des métaux* by Oscar Montelius]. The Oscar Montelius Archive (Folder F1a, vol. 133), Antikvarisk-topografiska arkivet (ATA), Riksantikvarieämbetet (RAÄ), Stockholm, Sweden.

Secondary Sources

Allsebrook, M. (2002). *Born to rebel: The life of Harriet Boyd Hawes* (Repr. with corrections and a postscript). Oxford: Oxbow Books.

Arwill-Nordbladh, E. (1989). Oscar Montelius and the liberation of women: An example of archaeology, ideology and the early Swedish women's movement. In Th. B. Larsson, & H. Lundmark (Eds.), *Approaches to Swedish prehistory: A spectrum of problems and perspectives in contemporary research* (British Archaeological Reports, International Series, 500) (pp. 131–142). Oxford: British Archaeological Reports.

Arwill-Nordbladh, E. (2018). *Agda Georgina Dorothea Montelius. In Svenskt kvinnobiografiskt lexikon.* https://www.skbl.se/en/article/AgdaMontelius

Baudou, E. (2012a). *Oscar Montelius. Om tidens återkomst och kulturens vandringar.* Stockholm: Kungl. Vitterhets historie och antikvitets akademien i samverkan med Atlantis.

Baudou, E. (2012b). Forskarbiografi och vetenskapshistoria: Montelius och Hildebrands liv och verk. In B. Petersson, & P. Nicklasson (Eds.), *Att återupptäcka det glömda: Aktuell forskning om forntidens förflutna i Norden (Acta Archaeologica Lundensia, Series in 4°, 32)* (pp. 179–197). Lund: Institutionen för arkeologi och antikens historia, Lunds universitet.

Bokholm, S. (2000). *En kvinnoröst i manssamhället: Agda Montelius 1850–1920.* Stockholm: Stockholmia.

Claassen, C. (Ed.). (1994). *Women in archaeology.* Philadelphia: University of Pennsylvania Press.

Coltofean, L. (2014). Unveiling Zsófia Torma: The diary of a woman, an archaeologist and a visionary. In J. Marler (Ed.), *Fifty years of Tărtăria excavations. Festschrift in honor of Gheorghe Lazarovici* (pp. 258–273). Sebastopol, CA and Suceava: Institute of Archaeomythology.

Coltofean, L. (2017). Zsófia Torma: A pioneer of prehistoric archaeology in nineteenth-century Transylvania. In Cs. Szabó, V. Rusu-Bolindeț, G. T. Rustoiu, & M. Gligor (Eds.), *Adalbert Cserni and his contemporaries. The pioneers of archaeology in Alba Iulia and beyond* (pp. 327–354). Cluj-Napoca: Mega Publishing House.

Delsalle, P. (1998). *Une histoire de l'archivistique.* Sainte-Foy: Presses de l'Université du Québec.

Delsalle, P. (2018). *A history of archival practice* (M. Procter, Trans. & Rev.). Abingdon: Routledge. (Original work published 1998)

Derrida, J. (1996). *Archive fever: A Freudian impression.* Chicago: University of Chicago Press.

Díaz-Andreu, M., & Sørensen, M. L. S. (Eds.). (1998). *Excavating women: A history of women in European archaeology.* London and New York: Routledge.

Douglas, J., & Mills, A. (2018). From the sidelines to the center: Reconsidering the potential of the personal in archives. *Archival Science, 18*(3), 257–277. https://doi.org/10.1007/s10502-018-9295-6

Drower, M. S. (1985). *Flinders Petrie: A life in archaeology.* London: Victor Gollancz.

Evans, J. (1943). *Time and chance: The story of Arthur Evans and his forebears.* London: Longmans, Green & Co.

Gustavsson, A. (2018). "Sveriges mest beresta kvinna". Glimtar ur Agda Montelius liv. In M. Sjöberg (Ed.), *Allvarligt talat. Berättelser om livet* (pp. 218–229). Göteborg: Makadam.

Gustavsson, A. (2020). Geographies of networks and knowledge production: The case of Oscar Montelius and Italy. In J. Roberts, K. Sheppard, H. R. Hansson, & J. R. Trigg (Eds.), *Communities and knowledge production in archaeology* (pp. 109–127). Manchester: Manchester University Press.

Johnsen, B., & Welinder, S. (2002). Johanna Mestorf. The link between Swedish and continental archaeology in the Golden Age. In J. K. Koch, & A. Mertens (Eds.), *Eine Dame zwischen 500 Herren: Johanna Mestorf—Werk und Wirkung* (pp. 71–101). Münster, New York, München and Berlin: Waxmann.

Lagerberg, M. (1872). *Bref från söderns länder: Dagboksanteckningar från en resa till Italien för deltagande i arkäologiska kongressen i Bologna 1871.* Göteborg: N. Zetterström.

Livingstone, D. N. (2003). *Putting science in its place: Geographies of scientific knowledge.* Chicago: University of Chicago Press.

Livingstone, D. N., & Withers, C. W. J. (Eds.). (2011). *Geographies of nineteenth-century science.* Chicago: University of Chicago Press.

Montelius, O. (1869a). *Från jernåldern.* Stockholm: Häggström.

Montelius, O. (1869b). *Remains of the Iron Age in Scandinavia.* Stockholm: Häggström.

Montelius, O. (1880–1882). *Spännen från bronsåldern och ur dem närmast utvecklade former: Typologisk studie.* Stockholm: Kongl. Vitterhets-, historie- och antiqvitetsakademien.

Montelius, O. (1883b). Några minnen från Sardinien. *Ymer, 1883*, 15–39.

Montelius, O. (1885). *Om tidsbestämning inom bronsåldern med särskildt afseende på Skandinavien.* Stockholm: Kongl. Vitterhets-, historie- och antiqvitetsakademien.

Montelius, O. (1986). *Dating in the Bronze Age: With special reference to Scandinavia* (H. Clarke, Trans.). Stockholm: Kungl. Vitterhets-, historie- och antikvitetsakademien. (Original work published 1885)

Montelius, O. (1895–1910). *La civilisation primitive en Italie depuis l'introduction des métaux* (Vols. 1–5). Stockholm: Imprimerie Royal.

Montelius, O. (1912). *Die vorklassische Chronologie Italiens.* Stockholm: Häggström.

Montelius, O. (1924–1928). *La Grèce préclassique (Vols. 1–2).* Stockholm: Kungl. Vitterhets-, historie- och antikvitets akademien, Häggström.

Nicklasson, P. (2021). Ida Nilsson vid arkeologkongressen i Köpenhamn 1869. *Fornvännen. Journal of Swedish Antiquarian Research, 116*(2), 129–144.

Nicotra, L. (2004). *Archeologia al femminile: Il cammino delle donne nella disciplina archeologica attraverso le figure di otto archeologhe classiche vissute dalla metà dell'Ottocento ad oggi (Studia Archaeologica, 129).* Roma: "L'Erma" di Bretschneider.

Nordström, P. (2014). *Arkeologin & livet: Ett dubbelporträtt av paret Agda och Oscar Montelius genom deras brevväxling 1870–1907.* Stockholm: Atlantis.

Roberts, J., & Sheppard, K. (2020). Introduction: Clusters of knowledge. In J. Roberts, K. Sheppard, U. R. Hansson, & J. R. Trigg, (Eds.), *Communities and knowledge production in archaeology* (pp. 1–13). Manchester: Manchester University Press.

Schlanger, N., & Nordbladh, J. (Eds.). (2008). *Archives, ancestors, practices: Archaeology in the light of its history.* New York and Oxford: Berghahn Books.

Smith, B. G. (1998). *The gender of history: Men, women, and historical practice.* Cambridge, MA: Harvard University Press.

Sørensen, M. L. S. (1998). Rescue and recovery: On historiographies of female archaeologists. In M. Díaz-Andreu, & M. L. S. Sørensen (Eds.), *Excavating women: A history of women in European archaeology* (pp. 31–60). London and New York: Routledge.

Stoler, A. L. (2009). *Along the archival grain: Epistemic anxieties and colonial common sense.* Princeton and Oxford: Princeton University Press.

Caroline Ransom Williams and the Foundations of Egyptian Archaeology in the United States

3

Kathleen L. Sheppard

Abstract

In the early days of professionalizing archaeology, women were able to participate in the discipline more easily than they could in a number of other scientific practices. Caroline Ransom was the first university-trained woman practicing as an Egyptologist in the United States in 1905. Contrary to what her male counterparts did, Ransom's most influential work was done on the administrative side of the discipline, and not on the field site. It was through this gendered domestic academic work that she established her wide ranging and influential professional networks. The focus of this chapter is Ransom's career as an assistant curator at the Metropolitan Museum of Art in New York, where she had constant contact with patrons and archaeologists from all over the world. Ransom, and others like her, organized and maintained the foundational universities, museums, societies, and collections forming the center of the work. Using Ransom's career as a case study, this chapter argues that we must study women in the early days of the discipline who were doing institution-building work. They were central to the narrative of Egyptology as network nodes, and they must be seen as such in order to understand the reality of their undeniable disciplinary impact.

Keywords

Caroline Ransom Williams · Metropolitan museum of art · James Breasted · Albert Lythgoe · Curatorial work · Women in archaeology · Egyptology

3.1 Introduction

Her work at Bryn Mawr, a liberal arts women's college in Pennsylvania, had become a burden. Within five years of earning the first PhD granted to a woman in the field of Art History and Egyptology from the University of Chicago in 1905, Caroline Ransom (1872–1952) was a professor and chair of the Department of the History of Art and Classical Archaeology at Bryn Mawr (Univ. Chicago Registrar n.d.; Ransom 1905). She had a heavy teaching load and too many administrative duties. She felt pulled in all directions and could not focus on the research she knew she could excel at if given the time. When, in the spring of 1910, it was time to renew her contract at the women's college, they offered Ransom $2500 per year and a slightly reduced workload.[1] She considered the offer. She liked Bryn Mawr: her students showed promise and the pay was good. Despite her prospects within the department, she wanted to see if it was possible for her to do more research and less teaching and administrative work. To that end, in March of that year she wrote to Albert Lythgoe (1868–1934), the new curator of the newly-formed Egyptian Art department at the Metropolitan Museum of Art (MMA) in New York to enquire about a potential position. She told him the details of her new contract, especially the pay, and continued:

> At the same time the conditions for conducting archaeological work here are so difficult and my desire really to specialize so great, that I do not wish to sign [the new contract] without making sure that under the new arrangements in your department of the museum you have no use for me. (Ransom Williams 1909–1917, Ransom to Lythgoe, March 30, 1910)

[1] $2500 in 1910 is equivalent to around $83,000 today.

K. L. Sheppard (✉)
Department of History and Political Science, Missouri University of Science and Technology, Rolla, MO, USA
e-mail: sheppardka@mst.edu

L. Coltofean et al. (eds.), *Connecting People and Ideas*, Themes in Contemporary Archaeology,
https://doi.org/10.1007/978-3-031-81006-0_3

Lythgoe wrote back to her immediately, letting her know he definitely had work for her and would be grateful to have her expertise (Ransom Williams 1909–1917, Lythgoe to Ransom, April 5, 1910). By April 17, Ransom had officially accepted a position as an assistant curator in the Egyptian Art department, a position she held for the next six years (Ransom Williams 1909–1917, Ransom to Lythgoe, April 17, 1910).

Ransom clearly hoped to have more time for research, as she wrote to her mentor James Breasted (1865–1935): "I shall have the opportunity to devote all my strength (…) for me to do good work in the Egyptian field. (…) I shall have a position of great responsibility and opportunity in charge of the collections here" (Breasted 1895–1935, Ransom to Breasted, May 12, 1910). Her new position as assistant curator of the largest Egyptian collection in the United States dramatically shifted her career trajectory and what she could accomplish in the discipline. It also made her arguably one of the most important scholars in the discipline at the time and the central node in a world-wide network of Egyptologists. For the better part of a decade, most scholars and donors working in Egyptology such as Breasted, Lythgoe, Herbert Winlock (1884–1950), Theodore Davis (1838–1915), Howard Carter (1874–1939) and more would be dependent on her for acquisitioning and organizing artifacts as they arrived at the MMA. Ransom was the one who could grant or deny access to or information about those pieces. Her own original research on artifacts in the collections became the first, and sometimes only, analyses of these artifacts. Even after she left the MMA in 1916, due to budget cuts caused by impending war for the United States, Ransom still worked for them as a consultant. She continued her research into the collections, and she continued her role as the center of a network extending from the resources at the MMA to scholars the world over.

While Ransom was able to do more original research on artifacts in the collections and write much more as a curator, she also, as a matter of course, took on a lot of administrative duties. Lythgoe was officially the curator of the department, but the fact that he was in the field in Egypt for half the year meant that Ransom took on many of his daily responsibilities, including personnel issues, acquisitions, conservation, and most of all, correspondence. During Ransom's tenure at the MMA, she built their collection, and with it the discipline of museum Egyptology, almost from the ground up. We do not know a lot about her life or career relative to her male colleagues because she was an assistant curator who left the MMA after only six years, without any official promotion. Further, she was a woman who had domestic concerns and could not readily go into the field or establish herself in that privileged male space, so she did not leave behind the site reports that have been crucial for disciplinary memory.

This chapter will analyze the life and career of Caroline Ransom Williams through multiple lenses and is informed by many different theoretical foundations. The methodology and theoretical underpinnings of the history of science, as William Carruthers and Stéphane Van Damme argue, "can offer an oblique and novel perspective on the ways in which archeological objects and the archaeological past have been formed. (…) Thinking through the history of archeology with the history of science enables us to draw new conclusions" (Carruthers and Van Damme 2017: 271). For example, in the history of archaeology, there has been work done on analyzing women in the field, mostly within biographies or autobiographies (e.g., Murray 1963; Caton-Thompson 1983; Díaz-Andreu and Sørensen 1998; Cohen and Joukowsky 2004; Carr 2012; Sheppard 2013). These pieces of scholarship tend to focus on the exciting field work that women did during their careers, because in the history of archaeology the field has been the privileged space of scientific practice (Wallach 1996; Kohler 2002; Livingstone 2003; Howell 2010; Livingstone and Withers 2011). However, archaeologist and art historian Margaret Cool Root notes that women like Ransom, who were not able to go into the field for one reason or another and whose careers "emerge[d] out of the depths of Victorian ideologies about the subordinate position of women (…) must ultimately be understood on her own terms as an intellect and as a human being" (Root 2004: 1). In order to do this, scholars have to contend with the evidence available. The best evidence we have for Ransom's life and career activities are two relatively small collections of her correspondence, although there are scattered letters in other institutions.[2] The main collection is located at the Institute for the Study of Ancient Cultures (ISAC) at the University of Chicago and dispersed within the correspondence collection of the main figure there: James Breasted. The other main collection of letters is within the MMA's Egyptian Art Department's archival collections. This chapter uses these two primary collections to show that Ransom—though today not remembered as a major figure in the story of early Egyptology—was, for a time, the central scholar in a vast network of Egyptologists. Grounding my analysis of her work within the history and philosophy of science, including actor-network theory and the importance of cognitive landscapes in understanding knowledge networks, I centralize the seemingly peripheral activities and work of Ransom in her position with the MMA, and by extension women like her, and highlight them as foundational roles in the discipline (Rudwick 1985; Latour 2005; see also Sheppard 2020).

[2]These include the University of Chicago's Institute for the Study of Ancient Cultures, the Egyptian Art Department at the MMA, the New York Historical Society, the Toledo Museum of Art, and possibly Bryn Mawr University and Mount Holyoke College. Both Bryn Mawr's and Mount Holyoke's archival collections have mentioned Ransom in the past, but their current holdings show no letters there. The Epigraphic Survey at ISAC also contains some photographs and other documentation about her time in the field in 1926–1927.

Ransom's career further demonstrates that we must shift the focus of the discipline's history from the field to the institution. She left no academic descendants. She did not establish a theoretical or practical school. She did not uncover a famous site or discover a golden artifact. Ransom's work was foundational because of the network she built at the MMA. Inheriting Lythgoe's position, *de facto*, and his correspondence, she established both herself and the MMA as centers of Egyptology in the early twentieth century; the MMA remains one. Ransom did the necessary domestic work in the museum that men like Lythgoe simply did not do (see Dyhouse 1995; Shteir 1996; Sheppard 2014). She accessioned objects, corresponded with scholars, and wrote the scholarship necessary to lay the foundation for men like Lythgoe to stand on. Over 20 years ago, Root posed the question: "Are there great women archaeologists?" (Root 2004: 8–12). She went on to answer that yes, there are, but we must look for them in the right places, which is to say, not necessarily in the field (Root 2004: 12–15). Root stated that we have to look for them "first in artifactual or detail analysis and then in the synthetic studies of iconography, stylistic valence, cultural transmission, social history and economics emerging out of these systematic and close readings" and, when we find them, we see their work is "truly pathbreaking" (Root 2004: 12). Plenty of scholarship has been done on Egyptology in museums (Moser 2006; Colla 2007; Reid 2015) but very little of it mentions women. Scholars like Rosalind Janssen (1992), Alice Stevenson (2019), and Amara Thornton (2018) position women as central to Egyptological collections in Britain, and it is within these contexts that I discuss Ransom in the United States.

The domestic museum that Ransom devoted her career to made her, not Lythgoe, the center of the Egyptological knowledge network based at the MMA. Despite her publications, her correspondents, and her clear success as a curator, we usually look to Lythgoe as a principal early figure in Egyptology in the United States, and not to Ransom, because of the excavation work he did. Women's work in Egyptology should be centralized. We know they did this germinal work, but who did it, where they did it, and why they did it, matters to the history of archaeology in general and Egyptology specifically.

3.2 Early Life and Education

Unlike many early Egyptologists and archaeologists, Caroline Louise Ransom was not wealthy, nor did she have connections to wealthy donors, so she did not have traditional access to the discipline. She came from a comfortably middle-class family from Toledo, Ohio (Sheppard 2018: 7–8). She received a consistent college preparatory education growing up, with the support of her parents, and went on to university at the age of 18. In a city such as Toledo in this period, it would be common for the daughters of prominent community members, like her parents, to be educated. Her mother and her mother's sister also had extensive educations, so Ransom was likely expected to follow suit. During her time as a student at the prestigious Mount Holyoke College in Massachusetts, she went on a two-year European Grand Tour with her aunt, Louise Fitz Randolph (1851–1932), who taught art history there (Higley 1937: 131; Sheppard 2018: 21). The tour included Cairo and a trip up the Nile (Coletta 2015). Like so many before her, after just a few weeks there, Ransom had fallen in love with the people, the place, and its history.

After earning her Bachelor's degree in 1896, Ransom wrote a letter to Egyptologist James Henry Breasted at the University of Chicago. In it, she outlined her education and interest in ancient art, and asked for his support in her application for a fellowship in either the Egyptology department or Classical Archaeology department for her to continue her studies. Although the letters from Breasted do not survive, we know from Ransom's letters that she arrived in Chicago with a fellowship to study with both Breasted and Classical archaeologist Frank Tarbell (1853–1920), in October of 1898. Because she was working with them on a daily basis, there are few letters from Ransom during the time she was a student. She earned her Master's degree in 1900 under Breasted and Tarbell (Sheppard 2018: 9). To continue her studies, Breasted then sent her, as he would any talented male student, to the University of Berlin to study with renowned linguist Adolf Erman (1854–1937) (Sheppard 2018: 9). She flourished in Berlin and returned to Chicago in 1903 to finish her PhD in History of Art and Egyptology by 1905. Her dissertation *Studies in Ancient Furniture* became the basis for her first book, *Studies in Ancient Furniture: Couches and Beds of the Greeks, Etruscans and Romans* (1905) and, once it was complete, her career could begin. It may have been around this time that her erstwhile suitor, Grant Williams (1865–1942), a real estate investor from her hometown of Toledo and 7 years her senior, began pursuing her. She decided to pursue a career instead, but Williams was willing to wait (Sheppard 2018: 11–12).

Like many prospective male archaeologists did at the time, Ransom connected herself to an established male figure from a well-known institution in order to establish her bona fides and a strong web of professional connections, which began with Breasted and Erman. However, unlike many men in the early days of Egyptology, Ransom had earned a PhD before she entered academia as one of the few women in the discipline at the time. She immediately took a position at Bryn Mawr and trained a cadre of students (Fig. 3.1). One of her star students, Mary Hamilton Swindler (1884–1967), later became a major figure in Classical Archaeology. Swindler was at Bryn Mawr for decades and

Fig. 3.1 Caroline Ransom, 1908. (Reproduced from Bryn Mawr College 1908, public domain)

trained dozens of her own students who then dispersed and built Classical Archaeology as a discipline in the United States (Thompson 1950; Mellink 1967; Leach 2021). In fact, in her magnum opus, Swindler (1929: v) included a dedication to "Caroline Ransom Williams in appreciation of her qualities as scholar, teacher, and friend." It is difficult to say what would have happened had Ransom stayed at Bryn Mawr and continued to train women students. It is probable that she could have, and would have, accomplished as much there as Swindler did. Instead, Ransom and her contributions to the field have been relegated to the margins of the history of Egyptology so much so that she is rarely even a footnote, although that situation is changing.

While at Bryn Mawr, Ransom added a number of new courses to the curriculum, including Egyptian art and archaeology (Navarro 2021). She also added to their teaching tools by acquiring over 800 lantern slides for the collection of slides of Egyptian sites and objects (Bryn Mawr College Egyptology Association 2019; Navarro 2021). At the same time, she was caring for her aging mother who had been living with her full-time. She missed researching, writing, and discovering, and needed a stable, less time-intensive position. In hopes of filling these needs, she arrived in New York in the fall of 1910 as one of the assistant curators of the Egyptian Art department. She negotiated a salary, likely lower than a man would have taken but that would still allow her to live in Manhattan, and the time for research. It is also here that she began the work that would make her the center of a network of Egyptologists and the world of Egyptology at a pivotal time for the discipline.

3.3 Central Park as a Central Node: Work at the Metropolitan Museum of Art

Ransom's new position as an assistant curator in the Egyptian Art department provided her with time to research more than at Bryn Mawr, at the start. Her museum workday began at around ten o'clock in the morning and lasted until five or six, so from eight to ten she would research on her own. Due to the rhythms of excavation and museum work at the time, Ransom was also given four to six weeks of leave in the summer for her research interests and language training as well as weekly time to take care of her mother (Breasted 1895–1935, Ransom to Breasted, October 16, 1910). Once the Egyptian department had been set up for display and officially opened in November 1911 with Ransom's accompanying handbook as a guide, Lythgoe left for the field in early 1912 (Metropolitan Museum of Art 1911; Breasted 1895–1935, Ransom to Breasted, November 19, 1911). He and his crew, including the other assistant curators at the MMA, Arthur Mace (1874–1928) and Herbert Winlock, were often in Egypt and Europe for more than half the year, from December to July, excavating sites, dispersing finds, and meeting colleagues. Ransom was tasked with administrative duties in his absence. Ransom effectively took on the curatorial role for most of the year when personnel decisions needed to be made, letters written, and exhibits organized. In addition, Lythgoe continued to send hundreds of boxes of artifacts from Egypt back to her at the MMA throughout each season. These finds, along with periodic donations and loans, formed the bedrock of the museum's Egyptian collections ("The Theodore M. Davis Bequest," 1931; Adams 2013: 313–14). In this, she was not unlike several other administrative women who stayed behind in the domestic museum spaces while their male counterparts were out in the field, like Margaret Murray (1863–1963) did for Egyptologist Flinders Petrie (1853–1942) at University College, London (Sheppard 2014). Importantly, once Lythgoe left for the winter, everyone who needed to work with him and the growing collections would have known and needed to work with Ransom, who was, by default, the central disciplinary gatekeeper for one of the nation's preeminent Egyptian collections.

As Ransom grew in her professional stature and expertise, she was often busy with publications, such as short articles and handbooks guiding visitors through the collections (Ransom 1911, 1912, 1913b, c, d, e, 1914a, b, 1915a, b, c). Many of these were in-depth analyses on the objects in the collections, such as the stela of a Middle Kingdom official, Mentuwoser, her first publication of a hieroglyphic translation of and commentary on an object (Ransom 1913a).[3] The stela of Mentuwoser is from the Twelfth Dynasty

[3]The stela is on display at the MMA in Gallery 110 (http://www.metmuseum.org/collection/the-collection-online/search/544320, accessed May 30, 2018). Today, the name is transliterated Mentuwoser, but Ransom's book is titled with the old transliteration.

(c. 1944 BCE), found at Abydos. A finely carved and painted rectangular stone, it shows Mentuwoser at his funeral banquet. The MMA was given the piece in 1912 by the wealthy philanthropist Edward S. Harkness (1874–1940) as a gift after he bought it from New York-based antiquities dealer Dikran G. Kelekian (1867–1951) (Ransom 1913a: 5). Other pieces came in that year, such as the masterfully carved New Kingdom stela of Senu, a New Kingdom royal scribe, but Ransom chose Mentuwoser out of all of them to research, translate, and describe in fine detail. She informed Breasted that "the publication is intended primarily to make the stela known to Egyptologists, [but] I have had to consider the fact that it is to be sold at the Museum entrance" and so she wrote for the public, as well (Breasted 1895–1935, Ransom to Breasted, May 2, 1913). While it is difficult to ascertain the sales figures of these kinds of small books, her research reached countless visitors to the museum and helped to bring a larger audience to the growing collections and establishing the MMA itself as a hub of Egyptology research and preservation in the eyes of public visitors.

After the Mentuwoser stela was published, one of Ransom's biggest responsibilities was the acquisition, organization, unpacking, and installation within the museum of an entire tomb from Saqqara, Egypt in 1913, known as the mastaba tomb of Perneb (Lythgoe and Ransom 1916). Perneb was a high official in the reign of Djedkare Isesi or Unas (at the end of the Fifth Dynasty, c. 2650 BCE). The tomb was plundered in antiquity, but then over the course of the millennia, as stones were taken from other mastaba tombs and repurposed for structures in the nearby villages (even as late as the 1840s and 1850s), Perneb's tomb remained undisturbed until the early twentieth century (Lythgoe and Ransom 1916: 24–27).

The tomb was inadvertently discovered in 1907 by British Egyptologist James Quibell (1867–1935) and his diggers while attempting to retrieve painted walls from a different tomb that the MMA, through Harkness's generosity, had already purchased. In 1913, Harkness also bought Perneb's tomb. In total, Perneb's tomb comprised 601 boxes of blocks which took two months to pack, ship, and move to New York. Ransom accessioned all of the boxes of blocks from the tomb upon their arrival. Then she had a brand-new problem on her hands. Egyptian limestone contains certain salts that are dormant in the dry desert air but cause major issues for the limestone and painted surfaces in the relatively humid New York climate. There were several options, the most effective being the method of soaking each stone in water until the salts had, essentially, dissolved. However, they could not do this without harming the painted surfaces. So instead, they sealed the stone, thereby locking out any other substances that could get in; they had to do this for every single block. The process was time consuming (Ransom Williams 1909–1917, Ransom to Lythgoe, May 7, 1914). During this process, the museum gallery on the northern end of the Fifth Avenue hall was prepped for installation, which began in August 1914 and was finished by early 1916 (Lythgoe and Ransom 1916: 40–42). After the preservation, and a long and arduous installation, the tomb was finally opened to the public on February 3, 1916 (Lythgoe and Ransom 1916: 38–45).

From the time it opened, the tomb became the focus of the Egyptian collection and remains one of the most popular exhibits to visit at the museum. The *New York Times* reported on January 30: "Perneb's Gay Tomb Ready for Visitors" (1916). Visitors clamored for the first, and at that time the only, display of its kind outside of Egypt. Ransom wrote to Breasted about the opening day:

> The opening of Perneb's tomb took place with great éclat and the interest of the public has not yet subsided. (…) People were formed in line two abreast all the way back to the Fifth Avenue entrance to get into the chambers. Glass positions electrically lighted illustrate the former position and the taking down of the tomb. There are two cases of the objects found in the course of the excavations including the greater part of Perneb's skull. A model of the entire tomb makes clear the position of the burial chamber. (Breasted 1895–1935, Ransom to Breasted, February 7, 1916)

Ransom spearheaded this initiative and produced, for those visiting the museum, a guidebook for the exhibit (Lythgoe and Ransom 1916). It was published around the same time as the tomb opened, and she told Breasted:

> It is only a popular thing, but in a way I think such pieces of writing are a more responsible undertaking than a scientific work. If you make mistakes in writing for your fellow Egyptologists, they soon set you straight, but the public is so trusting and rarely goes into the more technical books! (Breasted 1895–1935, Ransom to Breasted, January 18, 1916)

Here it is clear that Ransom used her scholarly influence to pioneer the MMA's methodology in displaying a fully articulated tomb. At the time, all other collections looked to them as an example. The public handbook was in fact so well-written and popular with visitors that Breasted used it in Chicago to persuade the Field Museum to install their own mastaba chambers (Breasted 1895–1935, Breasted to Ransom, February 10, 1916).

Lythgoe wrote the first chapter in the guidebook, detailing the discovery, purchase, packing, and shipping of the tomb. Ransom opened her chapter of the book by informing readers and viewers of Perneb's tomb that he had a family who would visit his mastaba (Fig. 3.2). Upon approaching the tomb, Ransom invited modern visitors to step into the place of Perneb's family. Both would have found "a startling reminder of the appearance of the departed nobleman as they had often seen him in life, issuing, staff in hand, from the door of his house" (Lythgoe and Ransom 1916: 49). The features that were portrayed on his likeness "while not a detailed portrait, were strongly reminiscent of the proud and noble countenance of the deceased grandee" (Lythgoe and Ransom

Fig. 3.2 Perneb on the left of the doorway to his tomb. (Reproduced from Lythgoe and Ransom 1916: 51)

1916: 49). But, as Ransom continued, visitors could be sure they were seeing "typical" Egyptian figures from which scholars could learn a lot about the Fifth Dynasty in which he lived. She described in detail what the hieroglyphs said about Perneb and his family, but that the scenes were incomplete because the tomb itself was never fully finished. As the leading scholar of Perneb's tomb, Ransom had closely examined every surface with a magnifying glass in order to study the various steps the original artists followed to get the almost-finished, colorful tomb in order. Her fascination was with the use of the Egyptian blue throughout the tomb, demonstrating, as she argued that "[f]ar from being crude and all too simple, as a superficial acquaintance might suggest, these painted reliefs reveal vividly conceived scenes, a facile technic, and a sophisticated color scheme," which demonstrated a long development of religious artwork and superior talent (Lythgoe and Ransom 1916: 79).

Ransom was an art historian and museum curator whose expertise was decorative elements and language, so the opportunity to design and install an exhibit like Perneb's tomb was a fitting job for her. Further, because she was the main curator-in-charge of the conservation, organization, and installation of the tomb, she had the most in-depth knowledge of these elements. In archaeology, this is not the exciting fieldwork that Lythgoe, Mace, and Winlock did; however, it is the necessary academic work that needs to be done to further the historical narrative. In addition, at this time in her life, it was the only work she could have done. She was a woman working in Egyptology and had her home life with her mother to consider, so her prospects for going to the excavation site—especially from the United States—were not good most years. Her written work was useful to the MMA, and patrons liked it, so Lythgoe asked Ransom to write two more public guidebooks for their other sets of mastaba reliefs (e.g., Ransom 1913c). In a second book about the tomb, *The Decoration of the Tomb of Per-Neb*, finally published in 1932, Ransom analyzed the techniques and color conventions of the wall paintings in the tomb, and thereby of the Old Kingdom; it became a useful reference work and is still used today (Ransom Williams 1932).[4] These books, and the public reception of them, made both the MMA and Ransom foci of the public audience that wandered through the displays and as such, she had considerable influence on what the public saw and what they thought about the objects. The ways in which she engaged and influenced the public's interactions with the museum collections made Ransom a central figure in the network of practitioners in major institutions, of which there were very few.

While she was busy with research and publication, Ransom was the main correspondent between Lythgoe and the activities at the Museum. In one of several long letters to Lythgoe, who was in Thebes (Luxor) at the time, she detailed several things that had happened during his long absence in the field. She opened the letter, however, by telling him that she had "so much to relate that I think I will confine myself to those things which you ought to perhaps hear before you return" (Ransom Williams 1909–1917, Ransom to Lythgoe, April 7, 1913). She had been working under the supervision of Edward Robinson (1858–1931), the director of the Museum, while Lythgoe was away and it seemed that Lythgoe wrote more letters to Robinson than he did to Ransom, likely so he did not have to repeat information to either of them. In return, she reported on personnel news such as vacation plans and who was writing which articles about which objects. Ransom supervised Bernice Cartland (b. 1887) and a Miss McCann who were both busy writing

[4]After Ransom married Grant Williams in mid-1916, she changed her name to Caroline Ransom Williams. I will use her published name in citations and the name she herself used at any given time in the text.

exhibition reports (Ransom Williams 1909–1917, Ransom to Lythgoe, March 28, 1914). Ransom also organized and opened new exhibits, such as the month-long show she had arranged in the Recent Accessions Room that displayed Egyptian furniture and musical instruments. She wrote the accompanying article for this exhibit as well (Ransom 1913b). She also had to deal with personnel problems, as Mr. Kent, the photographer at the Museum, changed the process by which pieces could be photographed. The new process took longer, the photos were smaller, and the details were not visible. To circumvent this, Ransom asked Lythgoe to have everything photographed in the field, as it might be cheaper and easier.

Ransom had inherited a network of correspondents from Lythgoe that she quickly made her own. She was writing letters to dozens of Egyptologists and archaeologists needing information and collaboration with the MMA, including Howard Carter, Flinders Petrie, John Pierpont Morgan (1837–1913), and more. People who had private collections that needed to be assessed for importance and monetary value came to her for valuations. One of them was Theodore Davis, an American millionaire whose excavations with his partner Emma Andrews (1837–1922) in the Valley of the Kings from 1902 to 1913 created massive personal collections. Ransom met with Davis (and likely Andrews) and went to their home in Rhode Island a few times to see and secure a number of pieces, ensuring a good relationship between Davis and the museum (Breasted 1895–1935, Ransom to Breasted, October 16, 1910). Upon Davis's death in 1915, his will bequeathed much of his collection to the MMA that was not already there; after over a decade of legal battles, the MMA officially claimed the objects as their own in 1931. Andrews had also left collections and money to the Museum in her will (Adams 2013: 313–15). Edward Harkness, the wealthy philanthropist whose money had funded the tomb of Perneb at the MMA, consulted with Ransom about possible funding of more excavations and museum building projects (Ransom Williams 1909–1917, Ransom to Lythgoe, March 28, 1914; Ransom to Lythgoe, April 11, 1914), but she and Robinson could not agree on what the building needed or how to agree on the use of space. Harkness also asked Ransom about a few pieces that he wanted to purchase and loan to the museum, and she reported to Lythgoe that the purchases were successful (Ransom Williams 1909–1917, Ransom to Lythgoe, March 28, 1914). Ransom was also a close friend of Mary Emma Harkness (1874–1950), Edward's wife, so she enjoyed some connections that many curators might not (Breasted 1895–1935, Ransom to Breasted, November 11, 1911), but it is unclear if she used that friendship for garnering more donations.

The letters between Ransom and Lythgoe in the early days of the Egyptian Art department clearly demonstrate the important discipline- and institution-building work that women like Ransom did in this period. She held the department together while the head curator was gone. More than that, she actively built up the collections within the department on her own, reporting to Lythgoe in a few long missives to the field about the tasks she had completed. Ransom, and other professional women in the discipline at the time, were needed "at home," which meant in the home institutions, so they rarely made it to the field (e.g., Sheppard 2014; Hill 2016).[5] In this way, in a brief six years at the helm, she had made the MMA an Egyptological center of work. In that time, she had also made herself a hub in a professional Egyptological network and she was able to take that influence with her wherever she went.

3.4 Love, Marriage, and Toledo

By 1916, the Great War was raging in Europe and, even though it had yet to touch the United States, the MMA was feeling the tightening purse strings of its donors. Budgets were cut, excavation seasons were cancelled (due also to safety), and Ransom ultimately left the museum as a full-time employee. Possibly an outcome of this change in her position, or perhaps the cause of it, Ransom finally agreed to marry Grant Williams, and the couple were married on June 28 of that year (Sheppard 2018: 11). She was 44 years old when she changed her name to Caroline Ransom Williams and people began to write to her not as Dr. Ransom, but as Mrs. Williams. She and her mother moved to Toledo with Williams, and Ransom Williams' domestic concerns doubled almost overnight, with both a husband and a mother to care for. The Williamses moved into an apartment hotel, and her mother took rooms there, too. While she yearned to set up her work in the first "really commodious and healthful study" she had since her faculty office at Bryn Mawr, she was constantly delayed in returning to research and publication (Breasted 1895–1935, Ransom Williams to Breasted, September 2, 1916).

Having taken so much time away from work, she worried she was beginning to disappear from the discipline. Her colleagues were also worried. For example, scholar Alan Gardiner (1879–1963) wrote to her that, now that she was married, he feared she would be giving up Egyptology for good (Ransom Williams 1917a). She told Gardiner that that would not happen. She also wrote to Breasted, who had been concerned for her productivity, that her "husband is

[5] There is older scholarship about these issues, the most influential being Margaret Rossiter's "The Matthew Matilda Effect in Science," from 1993. Rather than cite from these foundational texts, I chose to use more recent sources. However, I would like to thank early authors like Rossiter who started the practice of questioning where women were in the narrative and putting them back where they belonged—at the center of the story.

unselfishly interested in my plans and encourages me to take any opportunity for congenial work that comes my way" (Breasted 1895–1935, Ransom Williams to Breasted, December 2, 1916). Ransom Williams continued to work for the MMA by commuting to New York a few times a year for several weeks at a time. It helped that Toledo was on a main railroad line and the journey, though long, was easy. Much like Margaret Murray and other women had to do at the time, in addition to her work at the MMA, she began to piece small part-time jobs together to be able to continue the research she loved to do. She taught several courses at the University of Michigan and a few in Chicago, where she was also busy with research. She also catalogued the Egyptian collections at the Cleveland Museum of Art, the Detroit Museum of Art, and the Drexel Collection in Minnesota (Navarro 2021). The tasks themselves would keep her at least peripherally active in the discipline and her productivity remained relatively steady thanks to her continued connection to these collections. Not one to be relegated to the periphery, Ransom used her position as the middle of a vibrant scholarly network that she had established at the MMA to remain in the field and to continue to position herself as essential in a discipline where the immediate aftermath of marriage to a non-Egyptologist, for a woman, was usually expulsion (Pycior et al. 1996).

Not long after the Perneb display opened in 1916, Breasted came upon the Abbott collection at the New York Historical Society (NYHS). The collection, containing around 2000 objects, is named for British physician and antiquities collector Henry Abbott (1807–1859) and came to the NYHS in 1860 (Ransom Williams 1917b). It was the first Egyptian antiquities collection owned by an institution in the United States.[6] While it contained many important objects, it had been virtually untouched by academics and sat, decidedly unorganized, in the NYHS for over 50 years. Upon seeing the old catalogue in December of 1916, which Abbott himself had put together, Breasted wrote to Ransom Williams that: "(...) the unpublished things there made my mouth water. There is a chance for a large and beautiful monograph on these things" (Breasted 1895–1935, Breasted to Ransom Williams, December 21, 1916). He immediately thought Ransom Williams would be perfect to catalog, organize, and publish the collection, and Lythgoe agreed. Ransom Williams, as a woman, commanded a low salary despite her expertise, and she was arguably the most well-qualified scholar of any gender to do it. In January of 1917, she responded with thanks to Breasted for recommending her for the job, and said, "If I undertake it I shall endeavor to endure to the end!" (Breasted 1895–1935, Ransom Williams to Breasted, January 16, 1917). And she did. The Society announced her involvement in April 1917, believing it would take "a year or so" for her to do the "exhaustive scientific catalogue (...) to replace the catalogue made by Dr. Abbott at the time the Society secured his collection, which is still in use" (1917: 12). In the end, it seems, it took seven years to complete a partial catalogue and get it published (Ransom Williams 1924).

The catalogue took so long to complete mainly because she was commuting a very long distance by train in the early twentieth century, along with her other paying jobs and domestic tasks. She was unable to have access to the pieces all the time, which meant that she could not properly study or organize them full-time. She worked quickly, but carefully, when she was in New York, and many times she was able to get photographs of objects to take back to Toledo with her. In Toledo, she was still caring for her mother and husband, and, unlike her male counterparts in Egyptology and museum work, many of whom had wives to take care of things in the home and family, Ransom Williams was dealing with the double bind of domestic and professional duties at the same time. It is possible that someone without these concerns could have done the job more quickly, but she was exacting and detailed.

At a few points over those seven years in which she worked on the catalogue, Ransom Williams tried use her considerable influence to hand the task over to other scholars to give them some professional experience, but to no avail (e.g., Breasted 1895–1935, Ransom Williams to Breasted, December 9, 1921). In late November of 1920, she wrote to Breasted about "the medical papyrus of the Smith collection" owned by the NYHS but that she did not have the time to translate and publish (Breasted 1895–1935, Ransom Williams to Breasted, November 22, 1920). She further wrote to him:

> (...) I feel that it is imperative, if possible, to get work started soon on that medical papyrus by a competent authority, whose name on the titlepage would carry conviction that the papyrus had been well handled. I am authorized by a sub-committee of the Executive Committee to ask you whether you would care to prepare the publication and would have time for it. The papyrus is probably the most valuable one owned by the Society and I am ready to waive my interest in it, in the hope that it may be published sooner and better than I could do it. (Breasted 1895–1935, Ransom Williams to Breasted, November 22, 1920)

In the end, Breasted agreed to the work and by 1930 it was published as the Edwin Smith Surgical Papyrus, in two volumes (Breasted 1930). It was not that she thought Breasted could use the experience, but she knew how important the papyrus would be to Egyptology and wished to have it published as soon as possible. She had the expertise, she just did not have the time, so she used her influence with Breasted to complete the work. She also helped with editing his translations, marking up his page proofs, and communicating with the NYHS about the progress of his work to a degree that she was almost like a research assistant (Breasted 1895–1935,

[6] In 1937, the Brooklyn Museum took it on loan; they purchased it from the NYHS in 1948.

Ransom Williams to Breasted, July 21, 1930). While he acknowledged her in the volume, very few scholars note that Ransom Williams was the one who brought the papyrus to light in the first place (Breasted 1930: xix).

By 1922, she wished to take a break from the cataloging project, which had already been going on for almost five years. To the NYHS, she suggested Nathaniel J. Reich (1876–1943), an Egyptologist working in Austria-Hungary until the First World War ended, to replace her. As part of the periphery of each of their networks, both Breasted and Ransom Williams tried to get Reich work in the United States so he could emigrate and stay in the country (Breasted 1895–1935, Breasted to Ransom Williams, February 6, 1922; Ransom Williams to Breasted, February 9, 1922). She also tried to get him some work in Cleveland, with the same result. Simply put: neither museum could afford to hire a man to do the work they paid a woman to do. Due to the gender pay expectations at the time, the collections would have had to pay him far more than they were paying Ransom Williams; it was not tenable. In fact, by 1927, the post-war situation in Europe had turned so dire that both Breasted and Ransom Williams tried to find paying jobs for many of their colleagues. Despite the fact that there were so many Egyptologists they knew both in the United States and across Europe, there were not enough jobs and not enough funding to hire them (Breasted 1895–1935, Ransom Williams to Breasted, August 29, 1926). Ransom Williams had no choice but to continue the Abbott cataloging work. When it was completed in 1924, the catalogue was well-received among Egyptologists and jewelry historians and it is still used by scholars who study the collections today. Even though she did not complete the whole catalogue, her work began the long and difficult task. Again, as a woman in the field, she used her situation as an institution-building curator and built the foundation for the cataloging and academic study of the Abbott collection and other small collections in the Midwestern United States.

3.5 Conclusions

Most of Ransom Williams' career was made possible by the fact that, during the first ten years of it, she worked to establish herself as the hub of a wide Egyptological network. Because of her position, when she moved jobs, and moved towns, she was able to maintain her workflow with minimal interruptions. Her main roles were in artifacts and collections analysis and cataloging, conservation, and publication. It was in those positions that she made her mark on the discipline. She did not publish widely, but she wrote a lot of important early works on new and emerging collections in the United States. She was a museum administrator, and it was under her leadership that the MMA and the Abbott collection were curated and presented to the public. Much like other women in early Egyptology, Ransom Williams' work is not as recognized as that of her male counterparts such as Breasted, Lythgoe, Winlock, or Davis because her name is mainly associated with the domestic, institutional aspects of the discipline. However, various small correspondence collections demonstrate her importance in these domestic spaces. Alan Gardiner, Adolf Erman, James Breasted, Howard Carter, Albert Lythgoe, and Herbert Winlock considered her a valued and trusted member of the discipline and looked to her for guidance and expertise.

When we write about the careers of women in archaeology using the methodology based in history and philosophy of science, we can highlight the work that other methods miss. Ransom Williams is just one of many women, along with Kate Griffith (née Bradbury) (1854–1902), Emily Paterson (1861–1947), and Margaret Murray, who formed the backbone of Egyptology institutions around the world. She was central to the formation and curation of the Egyptian collection at the MMA and the fact it became a strong center of creating and curating knowledge about ancient Egypt. Although her work was not in excavation—instead curating those same objects back home—she had an equally foundational impact on the collections and their importance to the discipline. Ransom Williams' work with Lythgoe and Breasted was what made their own careers possible, by not only building the institutional foundations at home, but also in Egyptology's scholarly and public engagement in the United States.

If we focus only on site reports or traditionally male activities, we will miss the women who simply were not included in those documents but whose work was vital to Egyptology (Stevenson 2015: 102–5). Instead, we have to shift the narrative so we can see women's full lives to analyze what they really did and the impacts they made. For too long in archaeology we have been locating these women, only to try to fit them into the male heroic discovery narrative. We struggle to do that because they do not easily fit, and, since women's roles were relatively limited in that context, they end up being left out. We are left with the assumption that they simply were not present, or they were doing helpmate, assistant-type work.

We find the most evidence about active women such as Ransom Williams in archives, correspondence collections, or in footnotes of other biographies. Critically reading and then using this new-found material is important to writing the lives of these scientists. While this chapter has not told Ransom Williams' complete story (that has yet to be fully written), it has introduced some of her principal activities as being not only fundamental to the discipline of Egyptology, but also central to her own life. We must interrogate the archives, ask who the women are who may appear a few times in letters or once or twice in a memoir, figure out what their jobs were and what real impact they had, and in turn place them in the narrative where they belong.

Albert Lythgoe was well known for "his finest achievement" which was "the arrangement of the Egyptian Collection of the Metropolitan Museum of New York, which all who have seen it praise in almost unmeasured terms" ("Notes and news," 1934: 107). Yet, "he was often to be seen in Egypt, where he conducted the digs at Lisht, Thebes, and elsewhere" ("Notes and news," 1934: 107). How could he be in Egypt for so much of the year but also be responsible for the curation of the Collection at the MMA? The answer is simple—he was not really the one responsible; it was largely the work of Ransom Williams. Lythgoe was also remembered widely for "the prompt liberality with which he lent his staff to assist in the excavation of Tut'ankhamun's tomb" ("Notes and news," 1934: 107). But, as Winlock noted in a letter to Ransom Williams, while she was finishing yet another book, this one about Lisht, "Lythgoe died without doing a single thing on any of the excavations which he had ever conducted, and his notes remain locked up in a trunk" (Ransom Williams 1909–1917, Winlock to Ransom Williams, April 2, 1934). Ransom Williams' legacy has been locked up in a trunk as well, but by seeing her as a curator and scholar who made contributions in her own right as well as enabling the work of others, we will open the lid on her story, and she will finally be at the center of the discipline she helped to build.

Acknowledgements I would like to thank Julie McMaster, archivist at the Toledo Museum of Art, for her help on dating some activities in Ransom's life. Thanks also to Clare Lewis, Leila McNeill, Gabriel Moshenska, Anna Reser, Julia Roberts, the editors of this volume, and the anonymous reviewers for reading various iterations of this chapter and giving feedback. Parts of the research for this project were funded by grants from the University of Missouri System.

References

Primary Sources

Archival Material[7]

Breasted, J. H. (1895–1935). *Correspondence*. James Breasted Office Files, Archives of the Institute for the Study of Ancient Cultures, University of Chicago, Chicago, IL, United States.

Ransom Williams, C. (1909–1917). *Papers*. Egyptian Art Archives, Metropolitan Museum of Art, New York, NY, United States.

[7]A note on archival material: since I have consulted these archives, each of them has gone through significant restructuring. I have given the citation information as I found it.

Ransom Williams, C. (1917a, July 19). [Letter to Alan Gardiner]. Griffith Institute (Gardiner MSS), Oxford University, Oxford, United Kingdom.

Registrar's Office of the University of Chicago.

Articles, Books and Chapters

Breasted, J. (1930). *The Edwin Smith surgical papyrus: Hieroglyphic transliteration, translation, and commentary* (Vols. 1–2). Chicago: The University of Chicago Press.

Bryn Mawr College. Senior Class. (1908). *Bryn Mawr College yearbook. Class of 1908*. Bryn Mawr, PA. https://repository.brynmawr.edu/bmc_yearbooks/6

Higley, M. C. J. (Ed.). (1937). *One hundred year biographical directory of Mount Holyoke College, 1837–1937* (Bulletin Series 30, 5). South Hadley, MA: Alumnae Association of Mount Holyoke College.

Lythgoe, A., & Ransom, C. (1916). *The tomb of Perneb: With illustrations*. New York: Metropolitan Museum of Art.

Metropolitan Museum of Art. (1911). *A handbook of the Egyptian rooms*. New York: Gillis Press.

Murray, M. (1963). *My first hundred years*. London: William Kimber.

Notes: The Abbott Collection of Egyptian Antiquities. (1917). *New York Historical Society Quarterly Bulletin, 1*(1), 12.

Notes and news. (1934). *The Journal of Egyptian Archaeology, 20*, 107–109.

Ransom, C. (1905). *Studies in ancient furniture: Couches and beds of the Greeks, Etruscans, and Romans*. Chicago: University of Chicago Press.

Ransom, C. (1911). *A handbook of the Egyptian rooms, Metropolitan Museum of Art*. New York: Metropolitan Museum of Art.

Ransom, C. (1912). The value of photographs and transparencies as adjuncts to museum exhibits. *Metropolitan Museum of Art Bulletin, 7*(7), 132–134.

Ransom, C. (1913a). *The stela of Menthu-Weser*. New York: Metropolitan Museum of Art.

Ransom, C. (1913b). Egyptian furniture and musical instruments. *Metropolitan Museum of Art Bulletin, 8*(4), 72–79.

Ransom, C. (1913c). A model of the mastaba-tomb of Userkaf-Ankh. *Metropolitan Museum of Art Bulletin, 8*(6), 125–130.

Ransom, C. (1913d). Nubian objects acquired by the Egyptian Department. *Metropolitan Museum of Art Bulletin, 8*(9), 200–208.

Ransom, C. (1913e). The stela of Menthu-Weser. *Metropolitan Museum of Art Bulletin, 8*(10), 213, 216–218.

Ransom, C. (1914a). A Late Egyptian sarcophagus. *Metropolitan Museum of Art Bulletin, 9*(5), 112–120.

Ransom, C. (1914b). Pots with Hieratic inscriptions. *Metropolitan Museum of Art Bulletin, 9*(11), 236–243.

Ransom, C. (1915a). A commemorative scarab of Thutmose III. *Metropolitan Museum of Art Bulletin, 10*(3), 46–47.

Ransom, C. (1915b). Heart scarab of Queen Amenardis. *Metropolitan Museum of Art Bulletin, 10*(6), 116–117.

Ransom, C. (1915c). Three sets of Egyptian gold pendants. *Metropolitan Museum of Art Bulletin, 10*(6), 117–120.

Ransom Williams, C. (1917b). The Abbott collection. *New York Historical Society Quarterly Bulletin, 1*(2), 34–37.

Ransom Williams, C. (1924). *The New York Historical Society catalogue of Egyptian antiquities, Numbers 1–160: Gold and silver jewelry and related objects*. New York: *New York Historical Society*.

Ransom Williams, C. (1932). *The decoration of the tomb of Per-Neb: The technique and the color conventions*. New York: Museum of Modern Arts.

Swindler, M. H. (1929). *Ancient painting: From the earliest times to the period of Christian art*. New Haven: Yale University Press.

The Theodore M. Davis Bequest. (1931). *Metropolitan Museum of Art Bulletin*, *26*(3.2).

Newspaper and Magazine Articles

Perneb's Gay Tomb Ready for Visitors. (1916, January 30). *The New York Times*, 13.

Secondary Sources

Adams, J. M. (2013). *The millionaire and the mummies: Theodore Davis's Gilded Age in the Valley of the Kings*. New York, NY: St. Martin's Press.

The Bryn Mawr College Egyptology Association. (2019). *Bryn Mawr College's legacy in Egyptology: Past, present, and future*. https://brynmawrcollegeegyptologyassociation.digital.brynmawr.edu/on-campus-exhibit/

Carr, L. (2012). *Tessa Verney Wheeler: Women and archaeology before World War Two*. Oxford: Oxford University Press.

Carruthers, W., & Van Damme, S. (2017). Disassembling archeology, reassembling the modern world. *History of Science*, *55*(3), 255–272.

Caton-Thompson, G. (1983). *Mixed memoirs*. Gateshead: The Paradigm Press.

Cohen, G. M., & Joukowsky, M. S. (Eds.). (2004). *Breaking ground: Pioneering women archaeologists*. Ann Arbor, MI: University of Michigan Press.

Coletta, L. (Ed.). (2015). *The legacy of the Grand Tour: New essays on travel, literature, and culture*. Lanham, MD: Rowman & Littlefield.

Colla, E. (2007). *Conflicted antiquities: Egyptology, Egyptomania, Egyptian modernity*. Durham, NC: Duke University Press.

Díaz-Andreu, M., & Sørensen, M. L. S. (Eds.). (1998). *Excavating women: A history of women in European archaeology*. London: Routledge.

Dyhouse, C. (1995). *No distinction of sex? Women in British universities, 1870–1939*. London: University College London Press.

Hill, K. (2016). *Women and museums, 1850–1914: Modernity and the gendering of knowledge*. Manchester: Manchester University Press.

Howell, G. (2010). *Gertrude Bell: Queen of the desert, shaper of nations*. New York: Farrar, Straus and Giroux.

Janssen, R. M. (1992). *The first hundred years: Egyptology at UCL, 1892–1992*. London: University College London Press.

Kohler, R. (2002). *Landscapes and labscapes: Exploring the lab-field border in biology*. Chicago: University of Chicago Press.

Latour, B. (2005). *Reassembling the social: An introduction to actor-network theory*. Oxford: Oxford University Press.

Leach, E. W. (2021). *Mary Hamilton Swindler*. Breaking Ground: Women in Old World Archaeology. https://www.brown.edu/Research/Breaking_Ground/bios/Swindler_Mary%20Hamilton.pdf

Livingstone, D. (2003). *Putting science in its place: Geographies of scientific knowledge*. Chicago: University of Chicago Press.

Livingstone, D., & Withers, C. (2011). *Geographies of nineteenth-century science*. Chicago: University of Chicago Press.

Mellink, M. J. (1967, May). Mary Hamilton Swindler, January 1, 1884–January 16, 1967. *Expedition Magazine*, 9(3). https://www.penn.museum/sites/expedition/mary-hamilton-swindler

Moser, S. (2006). *Wondrous curiosities: Ancient Egypt at the British Museum*. Chicago: Chicago University Press.

Navarro, D. (2021). *America's First Woman Egyptologist—Caroline Ransom Williams*. Digital epigraphy: An epigraphic survey. https://www.digital-epigraphy.com/readings/americas-first-woman-egyptologist-caroline-ransom-williams

Pycior, H. M., Slack, N. G., & Abir-Am, P. (Eds.). (1996). *Creative couples in the sciences*. New Brunswick, NJ: Rutgers University Press.

Reid, D. M. (2015). *Contesting antiquity in Egypt: Archaeologies, museums, & the struggle for identities from World War I to Nasser*. Cairo: The American University in Cairo Press.

Root, M. C. (2004). Introduction: Women of the field, defining the gendered experience. In G. M. Cohen, & M. S. Joukowsky (Eds.), *Breaking ground: Pioneering women archaeologists* (pp. 1–33). Ann Arbor, MI: University of Michigan Press.

Rossiter, M. (1993). The Matthew Matilda effect in science. *Social Studies of Science*, *23*(2), 325–341.

Rudwick, M. J. S. (1985). *The great Devonian controversy: The shaping of scientific knowledge among gentlemanly specialists*. Chicago: University of Chicago Press.

Sheppard, K. L. (2013). *The life of Margaret Alice Murray: A woman's work in archaeology*. Lanham, MD: Lexington Books.

Sheppard, K. (2014). Margaret Alice Murray and archaeological training in the classroom: Preparing 'Petrie's pups'. In W. Carruthers (Ed.), *Histories of Egyptology: Interdisciplinary measures* (pp. 113–128). London: Routledge.

Sheppard, K. (Ed.). (2018). *'My dear Miss Ransom…': Letters between Caroline Ransom Williams and James Henry Breasted, 1898–1935*. Oxford: Archaeopress.

Sheppard, K. L. (2020). 'Trying desperately to make myself an Egyptologist': James Breasted's early scientific network. In J. Roberts, K. Sheppard, U. R. Hansson, & J. R. Trigg (Eds.), *Communities and knowledge production in archaeology* (pp. 174–187). Manchester: Manchester University Press.

Shteir, A. B. (1996). *Cultivating women, cultivating science: Flora's daughters and botany in England, 1760–1860*. Baltimore: The Johns Hopkins University Press.

Stevenson, A. (2015). 'To my wife, on whose toil most of my work has depended': Women on excavation. In A. Stevenson (Ed.), *Petrie Museum of Egyptian Archaeology: Characters and collections* (pp. 102–105). London: University College London Press.

Stevenson, A. (2019). *Scattered finds: Archaeology, Egyptology and museums*. London: University College London Press.

Thompson, D. B. (1950). Mary Hamilton Swindler. *American Journal of Archaeology*, *54*(4), 292–293.

Thornton, A. (2018). *Archaeologists in print: Publishing for the people*. London: University College London Press.

Wallach, J. (1996). *Desert queen: The extraordinary life of Gertrude Bell, adventurer, adviser to kings, ally of Lawrence of Arabia*. New York, NY: Doubleday.

The Dutch Years of F. W. Freiherr von Bissing (1922–1926): Networks, Egyptian Objects, and an Academic *Persona* on the Move

4

Martijn Eickhoff and Mathijs Smith

Abstract

Among Egyptologists, Friedrich Wilhelm Freiherr von Bissing (1873–1956) is considered one of the most controversial figures in the history of their research field. In this chapter we discuss the network, life and work of this German Egyptologist who left Germany for the Netherlands, together with his impressive private collection of Egyptian antiquities, a few years after the defeat of Germany in 1918. He was appointed as *bijzondere hoogleraar* (endowed professor) of Egyptian and Near Eastern Art History at Utrecht University in 1922, and his collection became part of the newly opened Museum Carnegielaan in The Hague. The following Dutch episode of his life ended in 1926, when he was forced to leave the country due to a scandal surrounding his covert homosexuality. Against the backdrop of this dramatic event, the *persona* of von Bissing will be reconstructed by looking at the interaction between his ideas and actions on an academic and private level. We show how this *persona*, together with his collection and transnational network, constitutes a crucial backdrop for understanding both the enthusiasm and problems von Bissing generated after acquiring the status of a professor in the Netherlands.

Keywords

Egyptology · Von Bissing · Scheurleer · National Socialism · First World War · Collections · Homosexuality

M. Eickhoff
NIOD Institute for War, Holocaust and Genocide Studies, Amsterdam, The Netherlands
e-mail: m.eickhoff@niod.knaw.nl

M. Smith (✉)
Independent Researcher, Leiden, The Netherlands

4.1 Introduction

On December 20, 1924, Museum Carnegielaan 12 in The Hague was opened with a great celebration. The museum, which became known as the Archeologisch Museum Scheurleer (Scheurleer Archaeological Museum) displayed the antiquities collections of the private collectors Constant Willem Lunsingh Scheurleer (1881–1941; Hoorn and Gelder 1941), Eduard Freiherr von Heydt (1882–1964; Illner 2013), and Friedrich Wilhelm Freiherr von Bissing (Fig. 4.1). The latter collection was slowly being acquired for the museum due to a deal between its owner and the Academie van Beeldende Kunsten (Academy of Art) on whose behalf von Bissing had become *bijzonder hoogleraar* (endowed professor) at Universiteit Utrecht (Utrecht University). The opening of the museum was the result of the tireless work of the banker and classical art historian Lunsingh Scheurleer—who was on this occasion knighted for his efforts (S.n. 1924j)—and attracted the cream of the crop of Dutch cultural high society. The guest list included the minister, secretary-general, and several department heads of the Ministerie van Onderwijs, Kunst en Wetenschappen (Ministry of Education, Art and Science); the administrator of the Ministerie van Buitenlandse Zaken (Ministry of Foreign Affairs); the director of the Rijksbureau voor Monumentenzorg (State Office for Monument Care); members of the local council and the mayor of The Hague; board members of the art society Pulchri and the Vereniging van Vrienden van Aziatische Kunst (Society for Friends of Asiatic Art); representatives of the council of Utrecht University; several professors; and no less than seven museum directors and curators. This was von Bissing's finest moment in the Netherlands. Since 1919 he had sought a safe haven for himself and his collection, and now he had found one. Despite outrage over his political writings published earlier that year (Bissing 1924b, c), his star was on the rise.

In this chapter we analyze the dynamic relationship between the *personae*, collection, and network of von

L. Coltofean et al. (eds.), *Connecting People and Ideas*, Themes in Contemporary Archaeology,
https://doi.org/10.1007/978-3-031-81006-0_4

Fig. 4.1 F. W. Freiherr von Bissing. (Photo: Theodor Hilsdorf [Atelier Müller-Hilsdorf]; © Münchner Stadtmuseum, Munich, Germany, Inv.-Nr. G 28/1379)

Bissing. The research for this chapter was carried out in the context of the collaborative international project, *The Bissing Link* (2020), whose Dutch partners include the Allard Pierson—the museum and knowledge institute for the heritage collections of the University of Amsterdam—and the NIOD Institute for War, Holocaust, and Genocide Studies. The aim of the project was to track down relevant archival material in the Netherlands associated with Friedrich von Bissing in order to shed a new light on the Dutch period in his life, in relation to his political stance and the colonial dimension of his collecting activities. Documentation concerning von Bissing is kept in seven different institutions: Allard Pierson (APAA: Allard Pierson Museum Archive), the Stadsarchief Amsterdam (AMS: City Archive Amsterdam), the Haags Gemeentearchief (DH: Municipal Archive The Hague), the Nederlands Instituut voor het Nabije Oosten (NINO: Netherlands Institute for the Near East),[1] the Rijksmuseum van Oudheden (RMO: National Museum of Antiquities Leiden), the Universiteitsbibliotheek Leiden (LUL: Leiden University Library) and the Utrechts Archief (UTR: Utrecht Archive). This documentation includes official, personal and professional correspondence, internal memos, contracts, bills of sale, photos, postcards, and books. In addition to this material, we consulted books and articles written by von Bissing during his stay in the Netherlands. Although additional research can still be done, the data collected for this project allows for a detailed reconstruction of von Bissing's time and networking in the Netherlands.

The life and work of von Bissing has previously been documented and described in obituaries (Babinger 1956; Brunner 1956; Müller 1965), as well as biographical (S.n. 1953; Garnett 1991/2020: 512–22; Grimm and Schoske 2010; Raulwing and Gertzen 2013; Haarlem 2014; Bierbrier 2015) and bibliographical publications (Wallert 1959). Some of these works are of clear hagiographical character. The Egyptologist Alfred Grimm's biographical sketch of von Bissing's life, for example, largely omits his dealings with the Nationalsozialistische Deutsche Arbeiterpartei (NSDAP: National Socialist German Workers' Party) and his anti-Semitism, instead focusing on his views of the NSDAP in the period after he was forced out of the party. Likewise, Egyptologist and curator Willem van Haarlem's portrayal of von Bissing attempts to soften his anti-Semitism by asserting that von Bissing showed concern for his Jewish colleagues before and during the Second World War (Haarlem 2014).[2] As such they connect the traditional approaches of the history of archaeology and Egyptology in which discoveries and scientific progress are the key themes (for a recent example, see Fagan 2014). In recent decades a more critical approach has developed in the history of archaeology and Egyptology. Key themes in this context are the involvement—and/or exclusion—of Egyptians (e.g., Quirke 2010),[3] the knowledge of ancient Egypt in Medieval Arabic writings (e.g., El-Daly 2005), the politics of Egyptology, in particular the connections between nationalism, (informal) colonialism and Egyptology (e.g., Reid 2002; Colla 2008), and the ethics of excavating and collecting (e.g., Merryman 2006; Renfrew 2000). Recent years have also witnessed the rise of studies devoted to National Socialist Egyptology (e.g., Schneider and Raulwing 2013). National Socialist and critical Egyptological history also relate to von Bissing, but only some of these topics have been taken up as central themes in the biographical studies dedicated to his life and work.

Inspired by some of these recent historiographical developments, while also reflecting on the existing biographical

[1] The authors regret that they were not able to include new archival material discovered in the NINO by Mathijs Smith in November 2023, which contains correspondence between C. W. Lunsingh Scheurleer and F. W. von Bissing from the period 1914–1925.

[2] Van Haarlem's main source is the so-called Steindorff list of Nazi-Egyptologists. Steindorff is lenient about von Bissing, who was in any case too old to play a significant role in politics after the War, because Friedrich visited him personally after a pogrom in 1938 (see e.g., Schneider and Raulwing 2013). However, in a letter from von Bissing to Franz de Liagre Böhl, the former remarks on Steindorff's situation: "personally these things are all very sad, but it is a strange nemesis [the Jews], which no one should forget, holds the power" (Bissing 1939b).

[3] The conference *Egyptian pioneers in Egyptology: Writings and archives*, of January 18, 2020 at the DAIK, was also dedicated to this topic.

Fig. 4.2 Top: Outside view of Museum Carnegielaan 12. Bottom: Interior of Museum Carnegielaan 12, likely room III after 1926. (Top: Photo by Dienst Stadsontwikkeling en Volkshuisvesting [Photographer C. J. de Gilde], collection Haags gemeentearchief 0.14912. Bottom: Allard Pierson, archief C. W. Lunsingh Scheurleer, University of Amsterdam, The Netherlands)

studies (for a thorough history of research, see Raulwing and Gertzen 2013), we will start this chapter by giving a brief biographical overview of von Bissing. We will focus on his Dutch years that began with so much promise in 1922. Von Bissing left Germany that year as a self-declared "academic refugee" and was appointed as *bijzonder hoogleraar* of Egyptian and Near Eastern Art History at Utrecht University in 1922 on behalf of the Academy of Art in The Hague. His collection became part of the newly opened Museum Carnegielaan in The Hague (Fig. 4.2). The following four Dutch years, which he described euphemistically as "four rich years",[4] ended in 1926 when he was forced to leave the country due to a scandal surrounding his covert homosexuality. Against the backdrop of these dramatic events, we reconstruct the academic *personae* (Daston and Sibum 2003: 1–8) of von Bissing, which include archetypes such as the 'true German', the passionate collector, the empathic Maecenas and the knowledgeable but eccentric Egyptologist. We will analyze the interaction between ideas and actions on both public and private levels (Paul 2016). In that context we focus on the relationship between his collection and his transnational network, which consisted of a mix of members of Dutch cultural elite, German nobility, Egyptologists, homosexuals, collectors and traders of antiquities, and supporters of the NSDAP and the Greater Netherlands movement. Taken together, the *personae*, collection, and network of von Bissing allow us to develop an understanding of both the enthusiasm and problems he generated after acquiring the status of a professor in the Netherlands.

4.2 The Life and Times of F. W. von Bissing

Friedrich Wilhelm Freiherr von Bissing (Fig. 4.1) was born on September 22, 1878 in Potsdam, Germany, as the eldest son of Moritz Ferdinand Freiherr von Bissing (1844–1917) and Myrrha Freifrau von Bissing *née* Wesendonck (1851–1888) who was the main source of the family's wealth and cultural capital. Like his mother, young "Fritz's" health was poor and necessitated frequent travels to Italy and Egypt where the climate was thought to alleviate his health problems. It was in these early years that he first encountered the art and culture of ancient Egypt. Due to his father's military career the family moved around often and Friedrich was home-schooled. In 1883 the family moved to Berlin due to Moritz's promotion to the Große Generalstab (German General Staff). Friedrich transferred to the Joachimsthalsches Gymnasium in Wilmersdorf, graduating in 1891. After his graduation, he went to study Classical philology, Archaeology, Egyptology, and Art History in Bonn am Rhein. He did this against his father's wishes, who wanted a military or diplomatic career for his son.

Von Bissing obtained his Ph.D. in 1896 under Egyptologist Alfred Wiedemann (1856–1936) and subsequently moved to Cairo to work on the *Catalogue Général des Antiquités Égyptiennes du Musée du Cairo* (General catalogue of Egyptian antiquities of the Cairo Museum). This was a momentous international project which aimed at documenting the entire collection of the Cairo Museum and included members from various countries. The director of the Kaiserlich Deutsches Institut für Ägyptische Altertumskunde in Kairo (The imperial German institute for the study of Egyptian antiquity in Cairo, later the DAIK), Ludwig Borchardt (1863–1938), was instrumental in this project. He was also the one who nominated von Bissing for the German position within the project. Von Bissing's association with the German Egyptologists in Cairo initially went well; his

[4] Von Bissing wrote this on a portrait picture of himself, kept in the Allard Pierson.

family fortune[5] allowed him to buy the library of German novelist and Egyptologist Georg Ebers (1837–1898) for the institute and finance excavations in the ancient Egyptian site of Abu Gurob (1899–1901), where he worked together with Borchardt and the Egyptologist Heinrich Schäfer (1868–1957) (Voss 2013: 71). Soon, however, the relationship between Borchardt and von Bissing turned sour. In 1898, von Bissing was elected to the *comité d'archéologie* (Archaeological Committee), which formerly consisted of British, French, and Egyptian dignitaries under the chairmanship of the French director of the Antiquities Service, but now expanded to include other nationalities. Borchardt had hoped for this position himself, but rumors about his 'anti-British sentiments' and political struggles had barred him from the position (Voss 2013: 51–52). Von Bissing's unauthorized actions in the Archaeological Committee in spring 1900, voting against the perceived German interest, led to tensions between Borchardt and von Bissing (Voss 2013: 74, 147). These were likely exacerbated by von Bissing's anti-Semitic worldview since Borchardt was Jewish (on this anti-semitism, see e.g., Raulwing and Gertzen 2013: 63f).

In 1904, von Bissing moved back to Germany where he was appointed to the chair of Egyptology at the Ludwig-Maximilian University in Munich, where he had obtained his habilitation in 1900. That same year he had married Elisabeth (Elsa) Freifrau von Carlowitz (1875–1961). Their daughter, Myrrha, was born four years later. After his move back to Germany, and even during the First World War, von Bissing kept up his feud with Borchardt (Voss 2013: 126, 147–50).

The First World War had a great impact on the life and work of von Bissing. On August 4, 1914, the German empire violated Belgian neutrality and invaded the kingdom. Occupied Belgium was soon divided into three distinct zones. Eastern and central Belgium became part of the Kaiserliches Deutsches Generalgouvernement Belgien (Imperial German General Government). Friedrich's father, *generaloberst* Moritz Ferdinand Freiherr von Bissing, was recruited from retirement and appointed as Governor-General. The German occupation government tried to divide Belgium by stirring up tensions between the French-speaking part of the Belgians: the Walloons, and the Dutch-speaking Belgians: the Flemish. By appealing to a sense of Pan-Germanism they hoped to enlist the help of the Flemish people. Many of them had for a long time felt underrepresented within the Belgium Kingdom. One of the key points of this *Flamenpolitik* was the transformation of the University of Ghent into a Dutch-speaking university, a long-held wish of Flemish activists in Belgium. This plan was enacted in 1916, and Moritz von Bissing had enlisted his academic son to help with the reorganization of the university. Some authors have noted that this gap in his Egyptological career almost appears to have been a penance exacted for having pursued an academic career against his father's wishes; Friedrich himself writes that he has now become his father's secretary (Gertzen 2016: 104–5). In Belgium, von Bissing junior applied himself to the German war effort by writing propaganda material (Gertzen 2016) and made contact with several Flemish scholars and students who had embraced the 'gift' of the German Occupational Government. Several of them would flee to the Netherlands after the German defeat branded them as collaborators. At the time, in the Netherlands there existed broad support for the idea that cultural ties between the Dutch and Dutch speaking populations in Belgium, France, and South-Africa should have political consequences (Eickhoff 2003: 127, 155–57). The transformation of Ghent University into a Dutch-speaking university—known as the Von Bissing University after the Governor-General—was generally approved of in Dutch society. Among students, the Algemeen Nederlands Verbond (General Dutch Alliance) and the more radical Dietsche Studentenverbond (Dutch/Flemish Student-alliance) publicly supported the 'Dutchification' of the university.

After the death of his father in 1917, Friedrich von Bissing moved back to Germany and once again took up teaching in Munich, even briefly becoming dean of the Faculty of Philosophy. It was in Munich that he was confronted with a Germany transformed after the defeat of 1918 and the events surrounding the short-lived Münchner Räterepublik (Bavarian Soviet-Republic; April 7, 1919–May 3, 1919). Von Bissing was arrested (Raulwing and Gertzen 2013: 86) and, according to his own autobiographical statements, his house was expropriated to serve as living quarters by Mayor Eduard Schmidt (1861–1933), who was appointed mayor after the end of the ill-fated republic. This period seems to have been highly traumatic for von Bissing and as such he mentioned it years later in his writings (see e.g., Bissing 1924d). On July 14, 1919, he wrote to Egyptologist Adolf Erman (1854–1937) in a private letter that he had given up on his plans to open a large public collection in Munich and that he owed no allegiance to this government (Raulwing and Gertzen 2013: 87). He felt that in such a situation he—as a 'true German'—could leave his beloved German fatherland, be it *à contre-coeur*. Contemporary newspapers speculate that the move may not have been informed by politics alone, but also by the devaluation of the German economy (S.n. 1922a, b, c). In addition, the discovery by his personnel of personal correspondence explicating his covert homosexuality (Bissing-Carlowitz 1925b)—then punishable by German law—likely played a role in his decision to leave Germany. Whatever the reasons, the same year von Bissing engaged in negotiations with the National Museum of Antiquities in Leiden about the sale of his collection. The Museum sent curator Pieter Boeser

[5]Arthur Weigall, who worked with von Bissing between 1902–1905, in 1902 estimated his fortune to be close to 7 million marks, the real wealth equivalent of c. 37,5 million pounds in 2019 (Weigall 1902).

(1858–1935) to Munich to inspect the collection. That von Bissing was serious about his wish to leave Germany can be deduced from a letter to the Dutch Ministry of Arts, Culture and Sciences from November 12, 1919 in which Boeser writes that: "hij komt in Holland wonen en heeft de wensch Nederlandsch staatsburger te worden naar de wet" (he [von Bissing] will live in Holland and expresses the wish to become a Dutch citizen by law) (Boeser 1919).

The museum expressed interest in buying part of the collection, but negotiations broke down at the moment when the Dutch government proved unwilling to pay. In 1920 new negotiations started, this time with the Academy of Art in The Hague through the intervention of its chairman Constant Willem Lunsingh Scheurleer. The Scheurleer family had made their money in banking, but C.W. Lunsingh Scheurleer's real passion was classical art history. Von Bissing and Scheurleer had recently met during a trip the latter took to acquire some marble fragments of classical sculpture from Munich (Gelder 1941: 5). Some scholars have suggested that the men first met in Egypt in 1902 when Scheurleer visited Egypt together with the Dutch maecenas Adriaan Eliza Herman Goekoop (1859–1914; Jonges 1990: 28). In 1920, Scheurleer managed to arrange for von Bissing to be appointed as *bijzonder hoogleraar* of Egyptian and Near Eastern Art History at Utrecht University on behalf of the Academy of Art in The Hague. He and von Bissing entered into an agreement in which Scheurleer bought the collection from von Bissing via his Bank Scheurleer & Zoonen on behalf of the Academy of Art in installments, which were paid to von Bissing in the form of a yearly *f*. 15,000 salary. Friedrich kept the exclusive right to publish the collection, with the exception of general catalogues of the new museum (Contract between Bissing and Scheurleer n.d.). This construction would later prove problematic since the collection was regarded as the personal property of Scheurleer when he filed for bankruptcy in 1932.

On June 14, 1922, von Bissing held his inaugural lecture *Die Bedeutung der orientalistischen Kunstgeschichte für die allgemeine Kunstgeschichte* (The significance of Oriental art history for general art history) at Utrecht University (Utrecht University 1916–1935: 135). Here he taught *Oude kunstgeschiedenis van Egypte en Voor-Azië* (Ancient art history of Egypt and the Near East; see Rijksuniversiteit Utrecht 1921–1926). In addition to his work for the university, he engaged in a range of public lectures on the ancient Near East and modern Egypt. Despite its distance from the university, von Bissing moved to a residence at Statenplein 16 in The Hague near the Academy of Art (Gemeente Den Haag 1922). Von Bissing's much heralded collection was first partially shown to the public in a temporary exhibition at the Museum van Kunstnijverheid (Museum of Applied Art) in The Hague in 1923, after which the collection was transferred to Scheurleer's newly-built Museum Carnegielaan in the same city (Academie van Beeldende Kunsten 1923: 8–9; Fig. 4.2). Although the Egyptian department remained unfinished, the museum opened with much fanfare on December 20, 1924. During his time in the Netherlands, von Bissing would continually refine the display of his collection.

During his stay in the Netherlands, von Bissing did his best to settle in Dutch society: he gave lecture series, worked on the display of his collection, wrote articles for Dutch magazines about the recently found tomb of Tutankhamun (Bissing 1924a) and served on the board of the Museum of Applied Art. In 1924, these efforts were marred by two articles in a weekly Dutch newspaper published in German, *Deutsche Wochenzeitung für die Niederlande*, which were of an explicitly political character. On June 14, 1924, the newspaper re-published an earlier article of von Bissing's for the German magazine *Tannenberg*, possibly without his knowledge (Bissing 1928). In it, he explained his views on the interaction between *Völkische* ideology and Christianity, and attacked the Catholics, Jews, communists, and social democrats. He accused Catholics of being traitors and called the Old Testament "jüdische schmutzige Phantasie" (filthy Jewish fantasy) (Bissing 1924b). The first criticism came from Germany. The *Berliner Tageblatt* daily newspaper (S.n. 1924a) strongly condemned von Bissing for what was viewed as a severe slight against Dutch hospitality. Then, several Dutch newspapers picked up on the reporting and the article received a lot of backlash from the Catholic and Jewish communities in the Netherlands (Cohen 1924; S.n. 1924b, c, d, e, g, h, i; Velders 1924). This dampened the relationship between Scheurleer and von Bissing (Bissing 1928). A response in the journal *Joodsche Wachter* shows that the problem was not only the subject-matter of the article, it was also considered unseemly for a professor to express himself politically, and to do so while mentioning his affiliation with Utrecht University (Cohen 1924). He thus violated the tradition of Dutch academia to refrain from political statements in the public sphere (for this tradition, see Dorsman and Knegtmans 2006). That same year von Bissing published a second article of political character, this time without mentioning his affiliation with Utrecht University, on the future of the Rhineland (Bissing 1924c). In it, von Bissing accused the Dutch writer and politician Frans Drion (1874–1948) of working for the French. The article earned von Bissing a libel suit, which was only dropped after he publicly withdrew his statement (S.n. 1924f, i). Despite slighting Drion, this publication caused considerably less outrage in comparison to his earlier article.

In 1925 the number of problems von Bissing was causing in the public sphere grew. That year he clashed with the staff members of the Museum Carnegielaan in an argument over people walking through the Egyptian Department while he was working (Bissing 1925). This was the latest in a series of clashes between von Bissing and the staff, who found him impossible to work with and were fed up with his incessant commands in a tone which Scheurleer described as a "Preussische[r] Feldwebelton" (Prussian Sergeant's tone; Lunsingh Scheurleer 1925a). The conflict was exacerbated by an apparent nervous breakdown. According to Scheurleer, von Bissing moved his collection around the museum seemingly at random, constantly talking to himself (Lunsingh Scheurleer 1925b, 1925f), and added Petrie excavation numbers to random objects in the collection (Lunsingh Scheurleer 1925e). He also noted that it was: "Es ist das Spielen eines Kindes mit Bausteinen, die er heute so, morgen anders legt, weil es ihm eben Spass macht" ([like] a kid playing with building blocks, which he today moves here, tomorrow somewhere else, because it pleases him for a while) (Lunsingh Scheurleer 1925c).

In November 1925 Scheurleer was confronted with a different aspect of von Bissing's private *persona*. He was approached by a young man who showed him letters sent by von Bissing with homosexual undertones (Lunsingh Scheurleer 1925d, e, f). Homosexuality was much frowned upon in Dutch society, and punishable by law if one of the partners was under 21 years of age. In Germany, homosexuality was punishable by law under section 175 of the German Criminal Code. After discussions between the family, Scheurleer, and von Bissing's wife, the case was quietly resolved. At the end of January 1926, Scheurleer was notified of another homosexual incident, this time in Utrecht. To avoid a scandal for both parties, von Bissing resigned from the university for "health reasons" (Board of the Academy of Art 1926).[6]

After these incidents von Bissing was diagnosed by the psychiatrist Oswald Bumke (1877–1950) as hypersensitive and hyperimpressible. According to von Bissing, Bumke made von Bissing's homosexuality into an aftereffect of a tropical disease, by explaining it as caused by a severe case of paratyphoid fever he had contracted in Egypt in 1901 (Bissing 1926). His family, on the other hand, viewed it in terms of a nervous disorder, even considering legal incapacitation. Von Bissing withdrew to his German estate near Agg in Oberaudorf am Inn, the "Agger Bichl", where he continued publishing and researching. During this time his Protestant faith became an increasingly important factor in his life (Bissing-Carlowitz 1956). In 1925 he had become a member of the Synod of the Evangelical Lutheran Church in Bavaria. In the same year he officially joined the NSDAP (Raulwing and Gertzen 2013: 99), which he had admired since 1923. As a member of the Nazi Party, he was keenly interested in the interaction between Nazi ideology and the church. His interest did not go unnoticed and he became a negotiator between the leadership of the NSDAP—Adolf Hitler (1889–1945) and Deputy Führer Rudolf Heß (1894–1987)—and the Vatican (Raulwing and Gertzen 2013: 101, f. n. 276).

Von Bissing's interest in the interplay between *Völkische* ideology and the church predated his official affiliation with the NSDAP as is clear from his much-lamented article in the *Deutsche Wochenzeitung für die Niederlande* (Bissing 1924b). Lutheran Protestantism ties into von Bissing's ideas of "true Germanness". In his view Christianity was the only viable religion; it was no use trying to revive the ancient rites of pagan religions since they had no persuasiveness. Protestantism was the only option for the 'true German' because it "renders unto Caesar the things that are Caesar's, and unto God the things that are God's" (Matthew 22: 15–22; Mark 12: 17), i.e., a separation of church and state. According to von Bissing, Catholics or Jews could never be 'true Germans' because their allegiance was to powers outside of Germany. For a long time, von Bissing believed that Hitler did not know of the increasing influence of the NSDAP on the German church (Bissing 1934). In a letter to Benedictine abbot Alban Schachleiter (1861–1937)—a supporter of National Socialism—he described Hitler as a blind man leading the blind (Raulwing and Gertzen 2013: 98). Unfortunately for von Bissing, the letter was intercepted by Schachleiter's housekeeper Gildis Engelhardt, who informed the NSDAP. Despite a personal apology to Hitler, von Bissing was expelled from the Party on May 24, 1937 (Raulwing and Gertzen 2013: 98). The sources on von Bissing's position during the Second World War are scarce, but it is clear that the realities of war were hard on him. His brothers and nephews were stationed at the eastern front, and in 1941 another nephew, who did not serve in the army, died. After the war, von Bissing's stepbrother was taken prisoner by the Russian army. The post-war reforms left von Bissing poor and disgruntled. By that time, he referred to the war as foolish and expressed his hope for rebuilding Germany (Bissing 1947). He died on January 12, 1956.

4.3 Von Bissing in the Netherlands: His Network and Collection

Von Bissing's personal and professional network played a crucial role in his entry into and acceptance in Dutch society. During his stay in the Netherlands in the 1920s, von Bissing could look back on a long international career spanning a quarter-century. His network was strongly rooted in the international world of Egyptology. It also included German nobility, collectors and traders of antiquities, and supporters of the Greater Netherlands movement. Throughout his network we find that the intellectual, political, and commercial

[6]It has been suggested that von Bissing was forced to leave due to his political opinions. There is no archival material to support this idea. The only documentary evidence is an anti-NSDAP article from 1932 (S.n. 1932a).

worlds collide. For example, one of his correspondents in the Netherlands was the Dutch-German Assyriologist Franz Marius Theodor de Liagre Böhl (1882–1976). Both men shared an interest in the ancient Near East and theology. But their paths also intersected in the political realm. Böhl was personally acquainted with the ex-emperor Wilhelm II (1859–1941), who had found refuge in Doorn in the Netherlands after the First World War. Later, he would become part of the Doorner Arbeits-Gemeinschaft (Doorner Workgroup; 1931–1938), an annual get-together of a select group of scholars under the auspices of the *Kaiser* at his residence-in-exile *Huize Doorn* to discuss scholarly questions in an informal setting (Liagre Böhl 1931; Franzen 2017). Von Bissing's admiration for the German monarchy is well-documented. He viewed it as the best form of polity (Bissing 1931, 1939a, c).[7]

The acceptance of von Bissing in Dutch society was prefaced by serious concern about his "hyper nationalism". Following a report by the *Vossische Zeitung* (S.n. 1922a), Dutch newspapers such as the local *Utrechtsch Nieuwsblad* (S.n. 1922b), and the Catholic daily paper *De Tijd* (S.n. 1922c) commented on his nationalism. Such reports were taken very seriously in the Netherlands. Scheurleer warned von Bissing not to enter his polemics into his inaugural lecture at Utrecht University (Bissing 1922). Yet, we will see that any hurdles for his appointment in the Netherlands were smoothed out by the influence of his network. In particular, three groups within his network played a crucial role in his entry into and acceptance in Dutch society: his Belgian network from the First World War, his reputation in his Egyptological network, and the network which revolved around his collection.

Von Bissing's position as a professor had to be formally approved by Utrecht University's college of curators. By law, they not only had to approve of the creation of a new chair in Egyptian and Near Eastern Art History, but also the academic status of the man who would fill this chair. Utrecht University was acutely aware of von Bissing's role in the Dutch-ification of Ghent University. However, his Belgian network vouched for him. Willem Vogelsang, professor of Art History (1875–1954), wrote in a letter to the secretary of the college of curators on June 10, 1921:

> In den grooten oorlog is hij [von Bissing] korten tijd belast geweest met de organisatie van het onderwijs in het bezette gebied van Belgie en werd daar, naar ons van betrouwbare zijde medegedeeld, bij uitzondering ook van Belgischen kant gewaardeerd.
> (During the Great War he [i.e., von Bissing] was briefly in charge of the organization of h[igher] education in occupied Belgium and, as we have heard from reliable sources, was one of the few who was also appreciated by the Belgians.) (Vogelsang 1921b)

These "reliable sources" were likely some of the collaborating Flemish activists who had fled to the Netherlands after the German defeat in 1918. Several professors of the 'Von Bissing University' had made their way to the Netherlands: Middle Dutch professor and librarian Willem Lodewijk de Vreese (1869–1938) had become the director of the Municipal library of Rotterdam (Gommers 2004), and Zoology professor Jan Versluys Janszoon (1873–1939) had settled in Hilversum (Bemmel 1939). Von Bissing knew these men personally (Bissing 1923). Several students also made their way to the Netherlands. For example, Robert van Genechten (1895–1945) continued his studies at Utrecht University, where he became the leader of the Dutch/Flemish Student-alliance. Some archaeologically-inclined students of the Von Bissing University later also made their way to the Netherlands: Augustus Remouchamps (1892–1927) became curator of Prehistoric Archaeology at the National Museum of Antiquities in Leiden in 1925 (Eickhoff 2003: 57–61, 330), and Ferdinand de Waele (1896–1977) became professor of Archaeology at the Catholic University of Nijmegen in 1932 (Janssen 1979). A similarly political and academic transfer was made by the Belgian archaeologist Jacques Breuer (1892–1971), who after 1918 temporarily worked for the National Museum of Antiquities in Leiden (Eickhoff 2003: 330). In addition, von Bissing had corresponded about his Belgium propaganda pamphlets with students of the Utrecht department of the General Dutch Alliance in 1915 (S.n. 1915).

The advocacy for von Bissing's professorship was helped by his focus on art history. Although he also published on philology and religious studies, von Bissing was at the time best known for his art historical approach, exemplified in his three-volume *Denkmäler ägyptischer Sculptur* (1906–1914) which was, partially, based on his own collection. This approach made him an attractive academic member of the Utrecht academic community as Utrecht University specialized in art history. In 1907 both Leiden and Utrecht University had vied for a permanent professorship in art history. Yet, the budget only allowed for one position. Utrecht succeeded in securing the ordinary professorship, while Leiden had to content itself with a part-time endowed professorship. Willem Vogelsang became the first ordinary professor of art history in the Netherlands, at Utrecht University (Hoogenboom 1998). Vogelsang was a proponent of the formalist school of art history that was based on the work of Swiss art historian Heinrich Wölfflin (1864–1945) and got his doctorate under art historian Berthold Riehl (1858–1911) and archaeologist Adolf Furtwängler (1853–1907). Von Bissing was steeped in the same school of art history as Vogelsang and had completed his habilitation under

[7] We do not know how Böhl and von Bissing's political views overlapped. Von Bissing talks openly about politics in his letters to Böhl. In Böhl's recent biography von Bissing is only mentioned once, as the buyer of finds from Sichem (De Liagre Böhl 2021).

Furtwängler. In a letter to the college of curators of June 8, 1921, Vogelsang wrote:

> Hij ziet daarin een welkome aanvulling van de studie van Kunstgeschiedenis aan ons Instituut, die zonder de kennis der Egyptische Kunst haar onderbouw mist, terwijl het niet aangaat, onze studenten in Kunsthistorie voor dit onderdeel naar Leiden te verwijzen, aangezien een waarlijk vruchtbare studie voortdurend contact met de Egyptische kunstsfeer veronderstelt.
> (He sees it [i.e., the chair in Near Eastern and Egyptian art history] as a welcome addition to the study of art history at our institute, which lacks a foundation without knowledge of Egyptian art. It would not be proper to send our art history students to Leiden for this part of their studies because a truly fruitful education presupposes prolonged contact with Egyptian art.)
> (Vogelsang 1921a)

In other words, von Bissing's appointment would enable Vogelsang to bolster the position of art history at Utrecht University in the Dutch academic landscape, especially in comparison to Leiden. Von Bissing's auxiliary teachings would also allow Vogelsang to ensure a certain amount of basic knowledge in the training of his students (compare Hoogenboom 1998: 31–31). In the same letter, Vogelsang revealed two further factors which spoke in Friedrich's favor: his Egyptological standing and his collection.

The fact that von Bissing was regarded as a lightweight by members of the Berlin School of Egyptology due to his (perceived) lack of philological expertise (e.g., Grapow 1973: 25–26; Raulwing and Gertzen 2013: 64, 126; Voss 2013: 147) was likely an open secret among Egyptologists. Yet, within the Dutch Egyptological tradition and institutional infrastructure other priorities were at play. Most of the early practitioners of Egyptology in the Netherlands were theologians and classicists with no formal training in Egyptology. Egyptologists like Willem Pleyte (1836–1903)—whose name is prominently displayed on the façade of the Egyptian Museum at Tahrir square in Cairo—had been largely self-taught. It was only in 1902 that the first university position in Egyptology was created in the form of a *privaatdocentschap* for Pieter Boeser at Leiden University (Kaper 2014: 40). The early twentieth century saw an increased professionalization of Egyptology in which German-speaking Egyptology played a major role. Boeser and his Groningen colleague Gerardus van der Leeuw (1890–1950) were among the first Dutch Egyptologists to have a formal education in Egyptology. Boeser had studied under Egyptologists Adolf Erman and Georg Steindorff (1861–1951) in Berlin and Leipzig; Van der Leeuw had studied Egyptology with the theologian William Brede Kristensen (1867–1953) in Leiden, the Egyptologists Adolf Erman and Georg Möller (1876–1921) in Berlin, and the Egyptologist Kurt Sethe (1869–1934) in Göttingen. German-speaking Egyptology therefore had great standing in the Netherlands (compare Raven 2021: 142).[8] In this context, despite his shortcomings as a philologist, von Bissing was regarded as an art historian who could strengthen the position of Egyptology as an academic specialization in the Netherlands. Vogelsang also highlights von Bissing's "excellent publications on the art of ancient Egypt" and thus propagates the (semi) public image of von Bissing as a knowledgeable scholar (Vogelsang 1921b; compare Akademie van Beeldende Kunsten 1923: 8–9; S.n. 1923).

Another, and possibly more important, reason for von Bissing's appeal was his immense collection, which had been formed over 30 years of collecting and contained about 15,000 objects. A large part of the collection was acquired on the art markets in Egypt, Italy, France, Germany, and Great Britain, but it also contained artifacts from formal excavations in Egypt. This included objects from his own excavations of the Sun Temple of Nyussere at Abu Gurob, pottery fragments that he collected at Naukratis, objects from the excavations of the British School of Archaeology in Egypt (BSAE) at Abydos, Deir el-Bahri, Ehnasiya, Gheyta, Saft el-Henna, Memphis, Hawara and Tarkhan, and objects from Egyptologist Auguste Mariette's (1821–1881) excavations at the Serapeum and the excavations of Said Khashaba *pasha* (fl. 1910–1950) (Bissing 1924d, 5; Essen 1927). In the Netherlands, Egyptology had traditionally centered around the city of Leiden, where in 1818 the National Museum of Antiquities was founded in conjunction with a university chair in archaeology. By the 1920s this museum had acquired an impressive array of Egyptian antiquities, which formed the basis for the study of ancient Egypt in Leiden (Keurs and Wirtz 2018). At the start of the twentieth century, there were two further major centers of academic Egyptology in the Netherlands: the university cities of Utrecht and Groningen. These cities did not possess impressive collections and were eager to acquire study collections of their own. Utrecht did in fact obtain a small study collection in 1923 after von Bissing became extraordinary professor, i.e., a professor without a chair. Another small collection became available to Groningen in 1925 with the *Semitistische werkkamer* (Semitistic Work-room) (S.n. 1925, November 8), whose Egyptian antiquities were described by von Bissing.

Von Bissing's move to the Netherlands would not have been possible without his collection and the network that had formed around it. In the Dutch context Scheurleer was a key player. He had long been a patron of the arts and had contributed several antiquities to the National Museum of Antiquities in Leiden. But although this museum was nearby and there

[8] This change in direction is also visible in the foreign language publications of Dutch Egyptologists, e.g., Pleyte's foreign language publications are all in French while Boeser preferred German.

Fig. 4.3 From left to right: C. W. Lunshing, E. F. Prins de Jong, an unknown boy, F. W. Freiherr von Bissing and an unknown man behind Scheurleer's house in the Hague on the site of the still-to-be-build Museum Carnagielaan, 1922. (Allard Pierson, archief C. W. Lunsingh Scheurleer, University of Amsterdam, The Netherlands)

was even already a museum containing an archaeological collection in The Hague (Museum Meermanno), Scheurleer dreamt of opening his own modern, public archaeological museum in The Hague. This museum was to display his own collection but would also allow visitors to engage in comparative critique with art from the entire ancient world. Von Bissing's collection allowed Scheurleer to bring this dream to fruition. Scheurleer was not the only one who benefitted from von Bissing's collection. His right-hand woman, the art historian Elisabeth Prins de Jong (1882–1957; Marcus-de Groot 2003: 105–7), wrote her doctoral thesis on pottery from Naukratis in von Bissing's collection (Prins de Jong 1925). Von Bissing himself also recognized that a large part of "his value" in the Netherlands was generated by his collection (Fig. 4.3). In a letter of March 14, 1921 to Lunsingh Scheurleer concerning the possibility of a position in the Netherlands, he wrote: "Welche Möglichkeit hat Holland bei meiner Berufung die Mitnahme meiner Sammlung und Bibliothek zu sichern? Ohne beide bin ich natürlich wertlos" (What options does Holland have to guarantee that I can take my collection and library with me when I am called to the Netherlands? Without both I am, of course, worthless) (Bissing 1921b). This feeling of dependence was part of a larger affective relationship with the objects that was deeply personal and emotional. In a letter from Elisabeth von Bissing to Scheurleer of April 3, 1925, she describes how von Bissing kept his favorite pieces in his bedroom where only he could enjoy them (Bissing-Carlowitz 1925a).

Despite von Bissing's strong political opinions, his political network was not a major force in the Netherlands. Several friends and tangential acquaintances such as the Assyriologist Franz de Liagre Böhl and the banker and collector Eduard Freiherr von der Heydt were closely acquainted with ex-Emperor Wilhelm II. There is evidence to suggest that von Bissing was also a frequent guest at *Huize Doorn*, his residence in the Netherlands after 1919, but the sources are ambiguous as to the precise period that these visits took place (Snijders 1931; UTR, archive 14). Although von Bissing was open about his personal political convictions in his letters to Scheurleer (Bissing 1920, 1921a), his NSDAP network seems to have been confined to Germany.[9] This is not to say that von Bissing's Dutch circle of acquaintances did not include reactionary thinkers and right-wing sympathizers; contributor to the Scheurleer Museum, von der Heydt became a NSDAP member in 1933 and former Von Bissing University student Robert van Genechten later became a prominent member of the Nationaal-Socialistische Beweging (NSB: Dutch National Socialist Party).

Despite von Bissing's attempts to integrate in Dutch society, we know very little about his private life in the Netherlands. His wife Elisabeth and daughter Myrrha did not

[9] It has been suggested that von Bissing syphoned money for the NSDAP through his Dutch bank account, but we have not been able to find any contemporary sources to substantiate this rumour. However, his name is mentioned in connection to this in a 1928 article in the *Bayerischer Kurier* (S.n. 1928). On February 11, 1932, a "professor of the university, named von Bissing" is mentioned in the minutes of the French Parliament by the member of parliament Paul Faure (1878–1960), based on an article in *Le Journal* (see S.n. 1932b: 575). We have not been able to locate this article.

live with him in The Hague for most of his time there. Elisabeth considered Friedrich's homosexuality harmful to Myrrha's development (Bissing-Carlowitz 1925b). His homosexual network itself remains unknown. The only sources which, except for his sexual orientation, reveal something of his private life are his letters to Willem Lodewijk de Vreese, whom he knew from his days in Belgium; Constant Willem Lunsingh Scheurleer, who became his surrogate family in the Netherlands; and Franz Marius Theodor de Liagre Böhl. In these sources we can glimpse the man behind the academic articles and the political polemics. He talks about his holidays, visits of Scheurleer's children, and his plans. But it is also in these sources that we find the clearest indication of von Bissing's rampant antisemitism and racism and the way he equated "Russian", "anti-Christian" and "Jewish" with "Liberalism" and "Berlin". In a letter to Böhl of January 26, 1939 we find:

> Jeder Fellah ist ein Kulturträger selbst gegenüber Tolstoi oder Dostojewski, die ich neben Nietzsche in tiefster Seele hasse und verachte. An diesen Kerlen ist die Welt krank geworden mehr als an irgendeiner Pest. Aber der moskowitische Ungeist-Nietzsche hatte slawisches Blut- steht immer wieder auf, in [Alfred] Rosenberg z.B. Wie soll das enden? Wir bekennen uns in Deutschland zum Christentum, aber immer wieder wird der Versuch gemacht Christus auszuschalten. Man hat jetzt den Begriff des „Gottgläubigen" erfunden, den ich mit „jüdisch" wiedergeben muss; denn er letzt Christentum ab. So treiben die Herrn jüdische Propaganda und merken es nicht (...).
> (Every *fellah* [Egyptian farmer] is a Kulturträger [culture bearer] compared to Tolstoi or Dostojewski whom, in addition to Nietzsche, I hate and despise with a passion. These guys have infested the world worse than any plague. But the Moscovian demon—Nietzsche had Slavic blood—always turns up again, e.g., in [Alfred] Rosenberg. How is it supposed to end? We profess Christianity in Germany, but again and again attempts are made to eliminate Christ. Now they have invented the term, "Gottgläubiger" [believer in God] which I must translate to "Jewish" since it turns away from Christianity. So, these gentlemen push Jewish propaganda and don't even notice.) (Bissing 1939a)

4.4 Conclusions

Friedrich von Bissing's four year stay in the Netherlands after the end of the First World War was inextricably tied in with his network, his collection of Egyptian objects, and his private and public academic *persona*. He entered Dutch society by the grace of his collection and his public *persona*. Behind the scenes, his network removed any obstacles for his successful move to the Netherlands, and justified parts of his personal history which did not fit his public *persona* and may have caused problems. His network crucially included Flemish scholars and activists who defended his previous work in Belgium, and Dutch scholars who emphasized his public *persona* as a knowledgeable art historian, while ignoring his reputation as a philologist. However, it was his private *persona* that soon caused much friction in the Netherlands, also within his network. His political convictions and activities affected his academic credibility because the Dutch academic tradition did not allow professors to explicitly express themselves politically in connection to their position. In addition, his homosexuality, once it became more public, led to his forced departure back to Germany. His academic network started to withdraw from him, yet allowed him—in part due to the cultural capital his collection afforded him—to make a quiet exit to save face. As a result, certain stereotypes within von Bissing's public *persona*—the knowledgeable (but eccentric) Egyptologist and the passionate collector—were allowed to persist.

In Dutch society, von Bissing is nowadays best remembered because of his collection (e.g., Lunsingh Scheurleer 2009: 14–15), which left an indelible mark on the landscape of Dutch Egyptological collections. After the bankruptcy of Bank Scheurleer & Zoonen in 1932, parts of von Bissing's collections found their way to the Municipal Museum The Hague, the Allard Pierson Museum in Amsterdam, the Art Historical Institute in Utrecht and the National Museum of Antiquities in Leiden. The collections of the Municipal Museum The Hague and the Art Historical Institute eventually found their way to the Allard Pierson. There, von Bissing's private *persona*, which had caused him so much trouble in the Netherlands, has recently gained new attention in the context of provenance research, initiated by a new post-colonial awareness of the politics of collecting and collections. As a result, von Bissing has gradually outgrown the traditional above-mentioned stereotypes that had been sanctioned by his former network, and his biography, including his formerly hidden homosexual life, has been contextualized within the larger framework of the intellectual and social history of Egyptology.

References

Primary Sources

Archival Material

Academie van Beeldende Kunsten (1923). [Annual report 1923]. Haags Gemeentearchief (archive 0058-01, inventory number 547.4), The Hague, The Netherlands.
Bissing, F. W. Freiherr von (1920, October 28). [Letter to Constant W. Lunsingh Scheurleer]. Allard Pierson Archive (inventory number 142), Amsterdam, The Netherlands.
Bissing, F. W. Freiherr von (1921a, March 8). [Letter to Constant W. Lunsingh Scheurleer]. Allard Pierson Archive (inventory number 142), Amsterdam, The Netherlands.
Bissing, F. W. Freiherr von (1921b, March 14). [Letter to Constant W. Lunsingh Scheurleer]. Allard Pierson Archive (inventory number 142), Amsterdam, The Netherlands.

Bissing, F. W. Freiherr von (1922, June 5). [Letter to Constant W. Lunsingh Scheurleer]. Allard Pierson Archive (inventory number 142), Amsterdam, The Netherlands.

Bissing, F. W. Freiherr von (1923, April 1). [Letter to Willem L. de Vreese]. Universiteitsbibliotheek Leiden (BLP 2998), Leiden, The Netherlands.

Bissing, F. W. Freiherr von (1925, March 7). [Letter to Constant W. Lunsingh Scheurleer]. Allard Pierson Archive (inventory number 143), Amsterdam, The Netherlands.

Bissing, F. W. Freiherr von (1926, February 26). [Letter to Constant W. Lunsingh Scheurleer]. Allard Pierson Archive (inventory number 142), Amsterdam, The Netherlands.

Bissing, F. W. Freiherr von (1928, January 3). [Letter to Franz M. T. de Liagre Böhl]. Universiteitsbibliotheek Leiden (BPL 3081), Leiden, The Netherlands.

Bissing, F. W. Freiherr von (1931, May 10). [Letter to Franz M. T. de Liagre Böhl]. Universiteitsbibliotheek Leiden (BPL 3081), Leiden, The Netherlands.

Bissing, F. W. Freiherr von (1934, August 27). [Letter to Franz M. T. de Liagre Böhl]. Universiteitsbibliotheek Leiden (BPL 3081), Leiden, The Netherlands.

Bissing, F. W. Freiherr von (1939a, January 26). [Letter to Franz M. T. de Liagre Böhl]. Universiteitsbibliotheek Leiden (BPL 3081), Leiden, The Netherlands.

Bissing, F. W. Freiherr von (1939b, March 2). [Letter to Franz M. T. de Liagre Böhl]. Universiteitsbibliotheek Leiden (BPL 3081), Leiden, The Netherlands.

Bissing, F. W. Freiherr von (1939c, March 3). [Letter to Franz M. T. de Liagre Böhl]. Universiteitsbibliotheek Leiden (BPL 3081), Leiden, The Netherlands.

Bissing, F. W. Freiherr von (1947, January 17). [Letter to Franz M. T. de Liagre Böhl]. Universiteitsbibliotheek Leiden (BPL 3081), Leiden, The Netherlands.

Bissing-Carlowitz, E. Freifrau von (1925a, April 3). [Letter to Constant W. Lunsingh Scheurleer]. Allard Pierson Archive (inventory number 147), Amsterdam, The Netherlands.

Bissing-Carlowitz, E. Freifrau von (1925b, November 28). [Letter to Constant W. Lunsingh Scheurleer]. Allard Pierson Archive (inventory number 147), Amsterdam, The Netherlands.

Bissing-Carlowitz, E. Freifrau von (1956, February 8). [Letter to Frans M. T. de Liagre Böhl]. Netherlands Institute for the Near Neast (collectie Böhl, (Collegae brieven (von Soden) (von Bissing)), Leiden, The Netherlands.

Board of the Academy of Art (1926, February 10). [Letter to the college of curators of Utrecht University], Utrechts Archief (archive 59, inventory number 8380, Utrecht, The Netherlands.

Boeser, P. A. A. (1919, December 10). [Report to the Ministry of Education, Arts and Sciences concerning the collection of F. W. Freiherr von Bissing]. Rijksmuseum van Oudheden (BNR 1-17-17-5-168), Leiden, The Netherlands.

[Contract between von Bissing and Scheurleer]. Stadsarchief Amsterdam (archive 30238, inventory number 2352), Amsterdam, The Netherlands.

Essen, C. C. van (c. 1927) [Copy of a list of von Bissing inventory numbers]. Rijksmuseum van Oudheden (inventory number 6.4.3/19 (Documentatie Egypte 19 Coll. Von Bissing), Leiden, The Netherlands.

Gemeente Den Haag (c. 1922). [Citizen Registration of F. W. Freiherr von Bissing]. Haags Gemeentearchief (archive 0354-01, inventory number 00984, scan 309). The Hague, The Netherlands.

Liagre Böhl, F. M. T. de (1931, October 15/16). [Letter to the Hofmarschallamt of the former Kaiser Wilhelm II]. Utrechts Archief (archive 14, inventory number 273, scan 299–301), Utrecht, The Netherlands.

Lunsingh Scheurleer, C. W. (1925a, March 11). [Letter to Elisabeth Freifrau von Bissing]. Allard Pierson Archive (inventory number 147), Amsterdam, The Netherlands.

Lunsingh Scheurleer, C. W. (1925b, May). [Letter to Elisabeth Freifrau von Bissing]. Allard Pierson Archive (inventory number 147), Amsterdam, The Netherlands.

Lunsingh Scheurleer, C. W. (1925c, May 8). [Letter to Elisabeth Freifrau von Bissing]. Allard Pierson Archive (inventory number 147), Amsterdam, The Netherlands.

Lunsingh Scheurleer, C. W. (1925d, November 3). [Letter to Elisabeth Freifrau von Bissing]. Allard Pierson Archive (inventory number 147), Amsterdam, The Netherlands.

Lunsingh Scheurleer, C. W. (1925e, November 11). [Letter to Elisabeth Freifrau von Bissing]. Allard Pierson Archive (inventory number 147), Amsterdam, The Netherlands.

Lunsingh Scheurleer, C. W. (1925f, November 25). [Letter to Elisabeth Freifrau von Bissing]. Allard Pierson Archive (inventory number 147), Amsterdam, The Netherlands.

Snijders, G. A. S. (1931, May 9). [Letter to A.H. van Karnebeek]. Stadsarchief Amsterdam (archive 30288, inventory number 2295), Amsterdam, The Netherlands.

Utrecht University (1916, September 1–1935, September 31). [Resolutions and minutes of the senate of Utrecht University], Utrechts Archief (archive 292–1, inventory number 1–9), Utrecht, The Netherlands.

Vogelsang, W. (1921a, June 8). [Letter on behalf of the senate of Utrecht University to the college of curators of Utrecht University]. Utrechts Archief (archive 59, inventory number 834), Utrecht, The Netherlands.

Vogelsang, W. (1921b, June 10). [Letter to Johan F. Hooft Graafland]. Utrechts Archief (archive 59, inventory number 834), Utrecht, The Netherlands.

Weigall, A. (1902, March 29). [Letter to Alice Craggs]. Metropolitan Museum of Art (Arthur Weigall papers, EAA-479), New York, United States.

Newspaper and Magazine Articles

Bissing, F. W. von (1924a, January-June). Om Toeot-Anch-amoen's graf: De reclame en het resultaat. *Haagsch Maandblad*, 75–81.

Bissing, F. W. von (1924b, June 14). Christentum, Völkische Bewegung u. Ultramontanismus. *Deutsche Wochenzeitung für die Niederlande*, 2–3.

Bissing, F. W. von (1924c, September 6). Preussenflucht. *Deutsche Wochenzeitung für die Niederland*, 2–3.

Bissing, F. W. von (1924d, December 19). De verzameling Von Bissing in het nieuwe Museum Scheurleer: Hoe de verzameling is ontstaan. *Het Vaderland: staat- en letterkundig nieuwsblad, avondblad*, 5.

Cohen, I. (1924, July 25). Het geval van Professor Freiherr von Bissing. *De Joodsche Wachter: Halfmaandelijksch orgaan voor Groot-Nederlands Jodendom*, 153.

Sine nomine. (1915, July 3). De Vlaamsche quaestie. *De Toekomst: Weekblad voor Nederland*, 277–278.

Sine nomine. (1922a, January 19). Der Ruf nach Utrecht. *Vossische Zeitung*, 2.

Sine nomine. (1922b, January 20). Prof von Bissing naar onze stad? *Utrechtsch Nieuwsblad*, 2.

Sine nomine. (1922c, February 10). Een woeste Al-Duitscher professor in Utrecht. *De Tijd*, 2.

Sine nomine. (1923, November 30). Het onderwijs van prof. von Bissing. *Algemeen Handelsblad, avondblad*, vierde blad, 13.

Sine nomine. (1924a, July 7). Deutschvölkische im Ausland: Holländische Beschwerden über Mißbrauch des Gastrechts. *Berliner Tageblatt*, 3.

Sine nomine. (1924b, July 10). Optreden van in Nederland gevestigde „Deutschvölkische". *Rotterdamsch nieuwsblad*, 13–14.

Sine nomine. (1924c, July 18). Geïmporteerd Anti-Semitisme. *Het Nnieuw Isrealitisch Weekblad*, 5.

Sine nomine. (1924d, July 19). Een ongewensche vreemdeling. *Het Limbursch dagblad*, 2.

Sine nomine. (1924e, July 25). Ludendorff in Nederland. *De Java-post*, 355–356.

Sine nomine. (1924f, September 19). Beleediging en Smaadschrift. *Het centrum*, 12.

Sine nomine. (1924g, September 26). Ludendorff in Nederland. *De Java-post*, 464.

Sine nomine. (1924h, November 13). Een Duitsch professor in Nederland. *Het nieuws van den dag voor Nederlandsch-Indie*, 14.

Sine nomine. (1924i, November 30). Drion-von Bissing. *Algemeen Handelsblad, ochtendblad*, tweede blad, 5.

Sine nomine. (1924j, December 20). Het Museum Carnegielaan 12: De officiële openingsplechtigheid. *Haagsche Courant*, vijfde blad, 2.

Sine nomine. (1925, November 7). Babylonische en Egyptische oudheden. *Nieuwsblad van het Noorden*, 4.

Sine nomine. (1928, January 5). Hitlers Geldgeber. *Bayerischer Kurier*, 3.

Sine nomine. (1932a, January 16). Naar aanleiding van den Duitschen Antisemitischen import. *Het Vaderland: staat- en letterkundig nieuwsblad, avondblad B*, 2.

Velders, W. J. J. (1924, August 3). Zending onder de joden. *Amsterdamse kerkbode*, 2–3.

Secondary Sources

Babinger, F. (1956). Friedrich-Wilhelm Freiherr von Bissing. *Jahrbuch der Bayerischen Akademie der Wissenschaften, 1956*, 190–202.

Bemmel, J. F. van (1939). Levensbericht J. Versluys Jzn. *Jaarboek van de Koninklijke Nederlandse Academie van Wetenschappen, 1938–1939*, 2–14.

Bierbrier, M. L. (Ed.). (2015). *Who was who in Egyptology* (5th rev. ed.). London: Egypt Exploration Society.

Brunner, H. (1956). Friedrich Wilhelm Freiherr von Bissing (22. April 1873 bis 12. Januar 1946). *Archiv für Orientforschung, 17*, 484–485.

Colla, E. (2008). *Conflicted antiquities: Egyptology, Egyptomania, Egyptian modernism*. Durham, NC: Duke University Press.

Daston, L., & Sibum, H. O. (2003). Scientific personae and their histories. *Science in context, 16*(1–2), 1–8.

De Liagre Böhl, H. (2021). *Bijbel en Babel: Franz de Liagre Böhl (1882–1976)*. Amsterdam: Prometheus.

Dorsman, L. J., & Knegtmans, P. J. (Eds.). (2006). *Stille wijkplaatsen. Politiek aan de Nederlandse universiteiten sedert 1876*. Hilversum: Verloren.

Eickhoff, M. (2003). *De oorsprong van het eigene: Nederlands vroegste verleden, archeologie en nationaal-socialisme*. Amsterdam: Boom.

El-Daly, O. (2005). *Egyptology, the missing millennium: Ancient Egypt in medieval Arabic writings*. London: Routledge.

Fagan, B. (Ed.). (2014). *The great archaeologists*. London: Thames & Hudson.

Franzen, C. J. (2017). Wilhelm II. und die „Doorner Arbeits-Gemeinschaft". In T. Beigel, & S. Mangold-Will (Eds.), *Wilhelm II. Archäologie und Politik um 1900* (pp. 101–121). Stuttgart: Franz Steiner Verlag.

Garnett, R. S. (2020). *Lion, eagle, and swastika: Bavarian monarchism in Weimar Germany 1918–1933*. London: Routledge. (Original work published 1991)

Gelder, H. E. van (1941). Constant Lunsingh Scheurleer en zijn Haagsche milieu. *Bulletin van de Vereeniging tot Bevordering der kennis van de Antieke Beschaving, XVI*(1), 4–8.

Gertzen, T. G. (2016). Der verlorene Sohn? Das politische Engagement des Ägyptologen Friedrich Wilhelm von Bissing im besetzten Belgien während des Ersten Weltkrieges. In S. Bischoff, C. Jahr, T. Mrowka, & J. Thiel (Eds.), *Belgica—terra incognita? Resultate und Perspektiven der Historischen Belgienforschung* (pp. 97–106). Münster: Waxmann Verlag.

Gommers, M. (2004). Vreese, W. L. de. In W. van Anrooij, I. Biesheuvel, K. van Dalen-Oskam, & J. Noordegraaf (Eds.), *Bio- en bibliografisch lexicon van de neerlandistiek*. https://www.dbnl.org/tekst/anro001bioe01_01/vree007.php#92

Grapow, H. (1973). *Meine Begegnung mit einigen Ägyptologen*. Berlin: Richard Seitz.

Grimm, A., & Schoske, S. (2010). *Friedrich Wilhelm Freiherr von Bissing: Ägyptologe, Mäzen, Sammler* (R.A.M.S.E.S., 5). München: Staatliches Museum Ägyptischer Kunst.

Haarlem, W. van (2014). Lotgevallen van een verzamelaar: Friedrich Wilhelm Freiherr von Bissing' (1873–1956). *Allard Pierson Mededelingen, 109*, 2–5.

Hoogenboom, A. (1998). Kunstgeschiedenis aan de universiteit: Willem Vogelsang (1875–1954) en Wilhelm Martin (1876–1954). In P. Hecht, A. Hoogenboom, & C. Stolwijk (Eds.), *Kunstgeschiedenis in Nederland: Negen opstellen* (pp. 25–44). Amsterdam: Prometheus.

Hoorn, G. van, & Gelder, H. E. van (1941). In memoriam Professor Dr. C.W. Lunsingh Scheurleer. *Bulletin van de Vereeniging tot Bevordering der Kennis van de Antieke Beschaving te 's-Gravenhage, 16*(1), 4–8.

Illner, E. (Ed.). (2013). *Eduard von der Heydt. Kunstsammler – Bankier – Mäzen*. München: Prestel Verlag.

Janssen, A. J. (1979). Waele, Ferdinand de (1896–1977), archeoloog, classicus. *Nationaal Biografisch Woordenboek, VIII*, 871–884.

Jonges, A. (1990). De ontstaansgeschiedenis van het Allard Pierson Museum: Archeologisch museum van de Universiteit van Amsterdam. *Tijdschrift voor Mediterrane Archeologie, 6*, 27–32.

Kaper, O. E. (2014). De geschiedenis van de egyptologie aan Nederlandse universiteiten. In O. E. Kaper, & J. G. Derksen (Eds.), *Waar de geschiedenis begon: Nederlandse onderzoekers in de ban van spijkerschrift, hiërogliefen en aardewerk* (pp. 39–62). Leiden: Nederlands Instituut voor het Nabije Oosten.

Keurs, P. ter, & Wirtz, W. (Eds.). (2018). *Rijksmuseum van Oudheden Leiden: Een geschiedenis van 200 jaar*. Zwolle: Waanders.

Lunsingh Scheurleer, R. (2009). *Toekomst voor het Verleden: 75 jaar Allard Pierson Museum, archeologisch museum van de Universiteit van Amsterdam 1934–2009*. Zwolle: Waanders.

Marcus-de Groot, Y. (2003). *Kunsthistorische vrouwen van weleer: De eerste generatie vóór 1921*. Hilversum: Uitgeverij Verloren.

Merryman, J. H. (Ed.). (2006). *Imperialism, art and restitution*. Cambridge: Cambridge University Press.

Müller, H. W. (1965). Friedrich Wilhelm Freiherr von Bissing 1983–1956. *Zeitschrift für ägyptische Sprache und Altertumskunde, 81*, X–XII.

Paul, H. (2016). Introduction: Repertoires and performances of academic identity. *Bijdragen en mededelingen betreffende de geschiedenis der Nederlanden – Low Countries Historical Review, 131*, 3–7.

Prins de Jong, E. F. (1925). *Scherben aus Naukratis: Aus der Sammlung des Universitäts-Professors Dr. Fr. W. Freih. von Bissing* [Unpublished Doctoral dissertation]. Utrecht University, The Netherlands.

Quirke, S. (2010). *Hidden hands: Egyptian workforces in Petrie excavation archives, 1880–1924*. London: Bristol Classical Press.

Raulwing, P., & Gertzen, T. L. (2013). Friedrich Wilhelm Freiherr von Bissing im Blickpunkt ägyptologischer und zeithistorischer Forschungen: Die Jahre 1914 bis 1926. In T. Schneider, & P. Raulwing (Eds.), *Egyptology from the First World War to the Third Reich* (pp. 34–119). Leiden: Brill.

Raven, M. J. (2021). The Netherlands. In A. Bednarksi, A. Dodson, & S. Ikram (Eds.), *A history of world Egyptology*. Cambridge: Cambridge University Press.

Reid, D. M. (2002). *Whose pharaohs? Archaeology, museums, and Egyptian national identity from Napoleon to World War I*. Cairo: The American University in Cairo Press.

Renfrew, C. (2000). *Loot, legitimacy and ownership*. London: Duckworth.

Rijksuniversiteit Utrecht (1921–1926). *Jaarboek der Rijks-universiteit te Utrecht* (Series 1921–1922–1925–1926). Utrecht: Universiteit Utrecht.

Schneider, T., & Raulwing, P. (Eds.). (2013). *Egyptology from the First World War to the Third Reich*. Leiden: Brill.

Sine nomine. (1932b, February 11). *Journal officiel de la République française: Débats parlementaires. Chambre des députés : compte rendu in-extenso*, 575.

Sine nomine. (1953). Grußwort an F. W. Frh. Von Bissing zum 22.4.1953, *Zeitschrift für Ägyptische Sprache und Altertumskunde*, *79*, 1–2.

Voss, S. (2013). *Geschichte der Abteilung Kairo des DAI im Spannungsfeld deutscher politischer Interessen* (Vol. I) (Menschen – Kulturen – Traditionen, Forschungscluster 5, Vol. 8, 1). Leidorf: Verlag Marie Leidorf.

Wallert, I. (1959). Friedrich Wilhelm von Bissing. Verzeichnis seiner Schriften (1895–1956). *Zeitschrift für ägyptische Sprache und Altertumskunde*, *84*, 1–16.

Antiquaries and Indian Agents: "Field Networks" of Ethnology and Archaeology in the North American Southwest, 1860–1880

5

James E. Snead

Abstract

The practice of ethnology and archaeology in the southwestern territories of the United States in the nineteenth century was poorly organized and chaotic. The presence of indigenous populations, along with extensive archaeological evidence, was attractive to the communities of interest concerned with these issues. However, difficulties of access and the lack of expertise proved a considerable hindrance. The "field network" concept, developed by historian of science Jeremy Vetter, provides means to explore how information from this region was "produced" by members of the local Euro-American population, was obtained/used by scholars, and reflected the interests of an array of actors. The field network created by Lewis Henry Morgan (1818–1881) for research in the United States Southwest between 1860 and 1880 provides a particularly useful example of how such connections developed in a critical era for the development of American anthropology.

Keywords

U.S. Southwest · Ethnology · Archaeology · Field networks · Lewis Henry Morgan · Adolph Bandelier

5.1 Introduction

In 1864 John N. Ward (1822–1875) produced a report regarding the status of the Native American "Pueblo" communities of New Mexico Territory—in the southwestern United States—for whom he served as Indian Agent. Addressed to the Commissioner on Indian Affairs, the account included population statistics but also general observations, acknowledging that most accounts of the Pueblo people to that date were "mere speculation." Ward also observed that the ethnographic and archaeological evidence available offered

> an ample field for research, but as I consider a task of this kind more adapted to the researches of the antiquary than to those of an Indian agent, I will simply present such facts as have come under my personal observation, together with the information I have been able to obtain from the Indians themselves. (1864: 187)

Ward's comments reflected the crisis of access and expertise that characterized American ethnology and archaeology in the nineteenth century. Scholars in nascent national institutions struggled with basic questions of documentation and access. Funding was scarce, and elements of practice were undefined. The continental scope of the issues under debate was exacerbated by the sparsely-settled landscape and underdeveloped transportation system. Members of the public engaged the material history of the indigenous population via their own cultural perspectives (cf. Hinsley 1981, 1996; Snead 2018; Stocking 1991).

The limited institutional and intellectual infrastructure meant that knowledge networks were the primary mechanism of anthropological scholarship. Complex lines of communication crossed barriers of geography, class, economics, and culture. Letters, telegrams, and reports of various formality were exchanged; groups, organized at local, regional, and national scales, assembled information that was disseminated through meetings and reports (Snead 2020).

The complex networks associated with archaeology and ethnology in the nineteenth-century United States remain poorly understood. Conceptual frameworks including "professionalism," such as the organization of the discipline over time (i.e., Christenson 2011), and "centralization," particularly the establishment of institutional control (i.e., Burns 2008), have provided useful guidelines. Another pertinent concept is what historian of science Jeremy Vetter calls the "field network," defined as "a mode of knowledge production in modern science that has linked together geographi-

J. E. Snead (✉)
Department of Anthropology, California State University, Northridge, CA, USA
e-mail: james.snead@csun.edu

L. Coltofean et al. (eds.), *Connecting People and Ideas*, Themes in Contemporary Archaeology,
https://doi.org/10.1007/978-3-031-81006-0_5

cally dispersed lay people whose activities are coordinated and directed from a central location" (2011: 259). Consideration of field networks is particularly appropriate for examining practice in ethnology and archaeology, characterized by—as archaeologists Julia Roberts and Kathleen Sheppard have noted—"ephemeral groups in which some members are permanent fixtures (…) others come and go, and still others appear only once, briefly, and then vanish into the dust of the site and archive" (2020: 7).

John Ward's reference to antiquaries and Indian Agents acknowledged disjunctions in the field network within which he found himself, highlighting the gulf between those asking questions, and those positioned to provide answers. Among the numerous associated variables are "access" and "expertise." In the nineteenth-century United States, distance alone was a considerable hindrance to fieldwork, but "access" also pertained to administrative and social barriers. Expertise was an equally fluid concept, given the pre-professional status of both ethnology and archaeology. Working with living peoples, both in their own cultural context and in the volatile cultural/political framework within which they were placed by ongoing colonialism, was deeply challenging. Students of antiquities faced difficult logistics, but also the absence of any common protocol (cf. Hinsley 1981).

A useful case for field networks in ethnology and archaeology in the post-Civil War (1861–1865) United States is provided by the experience of Lewis Henry Morgan (1818–1881). Morgan was perhaps the most prominent "anthropologist" of the era, well-connected in national intellectual and political circles (cf. Kuper 2005: 59–81; also see Service 1981). Of particular emphasis here is Morgan's effort to acquire information on contemporary/historical indigenous communities in the American Southwest. Active between 1860 and 1880, his field network reflected the specifics of the region, the nature of ethnological and archaeological information, and the evolving dynamics of intellectual inquiry itself.

The following discussion divides Morgan's use of the field network into three periods reflecting function and chronology. The first, beginning in 1860, addresses Morgan's efforts to collect ethnological information about indigenous southwestern communities for a planned volume on human kinship organization. It is followed by an evaluation of the field network developed in 1878, when Morgan traveled to the Southwest to learn more about archaeological discoveries. The final section concerns the experience of Morgan's protégé, Adolph Bandelier (1840–1914), whose 1880 journey to the Southwest came at Morgan's instigation and marked the transformation of the field network under very different circumstances. These episodes shed light on Morgan's southwestern projects, as well as providing insight into the means and motivation of individual actors that shaped the effort.

5.2 1860: Lewis Henry Morgan's "Circular"

In a candid aside deep within his *Systems of Consanguinity and Affinity of the Human Family* (1871), Lewis Henry Morgan expressed frustration with his inability to obtain reliable information about the indigenous peoples of the American Southwest. The region, he complained, was "hermetically sealed, so far as ethnological investigations are concerned. China and India are both more accessible" (1871: 260). This was all the more frustrating because these communities had long been recognized as of ethnological interest. "This oasis deserves notice," noted ethnologist and linguist Albert Gallatin (1761–1849), "because the population, which was more dense than usual, appears to have been eminently agricultural," describing the indigenous people of the Southwest as "humane, and civilized" (1845: 201). The region had been incorporated into the United States following the Mexican-American War (1846–1848), leading to expectations that ethnological study would follow rapidly (Squier 1848). These hopes were frustrated, however, and the Southwest remained remote and inaccessible for many years thereafter (Fig. 5.1).

Morgan was independently wealthy and had no institutional appointment, so the *Consanguinity* project was based at his home in Rochester, New York, which became the center of a global ethnological network. Requests for detailed information about kinship and social organization were solicited via a printed "circular," inviting anyone with relevant experience to contribute. Such questionnaires were a time-honored tool of American scholarship under conditions of limited resources (cf. Goldstein 1994; Strang 2017; Snead 2018: 21). As noted in a circular sent out by the American Philosophical Society at the end of the eighteenth century, "without the aid of gentlemen who have taste and opportunity for such researches, our means would be very confined" (American Philosophical Society 1799: xxxvii). Circulars had been used by the first generation of American ethnologists, including Gallatin and Lewis Cass (1782–1866) (Cass 1823; Gallatin 1836: 1). Morgan's version, printed in 1859, included a preamble explaining the issues involved, followed by a lengthy "schedule" of kinship questions. In early 1860, the circular was dispatched through various networks in the United States and abroad, including those of the Smithsonian Institution and U.S. Diplomatic agents (Henry and Cass 1860).

Gallatin, however, had received only one response to his circular, and Morgan's experience 30 years later was similar. "From the complicated nature of the subject," he admitted, "the results (…) were inconsiderable." Morgan attempted to rectify the situation by identifying other networks, such as missionary "boards," but the results remained difficult to synthesize. He made several research trips of his own, and carefully scrutinized the official reports available. As anthro-

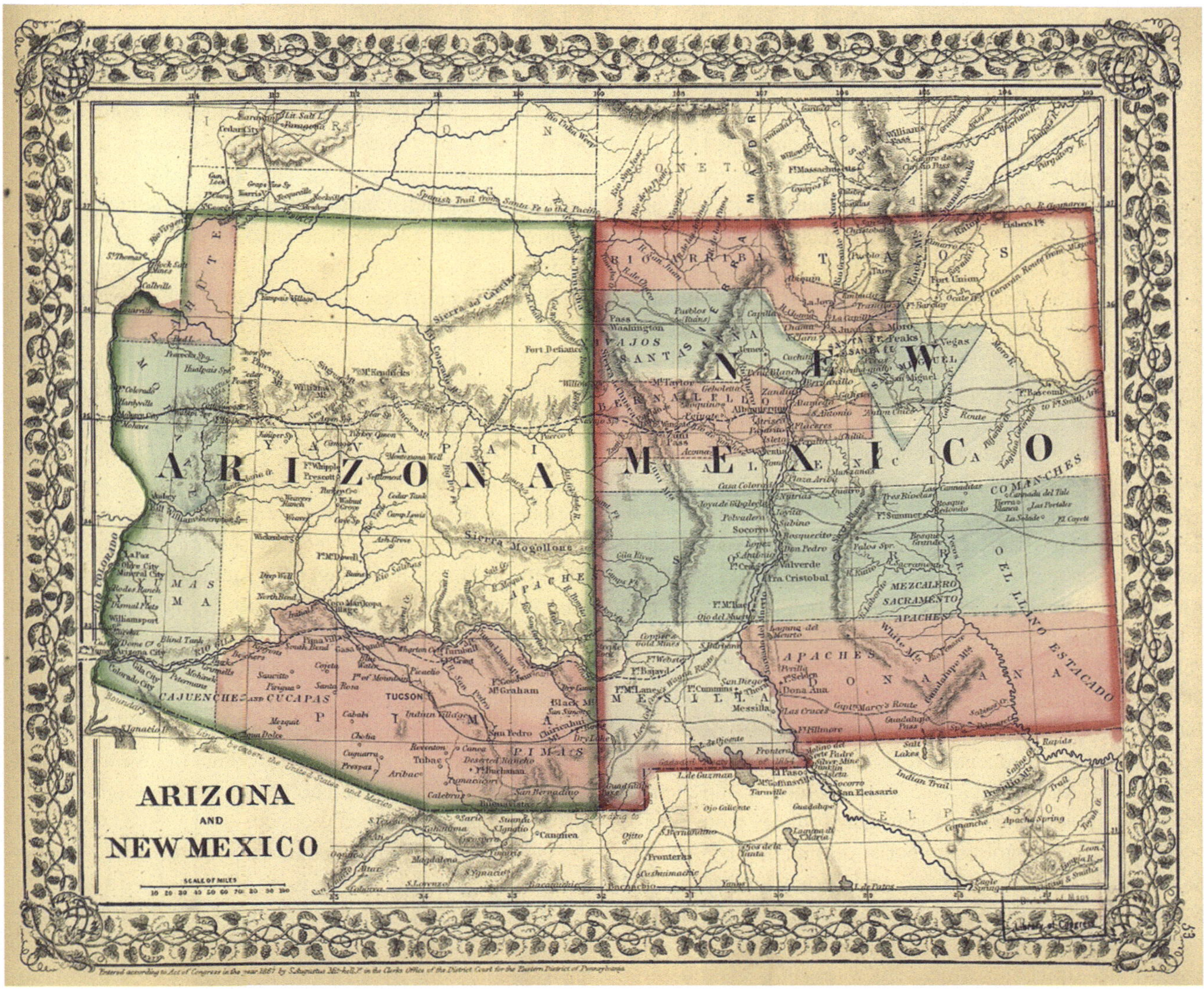

Fig. 5.1 Map of Arizona and New Mexico Territories, 1867. (Map: S. Augustus Mitchell; reproduced with the permission of the Library of Congress, Washington, D.C., United States)

pologist Elisabeth Tooker has noted, "Morgan collected data wherever he could" (1992: 365).

Morgan's circular traveled to the Southwest via seemingly random pathways. A copy was eventually received by the New Mexico Historical Society (Bloom 1943: 274–75). Founded in the territorial capital, Santa Fe, in 1859, the organization included many among the social/political leadership of both the Euro-American and Hispanic-American elite (Stensvaag 1980). After, perhaps, some discussion, the request was passed along members Dr. Michael Steck (1818–1880) and the Rev. Samuel Gorman (1816–1908), both of whom belonged to the society's section on "Indian Races" (Bloom 1943: 261, 273).

Steck and Gorman were practically archetypes of those on the frontiers of the United States expected to have ethnological expertise. After serving in the U. S. Army in the Southwest in the wake of the Mexican-American War, Steck was appointed Indian Agent to New Mexico in 1852 (Lester 1986: 30). Among those providing references for his appointment were the Secretary of the Smithsonian Institution, Joseph Henry (1797–1878), and the prominent ethnographer Henry Rowe Schoolcraft (1793–1864). Henry occupied the most prominent scientific post in the nation, but his connection with Steck is unclear. Schoolcraft had been an Indian Agent himself, and in the early 1850s was organizing a massive statistical survey of the indigenous population for the Indian Office (Bremer 1987: 293). He and Steck would have crossed paths in Washington. References from these two prominent scholars imply that Steck demonstrated at least the potential for intellectual inquiry.

Steck spent the 1850s negotiating the difficult political terrain faced by Indian agents, experiencing various changes in rank and responsibility. He was particularly focused on the indigenous Apache communities of the southern part of the state, where episodic warfare continued (Lester 1986). The expectation was that Steck could provide ethnological insights on Apache society for Morgan's project (Bloom 1943: 274).

Missionaries were the second profession tapped for ethnological information. Samuel Gorman attended the Granville Theological and Literary Institution in Ohio and thereafter led the life of an itinerant preacher.[1] The American Baptist Home Mission Society sent him to establish a school at Laguna Pueblo, New Mexico, and he arrived in the territory at about the same time as Dr. Steck (Gorman 1914: 317). Laguna was isolated from the Euro-American population centers, and only after months of extreme marginality did the Gorman family achieve some level of acceptance. A scornful report was made by Father Joseph P. Machebeuf (1812–1889), Vicar-General of the Archdiocese of New Mexico, who visited Laguna in 1858 and reported on Gorman's situation there, although rivalry with the Baptist minister clearly shaped his perspective (Steele and Weinberg 2005: 297–98). Another visitor noted that Gorman's young son James had achieved some fluency in the Keres language, serving as his guide through the community (Davis 1857: 394). Another for whom life at Laguna would have been a formative experience was Samuel Gorman's daughter Mary, who went on to a distinct career in education (e.g., Peard 2016).

Although these circumstances suggested opportunity, expectations that Steck and Gorman would be able to produce detailed ethnographic reports went unmet. By the time he received the circular, Gorman had already left Laguna and was ministering to a Baptist congregation in Santa Fe. He sent an "incomplete" schedule to Morgan in May 1860, but his opinions on Native Americans were sufficiently controversial to stir up (unspecified) debate in the historical society ("Lecture," 1861; Morgan 1871: 263, 290; Bloom 1943: 261). Whatever his contribution to ethnology may have been, Gorman departed Santa Fe ahead of the Confederate invasion of New Mexico in early 1862 and does not seem to have ever returned to Laguna.

Wartime conditions had a particularly drastic impact on Steck's response to Morgan's circular. He was at the center of numerous controversies associated with military campaigns against Native communities, including the forced removal of Navajo communities from their traditional homeland to Bosque Redondo, New Mexico (cf. Danziger 1970). Ethnological queries were obviously of low priority in such circumstances. Four years after receiving the circular, however, Steck responded to Morgan via the Commissioner of Indian Affairs in Washington, sending a packet of ethnographic material (Doyle 1864). Contrary to expectations, this information was derived from the indigenous pueblos of Taos and Tesuque, instead of the Apache communities, and at least in part was collected by another Indian Agent, John N. Ward.

Ward had been employed by the Indian Agency since at least 1851. Unlike Steck, he did not climb the administrative ladder, but was rotated to various posts over time (Abel 1941a, b; Hall 1984: 93–95). This involved constant travel, and over time much greater familiarity with indigenous communities. In 1863 Ward conducted a census of the Pueblo communities, providing more detailed information than had previously been available. His report, published in 1864, attracted some notice. British traveler A. W. Bell visited Santa Fe in 1867 and received similar information from Ward himself (1869: 224–25). Ward became a corresponding member of the American Ethnological Society and was noted by that organization as having "rendered a service to science" ("Complimentary," 1867).

Back in Rochester, Morgan attempted to process the information provided by Steck and Gorman. Regarding the kinship system of the Pueblo of Tesuque, he admitted that it was "impossible to form an opinion (…) upon the fragment given in the table," and apparently the information provided by Ward was not sufficiently relevant to be used. Information from the Pueblo of Laguna was also suspect. The global network of correspondents featured in Morgan's acknowledgements clearly did not extend to this corner of his own country (1871: viii, 263). With the publication of *Consanguinity*, the matter seems to have been dropped.

A dozen years after last hearing from Steck, Morgan received another letter from New Mexico. David Jefferson Miller (1827–1887) of Santa Fe, a clerk in the surveyor general's office, contacted Morgan out of general ethnological interest. The New Mexico Historical Society had not survived the Civil War, and there was no consistent local forum in which such issues could be discussed. Nonetheless, Miller was well-read, and empathized with Morgan regarding various anthropological controversies (Miller 1876; cf. Stern 1967: 118–19). Morgan, however, was less interested in scholarly exchange and more in the possibility that Miller could provide detailed ethnographic information than had been generated through his circular. He was in the final stages of preparing the comparative study that would become his most famous work, *Ancient Society* (Morgan 1877b). Ethnographic detail was still, however, a major goal, and he was particularly interested in Taos Pueblo, an indigenous community in northern New Mexico already famous for its "traditional" architectural style (Fig. 5.2). Morgan reasoned that Miller could provide the necessary information from Taos and proceeded to recruit him.

Miller was emblematic of the transformation of Euro-American society in the Southwest over the years since Morgan's first inquiries. Born in Alabama, he came west via Texas, securing a position as clerk to William Pelham (1803–1879), the newly-appointed Surveyor General of New

[1] Subsequently known as Denison University.

Fig. 5.2 Taos Pueblo, New Mexico, 1880. (Photo: Cunningham & Co.; reproduced with the permission of the Library of Congress, Washington, D.C., United States)

Mexico. In late 1845 he traveled with Pelham to Santa Fe in the company of the United States-Mexico Boundary Commission, led by Col. William Hensley Emory (1811–1887) ("Boundary Comission," 1854). Emory, a scholar in his own right, had been working in the region since the Mexican-American War and corresponded with Gallatin regarding Native American ethnology (Gallatin 1761–1849; cf. Emory 1847; Norris et al. 1998). Miller would thus have been introduced to the complex politics and history of the Southwest by one of the few Americans who knew it well.

Miller's work in the Surveyor General's office required him to quickly become familiar with complex legal and political issues associated with land claims in the southwestern territories ("Surveyor General," 1857). But he was also active in in the cultural organizations of Santa Fe's "American" community, such as the short-lived "Scientific Association of New Mexico" ("The Scientific Association," 1858). At first Miller found the "monotony of arid plains and bleak mountains" unappealing (Miller 1861). But he persisted, and after the Civil War kept his position in the office of the Surveyor General. Travel throughout the region and familiarity with the complex documentary record of associated land claims was essential: he was also sufficiently fluent in Spanish to serve as translator for a local newspaper ("Masthead," 1868; "Letter," 1875). Over time Miller acquired deep, personal familiarity with the Southwest, its history, and its people, and hence represented the access and expertise of most use to Morgan.

Morgan's instructions in the circular were detailed, covering 25 topics that included demographics, land ownership, and family organization, in addition to detailed architectural information. Miller dutifully copied these into a notebook, and in October 1877 made the 70-odd mile journey from Santa Fe to Taos. Accompanied by William [Will] M. Tipton (1857–1922), an associate in the Surveyor General's office who was later to become a well-known expert on land grants (Roeder 2009), Miller was well-prepared for the project.

Arrangements in Taos were facilitated by Louis Joseph Clothier and Frederick Muller. Miller described the two men as "merchants, partners in business, and clever and intelligent gentlemen" (Miller 1878). They were also brothers-in-law, married to daughters of Charles Beaubien (1800–1864), a politically-influential landowner who had come to Taos from Quebec 50 years earlier. These men were thus deeply embedded in the local community, circumstances that resembled those of Miller himself (U.S. Census 1870, Taos, New Mexico; cf. Gutierrez 2011; Beyreis 2019).

Once in Taos, Miller and Tipton began working "with the tape line [and] accurately measured the dimensions of the main building," included in the report, but a map was not made from this information (Miller 1877: 53). They also conversed with some of the Taos residents, recording their perspectives on matters of interest to Morgan. Back in Santa

Fe, Miller had the opportunity to interview two men from other indigenous communities, including this information as a postscript to the report. Morgan does not seem to have had any specific plan for this evidence, but—combined with the information originally collected by Ward—he found himself in possession of detailed measurements of the residential complex at Taos Pueblo, with additional ethnographic insights. Unlike his skepticism regarding Steck and Gorman, Morgan considered the work of his 1877 field network to be a success.

5.3 1878: Morgan and the "Cliff Houses"

Only six months after Miller dispatched the Taos notebook to Morgan, the senior ethnologist wrote to him again, announcing his intent to make a personal "visit to the Great West" (Miller 1878). This new plan was related to Morgan's evolving intellectual concerns. But it also reflected a new turn toward archaeology, requiring an expansion of his field network.

Morgan's interest in archaeology emerged from his conviction that southwestern antiquities were the key to central ethnological questions (cf. Trautmann and Kabelac 1994: 52; Moses 2009: 258). His thoughts on the matter appeared in an 1869 review that used the few archaeological reports available. Lt. James Simpson's report on the ruins of Chaco Canyon was cited at length, and was particularly praised for its architectural detail (Simpson 1852; Morgan 1869: 461–74). Interpretation of architecture was also a central part of his 1876 critique of historian Hubert H. Bancroft (1832–1918), particularly regarding what Morgan called the "communal character of Indian houses" (Morgan 1876: 286; cf. Resek 1960: 133; Stern 1967: 115; Trautmann 1987: 87). Specific references to both contemporary southwestern pueblos and the ruins of Chaco Canyon were made (Morgan 1876: 290). But if Morgan was to pursue this argument further, more detailed archaeological information was required.

In the 1870s, formal documentation of southwestern antiquities was almost entirely produced by survey teams dispatched by the Federal Government to map the western regions and document associated resources (Goetzmann 1966; cf. Worster 2000: 203–10). Science done "in the process" was very much a matter of opportunity and the aptitude of particular survey teams (cf. Kirsch 2002; McPherson and Neel 2016). Morgan was an avid consumer of information produced by the federal surveys, and his library was stocked with their reports (Trautmann and Kabelac 1994: 53).

How much "current" information was available to Morgan at the time is unclear. The different survey teams and their bureaucratic patrons represented a network that had limited overlap with the scholarly circles with which Morgan was integrated. Over time he developed better contacts with some of the senior members of the surveys, particularly John Wesley Powell (1834–1902), the noted western explorer and director of the Geographical and Geological Survey of the Rocky Mountain Region (cf. Worster 2000; Wilcox and Fowler 2002). Powell, having received a copy of *Ancient Society*, offered to collect kinship information on a planned trip to Hopi communities in Arizona (Morgan 1877a; Powell 1877).

The slow rate of publication of the survey reports meant that much of the public information about associated archaeological discoveries was derived from newspapers (cf. "The Far West," 1876). Government exhibits at the Centennial Exposition in Philadelphia featured photographs and models of these sites, and images were widely circulated (Fig. 5.3) (cf. "Ancient America," 1877). Morgan would thus have been quite aware of archaeological discoveries in southwestern Colorado, but little more.

The opportunity to resolve the questions raised by southwestern antiquities—and the stimulus for his letter to Miller—came in 1878. Morgan's 18-year-old grand-nephew, Columbia University student William Fellowes Morgan (1860–1943), "had a fondness for archaeology," and a western trip was planned for that summer (Resek 1960: 146). In addition to the senior Morgan the party would include another great-nephew, David Percy Morgan (1862–1932), with young friends William S. Sloan and Donald B. Toucey (White 1942: 3). It is clear that the senior Morgan took advantage of the enthusiasm and assistance of his younger relatives to add some personal experience to the scholarly products with which he was familiar.

Morgan's archaeological foray to the Southwest required an entirely different network than the one used to collect ethnographic information. The issues of access and expertise remained central, but the communities of interest involved were distinct. Southwestern Colorado had few centers of Euro-American settlement, and those present were ephemeral ranching/mining towns without the historical associations of those in New Mexico. The region was historically associated with the Pueblo people, but no inhabited communities were extant: the Ute and Navajo populations in the area had their own, complex interpretation of local antiquity. There were thus very few obvious sources of information to draw on. In the void of knowledge and scholarship, "access" became more of an issue of logistics than one of language or related cultural boundaries.

The party left the railroad in Canon City, Colorado, and traveled west. Their "wagon man" for much of the trip was Henry H. Waller, whose (unnamed) wife was an indigenous woman from the Northern Plains. Morgan's journal has long lists of Lakota vocabulary contributed by Waller, who also instructed the party in the art of assembling the indigenous lodging called a *tipi* (White 1942: 6). It was nearly three

Fig. 5.3 Archaeological site in Southwestern Colorado, 1872. (Photo: Timothy H. O'Sullivan; reproduced with the permission of the Library of Congress, Washington, D.C., United States)

weeks before Morgan saw his first antiquities, on a mesa near the town of Animas City in the southwestern part of the state.

Morgan soon discovered local antiquarian networks established in the region, and his own, personal connections with some of the participants. "At Animas City I met Gen'l Heffernan," he noted, "who lived in Albany" (White 1942: 16). "General Heffernan" was James J. Heffernan (1828–1886), a colorful mining entrepreneur who had recently been involved in various Utah smelting and transportation ventures (cf. "Personal," 1871). He was clearly well-connected among Colorado prospectors, and also reflected the collecting tradition increasingly evident in that landscape (cf. Blackburn 2006). Heffernan "gave us a large lot of pottery he had collected in the Mancos valley, in Montezuma valley and elsewhere," noted Morgan (White 1942: 16).

Heffernan also provided crucial references to ranchers who had first-hand familiarity with local ruins. These included John Gregor, who in the three years he had lived in the vicinity had assisted other antiquarian travelers, such as Samuel Beach Axtell (1819–1891), Territorial Governor of New Mexico (Forrester 1878). Gregor guided the visitors to a nearby "cliff house," where they made careful measurements (Morgan 1879).

The Morgan party devoted particular attention to the major archaeological complex subsequently designated "Aztec Ruin," now Aztec Ruins National Monument (Lister and Lister 1990). Aztec had first been documented by Federal survey teams almost 20 years earlier, although—given publication delays and geographic imprecision—Morgan may not have been aware of the previous work (i.e., Newberry 1876: 76–77). He put his young companions to work mapping the site, collecting more detailed information than was generally provided by such projects.

The next phase of the tour, in the McElmo Valley close to the border with Utah, required detailed consultation with the antiquarian network. Here they were guided by rancher Henry L. Mitchell, a recent arrival to Montezuma County, who had already acquired a detailed sense of the local landscape. For the review of local ruins that followed, the Morgan party was again joined by General Heffernan, along with his associate, "Captain" Ben Ford. Other locals, such as "Mr. Ward" and "Mr. White," visited the camp and took the opportunity to describe various antiquarian discoveries made in the vicinity (White 1942: 20).

Time pressure required the party to leave Colorado and rush southward, allowing Morgan to visit the scene of his traditional interest. He had written to Miller about the possibility of visiting Taos Pueblo during the trip, and Miller forwarded the names of Clothier and Muller (Miller 1878). Time was short, and while the rest of the party bought pottery, "Uncle Lewis (…) busied himself learning all he could about the customs, etc." with the elders of the community (Hollcroft 1953). With such limited time it was difficult to collect useful information, but Morgan clearly relished the opportunity. "At last I find myself at Taos," he wrote, and after 18 years of inquiry via others was at last able to ask a few questions himself (White 1942: 24).

5.4 1880: Bandelier and the Stevensons in Santa Fe

When the Morgan party left Colorado in August 1880, their first stop was St. Louis, Missouri, where the annual meeting of the American Association for the Advancement of Science was in progress. One of the scheduled sessions concerned the archaeology and ethnology of the Southwest. The details of the recent expedition were provided by the young William Fellowes Morgan, entitled "Description of a cliff-house on the Mancos River of Colorado, with a ground plan." He described the results of their modest fieldwork of less than a month previously (Morgan 1879). "I was now very frightened," Fellowes Morgan noted in his journal, "for this was my entrance into the arena of a scientific life" (Hollcroft 1953: 127).

After the session concluded, the Morgan party "went to a beer saloon," where they were joined by "(…) Major Powell (…) [and] Mr. Bandelier" (Hollcroft 1953: 127). Along with the senior Morgan, those at the table in St. Louis are listed among the most prominent American anthropologists of the late nineteenth century. "Major Powell" was John Wesley Powell, while "Mr. Bandelier" was Adolph Bandelier, a self-taught scholar from nearby Highland, Illinois, who had established himself as Morgan's premier protégé. The conversation in the beer garden was—like many such, in St. Louis and at innumerable meetings elsewhere—not recorded. But it clearly dealt with southwestern fieldwork.

The St. Louis encounter was emblematic of the rapidly changing field networks of southwestern ethnology and archaeology in the late 1870s. By that date the survey teams were being folded into various federal departments, one of which, the new Bureau of Ethnology, was specifically charged with collecting information about Native Americans.[2] The Bureau, as a component of the Smithsonian Institution, was formalized in Spring 1879, with Major Powell as director (cf. Noelke 1974; Hinsley 1981; Woodbury and Woodbury 1999).

Powell joined Morgan in advocating a systematic approach to fieldwork. The Bureau would have its own staff of "scholars and specialists," but would expand its mission "by inciting and guiding research immediately conducted by collaborators at work throughout the country" (Powell 1881: xiv). The expertise to apply such a centralizing agenda, however, remained scarce, particularly for the Southwest. The Smithsonian had dispatched circulars of its own—in 1863 to collect linguistic information (Gibbs 1863), and an archaeological circular in the Spring of 1878 (Snead 2020). The latter project received more than 300 responses: however, less than a dozen of these came from the southwest, and Santa Fe's lone respondent, the Rev. Henry Forrester, contributed little of interest (Forrester 1878). Powell was critical of such efforts, which he characterized as "filled with contradictions and absurdities" (1883: xxx). He had nonetheless developed his own survey form, in consultation with Morgan, although the number of responses is unclear (Noelke 1974: 34–37; Hinsley 1981: 161).

Morgan was thus aware of the centralizing federal agenda for anthropology but followed his own independent path, which he anticipated would lead him back to the Southwest. In early 1880, he was consulted by the new Archaeological Institute of America about ideas for a project and proposed an expedition to the Southwest under his own leadership (Lange and Riley 1996: 19–21, 32). The plan was

> to explore, delineate, and describe the house architecture of the Indian tribes as represented by the houses that are now to be found in ruins, or in actual occupation, in the region of the San Juan River, in New Mexico and Arizona, in Mexico and Central America, and to study such fictile wares, implements, and utensils that may be found therein, and may tend to illustrate the condition of the people. (Morgan 1880: 30)

That Morgan intended to take charge of the project is indicated by renewed contact with his southwestern field network. Miller responded with enthusiasm for "the visit of research among the ancient and modern Indian pueblos (…) you expect to make this spring [1880]" (Morgan 1880).

Miller's letter included suggestions for logistics and related support. "I am sure," he wrote, "that with one traveling companion you will be perfectly safe" (Miller 1880). He referred Morgan to men with the local experience and the linguistic skills that had been the traditional criteria for the southwestern field networks. One was "Capt. John Townsend (…) a man who takes much interest in Indian ethnology,"

[2] The Bureau of Ethnology was soon renamed the Bureau of American Ethnology, the name with which it is most closely associated today.

who was presently "serving as a clerk here in the quartermaster's office." He also recommended "Capt. Geo. H. Pradt (...) a civil engineer and adopted citizen of the Pueblo of Laguna" (Miller 1880). George H. Pradt (?–1927) was a surveyor who had worked with Miller in Santa Fe. In 1875, however, he had been sued by his superior, Surveyor-General James Kerr Proudfit (1831–1917), and moved to Laguna Pueblo. Once there, Pradt married into the community, gaining greater acceptance than Gorman had 20 years earlier, eventually taking a prominent role in local affairs (cf. Parsons 1923: 237; Walter 1927).

Morgan, however, had a different fieldwork plan in mind. In the winter of 1880, Bandelier was in the grip of a personal and professional crisis that threatened to derail his interests in ethnology. Morgan's solution was to invite him to join the southwestern project, as essentially his assistant (Lange and Riley 1966: 31). Bandelier would thus be the "one traveling companion" referred to by Miller; his familiarity with historical sources would provide critical expertise, and his enthusiasm and stamina would be essential under challenging local conditions.

Given his preparation and outreach, it is surprising that Morgan soon decided against making the southwest trip himself. The continued involvement of Bandelier, however, would be some consolation, so Morgan spent the spring of 1880 lobbying the Archaeological Institute to support the effort. After considerable negotiation, Bandelier was awarded a "fellowship" for fieldwork in the Southwest (Lange and Riley 1996: 36). If Morgan was unable to make the journey, he clearly anticipated oversight and guidance.

Bandelier arrived in Santa Fe in late August 1880 to find the town in transformation. The railroad had arrived in the previous year, making communication with the eastern states much more feasible. As the historic capital of the territory, it drew political figures, mercantile travelers, and tourists (Fig. 5.4). He immediately tapped into Morgan's field network in the person of Miller, who introduced him to the relevant members of the community and answered a stream of questions about southwestern ethnology and archaeology. Subsequently, Bandelier described Miller as "my most intimate acquaintance in New Mexico (...). He is very much respected & honored here by everybody, & to me is a warm friend & supporter" (Bandelier to Morgan, October 22, 1880, quoted in White 1940: 217).

With Miller's guidance Bandelier assembled a field network of considerable depth and diversity. Many he met in the fall of 1880 became long-term collaborators, including Territorial Secretary William G. Ritch (1830–1904), Territorial Librarian Samuel Ellison (1817–1889), and curio dealer Isaac Jacob ("Jake") Gold (1853–1905) (cf. Batkin 1998). Bandelier also made forays into the surrounding countryside, meeting with ranchers and others who had specific experience with the archaeological sites he hoped to study further. Besides Miller, however, those who had originally been in contact with Morgan—Ward, Gorman, Steck—had either died or moved on. Bandelier did, however, meet Pradt, finding him a useful source of information about Laguna Pueblo (Lange and Riley 1966: 174).

Perhaps the most significant "networking" encounter of Bandelier's first month in Santa Fe concerned national, rather local, connections. "Quite unexpectedly I met Colonel Stevenson and party," notes his journal for September 21, 1880 (Lange and Riley 1966: 89). James Stevenson (1840–1888) was appointed to lead the first of Major Powell's major fieldwork program under the auspices of the Bureau of Ethnology. "The object of the expedition," read the orders, were "(...) to make as careful a study as circumstances will permit of the Pueblo Ruins and caves of that district of country" (Powell 1879; cf. Parezo 1989: 12; Snead 2001: 9). He was accompanied by his wife Matilda Coxe Stevenson (1849–1915), whose reputation as an ethnologist would quickly eclipse that of her husband as an expedition leader (Parezo 1993; Miller 2007). Stevenson's own account was of the meeting only slightly more nuanced: "Mr. Bandalier [sic] is here," he wrote, remarking that "he (...) claims to have made some very important discoveries" (Stevenson 1888).

In theory the Stevenson party was highly prepared to conduct ethnological and archaeological research. In 1880 their party included photographer John K. Hillers (1843–1925), who had spent a decade with the survey teams, and two assistant "ethnologists," F. G. Galbraith and F. W. Morancey (cf. Hinsley 1992; for Hillers, see Hieb 2011). They also had funds to purchase collections and could draw on the logistical resources of the federal government (cf. Stevenson 1883: 211). Despite these apparent advantages, however, the Stevensons were as challenged by issues of expertise as was Bandelier. Despite their job titles, neither Galbraith nor Morancey had any substantive experience in the field, nor were they particularly prepared for the trip. "I found the gentlemen with me without means," complained Stevenson, "& had to let them have some time their vouchers are acknowledged" (Stevenson 1888). Galbraith had been recommended by Smithsonian Secretary Spencer Baird on the grounds that he would "be a very able assistant, & his services can be obtained at a cheap rate" (Baird 1880). James Stevenson described him as "not well enough educated to do scientific work," but thought that he had some potential: "it would be well to leave him at one of the pueblos. We might do so during the winter & see how he turns out" (Stevenson 1880).[3]

[3]Morancey was equally inexperienced. There are no further references to him in the Bureau of Ethnology's files: he may have been a short-term, political appointment. Galbraith did acquire a vocabulary of the Tewa language from Santa Clara Pueblo currently housed in the Smithsonian's National Anthropological Archives (MS 1016).

Fig. 5.4 Old Governor's Palace, Santa Fe, New Mexico, 1885. (Photo: Charles C. Pierce; reproduced with the permission of the Huntington Library, San Marino, California, United States)

When necessary, Stevenson recruited residents to assist in the effort. Excavations at the site of Cuyumungue north of Santa Fe were conducted by John W. Conway, whose occupation—listed in the 1880 Census as "teamster" (U.S. Census 1880)—indicates that he was hired as a manual laborer rather than someone with any specific expertise. Stevenson described the work as "quite successful," and the expedition catalog lists numerous lithics excavated from the site. Although Stevenson provides no further details, a paragraph contributed by Conway was published in a Smithsonian report shortly thereafter (Mason 1881: 442; Stevenson 1883: 433, 441).

Bandelier, meanwhile, kept up an active schedule, visiting sites and pueblos within reach of Santa Fe. He wrote reports on his field exploration and studied documents in local archives, including maps in the survey office provided by Miller. He successfully arranged to live in the Pueblo of Cochiti for a few weeks, conducting the most detailed ethnographic study in the region to that date (Lange and Riley 1996: 50). He kept in close touch with Morgan, providing updates on the project and suggesting their significance to broader ethnological issues (White 1940: 217). His relationship with the Bureau of Ethnology's expedition was increasingly competitive. Bandelier's journal for October 31 complained that the Stevensons had "bought nearly all the stone axes in town. Easy manner of collecting for the Smithsonian Institute!" (Lange and Riley 1966: 181).

By year's end both parties had completed their work. The Stevensons returned to Washington to organize the collections and prepare for the next field season. Bandelier visited Morgan at Rochester in January 1881, and they spent three days discussing the fieldwork (Lange and Riley 1996: 50). The two projects represented the increasingly complex use of field networks in the conduct of ethnological and archaeological research in New Mexico's territory: Morgan's death on December 17, 1881, along with Bandelier's subsequent cultivation of his own network, essentially brought his phase of exploration to an end.

5.5 Conclusions

The ultimate representation of Morgan's southwestern field network is his *Houses and House-Life of the American Aborigines* (1881), published just before his death by the Smithsonian. "I could not resist the invitation of my friend Major Powell," notes the preface, "(…) to put these chapters together as well as I might be able, that they might be published by the Bureau." Morgan acknowledged the limitations of the volume in light of his failing health, but presented it "cheerfully for the indulgence of [his] readers" (1881: viii).

Houses and House-Life is an amalgam of Morgan's previous writings connected with indigenous architecture, interlaced with newer material. The sections concerned with the Southwest refer to some of the recent survey reports, such as W. H. Jackson's account of Chaco Canyon (1879), but others were overlooked (cf. Holmes 1878). "Valuable as this information is to us," he observed, "it falls short of a full exposition" (1881: 135). Morgan did not spare himself criticism: a plan derived from his own 1878 fieldwork was included, "with some hesitation, because the sketch from which it [was] taken was executed in some haste, with no expectation of using it" (1881: 185).

Ironically, Bandelier's influence on *Houses and House-Life* was largely restricted to his observations about the history and kinship structure of the "Aztecs," rather than anything pertaining to the southwestern work. The early publication date made it difficult to include any substantive use of the 1880 results. A few footnotes suggest Morgan's awareness of the missed opportunity (1881: 41, 149).

In contrast, contributions of Morgan's original field network appear throughout *Houses and House-Life*. An entire section of the volume is derived from the Taos reports of Ward and Miller. Although architecture is emphasized, Morgan also makes considerable use of their observations on kinship and social organization. As for his own brief visit to Taos, Morgan ruefully noted his inability to acquire much useful information "from lack of sufficient time" (1881: 84). Those who enabled Morgan's 1878 archaeological work—Heffernan, Gregor, and Mitchell—were all acknowledged. Some of the published maps in *Houses and House-Life*, particularly of places visited with Mitchell, were derived from Morgan's 1878 field journal, while others were apparently reconstructed from measurements taken at the time (cf. Morgan 1878: Figs. 44–45, 1881: 189–90).

The interdigitation of the southwestern networks post-1878 is also evident in *Houses and House-Life*. One of the figures of Taos Pueblo used in the volume (Fig. 28) was drafted by F. C. Galbraith, probably obtained during the Stevenson party's visit to the community in September 1880 (Stevenson 1880; Morgan 1881: 149). James C. Piling (1846–1895), who supervised the Bureau of Ethnology's publication efforts, is acknowledged as playing a critical role in arranging for such figures (Morgan 1881: viii).

Of particular significance to the development of the southwestern field networks after Morgan was his use of Matilda Coxe Stevenson's description of Zuñi Pueblo (D. A. Miller 2007: 47). The citation is cryptic, attributed to James Stevenson's "wife," without additional source information (1881: 137). No correspondence between Morgan and either of the Stevensons is preserved in his archive (Melissa Meade, personal communication, August 2, 2022). The text, however, came directly from Matilda Coxe Stevenson's privately published *Zuñi and the Zuñians* (1881); she sent copies of the book to senior ethnologists,[4] and it may have come to Morgan directly. The fact that he devoted four pages to the quotation is evidence that Morgan perceived its significance.

There are dramatic differences between Morgan's southwestern field networks during the three phases of activity of the 1860s and those of the late 1870s. In 1860 he sought individuals who had access to indigenous communities and the expertise to provide relatively complex details of language and kinship. Those who responded were among the best available for the purpose, but were otherwise poorly equipped to assist. The disjunction between Morgan's "needs" and the available information were profound. The antiquarian network Morgan relied on to assist his 1878 fieldwork provided access to places that would otherwise be inaccessible or difficult to discover. They contributed little in the way of expertise, however, both because such skills had not yet been defined and because their lives provided little room for delving into ruins. The disjunction between the networks highlights not only geographic and social differences, but also the "gap" between ethnology and archaeology as fields of interest.

Bandelier's initial reliance on Morgan's contact Miller suggests the potential durability of these networks over time. However, the high mobility of frontier population, particularly in rural areas, made this a rare occurrence. As robust as the antiquarian scene of southwestern Colorado appeared in 1878, it had disappeared almost entirely within a few years despite continued interest in antiquities. The antiquarian entrepreneurs active in the region during the late 1880s had little overlap with those Morgan knew a decade earlier (cf. Blackburn 2006).

The growth of towns in the Southwestern territories created increasingly stable conditions. Local organizations and institutions had greater longevity and provided conditions for greater circulation of information. Miller's case is one example of how a more settled and integrated population allowed for the rise of field networks of greater reach and depth. The opportunity to acquire expertise was also enhanced under such circumstances, creating a substantial knowledge base. Increasingly, scholars from 'outside' used such communities of knowledge as a foundation from which to launch enquiries far more detailed than previously possible, framed by their own intellectual preparation.

[4] For example, a scanned copy of *Zuñi and the Zuñians* (1881), available via Google Books, came from the library of Edward Burnett Tylor (1832–1917) and is inscribed "with the compliments of Tilly E. Stevenson."

5.6 Coda: Bandelier at the New Mexico Historical Society

The transformation of the southwestern field networks, and the new opportunities they represented, was articulated by Bandelier on April 28, 1882. Having returned to Santa Fe, Bandelier addressed the New Mexico Historical Society—re-established in 1880—on the subject of "Kin and Clan." Members included not only his colleagues Miller, Ellison, and Secretary Ritch, but also former governor Axtell—now justice of the Territorial Supreme Court—and LeBaron Bradford Prince (1840–1922), who would himself be appointed governor a few years later (cf. Stensvaag 1980). Advance newspaper coverage touted association with Morgan and took particular note of "his archaeological researches [sic] in this territory" ("Kin and Clan," 1882).

Bandelier had learned the story of the original historical society from Miller, but it is unlikely that he understood the connection between that organization, Morgan, and the *Consanguinity* project. Nonetheless the introduction to his talk touched on issues of access and expertise similar to those that had existed 22 years earlier. "It is a difficult task," Bandelier acknowledged, "(…) to address an audience of persons, all of whom have the advantages of longer experience in New Mexico; consequently of greater practical knowledge" concerning the indigenous people of the territory. He persevered, however, expounding on the value of history, the opportunities of archaeology, and the pragmatic contributions of ethnography ("Kin and Clan," 1882).[5] Bandelier's talk was reviewed with satisfaction the next day. "He is not a mere theorist," noted the unnamed author, evoking sentiment similar to that expressed by Ward in 1864. "He has spent a good deal of time among the Indians of the southwestern states and territories, and has gathered his facts not from hearsay, but from living amongst them" ("Kin and Clan," 1882), incidentally providing a timely coda for Morgan's southwestern field network.

Acknowledgements The assistance of several archivists was critical in completing this chapter. Access to the Lewis Henry Morgan Papers was generously provided by Melissa S. Mead, John M. and Barbara Keil University Archivist and Rochester Collections Librarian, Department of Rare Books and Special Collections, University of Rochester. Tomas Jaehn, Director of Special Collections/CSWR at the University of New Mexico, provided access to the Michael Steck Papers. A scan of the letter from David J. Miller to his niece was generously provided by Heather McClure at the Fray Angelico Chavez History Library, New Mexico History Museum/Palace of the Governors.

[5] Bandelier provided the draft of his lecture to the *Santa Fe New Mexican*, allowing it to be printed in full (Lange and Riley 1966: 272).

References

Primary Sources

Archival Material

Baird, S. (1880, July 12). [Letter to John Wesley Powell]. Bureau of American Ethnology Papers, Letters Received, 1879–1888, National Anthropological Archives, Smithsonian Institution, Washington, D.C., United States.

Doyle, W. P. (1864, September 22). [Letter to Lewis Henry Morgan]. Lewis Henry Morgan Papers, Department of Rare Books and Special Collections, Rochester University Library, Rochester, NY, United States.

Emory, W. H. (1847, September 20). [Letter to Albert Gallatin]. Albert Gallatin Papers, New-York Historical Society, New York, NY, United States.

Forrester, H. (1878, May 14). [Letter to Joseph Henry]. Collected Letters on Ethnology (Record Unit 58), Smithsonian Institution Archives, Washington, D.C., United States.

Gallatin, A. (1761–1849). *Correspondence*. Albert Gallatin Papers, New-York Historical Society, New York, NY, United States.

Henry, J., & Cass, L. (1860). *Printed circular. Covering Letters (on same leaf) January 20, 1860 by Joseph Henry and January 5, 1860 by Lewis Cass*. Lewis Henry Morgan Papers (A.M85), Rare Books, Special Collections, and Preservation, River Campus Libraries, University of Rochester, Rochester, NY, United States. https://rbscpexhibits.lib.rochester.edu/items/show/6450

Miller, D. J. (1861, March 24). [Letter to "Niece"]. Fray Angélico Chávez History Library/New Mexico History Museum (AC 149-P), Santa Fe, NM, United States.

Miller, D. J. (1876, April 29). [Letter to Lewis Henry Morgan]. Lewis Henry Morgan Papers (A.M85), Rare Books, Special Collections, and Preservation, River Campus Libraries, University of Rochester, Rochester, NY, United States. https://rbscpexhibits.lib.rochester.edu/items/show/6440

Miller, D. J. (1877). *Taos (New Mexico) Memo Book, October-December 1877*. Lewis Henry Morgan Papers (A.M85, Box 24, Folder 127), Rare Books, Special Collections, and Preservation, River Campus Libraries, University of Rochester, Rochester, NY, United States. https://rbscpexhibits.lib.rochester.edu/items/show/6444

Miller, D. J. (1878, June 14). [Letter to Lewis Henry Morgan]. Lewis Henry Morgan Papers (A.M85), Rare Books, Special Collections, and Preservation, River Campus Libraries, University of Rochester, Rochester, NY, United States. https://rbscpexhibits.lib.rochester.edu/items/show/6442

Miller, D. J. (1880, February 12). [Letter to Lewis Henry Morgan]. Lewis Henry Morgan Papers (A.M85), Rare Books, Special Collections, and Preservation, River Campus Libraries, University of Rochester, Rochester, NY, United States. https://rbscpexhibits.lib.rochester.edu/items/show/6443

Morgan, L. H. (1877a, May 26). [Letter to John Wesley Powell]. Lewis Henry Morgan Papers, Department of Rare Books and Special Collections, Rochester University Library, Rochester, NY, United States.

Morgan, L. H. (1878). *Journal of a trip to New Mexico, June 21-August 7, 1878*. Lewis Henry Morgan Papers (A.M85, Series 3, Vol. 20), Rare Books, Special Collections, and Preservation, River Campus Libraries, University of Rochester, Rochester, NY, United States. https://rbbscpexhibits.lhester.edu/items/show/6445

Powell, J. W. (1877, July 18). [Letter to Lewis Henry Morgan]. Lewis Henry Morgan Papers, Department of Rare Books and Special

Collections, Rochester University Library, Rochester, NY, United States.
Powell, J. W. (1879, August 4). [Letter to James Stevenson]. Matilda Coxe Stevenson Papers, National Anthropological Archives, Smithsonian Institution, Washington, D.C., United States.
Stevenson, J. (1880, September 2). [Letter to John Wesley Powell]. Bureau of American Ethnology Papers, Letters Received, 1879–1888, National Anthropological Archives, Smithsonian Institution, Washington, D.C., United States.
Stevenson, J. (1888, September 22). [Letter to James C. Piling]. Bureau of American Ethnology Papers, Letters Received, 1879–1888, National Anthropological Archives, Smithsonian Institution, Washington, D.C., United States.

Newspaper and Magazine Articles

Ancient America. (1877, November 15). *Kansas Weekly Herald*, 6.
Boundary Commission. (1854, November 15). *Washington Sentinel*, 2.
Complimentary. (1867, June 15). *Weekly New Mexican*, 2.
The Far West. (1876, January 20). *St. Louis Republican*, 5.
Kin and Clan. (1882, April 29). *Santa Fe New Mexican*, 3.
Lecture. (1861, February 16). *Santa Fe New Mexican*, 2.
Letter from New Mexico. (1875, January 30). *Parsons Daily Sun*.
Masthead. (1868, August 26). *Santa Fe New Mexican*, 1.
Personal. (1871, April 22). *Corrine Daily Reporter*, 3.
The Scientific Association of New Mexico. (1858, February 20). *Santa Fe Weekly Gazette*, 2.
Surveyor General. (1857, October 21). *Santa Fe Weekly Gazette*, 3.

Secondary Sources

Abel, A. H. (1941a). Indian affairs in New Mexico under the administration of William Carr Lane from the journal of John Ward. *New Mexico Historical Review*, *16*(2), 206–232.
Abel, A. H. (1941b). Indian affairs in New Mexico under the administration of William Carr Lane from the journal of John Ward (concluded). *New Mexico Historical Review*, *16*(3), 328–358.
American Philosophical Society. (1799). Circular Letter. *Transactions of the American Philosophical Society*, *4*(1793–1798), 37–39.
Batkin, J. (1998). Some early curio dealers of New Mexico. *American Indian Art*, *23*(3), 68–81.
Bell, A. W. (1869). On the native races of New Mexico. *Journal of the Ethnological Society of London*, *1*(3), 222–274.
Beyreis, D. C. (2019). Dangerous alliances in the New Mexico borderlands: Charles Bent and the limits of family networks. *Journal of the Early Republic*, *39*(1), 57–80.
Blackburn, F. M. (2006). *The Wetherills: Friends of Mesa Verde*. Durango, CO: Durango Herald Small Press.
Bloom, L. B. (1943). Historical Society minutes, 1859–1863. *New Mexico Historical Review*, *18*(3), 247–311.
Bremer, R. G. (1987). *Indian agent and wilderness scholar: The life of Henry Rowe Schoolcraft*. Mount Pleasant, MI: Clark Historical Library, Central Michigan University.
Burns, J. C. (2008). Networking Ohio Valley archaeology in the 1880s: The social dynamics of Peabody and Smithsonian centralization. *History of Anthropology Annual*, *4*, 1–33.
Cass, L. (1823). *Inquiries, respecting the history, traditions, languages, manners, customs, religion, &c. of the Indians, living within the United States*. Detroit: Sheldon & Reed.
Christenson, A. (2011). Who were the professional North American archaeologists of 1900? Clues from the Work of Warren K. Moorehead. *Bulletin of the History of Archaeology*, *21*(1), 4–22.
Danziger, E. D., Jr. (1970). The Steck-Carleton controversy in Civil War New Mexico. *The Southwestern Historical Quarterly*, *74*(2), 189–203.
Davis, W. W. H. (1857). *El Gringo: Or, New Mexico and her people*. New York: Harper & Brothers.
Gallatin, A. (1836). A synopsis of the Indian tribes within the United States east of the Rocky Mountains, and in the British and Russian possessions of North America. *Archaeologia Americana: Transactions and Collections of the American Antiquarian Society*, *2*, 1–422.
Gallatin, A. (1845). Notes on the semi-civilized nations of Mexico, Yucatan, and Central America. *Transactions of the American Ethnological Society*, *1*, 1–354.
Gibbs, G. (1863). *Instructions for research relative to the ethnology and philology of America* (Smithsonian Miscellaneous Collections, 160). Washington, D.C.: Smithsonian Institution.
Goetzmann, W. H. (1966). *Exploration and empire: The explorer and the scientist in the winning of the American West*. New York: Alfred H. Knopf.
Goldstein, D. (1994). "Yours for science": The Smithsonian Institution's correspondents and the shape of scientific community in nineteenth-century America. *Isis*, *85*(4), 573–599.
Gorman, R. (1914). Samuel Gorman. *Old Santa Fe*, *1*(3), 308–331.
Gutierrez, P. S. (2011). More than a name on a land grant: Charles Hipolite Beaubien. *Wagon Tracks*, *25*(3), 5–11.
Hall, G. E. (1984). *Four leagues of Pecos: A legal history of the Pecos Grant, 1800–1933*. Albuquerque, NM: University of New Mexico Press.
Hieb, L. A. (2011). Collecting for the Centennial Indian Exhibit: The J. K. Hillers – Olin D. Wheeler expedition of 1876. *New Mexico Historical Review*, *86*(2), 157–195.
Hinsley, C. M. (1981). *The Smithsonian and the American Indian: Making a moral anthropology in Victorian America*. Washington, D.C.: Smithsonian Institution Press.
Hinsley, C. M. (1992). Collecting cultures and cultures of collecting: The lure of the American Southwest, 1880–1915. *Museum Anthropology*, *16*(1), 12–20.
Hinsley, C. M. (1996). Digging for identity: Reflections on the cultural background of collecting. *The American Indian Quarterly*, *20*, 180–196.
Hollcroft, T. R. (1953). The diary of William Fellowes Morgan. *The Scientific Monthly*, *77*(3), 119–128.
Holmes, W. H. (1878). Report on the ancient ruins of Southwestern Colorado, examined during the summers of 1875 and 1876. In *10th annual report of the United States geological and geographical survey of the territories* (pp. 383–408). Washington, D.C.: Government Printing Office.
Jackson, W. H. (1879). Report on the ancient ruins examined in 1875 and 1877. In *10th annual report of the United States geological and geographical survey of the territories* (pp. 411–430). Washington, D.C.: Government Printing Office.
Kirsch, S. (2002). John Wesley Powell and the mapping of the Colorado Plateau, 1869–1879: Survey science, geographical solutions, and the economy of environmental values. *Annals of the Association of American Geographers*, *92*(3), 548–572.
Kuper, A. (2005). *The reinvention of primitive society: Transformations of a myth*. Milton Park Abingdon: Routledge.
Lange, C. H., & Riley, C. L. (1966). *The Southwestern journals of Adolph F. Bandelier 1880–1882*. Albuquerque, NM: The University of New Mexico Press.
Lange, C. H., & Riley, C. L. (1996). *Bandelier: The life and times of Adolph Bandelier*. Salt Lake City, UT: University of Utah Press.

Lester, P. A. (1986). *Michael Steck and New Mexico Indian affairs, 1852–1865* [Unpublished Doctoral dissertation]. University of Oklahoma, United States.

Lister, R. H., & Lister, F. C. (1990). *Aztec Ruins National Monument: Administrative history of an archaeological preserve* (Southwest Cultural Resources Center, Professional Papers, 24). Sante Fe, NM: National Park Service.

Mason, O. T. (1881). Abstracts of the Smithsonian correspondence relative to Aboriginal remains in the United States. In *Annual report of the board of regents of the Smithsonian Institution for 1880* (pp. 441–448). Washington, D.C.: U.S. Government Printing Office.

McPherson, R. S., & Neel, S. R. (2016). *Mapping the four corners: Narrating the Hayden Survey of 1875*. Norman, OK: University of Oklahoma Press.

Miller, D. A. (2007). *Matilda Coxe Stevenson: Pioneering anthropologist*. Norman: University of Oklahoma Press.

Morgan, L. H. (1869). The seven cities of Cibola. *North American Review, 108*, 457–498.

Morgan, L. H. (1871). *Systems of consanguinity and affinity of the human family* (*Smithsonian Contributions to Knowledge*, 218). Washington, D.C.: Smithsonian Institution.

Morgan, L. H. (1876). Montezuma's dinner: An essay on the tribal society of North America. *North American Review*, *122*, 265–308.

Morgan, L. H. (1877b). *Ancient society: Or, researches in the line of human progress from savagery through barbarism to civilization*. New York: Henry Holt.

Morgan, L. H. (1880). A study of the houses of the American Aborigines; with a scheme of exploration of the ruins in New Mexico and elsewhere. In *Archaeological Institute of America. First annual report of the Executive Committee, with accompanying papers. 1879–80* (pp. 29–80). Cambridge, MA: John Wilson and Son.

Morgan, L. H. (1881). *Houses and house-life of the American Aborigines* (Contributions to North American Ethnology, 4). Washington, D.C.: U.S. Department of the Interior, Government Printing Office.

Morgan, W. F. (1879). Description of a cliff-house on the Mancos River of Colorado, with a ground plan. *Proceedings of the American Association for the Advancement of Science, 27th Meeting*, 300–306.

Moses, D. N. (2009). *The promise of progress: The life and work of Lewis Henry Morgan*. Columbia, MO: University of Missouri Press.

Newberry, J. W. (1876). *Report of the exploring expedition from Santa Fe, New Mexico, to the junction of the Grand and Green Rivers of the Great Colorado of the West, in 1859*. Washington, D.C.: U.S. Government Printing Office.

Noelke, V. H. (1974). *The origins and early history of the Bureau of American Ethnology, 1879–1910* [Unpublished Doctoral dissertation]. The University of Texas, United States.

Norris, L. D., Milligan, J. C., & Faulk, O. B. (1998). *William H. Emory: Soldier-scientist*. Tucson: University of Arizona Press.

Parezo, N. J. (1989). The formation of ethnographic collections: The Smithsonian Institution in the American Southwest. *Advances in Archaeological Method and Theory, 10*, 1–47.

Parezo, N. J. (1993). Matilda Cox Stevenson: Pioneer ethnologist. In N. Parezo (Ed.), *Hidden scholars: Women anthropologists and the Native American Southwest* (pp. 38–62). Albuquerque: University of New Mexico Press.

Parsons, E. C. (1923). *Laguna genealogies* (Anthropological Papers of the American Museum of Natural History, Vol. 21, Part 5). New York: Order of the Trustees.

Peard, J. G. (2016). *An American teacher in Argentina: Mary Gorman's nineteenth-century odyssey from New Mexico to the Pampas*. Lanham, MD: Bucknell University Press.

Powell, J. W. (1881). Introductory. In *First annual report of the Bureau of Ethnology* (pp. xi–xxxiii). Washington, D.C.: U.S. Government Printing Office.

Powell, J. W. (1883). Introductory. In *Second annual report of the Bureau of Ethnology to the Secretary of the Smithsonian Institution 1880–'81* (pp. xv–xxxvii). Washington, D.C.: U.S. Government Printing Office.

Resek, C. (1960). *Lewis Henry Morgan, American scholar*. Chicago: University of Chicago Press.

Roberts, J., & Sheppard, K. (2020). Introduction: Clusters of knowledge. In J. Roberts, K. Sheppard, U. R. Hansson, & J. R. Trigg (Eds.), *Communities and knowledge production in archaeology* (pp. 1–13). Manchester: Manchester University Press.

Roeder, F. (2009, April). Surveyor of the public domain—A portrait of William Pelham. *Lidar Magazine*. https://lidarmag.com/2009/04/11/surveyor-of-the-public-domain-a-portrait-of-william-pelham/

Service, E. R. (1981). The mind of Lewis Henry Morgan [and comments and reply]. *Current Anthropology*, *22*(1), 25–43.

Simpson, J. H. (1852). *Journal of a military reconnaissance from Santa Fe, New Mexico, to the Navajo Country*. Philadelphia: Lippincott, Grambo and Co.

Snead, J. E. (2001). *Ruins and rivals: The making of Southwest Archaeology*. Tucson: University of Arizona Press.

Snead, J. E. (2018). *Relic hunters: Archaeology and the public in nineteenth-century America*. Oxford: Oxford University Press.

Snead, J. E. (2020). Circular 316: Archaeology, networks, and the Smithsonian Institution, 1876–79. In J. Roberts, K. Sheppard, U. R. Hansson, & J. R. Trigg (Eds.), *Communities and knowledge production in archaeology* (pp. 34–46). Manchester: Manchester University Press.

Squier, E. G. (1848). New Mexico and California. *The American Review* (New Series), *2*(5), 503–528.

Steele, T. J., & Weinberg, F. B. (2005). By letter: Three years in the life of Vicar Machebeuf. *New Mexico Historical Review, 80*(3), 293–308.

Stensvaag, J. (1980). Clio on the frontier: The intellectual evolution of the Historical Society of New Mexico, 1859–1925. *New Mexico Historical Review*, *55*(4), 293–308.

Stern, B. J. (1967). *Lewis Henry Morgan, social evolutionist*. New York: Russell and Russell.

Stevenson, J. (1883). Illustrated catalog of the collections obtained from the Indians of New Mexico in 1880. In *Second annual report of the Bureau of Ethnology* (pp. 429–455). Washington, D.C.: U.S. Government Printing Office.

Stevenson, M. C. (1881). *Zuñi and the Zuñians*. Privately Published.

Stocking, G. (1991). *Victorian anthropology*. New York: Collier McMillan.

Strang, C. (2017). Scientific instructions and Native American linguistics in the imperial United States. *Journal of the Early Republic*, *37*(3), 399–427.

Tooker, E. (1992). Lewis H. Morgan and his contemporaries. *American Anthropologist*, *94*(2), 357–375.

Trautmann, T. R. (1987). *Lewis Henry Morgan and the invention of kinship*. Berkeley: University of California Press.

Trautmann, T. R., & Kabelac, K. S. (1994). The library of Lewis Henry Morgan and Mary Elizabeth Morgan. *Transactions of the American Philosophical Society* (New Series), *84*(6–7), 1–336.

Vetter, J. (2011). Lay observers, telegraph lines, and Kansas weather: The field network as a mode of knowledge production. *Science in Context*, *24*(2), 259–280.

Walter, P. A. F. (1927). Necrology: Major George H. Pradt. *New Mexico Historical Review*, *2*(2), 208–214.

Ward, J. N. (1864). Report of John Ward, agent for Pueblo Indians. In *Annual report of the commissioner of Indian affairs, for the year 1864* (pp. 187–199). Washington, D.C.: Government Printing Office.

White, L. A. (1940). *Pioneers in American anthropology: The Bandelier-Morgan letters 1873–1883* (Vol. II). Albuquerque, NM: University of New Mexico Press.

White, L. A. (1942). Lewis H. Morgan's journal of a trip to Southwestern Colorado and New Mexico, June 21 to August 7, 1878. *American Antiquity, 8*(1), 1–26.

Wilcox, D. R., & Fowler, D. D. (2002). The beginnings of anthropological archaeology in the American Southwest: From Thomas Jefferson to the Pecos Conference [Monograph]. *Journal of the Southwest, 44*(2), 121–234.

Woodbury, R. B., & Woodbury, N. F. S. (1999). The rise and fall of the Bureau of American Ethnology. *Journal of the Southwest, 41*(3), 283–296.

Worster, D. (2000). *A river runs west: The life of John Wesley Powell.* Oxford: Oxford University Press.

"Lieber Freund!" Nineteenth-Century Correspondence Networks and the Early History of the Milwaukee Public Museum

6

Bettina Arnold

Abstract

The first decades of the Milwaukee Public Museum (MPM), which dates back to 1851, are documented in the MPM Library and the Wisconsin Historical Society's archives mainly in German and often in a now-defunct traditional German script called Sütterlin, or Deutsche Schrift. The founding members of the institution were almost all German-born and maintained contacts with the German-speaking areas of Europe through extensive scholarly correspondence as well as occasional trips to the Old Country to obtain specimens and interact with like-minded colleagues. While the history of the MPM in the early twentieth century is well documented (it was officially chartered in 1882) the same is not true for the first 30 years of its existence. This chapter focuses on the hidden history represented by the correspondence of several important founding members of the MPM: its first director, Carl (Charles) Dörflinger; Adolph Meinecke, charged with overseeing the design and construction of the first museum building; William Frankfurth, whose early antiquarian research in southern Tirol has recently been acknowledged in a museum exhibition in Bolzano, and Carl Thal, first official MPM librarian.

Keywords

History of archaeology · Trans-Atlantic interactions · German-American studies · Scholarly networks · Thought-collectives · Milwaukee Public Museum · Smithsonian Institution

B. Arnold (✉)
Department of Anthropology, University of Wisconsin-Milwaukee, Milwaukee, WI, USA
e-mail: barnold@uwm.edu

6.1 Introduction

The epistolatory logorrhea of the nineteenth century in Europe has been a gift to scholars attempting to reconstruct the social history of that era. Literacy rates were at an unprecedented high by the end of the century and record-keeping at all scales, from individual diaries through household accounts to institutional ledgers, was accompanied by the output of a vast system of correspondence networks, much of it still accessible in libraries and archives today. The face-to-face interactions of early antiquarians in pubs like the Bear Tavern in London, where the Society of Antiquaries had its eighteenth-century beginnings (Sweet 2004: 84), were exponentially expanded by the addition of hundreds of corresponding members, a practice that continues today in many archaeological societies. The rise of print media, especially newspapers, at a time when archaeological investigations were just beginning to transition from amateur interest to professional practice, provided an additional outlet for reporting on local as well as global archaeological discoveries, whetting the public appetite for such news items. In some cases, members of these categories overlapped; the editors of local newspapers in the United States were frequently also members of antiquarian correspondence networks and active collectors (Snead 2002: 42). It was truly a Golden Age for the production of documentary source material, simultaneously a boon and a bane for scholars attempting to organize and systematically analyze this paper mountain today.

Scholarship on antiquarian and early archaeological networks in the United States has a long track record, with numerous volumes dedicated to institutional histories, especially museums and organizations like the Smithsonian (e.g., Hinsley 1994; Petraglia and Potts 2004; Ewing 2007) or Harvard's Peabody Museum (Watson 2001; Greis 2006) and their impact on the discipline, but few of these studies have systematically explored the impact of the nineteenth-century wave of immigrants from Europe, especially in the wake of

L. Coltofean et al. (eds.), *Connecting People and Ideas*, Themes in Contemporary Archaeology,
https://doi.org/10.1007/978-3-031-81006-0_6

the 1848 revolutionary movements, on United States archaeology. The fact that the Midwestern states in particular established so many antiquarian organizations and contributed disproportionately to requests by the Smithsonian for reports on local natural history and archaeology (Snead 2020: 42) has yet to be studied in light of the timing of this immigration phenomenon and the backgrounds of the individuals involved. States with visible above-ground monuments like those at Cahokia in Illinois (Kelly 2002) or the Effigy Mounds of Ohio and Wisconsin (Birmingham and Rosebrough 2017) were especially active in early archaeological investigations and the histories of their museums and other intellectual institutions reflect this. For various reasons, these were also the states with the largest influx of German-speaking immigrants in the aftermath of the 1848 revolutions and the close contacts many of these individuals maintained with their home countries in the later nineteenth century resulted in a dramatic expansion of the intellectual horizons of scholars on both sides of the Atlantic (Arnold 2019a; Kaufmann 2019). The early history of archaeology in the United States must therefore be viewed in the larger context of United States immigration history and the communication networks—what Ludwik Fleck (1935) has dubbed thought-collectives—that connected first generation European immigrants to existing communities of knowledge production in the Old Country (Milosavljević 2020; de Tomasi 2020). Many of those who were economically successful returned to the Old Country at least once after emigrating, creating a complex and multi-directional meshwork (Ingold 2015; Klenk 2018) of intellectual exchanges. Certain individuals served as high-profile bridges in this early information transfer, including the German polymath Alexander von Humboldt (1769–1859) (Wulf 2015; Penny 2019) and the Swiss naturalist Louis Agassiz (1807–1873) (Irmscher 2013) but many less well-known individuals trained in various scientific disciplines had a significant impact on thought-collectives in areas like the Midwest (Kelly 2002, among others). Scholarly investigation of the rich data sets represented by these networks in United States archives has been hampered by linguistic and epigraphic barriers and the German-language documentation is only now being slowly brought to light (Arnold 2013, 2019a, b; Arnold and Stadler 2014).

One of the major Midwestern hubs of this trans-Atlantic intellectual network is introduced here, represented by several individuals who were involved in founding and building the collections of the Milwaukee Public Museum (MPM), the oldest natural history museum in the Midwest (Lurie 1983). The MPM is well-known in museum circles as the place where famed taxidermist Carl Akeley (1864–1926) first developed the wildlife displays that came to feature prominently in United States natural history museums (Kirk 2010); the so-called muskrat colony case is the first complete museum habitat diorama in the world. What became known as the "Milwaukee style" of large and small-scale dioramas influenced many other United States museums in the nineteenth and early twentieth centuries. The focus of the institution has always been on research as well as on exhibits, public outreach and education, all of which were explicit goals of the German-American founders from its inception (Meinecke 1904; Lurie 1983).

6.2 Antiquarians in the German Athens

Although German immigrants came to dominate mid- to late-nineteenth century immigration to Wisconsin to such an extent that Milwaukee became known as the "German Athens" (Fig. 6.1) and boasted more German-language than English newspapers by the 1880s (Gurda 1999), the first serious study of the natural and pre-contact cultural history of Wisconsin was produced by a Yankee transplant by the name of Increase Lapham (1811–1875) (Bergland and Hayes 2014). Lapham arrived in 1836 in what was at that time still a trading post at the mouth of the Milwaukee River and was instrumental in establishing the initial land survey that eventually become the street grid of Milwaukee west of the river. This was a period of rapid growth and constantly changing borders in the Great Lakes region. The day Lapham stepped off the boat, Milwaukee was still in Michigan territory but three days later it became the Wisconsin Territory when Michigan became a state. By the time Wisconsin achieved statehood 12 years later in 1848, Lapham had become its foremost naturalist with the publication in 1844 of *A geographical and topographical description of Wisconsin with brief sketches of its history, geology, mineralogy, natural history, population, soil, productions, government, antiquities etc. etc.* Although archaeological sites are only mentioned at the very end of the title of this first work, they were a major feature of his 1855 monograph, *The antiquities of Wisconsin*, which included detailed surveys of mounds and mound groups that in many cases are the only remaining documentation of these monuments available to us today.

The significance of Lapham's work was recognized across the Atlantic as well as in United States antiquarian circles. Two years after publishing his *magnum opus* on Wisconsin's antiquities, Lapham received a letter from antiquarian Charles C. Rafn (1795–1864) in Copenhagen, Denmark, informing him that he had been elected a member of the Royal Society of Northern Antiquities, which had translated and published one of his articles on American Indian mounds and artifacts. The main motivation for initiating the correspondence on Rafn's part seems to have been a desire to facilitate the exchange of artifacts and publications, one of the primary mechanisms for scholarly information transfer at the time. Lapham would have been more impressed by the honor if his name had not been spelled incorrectly on the

Fig. 6.1 Panoramic view of Milwaukee c. 1898. (Photo: The Gugler Lithographic Co.; copyright free, available through the website of the Library of Congress, Washington, D.C., United States, https://www.loc.gov/resource/g4124m.pm010472/)

diploma (Bergland and Hayes 2014: 228–29). Clearly by the mid-nineteenth century there was growing awareness on the part of antiquarian networks in Europe of the benefits of establishing links to scholars in the Americas.

By the 1850s Lapham's interest in geology had brought him into the orbit of Louis Agassiz, recently established at Harvard after emigrating to the United States from Switzerland. The archives contain a lively correspondence between Lapham and Agassiz's assistant (and eventual detractor) Édouard Desor (1811–1882) after the latter's return to Europe (Bergland and Hayes 2014: 284). Desor's *carte de visite* is still in the Wisconsin Historical Society archives and his travels across the United States appear to have sparked a lifelong interest in the natural and cultural history of the Americas in this preeminent German-Swiss naturalist, who was elected a member of the American Antiquarian Society in 1871 (Kaeser 2004: 92–101). In the aftermath of the American Civil War, Lapham became involved in the newly founded German Natural History Society of Wisconsin, which eventually evolved into the Milwaukee Public Museum. In 1869 Carl H. Dörflinger (1843–1911), an active member of the Society at the time, gave Lapham a section of a meteorite discovered by a farmer plowing his field near West Bend and the stereograph of Lapham examining the specimen has become the iconic image of this preeminent early scientist (Fig. 6.2; Bergland and Hayes 2014: cover photo). The meteorite was part of the early collections of the German Natural History Society, which was reorganized and incorporated under a state charter as the Natural History Society of Wisconsin in 1879. This society was originally founded in 1857 by Peter Engelmann (1823–1874), whose impressive academic credentials outstripped those of most of his fellow 1848ers, mostly young radicals with progressive views who had fled Europe when the hoped-for reforms failed to materialize (Zucker 1950; Wittke 1952). In his brief biographical sketch of Peter Engelmann, Dörflinger describes the motivations of this generation of German intellectuals as follows:

> Like most students and many thousands of professors and other cultured men in Germany and Austria, young Engelmann in 1848 was drawn into the vortex of the idealistic uprising against the German specimen of secession, the disintegration of the German Empire, and the unbearable oppression of thirty-two (more or less) reactionary governments. (Dörflinger 1907: 3)

Educated at the Universities of Heidelberg and Berlin, Engelmann was the obvious choice of the German community in Milwaukee to lead the new German-English Academy, later to become the University School of Milwaukee. The school was established by mainly German immigrants dissatisfied with the public schools available in the city at the time, who wanted their children to have the benefit of a bilingual and progressive education. Most of the board members

Fig. 6.2 Increase A. Lapham inspecting a piece of the Trenton meteorite. (Photo: Bettina Arnold of cover of *Studying Wisconsin: The Life of Increase Lapham* [Bergland and Hayes 2014])

and trustees of what was to become the MPM were educated at the Academy or had been instrumental in its establishment (Lurie 1983: 5). Engelmann's pedagogical focus on learning by directly experiencing nature and objects, including archaeological artifacts, was part of the impetus for the accumulation of natural and cultural specimens in the early collections of the future MPM. The school served as a direct link between progressive pedagogical movements in Germany (Marchand 1996; Nyhart 2009) and the United States and had a major impact on the interest of the many Wisconsin residents who attended the Engelmann School, as it was also known, in the natural and cultural history of their state as well as other areas of the world. By the late nineteenth century, the German-English Academy, the Milwaukee Turners Society, which provided the funds for the construction of the Academy's gymnasium, and the MPM were at the center of cultural, intellectual, and political life in the city. Many of the same names appear on the founding documents of all three institutions, whose continued existence and importance in Milwaukee today are a testament to the energy and enthusiasm these immigrants were willing to commit to the natural and cultural history of their new home.

In his brief history of the Wisconsin Natural History Society, Dörflinger specifically singles out individuals listed as founding members of this organization, by indicating whether they were "Forty-eighters" (referring to the 1848 European revolutionary movement) or had served in the Union Army during the Civil War (1861–1865) (Dörflinger 1907: 3). Ten of the names on the list have both symbols after their names, including Dörflinger himself, whose ankle was shattered by a Minié ball at the Battle of Chancellorsville (1863) during the Civil War, when he was just 19 years of age (Dörflinger 1907: 10–11). By 1866 the list of Honorary Members included Andrew Kerr of Scotland and Dr. Emil Höpfel, a Forty-eighter who returned to Germany and became mayor of his hometown there, a good example of the return-migration that kept the links between the Old and New World communities of Germans strong (Dörflinger 1907: 12).

While the history of the MPM in the early twentieth century is well documented (it was officially chartered in 1882) the same is not true for the first 30 years of its existence, due to the dual barriers of language and the esoteric looking Sütterlin script used by those educated in German-speaking school systems in the nineteenth and early twentieth centuries. Apparently the first records of the Proceedings of the Wisconsin Natural History Society were lost (Dörflinger 1907: 11), so the first official record in the MPM archives is Dörflinger's 1866 annual report (Dörflinger 1907: 9). Minutes of meetings in the early decades of the MPM are documented in the MPM Library and the Wisconsin Historical Society archives entirely in German in both standard and Kurrent/Sütterlin script. Also referred to as *Deutsche Schrift*, this cursive form was designed in 1911 by graphic artist Ludwig Sütterlin (1865–1917) as a commission funded by the Prussian Ministry of Science, Art and Culture (Sütterlin 2023). The script was taught in all German schools between 1915 and 1941, when the Nazi party banned its use. It is virtually incomprehensible without special training (Ulmschneider 2011) and poses a significant barrier to research into the early history of institutions in areas of the United States where German-speaking immigrants were often in the majority.

The founding members of the MPM, whether first- or second-generation immigrants, were trained in the use of this script as well as standard cursive (Fig. 6.3). Their letters to one another while traveling as well as to colleagues in Europe were frequently written in Sütterlin, with dates, place names and personal names in standard script. The standard script sections of such documents provide at least an indication of the likely subject matter, but specialist transcription is usually required for complete comprehension. The earliest MPM ledgers contain the minutes of meetings and reports of members that shift between Sütterlin and standard cursive until 1899, when the official record made the transition to

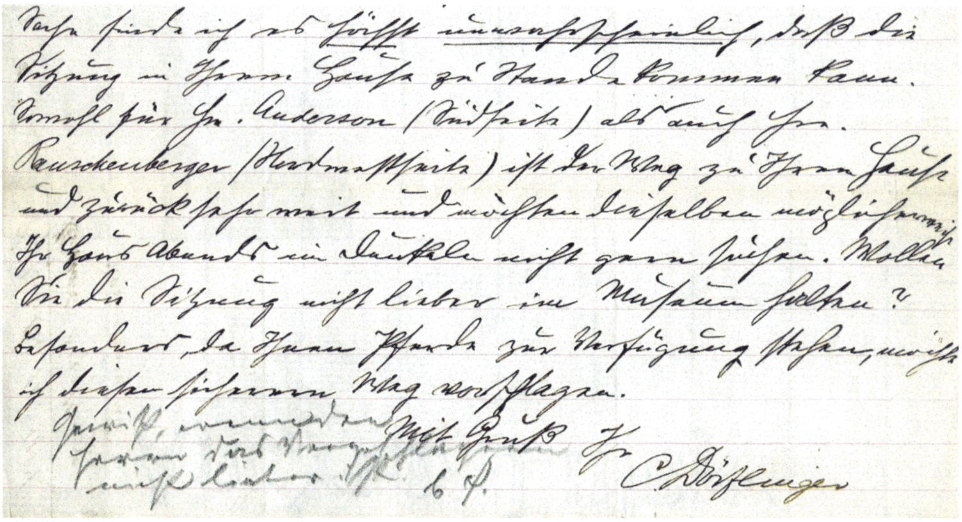

Fig. 6.3 Excerpt of letter by Carl Dörflinger in Sütterlin. (MPM Directors' Correspondence Files; Photo: Bettina Arnold)

English and the English-language *Bulletin* replaced the *Jahres-Berichte* that had chronicled the early years of the museum (Lurie 1983: 6). Some early collections, including specimens from the important Neolithic and Bronze Age lake-dwelling site of Robenhausen near Zürich, contained objects labeled using this script, which contributed to the slow sinking into obscurity of this material. The MPM directors, curators, and museum professionals after 1918 rarely spoke, read, or wrote German and the script would have exacerbated this gradual process of alienation.

To understand the connections between the various figures associated with the early history of the MPM and their colleagues in Europe, their biographies must be viewed generationally as well as relationally (Fig. 6.4). The additional filters of educational and socio-economic background and life course trajectories played a determining role in the influence of each of these individuals on developments in the natural sciences and archaeology, both in the United States and Europe. The founder generation in Wisconsin is represented by Lapham, an East Coast transplant of English descent trained as a surveyor and geologist with an active interest in Native American cultures, and first-generation German immigrant Engelmann, the only member of the group with a traditional academic background, described as the "spiritual father of the MPM" (Dörflinger 1907: 6).

The relationship between Lapham and Engelmann appears to have been an unusually fruitful meeting of minds, based on Dörflinger's assessment of the affinity between them:

> Thus the two friends of like idealistic aspirations (…) may go into history as the first two exchange professors between America and Germany, i.e. the Female College [the Downer Women's College, of which Lapham was a trustee and where he taught science classes] and the German English Academy [run by Engelmann, who occasionally taught chemistry at the Female College], only that they were not separated by an ocean but only by one block of Division Street (now Juneau Avenue), between Main Street (now Broadway) and Milwaukee Street. (Dörflinger 1907: 14)

The Honorary Members of the Wisconsin Natural History Society/MPM between 1857 and 1882 included five academics in Germany and one in France (François Daleau [1845–1927]) (Dörflinger 1907: 15), reflecting the expanding influence of the organization outside the region and in Europe. The French connection represented by Daleau, considered the founder of French prehistoric archaeology (Groenen 2021), was the direct result of Dörflinger's return trip to Europe in 1892–1893. Among other locations in France and Switzerland, Dörflinger visited Daleau's excavations at the Upper Paleolithic cave site of Pair-non-Pair near Bourg sur Gironde, where he acquired numerous lithic specimens and other artifacts "all authenticated by Dr. François Daleau, Bourg sur Gironde, the discoverer and explorer of Pair-non-Pair" (Dörflinger 1893). The recognition that parietal or cave art on the walls of Upper Paleolithic caves had been produced by the same ancient humans who made the stone tools found at such sites was still contested at the time but the antiquity of the engravings at Pair-non-Pair was accepted by 1896, shortly after Dörflinger's visit. It is therefore one of the most important sites in the history of Upper Paleolithic art in Europe (Lenoir 2006). Dörflinger also visited and acquired artifacts from the site of Robenhausen in Switzerland, which contributed to *Pfahlbaufieber* (lake-dwelling fever), a collecting frenzy that led to a diaspora of lake-dwelling artifacts from Switzerland to museums across the globe (B. Arnold 2013).

6.3 The Merchant Archaeologists

The next group chronologically speaking is represented by first-generation German immigrants Adolph Meinecke (1830–1905) and William (Wilhelm) Frankfurth

Fig. 6.4 Generational phylogeny of Milwaukee Public Museum's founders, from bottom to top. (Photo: Bettina Arnold)

(1829–1891), both successful businessmen who were actively involved in establishing various civic and cultural organizations in Milwaukee, including the Engelmann School, the Milwaukee Turners and the MPM.

Meinecke's wicker toy and furniture business was successful enough to allow him to travel to Europe on collecting and museum trips in 1884 and 1885 to acquire specimens for the ornithological collections of the MPM, his particular interest, as well as geological and archaeological material. Among other things he carried with him *papier mâché* models of Lapham's Effigy Mound display (Meinecke 1884), commissioned by the Smithsonian and designed by Lapham in 1875 for the 1876 Centennial Exhibition in Philadelphia. The Effigy Mounds are a unique cultural phenomenon restricted to the southern part of Wisconsin with some overlap into the bordering states of Illinois, Iowa, and Minnesota (Birmingham and Rosebrough 2017; Rosebrough 2017). Meinecke's delivery of the models based on Lapham's research would have exposed European museum visitors to the prehistory of this area of the United States for the first time. He in turn brought back to Milwaukee the first specimens of Swiss lake-dwelling material exhibited in the Midwest, part of the lake-dwelling collecting "fever" sparked by the publication of the English translation of prehistorian Ferdinand Keller's (1800–1881) *The lake dwellings of Switzerland* in 1866 (Keller 1866; Altorfer 2004; Arnold 2013).

Meinecke's dedication to the acquisition of natural and cultural specimens for the MPM is reflected not just in his letters to Dörflinger, who was still the de facto director of the museum at the time, but in an editorial published in the *Milwaukee Herald* on October 9, 1904. Entitled "Unser Museum", the history of the founding of the MPM reflects Meinecke's pride in his adopted city and its citizens, whose hard work and financial sacrifices had made possible a first-class institution dedicated to the education of the general public and free to all (Meinecke 1904). He reports on the visit of the famous German zoologist and author Alfred Brehm (1829–1884), who was on a lecture tour of the United States in the year before his death and was pleasantly surprised to find such an impressive institution in the "distant West". Meinecke also was instrumental in drawing up the plans and raising the $500,000 needed for the new museum building, which by 1904 contained more than 250,000 specimens and objects of archaeological and ethnographic importance. These efforts were directed at bringing the world to Milwaukee in order to educate and entertain the local population, a motivating factor in the founding of many such institutions in the United States and Europe at the time. Based on the documentary record available, his 1884–1885 trip to Europe was partly to gather intel on the latest European approaches to museum design and displays, which contributed to the innovative ideas he and Klas Linderfelt (1847–1900), the Head Librarian of the Milwaukee Public Library, developed in the plans for the new building (Linderfelt and Meinecke 1890).

The connections Meinecke made at museums and scientific societies in Lübeck, Berlin, Hamburg, Munich, Bolzano, Meran, Naples, Konstanz, and London during his travels in 1884–1885, may have partly influenced the itinerary of industrialist William Frankfurth. The latter's activities in

Germany and Austria during his trip to Europe between 1889 and 1891 included several of the localities in southern Tirol previously visited by Meinecke, his colleague on the MPM Board. Frankfurth's interest in all things historical and archaeological appears to have been established early in his life but he was unable to indulge it until 1875, after his hardware and iron monger's business in Milwaukee had been securely established and he was well over 40 years of age. In a different time and under more favorable circumstances he would very likely have received a university education and been trained as a professional archaeologist. Born in Gudensberg near Cassel in Hessen, we know little about his formative years other than that his father, a Swiss national with a civil service position in the *Hessische Regierung* (the government of Hessen an independent member of the German Confederation at the time) died when he was three years old. After receiving a primary school education, he moved to Arnstadt to be apprenticed to an uncle who owned a grocery store there. By the time he emigrated to the United States at the age of 19 he was a self-described Free Thinker, which appears to have influenced his interest in cross-cultural comparisons of ancient technology and settlement structures as well as his concern with education, civic engagement, and service (Arnold 2019a: 290–92).

That he was politically progressive is clear from a number of documentary sources that describe his activities after arriving in Milwaukee, where he quickly established a prominent position among his fellow Forty-eighters. In addition to serving as the director of the Milwaukee Exposition Association and the Engelmann German-English Academy, Frankfurth was a trustee of the MPM and first speaker of the Milwaukee Turnverein ("Died away from home," 1891). He was in regular contact with Engelmann and Lapham and based on locally acquired artifacts donated to the MPM, he engaged in archaeological investigations at various sites in and around Milwaukee, including the Mississippian outlier of Aztalan, the northernmost site associated with the major pre-contact center of Cahokia in Illinois. The platform mounds characteristic of Mississippian sites (dating from approximately AD 800–1600) captured Frankfurth's interest and attention. At a meeting of the Natural History Society of Wisconsin on November 14, 1885, Frankfurth is recorded in the Minutes as having moved

> that a committee be appointed to confer with the Wisconsin Academy of Arts, Sciences and Letters, the State Historical Society, and board of the Trustees of the Milwaukee Public Museum relative to the preservation of the more characteristic Mound-builders works located in different parts of Wisconsin. The motion was carried (Wisconsin Natural History Society Bulletin 1884–1888: 47)

This brief passage indicates that Frankfurth was concerned with preserving as well as exploring local archaeological sites. He also seems to have developed an eye for identifying artificial subsurface monumental constructions such as earthworks and tumuli during his exploration of pre-contact sites in the Midwest. This ability to read cultural landscapes obscured by vegetation was applied to great effect during Frankfurth's field surveys in Germany, the Innsbruck area and southern Tirol in 1889–1891, where antiquarian efforts were still mainly focused on above-ground monuments made of stone and dating to the Roman period or later (Arnold and Stadler 2014: 268–69; Kaufmann 2019: 306–07). His activities spurred local antiquarians on to follow-up excavations at sites that are now recognized as especially important, such as the Hochbichl (Kaufmann 2019: 306–13).

The trip to Europe in 1889 appears to have been mainly motivated by the ill health of both Magdalena, Frankfurth's wife, and one of his sons (it is unclear whether Hans or Lorenz was the sufferer). After an initial period of several months following their arrival in Vienna in December 1889 or January 1890 (the precise departure and arrival dates are also unclear), Frankfurth settled in for a period of serious antiquarian exploration in southern Tirol. The Alpine climate of this region was considered efficacious for sufferers of various ailments in the nineteenth century. Frankfurth is first mentioned in the local newspapers, especially the *Meraner Zeitung*, in the spring of 1890, after he had secured the services of a local expert (and dealer) in antiquities by the name of Fridolin Plant (1838–1911). Newspaper accounts describe their trips to several sites in and around Meran between February and May of 1890, including the Hochbühl, the Grumserbühl, the Sinichkopf, the Fachegg, the Glurnserköpfl, and the Tartscherbühl (Table 6.1; "Prähistorische Funde"; Arnold and Stadler 2014). These sites include several strategically located Iron Age settlements and ritual places that were important during the initial Roman forays across the Alps in the first century BCE (Kaufmann 2019) whose significance had not yet been recognized at the time of Frankfurth's investigations.

Handwritten notes that are obviously part of a diary kept by Frankfurth during these expeditions include descriptions of the find contexts of some of this material. In at least one case (the Sinichkopf), he refers to a sketch map with numbered features, which has unfortunately not yet turned up in the MPM archives but indicates that he conducted relatively extensive excavations at some of these sites. Plant and Frankfurth seem to have had a falling-out sometime in May of 1890 regarding which of them had first identified the prehistoric nature and significance of the Segenbühl (*Meraner Zeitung* 1890 May 21). In an article in *Der Burggräfler* on June 7, 1890, Frankfurth claimed, rather sarcastically, that if Plant had really recognized the site's importance years ago, as he had claimed in a previous newspaper piece ("Ausgrabungen in Tirol," 1890), he would surely have trumpeted the discovery far and wide, "denn ich halte ihn nicht für den Mann, der sein Licht unter den Scheffel stellt"

Table 6.1 Milwaukee Public Museum Accession 213 (Frankfurth Collection)

Provisional itinerary for William Frankfurth's European trip 1889–1891	
Dates	*Location/Activities*
November 1889	Frankfurth travels to Europe from Milwaukee with his wife and two sons
February–April 1890	Frankfurth meets Fridolin Plant in Meran; they visit the Hochbühl among other sites
March–April 1890	Frankfurth and his sons carry out excavations at the Grumserbühl with Fridolin Plant and Peter Reuter
April–May 1890	Frankfurth and Plant excavate at the Hochbühl; Frankfurth and his sons excavate at the Sinichkopf
May 1890	Frankfurth and his sons excavate at the Glurnserköpfl, the Tartscherbühl and the Fachegg
June–August 1890	Frankfurth excavates at the Sonnenburgerhügel near Innsbruck as well as in Natters und Hötting
September 1890	Frankfurth excavates at Ampaß (Adolf von Pichler)
October 1890	Frankfurth investigates the Flatschhügel (Adolf von Pichler)
February–September 1891	Frankfurth and his family reside in the Pension Hedwig, Wilhelmshöhe, Gudensberg, Hessen
February 1891	Frankfurth receives a letter from Ludwig Lindenschmidt in Cassel
July 21, 1891	Frankfurth identifies the Baunsberger hillfort as an Iron Age hillfort
September–October 1891	Frankfurth and his family return to Innsbruck; Lorenz travels back to Milwaukee
December 1, 1891	Frankfurth dies in Vienna of an unidentified stomach ailment (probably a bowel obstruction or appendicitis)

(because he isn't the type to hide his light under a bushel) (Frankfurth 1890). In the same article he refutes Plant's assertion that the finds had all been sent to America; rather, he (Frankfurth) had taken a third of the excavated finds from the site, Plant had also taken a third and the remainder was given to the town of Meran (*BG*, 1890, June 7). The highly competitive and frequently combative atmosphere of nineteenth-century antiquarian and scientific networks was in part due to the prestige associated with new discoveries and disagreements between scholars were often fought out in the newspapers of the day. Such fallings-out have been chronicled in other contexts around this time, famously in the previously mentioned case of Agassiz and his protégé Desor (Kaeser 2004; Irmscher 2013). The latter spent some time in the Midwest and Frankfurth is likely to have met him during his visit to Wisconsin.

By the beginning of June 1890 Frankfurth appears to have moved his family to Innsbruck, based on reports of his activities submitted to the *Meraner Zeitung*. During this period, he continued his explorations at sites in or near Natters and Hötting and secured official permission to engage in excavations at the important Iron Age site of Ampaß near Innsbruck from the local antiquarian and well-known Forty-eighter Adolf Pichler, Ritter von Rautenkar (1819–1900). The permits are part of the archival collection at the MPM, which includes an extensive correspondence between the two men. The official sanctioning of Frankfurth's activities does not appear to have mitigated the local sense of outrage at what was increasingly viewed as the depredations of an outsider, culminating in a temporary stay of Frankfurth's excavations (Fischnaler 1890). Frankfurth countered with a comment published in the *Meraner Zeitung* on October 8, 1890:

> Ich prätendire durchaus nichts, und nehme nur für mich das Recht in Anspruch, gerade so, wie ich schon in Amerika archäologische Studien und Forschungen machte, auch hier weiter zu führen. Dazu bietet Tirol, das ich durch einen fast einjährigen Aufenthalt recht lieb gewonnen habe, einen sehr günstigen Boden. Es mag in Europa wohl wenig Länder geben, die noch so viele prähistorische Schätze in der Erde bergen als Tirol. Je früher diese gehoben werden, um so besser. Hier wäre Arbeit für das Landesmuseum, hier ließen sich Lorbeeren erringen, nicht dadurch, daß man Leute, welche forschen und finden, benörgelt und verdächtigt. Wenn es dem Ferdinandeum daran liegt, Funde zu machen, so darf es z.B. nur da weiter nachgraben, wo ich mit meinen Söhnen bereits Entdeckungen gemacht habe und wir es das auch jetzt in der Nähe von Innsbruck gethan hat.
> (I make no claims and am only exercising my right to continue to conduct archaeological investigations and research just as I did in America. Tirol, of which I have grown very fond during my year-long sojourn, is a rich location in which to conduct such investigations. There are surely few countries in Europe that still retain so many treasures in their soil as does Tirol. The sooner these are recovered, the better. This would be a worthy task for the *Landesmuseum*, this would be a way to earn laurels, not by making complaints and casting suspicion against those who do seek and find. If the Ferdinandeum is interested in making discoveries, all they need to do is continue to excavate at those locations where my sons and I have already made finds, as we have now also done in the vicinity of Innsbruck.) ("Prähistorische Forschungen in Tirol," 1890: 3)

Conrad Fischnaler (1855–1941), curator at the Ferdinandeum in Innsbruck at the time, was even more vituperative in his newspaper campaign against Frankfurth than Fridolin Plant had been, trading insults with Frankfurth in a series of articles through October 17, 1890. Eventually Dr. Franz von Wieser (1848–1923), head conservator and director of the Ferdinandeum, weighed in by issuing a general interdict against "archäologische Raubgräberei" (archaeological looting), described in terms that made it clear that Frankfurth was its main target ("Behördliche Maßnahmen gegen archäologische Raubgräberei," 1890). A review of this feud was published in the *Augsburger Abendzeitung* by Arnold von der Passer (1851–1917), a pseudonym of the writer Franz L. Hoffmann, sometime in October 1890, in which he admits that Frankfurth's claim that the antiquarian elite of Tirol had neglected their prehistoric patrimony was justified: "da man sich tatsächlich bisher wenig um diese interessanten Ueberreste gekümmert hat. Das gebildete Publikum hat den Vortheil, daß es bei Gelegenheit dieses Streites über alle Fundstellen genau unterrichtet wird" (because there has in fact been little attention paid to date to these interesting remains. The educated public has the advantage due to this feud being fought in print of being informed as to the precise nature of all these sites) (von der Passer 1890).

The documentary evidence currently available is silent on the whereabouts of the Frankfurth family between October 1890 and the spring and summer of 1891, when he reappears on the scene in his hometown of Gudensberg in Hessen and joins the local Verein für hessische Geschichte und Landeskunde (Hessian Historical Society) there, which reported on his induction and death in the same year (*Mitteilungen*, 1891). Frankfurth led a Historical Society tour to the summit of the Baunsberg oppidum, which he correctly identified as a prehistoric hillfort site; he is credited with being the first person to recognize the significance of this fortified settlement (Roßner 2012). It is likely that he visited other sites in the area but until the archives of the local ministries and institutions have been investigated further, the only source of information is a map in the MPM collections that bears pencil marks showing possible routes he may have taken on foot to survey various likely localities near the Wilhelmshöhe Pension (similar to a long-stay bed and breakfast) that was the family's home base during that time (Arnold 2019a: Fig. 5).

Accession 213 at the MPM consists of material donated by William Frankfurth's descendants after his death as well as donations he made prior to departing for Europe in 1889, including specimens collected in Wisconsin (Table 6.2). What remains to be determined is whether the European material, which includes a large collection of books and maps, was acquired during the 1889–1891 trip or whether some of it may have been collected during earlier, shorter trips to Europe. The bulk of the collection is undoubtedly represented by the material excavated and collected by

Table 6.2 William Frankfurth's European itinerary 1889–1891

Estimated Milwaukee Public Museum Accession 213 and 29,433 (Frankfurth Collection)

Object category	*Object quantity*
Prehistoric bronze and iron objects	350
Terra sigillata	722
Other Greek/Roman pottery	76
Prehistoric pottery	171
Lake-dwelling material	107
Medieval or historic finds	225
Antiquarian books and maps	30
Total	**1681**

Frankfurth during the 1889–1891 Europe trip, which is first mentioned in his obituary notice in the *Milwaukee Sentinel*:

> While in Germany [sic], at his own expense, he superintended explorations of different districts where he made a large number of valuable prehistoric finds. Part of them he was compelled to deliver to the German [sic] government, but the balance he stipulated in his will were to be sent to the Public Museum in Milwaukee. There are seven large boxes, filled with these finds, stored away at Bremen at present, and will be sent to America as soon as Mrs. Frankfurth is able to carry out the details of her husband's wishes. ("An honored citizen," 1891: 3, col. A)

An entry in *A short history of the German-English Academy*, published on the occasion of the jubilee of the Engelmann School in May 1901, indicates that Frankfurth had planned to work on this material upon his return to the United States ("William Frankfurth," 1901: 71). In one of his letters to the *Meraner Zeitung* in response to complaints by Fischnaler he clearly stated his intention of publishing his discoveries after returning to the United States ("Prähistorische Ausgrabungen," 1890). Unfortunately, his premature death made that impossible and contributed to the gradual erosion of context information associated with the excavated material. When the family eventually delivered the seven crates of material to the MPM, there was little likelihood that their significance would be recognized. The entry for April 17, 1906, of the *Milwaukee Public Museum Monthly Reports* from 1905–1908 reads as follows:

> We have received from Mr. Lawrence [sic] Frankfurth, in the name of his father the late Wm. Frankfurth, two wagon loads of archaeological specimens, minerals and books. How much of this material will be of use to the museum can only be determined after it has been examined critically. (Ward 1906b: 54)

In a letter to Frankfurth's son Lorenz, dated September 26, 1906, then-MPM president Henry L. Ward (1863–1943) was less than enthusiastic about the research potential of the collection, which by that time had been subjected to a preliminary examination:

> Dear Sir, I have recently looked over the specimens donated to the museum and have picked out all of the cameos that were contained in it which, together with a number of photographs that may have some family value, I herewith return to you. [It is likely that some of the photographs included images of the sites where Frankfurth and his sons conducted their excavations, but Ward obviously was unaware of that.] I am sorry to find that so few of the specimens have adequate data preserved with them to make them of scientific value. Most of the pottery I have set aside until such time as I can secure the services of some expert who may be able to supply the requisite information concerning it. (Ward 1906a, September 26)

It is worth noting that Ward was the first director of the MPM who was not German-born or of German descent, although he evidently was able to read and write the language to some extent. He had little or no expertise in European pre- or early history and was unaware of the significance of much of what the collection contained. Based on Ward's comment about the pottery, the MPM appears to have conducted an initial sorting of the artifacts by material, which destroyed the potential site-based context of the assemblages. This is the main reason it is so difficult now to re-associate the material with locations. Matters were not improved by a subsequent attempt to catalog the collection detailed in the MPM *Monthly Report* for June 13, 1917, which implies that the material that was put on display was separated based on presumed cultural phase:

> A considerable amount of Dr. Hawkes' time during the month has been spent on working over the European archaeological collection and installing a large part of the same. This collection was originally catalogued [unclear if this refers to the MPM or to Frankfurth] by locality only, but inasmuch as various localities contained two or more distinct cultures it was necessary to make a careful examination of these specimens *in order to separate them according to the different culture periods represented.* We were fortunate in having a visit from Mr. Lytton of the American Museum of Natural History, New York, who had made a considerable study of this subject and who worked for several days gratuitously in connection with Dr. Hawkes thereby affording us the benefit of his experience in this particular line of research. (Ward 1917: 247; emphasis added)

Ernest William Hawkes (1883–1957) was an American anthropologist specializing in the indigenous peoples of Alaska and northern Canada, so his qualifications for cataloging the material Frankfurth had collected in southern Tirol were questionable to say the least and nothing further is known about Mr. Lytton, not even his first name. Most of the bronze and iron objects are late Bronze Age, Iron Age or provincial Roman but they were stored in drawers labeled "Greek and Roman Metalwork." The pottery that was not obviously Roman was never cataloged at all until it was recognized as potentially part of the Frankfurth assemblage (Arnold and Stadler 2014). Fragments of Fritzens-Sanzeno type ceramic vessels and a piece of a soapstone (Lavezstein) cooking vessel, all of which are characteristic of southern Tirol, make it clear that this material belongs with the remainder of Accession 213. However, because no documentation exists to confirm this attribution a different accession number was assigned to this material (Table 6.2).

By 2014, when the additional documentation and maps were discovered in the MPM Library, two Master's theses on portions of the Frankfurth collection had been completed under the author's supervision in the Department of Anthropology at the University of Wisconsin-Milwaukee (UWM). One focused on the *terra sigillata* material (Caywood 2011) and the other on the metal objects from sites in southern Tirol that corresponded to descriptions of finds in contemporary newspaper accounts (Arnold 2014). A cache of iron axes was attributed to Frankfurth's explorations on the Tartscherbühl (also known as the Tartscher Bichl or Tartscher Bühel) in a publication by regional archaeologist Reimo Lunz (2006), who had come across a description of the find in the *Meraner Zeitung*. Family members still living in Wisconsin have provided more documentary evidence (Arnold and Stadler 2014: 272–73) and there are plans for additional archival research in Innsbruck and Gudensberg that will, we hope, further fill in the gaps in the story of Frankfurth's archaeological explorations in Germany, Austria, and Italy. That would make it possible to position this biographical and museological narrative within the larger socio-political and intellectual developments in natural history, public education and science in Austria, Germany and Italy in the late nineteenth century (Applegate 1990; Hare 2015; Maner 2018).

Frankfurth's unexpected death in Vienna in 1891 partly explains why his antiquarian activities had more of an impact in Tirol and his home state of Hessen in Germany than in Milwaukee, where the enormous collection of material he excavated at various localities in those regions languished in near total obscurity at the MPM until recently (Arnold 2019a). In spite of the aspersions cast on Frankfurth's explorations by the well-known Tirolean physician and antiquarian Franz Tappeiner (1816–1902) and Fischnaler, Frankfurth's discovery of several important sites in the region stung local antiquarians into action. The founding of several museums in southern Tirol in the years after Frankfurth's death can be directly attributed to his agitation for the creation of such institutions during his travels and excavations in the region, a particularly good illustration of the impact of return migration on the development of professional archaeology in Europe (Kaufmann 2019: 311). Frankfurth's activities and a selection of the artifacts from his excavations in the MPM's collections were included in a 2019 exhibition in the Museo Archeologico dell'Alto Adige (Southern Tirol Archaeological Museum) in Bolzano featuring the early archaeology of the region (Fig. 6.5; Kaufmann and Putzer 2019). The MPM loan was made possible by connections established between the author and archaeologists in the region following the publication of an account of

Fig. 6.5 Cover of exhibition catalog for *Lost and found: Archaeology in Southern Tirol before 1919* (Kaufmann and Putzer 2019). (Photo: Bettina Arnold)

Frankfurth's activities in Austria and southern Tirol (Arnold and Stadler 2014). His important contributions to cultural and political life in Milwaukee, on the other hand, have been overshadowed by contemporaries like Meinecke and Dörflinger, both of whom outlived him by several decades. Although Accession 213 contains prehistoric and early historic European material not found in any other museum in the United States, until recently the MPM was unaware of Frankfurth's activities during his European sojourn and had not realized the importance of the material or his role in early archaeology in Austria, Italy and Germany.

6.4 Passing the Torch

Unlike Engelmann, Meinecke and Frankfurth, Dörflinger and Carl Thal (1854–1941), though also first-generation immigrants, were too young to have participated directly in the failed 1848 political reforms in Europe. However, they were equally committed to raising the cultural and intellectual profile of the city that they viewed as a second home. The fact that Dörflinger listed himself as a Forty-eighter in his 1907 history of the Wisconsin Natural History Society indicates that he was fully in sympathy with the civic upheavals that brought his parents across the Atlantic when he was seven years old. Dörflinger was the next of the MPM founders to travel back to the Old Country in 1892, following in Frankfurth's footsteps after retiring from his MPM position due to failing health exacerbated by his war injury. He visited many of the same localities Meinecke and Frankfurth had scouted out before him; in particular, all three men acquired lake-dwelling material from sites on or near Lake Constance that represents one of the largest assemblages of lacustrine material in the United States (Dernehl 1912: 11; Arnold 2013). Based on his correspondence with MPM colleagues during this trip, Dörflinger made the acquaintance of Albert Heim (1849–1937), a Swiss geologist whose studies of the Alps greatly advanced knowledge of the dynamics of mountain building and of glacial effects on topography and geology. In a letter dated

April 18, 1892, to Henry Nehrling (1853–1929), the MPM director at the time, Dörflinger recommends the adoption of Heim's classificatory scheme for specimen displays, concluding that

> probably no other museum in the US or Europe either, is as yet in possession of a similar special series, which will not only prove of interest to the student but will also attract the attention of the people generally by its many curious illustrations of natural phenomena. (Dörflinger 1892)

In the same letter he provides a spirited defense of the use of casts in museum displays, which had evidently been viewed by some MPM Board members as a swindle, noting that such "moulages" were used to great effect in museums in Europe. Dörflinger's particular interest in stone tool industries also led him to the Dordogne region of France, where various caves were being excavated by Daleau, including Les Eyzies, Laugerie-Haute, Laugerie-Basse, la Gorge d'Enfer and Pair-non-Pair. Dörflinger reported on his interactions with Daleau and his visits to these sites in the April 1899 meeting of the Wisconsin Natural History Society, at which he also circulated a pamphlet by Daleau with photographs of the animal images at Pair-non-Pair. The MPM's collections of Upper Paleolithic stone tools, animal bones and casts of Paleolithic cave art are unparalleled in the Midwest (Dybowski 2011) and are a legacy of Dörflinger's close personal relationships with major scholarly figures in Europe in the late nineteenth century.

One of the last examples of the MPM's direct links to a European scholarly network is represented by a letter sent in August 1904 by MPM Secretary Thal, autodidact and MPM librarian (Lurie 1983: 10), to Ward, who had taken on the position of director two years previously. The letter was written shortly before Thal's return to the United States after his first visit to Germany since emigrating 21 years earlier. In addition to descriptions of his family in Germany and hikes in the Harz Mountains, Thal reports on a visit to the Natural History Museum in Berlin and provides a detailed account of the displays there, describing it as "a great educational institution" (Thal 1904). He was less complimentary about the Braunschweig Natural History Museum, which "has splendid and much material, but very little care is given to it; it is so much neglected that I could not help but express it to the preparator, who showed me through it" (Thal 1904). In Berlin he asked for and was given a selection of plates, sample labels and instructions for labeling that he thought would be useful for the MPM and specifically mentions meeting the ornithologist Anton Reichenow (1847–1941), then assistant director of the Berlin Museum, who had been one of Meinecke's particular contacts during his trip to Berlin 20 years before.

The role played by the Smithsonian Institution (SI) in facilitating interactions between museums and collectors across the Atlantic must also be mentioned here. In addition to amassing extensive collections of European material via the efforts of two SI curators, Carl Rau (1826–1887) and Thomas Wilson (1832–1902), whose activities have been partly documented but would warrant further study (Kelly 2002; Maxwell 2013; Murphy 2016), the SI used government contacts to enable freight transport of artifacts for a number of early collectors in Europe and the United States. A letter from Ward dated June 19, 1902, to the secretary of the SI illustrates its role as a conduit of material and information, connecting thought-collectives on both sides of the pond:

> Dear Sir, I wish to inquire whether through your Exchange Department it would be possible for us to send to Germany certain specimens from the Swiss Lake Dwellings for authoritative determination and to be returned to us later. I am unaware whether you accommodate American museums in this manner but was informed that you did so. If this is a fact, kindly give me information as to how they should be sent. (Ward 1902)

6.5 Conclusions

Ten years after Thal's trip to Germany, the outbreak of the First World War severed the close connections between German-Americans and their homeland as anti-German sentiment swept across the United States. The MPM's directors in the years after 1918 were mostly Yankees of English descent. For example, Ward's father Henry A. Ward (1834–1906) was known as the King of the Museum Builders (Fairchild 1918: 241) and founded Ward's Natural Science Establishment in Rochester, New York. The next generation of directors and curators had little or no training in Old World archaeology and no particular interest or expertise in European prehistory. Gradually the pivotal role played by this cadre of German-Americans in the founding of the MPM was forgotten. The last MPM exhibit featuring prehistoric European artifacts was dismantled in 1985 when the Lake-Dwellers case was emptied to make way for new displays deemed more relevant to contemporary museum audiences (Anthropology Department Exhibit Archives). When the author was hired by the Anthropology Department at the UWM in 1996, there was no institutional memory of the extensive prehistoric European holdings at the MPM that had been acquired through excavation and travel by the three founding generations of the museum. In the course of searching for material suitable for graduate student thesis projects, the author and MPM Anthropology curator Dawn Scher Thomae, co-coordinators of the UWM Museum Studies Graduate Certificate at the time, gradually re-discovered these collections, beginning with the Swiss lake-dwelling material from the site of Robenhausen near Zürich. In the two decades since then more than a dozen graduate student theses and academic publications have rescued this material from obscurity, re-establishing academic connections between the MPM, UWM and various universities and museums in Switzerland, Germany, Austria, France, and Italy. The hidden history of this institution has been brought back to life because the artifacts themselves have retained their ability to communicate with us in the present in spite of limited documentary sources. It was these objects that provided the first clue to the rich vein of archival sources buried below the linguistic and epigraphic layers of neglect that had accumulated over the past century. Mining the existing sources until the scholarly networks they document have been fully revealed, and new links have been established, will reconnect the MPM to European scholarly thought-collectives and contribute to the history of archaeological knowledge transfer in the process.

References

Primary Sources

Archival Material

Anthropology Department Exhibit Archives, Milwaukee Public Museum, Milwaukee, WI, United States.

Dörflinger, C. H. (1892, April 18). [Letter to Henry Nehrling]. Milwaukee Public Museum Archives (Directors' Correspondence Files), Milwaukee, WI, United States.

Dörflinger, C. H. (1893). Loan Checklist 1893. Milwaukee Public Museum Archives (Directors' Correspondence Files), Milwaukee, WI, United States.

Linderfelt, K. A., & Meinecke, A. (1890). *Reports on the proposed library and museum building for the city of Milwaukee*. Milwaukee, WI, Trustees of the Public Library and Public Museum.

Thal, C. (1904, August 10). [Letter to H. L. Ward]. Milwaukee Public Museum Archives (Directors' Correspondence Files), Milwaukee, WI, United States.

Ward, H. L. (1902, June 19). [Letter to Smithsonian Institution Secretary Samuel P. Langley]. Milwaukee Public Museum Archives (Directors' Correspondence Files), Milwaukee, WI, United States.

Ward, H. L. (1906a, September 26). [Letter to Lorenz Frankfurth]. Milwaukee Public Museum Archives (Directors' Correspondence Files), Milwaukee, WI, United States.

Ward, H. L. (1906b). MPM Director's Monthly Report, 1906. Milwaukee Public Museum Archives (Monthly Reports 1905–1908), Milwaukee, WI, United States.

Ward, H. L. (1917). MPM Director's Monthly Report, 1917. Milwaukee Public Museum Archives (Monthly Reports), Milwaukee, WI, United States.

William Frankfurth. (1901, May). In *A short history of the German-English Academy, Jubilee of the Engelmann School* (p. 71). Writings and Research Notes 1935–1942, Farill-Grignon (Box 19 of 43, p. 71). Wisconsin Historical Society Archives, Writer's Program, Madison, WI, United States.

Wisconsin Natural History Society Bulletin. (1884–1888).

Newspaper and Magazine Articles

An honored citizen. Remains of William Frankfurth to arrive this morning. (1891, December 24). *Milwaukee Sentinel*, 3, col. A.

Ausgrabungen in Tirol. (1890, May 21). *Meraner Zeitung*.

Behördliche Maßnahmen gegen archäologische Raubgräberei. (1890, November 22). *Burggräfler*.

Died away from home. One of Milwaukee's leading merchants expires at Vienna. William Frankfurth no more. (1891, December 2). *Wisconsin Evening News*.

Fischnaler, C. (1890, October 1). Ausgrabungen in Tirol. *Innsbrucker Tagesbericht*.

Frankfurth, W. (1890, June 7). Eingesendet. *Der Burggräfler*, 469.

Meinecke, A. (1904, October 9). Unser Museum. *Milwaukee Herald Sunday*.

Meraner Zeitung. (1890, May 21).

Prähistorische Ausgrabungen. (1890, October 17). *Meraner Zeitung*.

Prähistorische Forschungen in Tirol. (1890, October 8). *Meraner Zeitung*, 3.

Prähistorische Funde. (1890, April 8). *Meraner Tagesbericht*.

von der Passer, A. (1890, n.d., probably October). *Augsburger Abendzeitung*.

Secondary Sources

Altorfer, K. (2004). Pfahlbautourismus und Pfahlbauentdeckungen in Ausland. In M. Flueler-Grauwiler, & J. Gisler (Eds.), *Pfahlbaufieber: Von Antiquaren, Pfahlbaufischern, Altertumerhändlern und Pfahlbaumythen* (Mitteilungen der Antiquarischen Gesellschaft in Zürich, 71) (pp. 91–101). Zürich: Chronos.

Applegate, C. (1990). *A nation of provincials*. Oakland, CA: University of California Press.

Arnold, B. (2013). The lake dwelling diaspora and natural history museums: Identity, collecting and ethics. In F. Menotti, & A. O'Sullivan (Eds.), *The Oxford handbook of wetland archaeology* (pp. 875–891). Oxford: Oxford University Press.

Arnold, B. (2019a). "Der stete Antiquar": William Frankfurth (1829–1891) aus Milwaukee, WI, USA. In G. Kaufmann, & A. Putzer (Eds.), *Lost and found: Die Anfänge der Archäologie im Süden Tirols: Forschungsgeschichte bis 1919* (Schriften des Südtiroler Archäologiemuseums, 6) (pp. 289–300). Bozen: Athesia.

Arnold, B. (2019b). Von Bozen nach Wisconsin und zurück – die Entdeckungen des William Frankfurth. *Archäologie in Deutschland, 5*, 44–45.

Arnold, B., & Stadler, H. (2014). Das Geheimnis von Sammlung 213: Milwaukee, USA. Der Hobbyarchäologe William Frankfurth in den Alpen. In F. Müller (Ed.), *Graben, Entdecken, Sammeln: Laienforscher in der Geschichte der Archäologie Österreichs* (pp. 259–276). Vienna: LIT-Verlag.

Arnold, W. B. (2014). *Fire on the mountain: The bronze and iron ash altar material in the Frankfurth collection at the Milwaukee Public Museum* [Master's thesis, University of Wisconsin-Milwaukee, United States]. https://dc.uwm.edu/etd/352/

Bergland, M., & Hayes, P. G. (2014). *Studying Wisconsin: The life of Increase Lapham*. Madison, WI: Wisconsin Historical Society Press.

Birmingham, R., & Rosebrough, A. (2017). *Indian mounds of Wisconsin*. Madison, WI: University of Wisconsin Press.

Caywood, A. M. (2011). *The choice of legions: The terra sigillata collection at the Milwaukee Public Museum* [Master's thesis, University of Wisconsin-Milwaukee, United States].

Dernehl, P. H. (1912). Karl Herman Dörflinger: In memoriam. *Bulletin of the Wisconsin Natural History Society, 10*(1–2), 8–12.

Dörflinger, C. H. (1907). *The genesis and early history of the Wisconsin Natural History Society At Milwaukee*. Milwaukee, WI: Milwaukee Public Museum.

Dybowski, D. G. (2011). *The analytical limits of nineteenth-century lithic collections in identifying variation and lithic variability in thirteen French Paleolithic assemblages from the Milwaukee Public Museum* [Master's thesis, University of Wisconsin-Milwaukee, United States].

Ewing, H. (2007). *The lost world of James Smithson: Science, revolution, and the birth of the Smithsonian*. London: Bloomsbury.

Fairchild, H. L. (1918). Biographic memoirs of deceased fellows: Henry Augustus Ward. *Proceedings of the Rochester Academy of Science, 5*, 240–251.

Fleck, L. (1935). *Entstehung und Entwicklung einer wissenschaftlichen Tatsache: Einführung in die Lehre vom Denkstil und Denkkollectiv*. Basel: B. Schwabe.

Greis, G. P. (2006). *A noble pursuit: The Duchess of Mecklenburg collection from Iron Age Slovenia*. Cambridge, MA: Peabody Museum Press.

Groenen, M. (2021). *François Daleau, fondateur de l'archéologie préhistorique*. New York: Millon.

Gurda, J. (1999). *The making of Milwaukee*. Milwaukee, WI: Milwaukee County Historical Society.

Hare, J. L. (2015). *Excavating nations*. Toronto, Canada: University of Toronto Press.

Hinsley, C. M. (1994). *The Smithsonian and the American Indian: Making a moral anthropology in Victorian America*. Washington, D.C.: Smithsonian Institution Press.

Ingold, T. (2015). *The life of lines*. London and New York: Routledge.

Irmscher, C. (2013). *Louis Agassiz: Creator of American science*. Boston, MA: Harcourt.

Kaeser, M.-A. (2004). *L'univers du préhistorien: Science, foi et politque dans l'œvre et la vie d'Édouard Desor (1811–1882)*. Paris: L'Harmattan.

Kaufmann, G. (2019). Der Beginn der Ringwallforschung im südlichen Tirol. In G. Kaufmann, & A. Putzer (Eds.), *Lost and found: Die Anfänge der Archäologie im Süden Tirols: Forschungsgeschichte bis 1919* (Schriften des Südtiroler Archäologiemuseums, 6) (pp. 301–345). Bozen: Athesia.

Kaufmann, G., & Putzer, A. (Eds.). (2019). *Lost and found: Die Anfänge der Archäologie im Süden Tirols: Forschungsgeschichte bis 1919* (Schriften des Südtiroler Archäologiemuseums, 6). Bozen: Athesia.

Keller, F. (1866). *The lake dwellings of Switzerland and other parts of Europe* (J. Lee, Trans.). London: Longmans.

Kelly, J. (2002). Charles Rau: Developments in the career of a nineteenth-century German pan-American archaeologist. In D. L. Browman, & S. Williams (Eds.), *New perspectives on the origins of Americanist archaeology* (pp. 117–132). Tuscaloosa, AL: University of Alabama Press.

Kirk, J. (2010). *Kingdom under glass*. New York: Henry Holt & Co.

Klenk, N. (2018). From network to meshwork: Becoming attuned to difference in transdisciplinary research encounters. *Environmental Science and Policy, 89*, 315–321.

Lenoir, M. (Ed.). (2006). *La grotte de Pair-non-Pair à Prignac-et-Marcamps (Gironde)*. Bordeaux: Société archéologique de Bordeaux.

Lunz, R. (2006). *Archäologische Streifzüge durch Südtirol* (Vol. 2). Bozen: Athesia.

Lurie, N. O. (1983). *A special style: The Milwaukee Public Museum, 1882–1982*. Milwaukee, WI: Milwaukee Public Museum.

Maner, B. (2018). *Germany's ancient pasts*. Chicago: University of Chicago Press.

Marchand, S. (1996). *Down from Olympus: Archaeology and philhellenism in Germany, 1750–1970*. Princeton, NJ: Princeton University Press.

Maxwell, K. G. (2013). *Historic museum collections as primary sources: Thomas Wilson's Robenhausen material at the Smithsonian Institution's National Museum of Natural History* [Master's thesis, University of Wisconsin-Milwaukee, United States].

Milosavljević, M. (2020). How archaeological communities think: Re-thinking Ludwik Fleck's concept of the thought-collective according to the case of Serbian archaeology. In J. Roberts, K. Sheppard, U. R. Hansson, & J. R. Trigg (Eds.), *Communities and knowledge production in archaeology* (pp. 14–33). Manchester, UK: Manchester University Press.

Murphy, L. (2016). *The unsung evolutionist: Charles Rau's Swiss Lake Dwelling collection at the Smithsonian Institution* [Master's thesis, University of Wisconsin-Milwaukee, United States].

Nyhart, L. K. (2009). *Modern nature: The rise of the biological perspective in Germany*. Chicago: University of Chicago Press.

Penny, H. G. (2019). *Im Schatten Humboldts: Eine tragische Geschichte der deutschen Ethnologie. München*. Munich: CH Beck.

Petraglia, M., & Potts, R. (2004). *The Old World Paleolithic and the development of a national collection* (Smithsonian Contributions to Anthropology, 48). Washington, D.C.: Smithsonian Institution Press.

Rosebrough, A. (2017). Effigy mounds: Monumente aus der Glaubenswelt der Ureinwohner. *Archäologie in Deutschland, 6*, 14–19.

Roßner, K.-M. (2012). *Führung zum Baunsberg und zum Ohlsborn*. Kassel: Stadtteilkultur Brasselsberg. http://www.stadtteilkultur-brasselsberg.de/PDFs/12-11-03_Baunsberg-Ohlsborn-02.pdf

Snead, J. (2002). The "Western idea": Local societies and American archaeology. In S. H. Allen (Ed.), *Excavating our past: Perspectives on the history of the Archaeological Institute of America* (Colloquia and Conference Papers, 5) (pp. 123–140). Boston, MA: Archaeological Institute of America.

Snead, J. (2020). Circular 316: Archaeology, networks, and the Smithsonian Institution, 1876–1879. In J. Roberts, K. Sheppard, U. R. Hansson, & J. R. Trigg (Eds.), *Communities and knowledge production in archaeology* (pp. 34–46). Manchester, UK: Manchester University Press.

Sütterlin. (2023). Wikipedia. Retrieved August 15, 2023, from https://en.wikipedia.org/wiki/S%C3%BCtterlin

Sweet, R. (2004). *Antiquaries: The discovery of the past in 18th-century Britain*. London: Hambledon and London.

de Tomasi, F. (2020). 'More for beauty than for rarity': The key role of the Italian antiquarian market in the inception of American Classical art collections during the late-nineteenth century. In J. Roberts, K. Sheppard, U. R. Hansson, & J. R. Trigg (Eds.), *Communities and knowledge production in archaeology* (pp. 47–65). Manchester: Manchester University Press.

Ulmschneider, H. (2011, August 10). Deciphering Sütterlin. *Archaeology Archives Oxford*. https://archaeologyarchivesoxford.wordpress.com/2011/08/10/deciphering-sutterlin/

Watson, R. (2001). *Opening the museum: The Peabody Museum of Archaeology and Ethnology* (Occasional Papers, 1). Cambridge, MA: Peabody Museum of Archaeology and Ethnology, Harvard University.

Wittke, C. (1952). *Refugees of revolution: The German Forty-eighters in America*. Philadelphia: University of Pennsylvania Press.

Wulf, A. (2015). *The invention of nature: Alexander von Humboldt's New World*. New York: Vintage.

Zucker, A. E. (1950). *The Forty-eighters: Political refugees of the German revolution of 1848*. New York: Columbia University Press.

Gábor Szinte and the Role of Networks in Early Research on Medieval Monuments in Nineteenth-Century Transylvania

7

Ionuț-Cosmin Codrea and Cristina Bodó

Abstract

This chapter examines how Hungarian illustrator and photographer Gábor Szinte (1855–1919)—through his network of archaeologists, historians, and other heritage enthusiasts—contributed to the incorporation of professional illustration and photography in nineteenth-century Transylvanian archaeological and built heritage research. The chapter integrates Szinte's activity in the cultural milieu of the Kingdom of Hungary after the Austro-Hungarian Compromise of 1867. Szinte graduated in 1879 from the Royal Hungarian College of Art in Budapest and at the beginning of the academic year 1883–1884 he was appointed to a position as an art teacher at the State Central Secondary School for Technology and Science in Deva. He almost immediately became a member of the History and Archaeology Society of Hunedoara County, to which he contributed his talents in drawing and photography until 1897, when he departed for Budapest. As an illustrator, Szinte contributed to an increase in knowledge of ancient and medieval monuments, capturing details which are no longer preserved today. Through the History and Archaeology Society of Hunedoara County, he participated in the programme organized by the National Monuments Commission in Hungary, which focused on the discovery and rescue of medieval frescoes in the country. Beginning with this project, his interest in medieval monuments increased and resulted in further research on ecclesiastical buildings in Hunedoara County.

Keywords

Networks · Medieval monuments · Frescoes · Archaeological societies · Transylvania · Hungary

I.-C. Codrea (✉) · C. Bodó
Museum of Dacian and Roman Civilization, Deva, Romania
e-mail: ionut.codrea@yahoo.com; emesbod@yahoo.com

7.1 Introduction

This chapter presents the work of Hungarian illustrator and photographer Gábor Szinte (Fig. 7.1) between 1883 and 1897 in collaboration with the National Monuments Commission in Budapest and the History and Archaeology Society of Hunedoara County, Hungary. The chapter also examines Szinte's contributions to the research and conservation of built heritage in Hunedoara County. As a member of the National Monuments Commission, Szinte was part of a network of art teachers who created survey drawings and sketches of artistic monuments at the Commission's request. In 1883 and 1884 he was commissioned to survey and draw the frescoes in the churches in Bârsău, Streisângeorgiu, and Deva in Hunedoara County, an administrative territory on the south-western border of historical Transylvania.[1] Beginning with this project his interest in medieval monuments increased and resulted in further research on additional historic buildings in Hunedoara County, together with other members of the Hunedoara County History and Archaeology Society.

Szinte's research activity during his stay in Deva (Hungarian: Déva) between 1883 and 1897 was reconstructed for this chapter based on archival records kept in the History and Archaeology Society of Hunedoara County repository of the Hunedoara County Service of the National Archives in Deva. These records include official documents, annual reports and minutes that mention Szinte's activity within the Society and are supplemented by information published in the Society's yearbook, *A Hunyadmegyei Történelmi és Régészeti Társulat Évkönyve*. In addition, Szinte's research on medieval monuments was published in several local and national publications of the time. The results of his collabora-

[1] During the period discussed in this chapter Transylvania was part of Hungary, which in turn was part of the Austro-Hungarian Empire. After the dissolution of the Empire at the end of the First World War, Transylvania became part of the Kingdom of Romania. This was confirmed by the Treaty of Trianon in 1920.

L. Coltofean et al. (eds.), *Connecting People and Ideas*, Themes in Contemporary Archaeology,
https://doi.org/10.1007/978-3-031-81006-0_7

SZINTE GÁBOR
(1855—1914.)

Fig. 7.1 Portrait of Gábor Szinte (1855–1914). Budapest, c. 1898. (Reproduced from *Értesítő* 1914: 9–10)

tion with the National Monuments Commission between 1883 and 1886 are now preserved in the Hungarian Museum of Architecture and Monument Protection Documentation Centre, in two separate repositories: the Plan Repository and the Scientific Archives of the National Monuments Commission. Information about Szinte's cooperation with the National Monuments Commission was also published in *Archaeologiai Értesítő*—one of the main Hungarian archaeology journals—in the 1880s.

Gábor Szinte was born in 1855 to a family of small Unitarian landlords in Valea Crişului (Hungarian: Sepsikőröspatak) in the Grand Principality of Transylvania. He completed his pre-college education in Transylvania at Cristuru Secuiesc (Hungarian: Székelykeresztúr) and Cluj (Hungarian: Kolozsvár). In 1875, he began his studies at the Magyar Királyi Mintarajztanoda és Rajztanárképezde (Hungarian Royal Drawing School and Art Teacher's College in Budapest),[2] where he obtained his diploma as a secondary school art teacher in 1879. His career began at the Váczi Királyi Magyar Siketnéma-Intézet (Vácz Royal Institute for the Deaf and Dumb) in Pest County, Hungary. Starting with the 1883–1884 school year, Szinte taught at the Dévai Magyar Királyi Állami Főreáliskola (State Central Secondary School for Technology and Science in Deva; Fig. 7.2)[3] in Hunedoara County. In September 1897, he was transferred to Budapest, where he worked until his death in 1914 (Bata and Tasnádi 2015: 6).

During the first half of the nineteenth century, two prestigious cultural institutions were established in Hungary: the Magyar Nemzeti Múzeum (Hungarian National Museum) in 1802, courtesy of Count Ferencz Széchényi's (1754–1820) donation; two decades later, his son István Széchenyi (1791–1860) laid the foundations of the Magyar Tudományos Akadémia (Hungarian Academy of Sciences) (Gáll 2014: 221). The establishment of these two institutions was critically important for the formation of archaeology and art history as scholarly disciplines in Hungary (Marosi and Cennerné Wilhelmb 1983: 8). Towards the middle of the century a call for the protection of artistic monuments was published by Imre Henszlmann (1813–1888), art historian and archaeologist, and Ferenc Toldy (1805–1875), Secretary of the Hungarian Academy of Sciences. However, any initiative to institutionalize this desideratum was crushed by the start of the 1848 Revolution. A year later, after the Revolution had failed, a neo-absolutist regime was established that split Hungary into five military districts. The situation was similar in Transylvania, which was however under direct control from Vienna. In 1850 the Kaiserliche-königliche Zentral-Kommission zur Erforschung und Erhaltung der Baudenkmale (Royal and Imperial Central Commission for Research and Preservation of Architectural Monuments) was established by a decree of Emperor Franz Joseph I. This institution was responsible for investigating and protecting monuments in Hungary, Transylvania, Croatia, and all the southern territories of the Empire (Marosi and Cennerné Wilhelmb 1983: 13–14; Sisa 2020: 191–92). The Commission's goal was to collect written and graphic information regarding as many monuments as possible, especially medieval ones. This goal was achieved through a network of correspondents who were active throughout the Empire's territory and who regularly submitted data or drawings of monuments with historical and artistic value (Sisa 2020: 192–93).

Szinte began his scientific activity at a favourable time, not only politically and economically but also culturally. After the Austro-Prussian War of 1866, Europe experienced a period of relative calm and stability that resulted in

[2]The Hungarian Royal Drawing School and Art Teacher's College in Budapest was established in 1871, its first director being Gusztáv Keleti (1834–1902). This institution provided preparatory education for students planning to go on to careers in arts, industry, or other technical occupations. The students received training in artistic drawing, architectural drawing, perspective, industrial art and design, artistic anatomy, modeling, and painting. In 1880, within the framework of the School of Drawing, the National Hungarian School of Applied Arts was launched. The activities of the school were influenced by the English Arts and Crafts movement with a tendency towards folk art (Sisa 2002: 133–34; Tasnádi 2006a: 76).

[3]In Hungary, the State Central Secondary School for Technology and Science was part of the secondary school system (together with the *Gymnasium*); it consisted of an eight-year course that was divided into two four-year cycles.

Fig. 7.2 The teaching staff of the State Central Secondary School for Technology and Science in Deva before 1898. Standing on the left is Gábor Szinte; Gábor Téglás, the School's director, is at the head of the table. (Reproduced with the permission of the Hungarian National Museum, Budapest Hungary; MNM: 561.1956)

economic growth. Moreover, the Compromise of 1867—which led to the creation of a semiautonomous state within the framework of the Austro-Hungarian Monarchy—allowed Hungary to develop a sense of national identity and cultural and political independence (Molnár 2001: 208–9; Sisa 2020: 197, 200). The Compromise gave Hungary almost complete power over its internal affairs, including establishing its own political system and cultural institutions. The Liberal government pursued the creation of a national liberal state, reminiscent of Republican France (Molnár 2001: 211).

7.2 Gábor Szinte's Activity and Network Within the National Monuments Commission in Budapest

After the Austro-Hungarian Compromise of 1867, it was necessary to reorganize the administrative and judicial institutions in Hungary as well as the cultural and educational ones. The founding of an institution to deal with the preservation of historical monuments in Hungary was delayed for a rather long time. For a while, monument protection remained the responsibility of the Central Commission in Vienna, while in 1858 the Archaeológiai Bizottság (Archaeological Committee) was housed in the Hungarian Academy of Sciences. It was only in 1872 that the Műemlékek Országos Bizottsága (National Monuments Commission)[4] was founded by a decree issued by the Vallás- és Közoktatásügyi Minisztérium (Ministry of Religious Affairs and Public Education). This new Commission was structured around active members as well as non-paying, honorary members: aristocrats, archaeologists, architects, art historians, historians, and artists (Lővei 2016: 555–56). The efforts of Henszlmann were largely responsible for the founding of this Commission. Once the Commission was created, Henszlmann also acted as a rapporteur, responsible for coordinating efforts focused on the study and preservation of architectural monuments in the Kingdom of Hungary.[5] His main remit was to inventory the artistic monuments. For this, Henszlmann relied on a network of 'associates' spread throughout the kingdom's territory: art teachers, local archaeologists, members of the clergy, and landowners who filled in the distributed record sheets and sent in drawings and photographs. All this work was financially supported by the Monuments Commission (Horváth 1990: 156–57; Bata and Tasnádi 2015: 13; Sisa 2020: 200).

The art teachers, part of the Royal Hungarian College of Art and found throughout the country, played an important and active role in the social and artistic movements of the second half of the nineteenth century. In addition to educating the young, they were engaged in historical, archaeological, and ethnographic projects. Furthermore, they contributed to the promotion of Hungarian culture and national identity within the local communities where they worked (Tasnádi 2006a: 76–77, 2016: 117). The network of art teachers that

[4]The institution's initial name was Magyarországi Műemlékek Ideiglenes Bizottsága (Hungarian Provisional Monuments Commission). Its name was changed to National Monuments Commission in 1881.

[5]Initially, only medieval buildings were considered historical monuments; later, Renaissance architecture was also included in this category, while the Baroque was totally neglected (Lővei 2016: 557).

collaborated with the National Monuments Commission in Budapest was widely distributed across the Kingdom of Hungary, even in some of the most remote places. In addition to Szinte, the network included József Könyöki (1829–1900), Viktor Myskovszky (1838–1909), Gyula Szabó, Lipót Ács (1868–1945), Antal Streitmann (1850–1918), László Ényi Dömötör (1850–1903), József Huszka (1854–1934), Ödön Nemes (1847–1902), and Adolf Csehély (1854–1904). They were all involved in researching and protecting historical monuments, and periodically submitted drawings and documentation to the National Monuments Commission. Many art teachers were also interested in ethnography, gathering and analyzing the decorative motifs of Hungarian culture (Tasnádi 2006a: 77).

One of the abovementioned art teachers, József Huszka, had a career path very similar to that of Szinte. As an art teacher in Sfântu Gheorghe (Hungarian: Sepsiszentgyörgy), beginning in 1881, Huszka traveled through the villages of the Székely Land (Hungarian: Székelyföld; Romanian: Ținutul Secuiesc[6]). In 1887 he began taking photos of the villages he visited. In 1891, he moved to Budapest but continued his research work during school breaks. At the National Monuments Commission's request Huszka investigated and documented the murals in the medieval churches of the Székely Land; between 1882 and 1905, he identified and recorded 23 church murals in the Székely Land and northern Hungary. According to the current practice, Huszka took photographs and made architectural drawings in addition to recording detailed descriptions of the churches. The reports together with his photographs and drawings (in color and scale reproductions in India ink on tracing paper) were sent to the Commission (Tasnádi 2006b: 35–37; Bata 2007: 127). During his research trips Huszka also discovered the textiles with folk-style motifs that decorated the churches, the richly adorned Székely gates, and other forms of Hungarian folk motifs found in the use of the peasantry—e.g., painted furniture, ceramics, clothing items—which he began to photograph, commissioned by the Commission, in 1887. Huszka used photography in addition to drawings to reproduce the frescoes as faithfully as possible. Later, the camera proved to be an important tool for studies on the Székely gates (Bata 2007: 127).

Szinte followed the same path as Huszka. Besides the fact that they both attended the Royal Hungarian College of Art at almost the same time, Szinte also began his carrier as a drawing teacher, started documenting medieval frescos for the National Monuments Commission, and was later interested in similar ethnographic topics. Even if Szinte did not use the camera in his research on medieval frescoes, he employed it intensively in documenting the Székely gates, homes, and wooden churches (Bata 2021: 90).

Szinte was drawn into the National Monuments Commission's network of collaborating intellectuals from the beginning of his teaching career in Deva in 1883. The first project he worked on was the documentation of the frescoes in the church at Bârsău (Hungarian: Berekszó). The church came to the attention of specialists due to the research of classical archaeologist Károly Torma (1829–1897), who underlined the importance and age of the monument, which he had visited in 1877 (Torma 1879). Later, on June 7, 1882, during the Executive Committee meeting of the History and Archaeology Society of Hunedoara County, local teacher and archaeologist Gábor Téglás (1848–1916) announced that the church at Bârsău, which housed old mural paintings and inscribed funerary stones, was threatened with demolition because of its advanced state of degradation (Executive Committee 1882a: 66). At the Society's meeting on July 5, 1882, the visual recording of the frescoes as a preservation measure was discussed (Executive Committee 1882b: 67v). Téglás recalled that, during his investigative excursion north of the Mureș River that November, he had visited the Eastern Orthodox Church in Bârsău. During this visit, Miklós Horváth, the art teacher at the State Central Secondary School for Technology and Science in Deva, made sketches of the valuable frescoes there. Téglás also drew attention to the fact that detailed drawings would require a more serious time commitment and financial investment (Kun 1884: 90). Horváth therefore began collaborating with the National Monuments Commission, for which he made seven drawings of the Bârsău church in 1882–1883 (Gerecze 1905: 268). In 1883, Horváth was compelled for personal reasons to retire from his teaching job and by association from his collaboration with the National Monuments Commission. He requested to be relieved of his tasks and proposed Szinte as his successor, transferring the money needed for the assignment to his replacement (Henszlmann 1883: IX; Zichy and Henszlmann 1883: VII). Téglás supported the recommendation as well; he forwarded Szinte's address and requested the transfer of the 50 forints allocated to Horváth for copying the Bârsău murals to Szinte (Zichy and Henszlmann: VII).

During the meeting of the National Monuments Commission on December 18, 1883, an announcement was made that the Ministry of Religious Affairs and Public Education had commissioned Szinte to copy the Bârsău church frescoes (Zichy and Henszlmann 1883: VIII). More details about this task are described in a letter sent by Szinte to the National Monuments Commission, dated January 23, 1886:

> A nagytekintetü műemlékek orsz. bizottsága még az 1883ik évben kegyes volt rám ruházni a Horvát Miklós tanár által be nem fejezett berekszói templom falfestményeinek lemásolását vázlatosan, csak körrajzban; továbbá a chorusban levő foga-

[6]Székely Land is an historic and ethnographic area located in the eastern part of Transylvania (central area of present-day Romania), corresponding mostly to the present-day counties of Harghita and Covasna and central and eastern Mureș County.

dalmi kép szines rajzát s a rajta elöforduló föliratok pauszirozását. Tizennégy táblán nyujtom e föladat megfejtését a Nagytekintetű bizottság elé, kegyes elnézésért esdve a huzamos idei várakoztatás miatt, nem mintha munkálatom nem lett volna már régen kész, hanem mert az itt előforduló számtalan kérdések földeritésére az összes régi templomokat szerettem volna átkutatni a megyében s határozott kész munkát szolgáltatni a Nagytekintetű bizottságnak; de mivel 1885/62 számu sürgetését kaptam a Nagyt. bizottságnak, nehogy hanyagság gyanujába ejtessem bátor vagyok fe dolgozatlanul, csak néhány szerény kiegészitő szóval beterjesztésemet megtenni.
(The venerable National Monuments Commission kindly entrusted me to sketch the outlines of the paintings on the walls of the Berekszó church, which had not been completed by the teacher Miklós Horvát; and to make color drawings of the votive panel and to copy on tracing paper the inscriptions on it. I hereby forward fourteen plates before the venerable Commission and I apologize for the delay, not because the work could not have been completed earlier, but because I would have preferred to study all the old churches in the county in order to solve the numerous issues here and to present a comprehensive report to the venerable Commission; yet because I was asked by the venerable Commission through the address no. 62 from 1885 to rush the completion and since I do not want to be accused of negligence, I take the liberty of sending the draft material, with some clarifying text.) (Szinte 1886a: 14)

In the same letter, Szinte described the architecture of the church and the stylistic elements of the frescoes' iconography (Szinte 1886a: 15–16). When Szinte visited the church, the bell tower had shifted well away from the nave, pulling the nave's western seal with it and causing immense cracks in the long walls. To prevent collapse, the tower had been supported by a large buttress. The old bell tower was demolished and replaced during the repair works finished in 1888 and marked by a plaque placed above the entrance under the new-built bell tower. This last restoration radically altered the western side of the church, such that Szinte's descriptions are the only written testimonies we have regarding its initial shape:

Bejárata délen tágas és renaissance stylú és tiszta, szép kivitelű kőből faragott ajtó rámázattal, melyet helyéből a későbbi profán alakitás kiemelt és a hajót átszelő ujabban készült falazatba rakott, hasonlókép ajtó szemöldöknek. A jóval kisebb bejárathoz alig fele használtatott e szép profiloknak, a többi rész ott hever az előcsarnokban ülő padként szolgálva a roman szokásként a főhajóból kizárt asszony népnek. Tagozata tojás és levél füzérrel rendkivül ügyesen van díszitve, a vizszintes felső orom népies motivumokat tár elénk. (Rajzát Horvát adta.)
(The southern side's entrance is wide, in a Renaissance style, with a beautiful decorative border of carved stone which was later removed and placed with a similar purpose on the newer [interior] wall that traverses the nave. Only half of it was used for the much smaller entry the remaining pieces were lying in the narthex and used as a sitting place for women, who are not allowed in the nave according to the Romanian custom. Its elements are beautifully decorated with egg-and-darts, and the horizontal upper piece has popular motifs. [The drawing of this section was forwarded by Horvát.]) (Szinte 1886a: 15–16)

The series of plates drawn by Szinte begins with the sanctuary: the altar vault (the scene of Mary enthroned and holding the Child) (Szinte 1886b: FM 42; also see Szinte 1886a: 16), followed by an essential theme of the altar's iconography: the Apostles' Eucharist, situated in the uppermost register of the four walls. Jesus appears twice, offering bread and wine to the Apostles in groups of six. The same plates copy the decorative floral borders, "vörös, fehér és zöld szinekkel" (painted in red, white and green), used to separate the horizontal fields; this motif is used throughout the church (Szinte 1886b: FM 43, FM 44). The next four plates copy the lower register of the altar's walls. The plates show the Holy Hierarchs (on the lower part of the wall), posed frontally, bearing censers or unwrapped frontlets; above are medallions with saints offering the blessing with their right hands and holding Bibles with their left (Szinte 1886b: FM 45, FM 46, FM 47, FM 48). The following two plates show an image of the infant Jesus offering a blessing (Szinte 1886b: FM 49), and one with the Archangel Gabriel as he tells the Virgin Mary of the miracle that is to come (Szinte 1886b: FM 50)—these are images that we find on the triumphal arch between the sanctuary and the nave.

On the southern wall of the nave, Gábor Szinte copied the votive painting (Plate XI), and in the caption he refers to Plates XII, XIII, and XIV (where the inscriptions are copied), which are only suggested in the copy of the main painting (Szinte 1886b: FM 56). In his letter to the Commission, Szinte offers a description of this painting including Saint George that occupies the entire wall right up to the base of the vaulted ceiling (Szinte 1886a: 16). Plate XI, the only colored one, depicts two women presenting a model of the church to its patron saint,[7] shown in the upper part, with an open book on his chest, providing a blessing with both hands. The inscriptions are reproduced in life-size in Plates XII, XIII, and XIV, and the Slavonic texts (written in Cyrillic characters) were translated into Hungarian by numismatist and ethnographer László Réthy[8] (1851–1914) (Szinte 1886b: FM 57, FM 58, FM 51). The first two inscriptions are very important because they identify the patrons of the murals: Mara, widow of Nicolae Cherepovici, and her daughter Catarina, widow of Blasius Török (Burnichioiu 2003: 107–8).

Réthy was one of Szinte's collaborators in researching the Bârsău church. Unfortunately, there is no information about any correspondence between them regarding the mural paintings from Bârsău. Still, the existing documents indicate the

[7]Szinte misidentified the saint above the votive painting as Saint George; the saint to whom the two women present the church is Saint Nicolas.

[8]László Réthy was a corresponding member of the Hungarian Academy of Sciences since 1892. He studied Slavic languages, as well as Sanskrit and Armenian, in Vienna and Cracow. He became a doctor of numismatics in 1880 and began to collaborate with the Hungarian National Museum in 1881, where in 1901 he was appointed Keeper of the Department of Numismatics and Antiquities (Harsányi 1915: 22–28).

two knew each other. These passionate men, interested in the history of the places they inhabited, worked through local cultural associations as well as regional and national ones, such as the Országos Régészeti és Embertani Társulat (National Archaeology and Anthropology Society), the Magyar Néprajzi Társaság (Hungarian Ethnographical Society), and the Erdélyi Magyar Közművelődési Egyesület (Hungarian Cultural Society of Transylvania). These were all cultural organizations in which intellectuals with related interests met and developed professional relationships. Both Réthy and Szinte were members of the National Archaeology and Anthropology Society,[9] the former beginning in 1880, and the latter from 1886 onwards ("Az Országos," 1889: 67–76). As a result of this existing, albeit long-distance relationship, initially maintained via correspondence, Szinte was able to turn to Réthy for the translation of the Bârsău inscriptions[10] ("A társulat igazgatása," 1886: 66). The two men's collaboration continued, as reported in a short newspaper article from 1886. Szinte sent Réthy two old Romanian religious books found in Hunedoara County for analysis that were written in Slavonic in the sixteenth century; Réthy's analysis of these books was later published ("Régi oláh," 1886: 3).

Regarding the Bârsău church, Szinte concluded that "első tekintetre arra a gondolatra vezetett, hogy tán többször is volt a templom festve, de gondos kutatásaim arra vezettek, hogy több réteg nincs" (at first glance, I thought that the church may have been painted more than once, but my careful investigations led me to the conclusion that there were no more layers). He therefore dated the frescoe to 1770, the year he had seen painted on a ribbon at the feet of the Virgin Mary, on the vault of the altar (Szinte 1886a: 17). Research over the past few decades has revealed that the paintings in the Saint Nicholas Orthodox Church in Bârsău, dating most likely to 1562–1572 (Burnichioiu 2003: 111), were repainted, color on color, in the second half of the eighteenth century, more precisely in 1770. The signature of the man who performed this 'refreshing' appears next to the year: "Stan Zugravu sin popii Radu" (Stan the painter, son of Radu the priest) (Burnichioiu 2003: 102). Despite this error, and considering the degraded state of the frescoes, especially the scene on the altar vault—which is almost completely gone—Szinte's drawings remain an important record of the interpretation of this monument. The date of the Bârsău church, which is older than the painted layer, is not fully settled even today; the construction method of the altar, with its axial closing of the apsis, is not often found in Transylvania. Even though the archaeological investigations in 1976 failed to identify chronological elements prior to the sixteenth century (Dumitriu 2000), the architectural features indicate that the church was built at least a century earlier (Greceanu 1971; Burnichioiu 1999, 2003). At the end of his report on the Bârsău church, Szinte expressed his uncertainty regarding the identity of the fresco painter: "E kérdés földeritésére minden utamba eső festett templomot megnéztem s végre meg is találtam a sytrigy Szt. Györgyi templomban, melyet kapcsolatosan a berekszoival bátor vagyok a Nagyt. bizottság elé terjeszteni" (To answer these questions, I have looked at all the painted churches on my route and finally found [the answer] in the church of St. George Church in Streisângeorgiu, which I have the courage to present before the Committee in connection with the one in Bârsău) (Szinte 1886a: 17).

The St. George Orthodox Church in Streisângeorgiu (Hungarian: Sztrigyszentgyörgy; a village that today has been incorporated by the town of Călan) was investigated by Szinte a year after his work at Bârsău. In a letter addressed to the National Monuments Commission, dated June 30, 1884, he requested 150 forints to defray the costs of travel expenses incurred while researching the frescoes. In this letter, he mentioned not only the church in Streisângeorgiu, but the whole valley of the Strei River, where "több ily régi templom volna frescokkal és a lakok által ismeretlen nyelvü szerb föliratttal ellátva, melyeknek mind átkutatása az egészre megállapitandó vélemény szempontjából is fölötte szükséges, mind a mellett hogy mű becsét biró részleteket tartalmaznak" (we might find more such old, frescoed churches with inscriptions in the Serbian language unknown to the locals, and whose investigation will be necessary to obtain a general view of them, in addition to the artistic details they preserve) (Szinte 1884: 191). The response of the Commission, dated October 24, 1884, was negative. This decision was based on the fact that regulations did not allow the reimbursement of expenses incurred in the course of research with no clear objective (Henszlmann 1884: 190–190v).

Along with the letter from January 1886, Szinte sent three plates of illustrations of the church in Streisângeorgiu to support his statements before the National Monuments Commission: Plate I. The Annunciation scene (the only one preserved today; see Szinte 1886c: FM 574); Plate II. The votive painting of the church with copies of the Cyrillic inscriptions and their decipherment; Plate III. The ground plan and longitudinal cross section of the church, including the inner and outer dimensions (scale 1:100) (Gerecze 1905: 271). This little church caught Szinte's attention due to its unusual architecture, particularly the position of the small bell tower, located on the western side of the nave, and the rectangular shape of the sanctuary. He considered it even more important than the Bârsău church. His letter to the National Monuments Commission laid out the arguments for

[9]The members of the Hunedoara County History and Archaeology Society also include Zsófia Torma, Gábor Téglás, and Ferencz Sólyom-Fekete beginning with 1879, Géza Kuun since 1880, and Pál Király since 1885 ("Az Országos," 1889: 67–76).

[10]It is possible that Gábor Téglás served as a mediator between Szinte and Réthy. Téglás was on the Society's board in 1885, together with Réthy.

this assessment (Szinte 1886a: 18). Szinte presented three inscriptions: the Slavonic dedication from the votive painting "mely elárulja hogy a templom 1409-ben épült és Szent Györgynek volt szentelve" (which tells us that the church was built in 1409, under the patronage of Saint George); an inscription from 1743 that mentions the painter Gheorghe Sandor from Făgăraş; and one dated to 1843 that mentions repair work (Szinte 1886a: 19). Also, comparing the Streisângeorgiu frescoes with the ones from Bârsău, Szinte found many similarities "redözet, távlat és architechtonica tekintetből is" (from the point of view of the draping of the clothes, the perspective, technique, and background) (Szinte 1886a: 19).

In the final part of the letter sent to the National Monuments Commission in January 1886, Szinte mentioned that, in addition to the drawings from Bârsău and Streisângeorgiu, he sent two color plates from the Calvinist Reformed Church in Deva. He also drew attention to the St. Michael the Archangel Orthodox Church in Gurasada (Hungarian: Guraszáda). Szinte noted that he made drawings of the respective church with precise measurements but did not include them because he was still pursuing his research on the monument. Based on his observations and research, he considered the first two churches as built under Serbian influence, while the ground plan of the Gurasada Church reminded him of the Balkans. In addition, Szinte suggested that the frescoes could be of interest to those studying religious mural paintings and dated them to 1760 (Szinte 1886a: 21).

The last plates sent by Szinte to the National Monuments Commission were survey drawings of the frescoes from the Calvinist Reformed Church of Deva. They were drawn at the Commission's request, as is recorded in the notes of the meeting held on November 24, 1885 (Szinte 1886a: 20). Szinte produced three life size plates based on the frescoes drawn in India ink on tracing paper and representing: "női szent" (a female saint), "szent király" (a holy king) and "sarkantyús, csizmás félalak" (a half figure with spurred boots). Although his interpretation of the frescoes was wrong (the figures in fact represent an apostle and two magi), Szinte's description remains to this day the most complete and exact record of the frescoes on the walls of the altar from the now-vanished church in Deva (see Sarkadi Nagy 2010: 944). The report and the attached plates were discussed in the Commission's above-mentioned meeting, at which it was also decided that Szinte should make color copies of the paintings, which he submitted on January 23, 1886.

Szinte's correspondence with the National Monuments Commission was mediated by Téglás, who had represented his interests on the Commission. The 19 plates from Bârsău, Streisângeorgiu and Deva (four in color and 15 in Indian ink on tracing paper), as well as Szinte's explanatory notes, were sent together with an address signed by Téglás. Based on the available documents, it is unclear why Téglás had to forward the results of Szinte's work to the Commission in Budapest; but his name certainly carried greater weight in the scientific community than that of his younger colleague. At that time, Téglás was already a member of several Hungarian and Transylvanian cultural associations and was in contact with various scientific institutions and personalities throughout Europe. Besides being the director of the State Central Secondary School for Technology and Science in Deva, he also was the director of the History Museum of the History and Archaeology Society of Hunedoara County from its beginnings in 1880 until his retirement in 1903. As the Society's director, he was responsible for increasing its collections through donations, acquisitions, and archaeological research. In recognition of his scientific activities, Téglás was named a corresponding member of the Hungarian Academy of Science in 1888 (Gáll 2014: 228–29, 231; Bodó 2020: 180).

Szinte's letters were addressed to the National Monuments Commission care of Ágoston Trefort (1817–1888), the Minister of Religious Affairs and Public Education. After a thorough analysis of the plates during the meeting of February 11, 1886, the Commission agreed to pay Szinte 120 forints, of which 50 forints had been sent in advance in 1883 (Henszlmann 1886: 12–13). The letter approving payment for his drawings, like all letters from the Commission, was signed by Imre Henszlmann. Acquiring such plates had been one of the National Monuments Commission's practices ever since its institutional beginnings. The survey drawings of mural paintings found in medieval churches carried out by art teachers were considered the most important tools for the investigation and protection of historical monuments (Lővei 2016: 558).

7.3 Gábor Szinte's Activity and Network Within the History and Archaeology Society of Hunedoara County

Throughout his career, Gábor Szinte maintained professional relations with scholars from all the cities where he had worked as a drawing teacher. During his time in Deva, in addition to his teaching activity at the State Central Secondary School for Technology and Science, he was also a member of the Hunyadmegyei Történelmi és Régészeti Társulat (History and Archaeology Society of Hunedoara County). Within this Society, he conducted field research with local scholars like Gábor Téglás, Pál Király (1853–1927), Sándor Szőts (1853–1905), Ferencz Sólyom-Fekete (1839–1906), and Oszkár Mailand (1858–1924).

The Society was established in 1880, after a call to research and preserve the remains of the past was published by Sólyom-Fekete[11] in the local newspaper *Hunyad* in 1879 (Sólyom-Fekete 1879: 1). The arguments presented in the article became the foundation of the new cultural association. However, this initiative was not new in Hunedoara County, as proposals to establish a cultural society had also been made in 1844 and 1856. Those who supported these proposals were members of the local elite, which was well integrated into the cultural life of Hungary (Bodó 2021: 11–13). Therefore, the proposal to establish a cultural association appeared as a natural step in a period of strong cultural effervescence in the country. Following the spirit of the era, the elites from Transylvania began to develop their own collections and establish cultural associations. The aim of these associations was to bring together those with shared interests and establish museums displaying the collections for the public, thus contributing to their education.

Sólyom-Fekete's call was successful and those who responded to it included the vice-lord-lieutenant of Hunedoara County, Kálmán Barcsay (1839–1903), representatives of local noble families, and almost all the teachers at the State Central Secondary School for Technology and Science and the Pedagogical School in Deva. They decided to establish a cultural association, and a committee was formed to draw up the Statute. After the final version of Statute was approved on May 13, 1880, the Constituent Assembly of the History and Archaeology Society of Hunedoara County met for the first time. The third paragraph of the Statute stipulated that to meet its objectives "a társulat a szakképességek és a szakkedvelők egyesitése (...) a műemlékek, a régiségek, nemkülönben a természeti nevezetességek tanyulmányozása, valamint ösmertetése által törekszik" (the Society would have to bring specialisms and specialists together, (…) it aims to study monuments, antiquities, and natural monuments and make these known). Another point in the Statute established the structure of the society's leadership: the General Assembly was held once a year, and the Executive Committee had monthly meetings to plan the Society's activities (Bodó 2012: 371–73). On the first meeting of the Constituent Assembly, the Society elected its leadership which consisted of Géza Kuun[12] (1838–1905) as president, Ferencz Sólyom-Fekete as vice president, Róbert Kun as secretary, Pál Király as librarian, Gergely Lengyel as cashier, and Petco Lazăr as jurist (General Assembly 1880: 9). At the time of this meeting, the Society already had 517 members, of which 162 ordinary members, 318 supporting members, and 37 founding (Bodó 2012: 372–74).

The members of the Society were driven in their work by passion and received no remuneration. Field investigations were usually conducted by groups of members with diverse scientific ambitions and interests. Their collaboration advanced the preservation of local cultural heritage and contributed to each member's personal and professional development. At the same time, the members of the Society—who came from various parts of Hunedoara County—formed a network of people who had a mutual interest in protecting the remains of the past.

One of the common efforts of the Society's members was the attempt to salvage the Calvinist church in Deva, Hunedoara County's administrative center. The frescoes of the old parish church were discovered in 1885, when the plastering on the northern wall of the sanctuary disintegrated. The Executive Committee of the Society was informed of this event by Sándor Szőts, Calvinist minister and member of the Executive Committee. Szinte and Téglás, together with Szőts, were sent to investigate the frescoes and to write a report. If the mural paintings had proved to be important, the discovery would have triggered the attention of the National Monuments Commission (Executive Committee 1885a:121v). One of the National Monuments Commission's first tasks was to establish rules and regulations for the registration and classification of historical monuments. As financial resources were limited, it was necessary to clarify criteria to decide between which monuments would be restored, preserved, or documented and which would be left to their fate. These criteria were based on a monument's artistic value and local importance (Lővei 2016: 557).

As far as the Society's members were concerned, the old parish church in Deva was one of high importance. The fres-

[11] Ferencz Sólyom-Fekete studied law in Pest, and in 1875 became President of the Royal Tribunal in Hunedoara County, located in Deva. Between 1880 and 1903, he was vice president of the History and Archaeology Society of Hunedoara County. He collected and studied historical documents referring to the county's history and occasionally even conducted investigations in the field (including archaeological digs) ("Tárca," 1906: 127–28).

[12] Count Géza Kuun, known as an epoch-relevant scientist, linguist, orientalist, and philologist, was the president of the History and Archaeology Society of Hunedoara County between 1880 and 1905. He became a corresponding member of the Hungarian Academy of Sciences in 1867 and an honorary member beginning in 1883. Between 1901 and 1904, he was vice president of the Hungarian Academy of Sciences. He left his personal library and archive to the Erdélyi Múzeum Egyesület (Transylvanian Museum Society) and the stone monuments on his property at Mintia to the museum of the History and Archaeology Society of Hunedoara County (Szádeczky 1905: 401–31).

coes' appearance supplied an additional argument to bring before the central authorities to justify the preservation of this critically threatened monument. Szinte also drafted a restoration design for the church (Szőts 1893), but despite his efforts, along with those of other members of the Society, the nave of the church, the most degraded part, was demolished in 1899, followed by the sanctuary in 1907 (Codrea 2010: 59–60). In September 1907, just before the altar was pulled down, Oszkár Mailand, at that time director of the Society's History Museum, had five of the church's frescoes photographed (Executive Committee 1907: 535).

The Society's vast network of members made it possible for the news about archaeological finds to quickly reach its leadership. During the November 1885 meeting, Sólyom-Fekete announced that, thanks to information received from Samuel Nemes, a member of the Executive Committee of the Society, a series of ruins, believed to belong to the medieval *castellum* of Haţeg (Hungarian: Hátszeg), erected over a Roman construction, were uncovered on the land of Miklós Bernáth[13] (Executive Committee 1885b: 127). In the spring of 1886, at the recommendation of Sólyom-Fekete, a group formed of Téglás, Szinte, and Pál Király[14] undertook a field trip that resulted in the discovery of the ruins of the Franciscan monastery at the site (Téglás and König 1886: 53). They presented the results of their trip at the Society's meeting on May 5 and interpreted the ruins as belonging to a large Gothic church built between the fourteenth and sixteenth centuries using numerous Roman architectural materials or *spolia*. According to their report, the ground plan was similar to that of the Calvinist Reformed Church in Deva (Executive Committee 1886: 137v; Téglás and König 1886: 53).[15] Archaeological excavations were not conducted at the site until the summer of 1888, when Téglás and Szinte engaged in a preliminary investigation. Both scholars signed the resulting archaeological report. Szinte's survey of the ruins was approximate and featured no scale (nevertheless, the dimensions of the constructions were noted) (see Rusu 1993: 139–41).

Based on the minutes of the Society, Szinte accompanied its members (e.g., Téglás, Király, and Mailand) on various archaeological excavations and documentary field trips (Bodó 2017: 240; 2021: 207–8, 212, 223–25) during which he was mainly in charge of the drawings and photographs. Among the Society's activities, he participated in the unearthing of the amphitheater of Sarmizegetusa—the capital of Roman Dacia—about which he published an initial report (accompanied by a survey of the amphitheater and some drawings of the findings) in the yearbook of the Society (Szinte 1897e). Archaeological excavations at Sarmizegetusa provided research opportunities for all members of the Society. A talented artist, Szinte created numerous archaeological field drawings, as well as images of artifacts. Apart from illustrating his own publications, he also executed archaeological record drawings for his Society colleagues. For instance, he created hundreds of drawings for archaeologist Zsófia Torma (1832–1899),[16] only some of which have been published (Coltofean 2017: 345); he served in the same capacity for Téglás, to illustrate the monograph of Hunedoara County, *Hunyadvármegye története*, together with László Dömötör, drawing teacher from Arad (Kuun et al. 1902: XIV, XVII, 65, 110, 181–82; Bata and Tasnádi 2015: 15).

The field trips organized by the Society facilitated the completion as well as initiation of various investigative projects. Archaeology and history (identifying, gathering, and investigating documents) were initially the main areas of interest, but ethnography and natural sciences followed relatively soon thereafter. Szinte's scientific progress followed a similar disciplinary ontology. In addition to participating in archaeological investigations, his interest in artistic monuments is clear, be they medieval or Roman, religious (Eastern Orthodox, Reformed or Catholic), military or civil constructions. At the same time, he had an interest in ethnographic studies, research that captured his full attention after his departure from Deva, and for which he is much better known today.

Téglás, Mailand, and Szinte investigated the southern part of Hunedoara County between July 15 and August 14, 1884 (Téglás 1884: 32). On that occasion, they visited the Crivadia Tower in the southernmost part of the Land of Haţeg (Hungarian: Hátszegvidék; Romanian: Ţara Haţegului). The

[13] Miklós Bernáth was a notary public from Haţeg and a founding member of the History and Archaeology Society of Hunedoara County.

[14] Pál Király was a history and geography professor at the State Central Secondary School for Technology and Science of Deva starting in 1877, a Society member and librarian from 1880 until he moved away in 1891. He actively participated in investigations in the field, especially the Sarmizegetusa archeological excavations (with Gábor Téglás), and he published many works on Roman civilization, making a name for himself especially with studies dedicated to the Sarmizegetusa *mithraeum* and the cult of Mithras (Deju and Bodó 2006: 634–39).

[15] Today we know that the ruins belonged to a Franciscan monastery, probably built in the middle of the fourteenth century during the reign of the Hungarian King Louis I 'the Great' of Anjou. The monastery was entirely destroyed in the course of the Ottoman incursions of 1479 (Rusu 1993: 137).

[16] Zsófia Torma, considered the first documented female Hungarian archaeologist, investigated the prehistoric site at Turdaş (Hungarian: Tordos) in Hunedoara County and presented her discoveries in international conferences. She made an important contribution to the development of prehistoric archaeology in Transylvania. She became a member of the History and Archaeology Society of Hunedoara County in 1880 (Coltofean 2017: 327–54).

investigation of this monument represents another example of the productive collaboration between Szinte and Téglás. The results were presented by Szinte during the Society's Executive Committee meeting on March 4, 1894 (Executive Committee 1894: 247v) and were published in the same year in *Archaeologiai Értesítő*. The watchtower was believed to be of Roman origin. Despite the incorrect chronological interpretation (the tower was built in the sixteenth century), Szinte's description and drawings of the site remain unparalleled to this day (Szinte 1894: 110). They are also as accurate as they can be, considering that the monuments' preservation state has not changed during the last century. Supported by Szinte's illustrations, Téglás' published description drew attention to the threatened situation of the ruins. Although their state of preservation was good, there was the risk that the tower could be damaged by robbing of stone for building material, as had happened to the nearby Bănița Dacian fortress whose stone blocks were re-used for infrastructure works. Starting in 1891, Téglás submitted several proposals to protect the Crivadia Tower from destruction. These included property transfer to the State of the land on which the tower was located, of the path leading to it and a 15 m radius around it (following the administrative course, expropriation or land exchange); supervisory oversight of the tower by the Hungarian Railway; and submitting a petition to declare the watchtower a historical monument by the Ministry of Religious Affairs and Public Education or by the Minister of the Interior (Téglás 1891: 42v).

Several years earlier, in 1886, another project had caught Szinte's attention: the ruins of the medieval monuments at Suseni-Colț (Hungarian: Kolc). The results of the research carried out at these monuments were published in the Society's yearbook in 1893. Located on a small plateau, the chapel (as Szinte called it) had been for a long time in an advanced state of decay. However, the sanctuary, the nave, and large sections of a mural painting—described by Szinte as "naiv festészetét" (naïve painting) and dated by him to the seventeenth century—were reasonably well preserved. Today most of the murals are lost. The nave remained uncovered for a long period and repeated rains washed away the frescoes, while in the altar area a regrettable restoration at the end of the twentieth century irremediably altered the valuable wall paintings. Szinte also drew attention to the destruction caused by the numerous scratched inscriptions on the church walls left behind by various visitors, some of which were incised into the frescoes, contributing to their disintegration. He concluded that if this habit persisted, "egy évtized alatt az érdekes falfestés eltűnik az értéktelen bekarczolások alatt" (within a decade, the interesting paintings would disappear beneath worthless scratches) (Szinte 1893: 70). He did not produce any drawings of the interior (or none that have come down to us), but he analyzed the frescoes' iconography very precisely (see Szinte 1893: 73–74). Across the valley from the church were the ruins of an aristocratic residence, above them, the fortification of the Kendeffy noble family rose on a crag. Szinte described the fortress' ground plan with much precision, along with the conservation state of the walls and some considerations on the chronology of the monument (see Szinte 1893: 74–77).

The two ruins had attracted visitors even before Szinte's report on the site. The robustly built little church had already been noted by Albert Nemes (1867), Ștefan Moldovan (1853: 284) and András Fodor (1844b: 16). The professional and cultural activity carried out by these individuals reflects the interest in cultural heritage of researchers from various disciplines in the mid-nineteenth century. Nemes was a music teacher who published many articles in the Hungarian newspapers of the time (e.g., *Vasárnapi Ujság*, *Fővárosi Lapok*), with short descriptions of historical monuments and the natural beauty of Alba and Hunedoara Counties, sometimes with his own drawings, especially between 1860 and 1875. He was one of the contributors to the *Hunyad* newspaper from its beginnings. We find him among the members of the History and Archaeology Society of Hunedoara County in 1880 ("Uj lap," 1877; Szinnyei 1903: 881–82). Moldovan was the Greek-Catholic vicar of Hațeg[17] between 1852 and 1857, when he developed an interest in the history of Transylvanian Romanians. The information he collected was published in multiple issues of the Romanian newspaper *Foaie pentru minte, inimă și literatură*[18] (Naghi 1979: 421–22). Fodor was the chief medical officer of Hunedoara County, a collector with a passion for the county's history. He wrote a series of articles about the areas of historical and archaeological interest in Hunedoara. In addition, he was among the first to propose the creation of a cultural association in the county that would be dedicated to collecting and studying the vestiges of the past (Fodor 1844a, b; Szinnyei 1894: 568–69; Bodó 2021: 11–12).

The results of historical investigations were also disseminated through periodicals with wide coverage among the population. While central newspapers were used, among other things, as platforms to make the authorities aware of the problems associated with preserving built heritage, local newspapers catered to the general public's taste for history, culture, and the protection of artistic monuments. *Hunyad* was a weekly newspaper founded in Deva in 1877. Over the years, many members of the Society were involved in its creation and maintenance. For example, Lajos Réthi

[17] Hațeg is the eponymous administrative center of the district, situated in southern Hunedoara County in an area populated predominantly by Romanians.

[18] This study, entitled *Annotatiuni despre tiéra Hatiegului*, involved many historical monuments within the Land of Hațeg. It appeared in 11 issues, between August 24, 1853, and September 26, 1854.

(1840–1912)[19] was the newspaper's founder, owner, and collaborator, while Róbert Kun (1842–1897)[20] was its editor in chief (1887–1897). This explains why the Society's activities were a major feature of the newspaper's content, which also published detailed accounts of its meetings. A series of scientific articles were published as well, including those by Szinte about the Deva Fortress.

Szinte's last major project in the period when he was an art teacher at in Deva was a report on the fortress of Deva itself. Built in the second half of the thirteenth century, with important restorations and extensions in the sixteenth, seventeenth and nineteenth centuries, the first ground plan of the fortifications produced for scholarly purposes is linked with Szinte's name. At the Society's meetings, Szinte gave several lectures on the history of the citadel. Although he was asked to publish it in the yearbook of the Society, it ultimately appeared in the local newspaper *Hunyad*, in 1897,[21] to which he was a regular contributor. Szinte only published a selection of the research results associated with the ground plan of the fortification in 1910 in the commemorative issue of *Unitárius Közlöny* (see Szinte 1910), dedicated to Ferenc Dávid (1520–1579, the founder of the Unitarian Church in Transylvania), which included a guide for visitors.

Szinte's interest in the fortress of Deva was not exclusively academic. He also engaged in efforts aimed at the conservation, protection, and promotion of historic buildings. On February 3, 1897, at his initiative, the Society's Executive Committee asked Szinte and two other members to submit proposals for clearing the rubble from the first curtain wall of the fortress in the cheapest way possible in order to facilitate the production of an architectural survey of the citadel (Executive Committee 1897a: 284v–285). The aim was to reveal some chambers, but also to remove large quantities of debris, which could have caused the collapse of the walls. During the Executive Committee's meeting on September 8, 1897, Szinte presented an estimate of the work undertaken at the fortress of Deva and reported on the artifacts discovered during the clearing of the inner court of the citadel (Executive Committee 1897c: 300).

In tandem with the Deva fortress project, Szinte submitted a proposal for the restoration of the main gate of the *Magna Curia* Castle (also known as the Bethlen Gate[22]). The castle housed the administration of one of the largest princely domains from its beginnings in the sixteenth century, and it was occupied by the administration of Hunedoara County and the royal financial authorities in the nineteenth century (Dévai 1880). On March 10, 1897, Szinte presented a photo and a reconstruction proposal for the main gate. He suggested solutions for reconstructing this historical monument and provided details of the financial support needed for the project to the Society's Executive Committee. At the urging of Téglás, it was decided that the gate would be restored at the expense of the Society under the direct supervision of Szinte (Executive Committee 1897b: 287–287v). Szinte seized the opportunity to propose this project when the main entrance into the castle grounds was demolished to make room for the new Royal Tribunal of Hunedoara County. The photograph taken by Szinte, which he donated to the Society's library, is the only image of the Gate where one can see the inscription as well as the Bethlen family's coat of arms and associated artistic elements (Codrea 2009: 38, Fig. 2; Bodó 2021: 269–70).

Before the demolition of the entrance, there had been repeated discussions about restoring the Bethlen Gate in the Society's Executive Committee meetings. Salvaging the gate became an urgent matter when proposals to construct the Tribunal in its place began to be circulated. As a last resort, on August 8, 1896, Gábor Téglás addressed the central Hungarian authorities (specifically László Lukács [1850–1932], the Minister of Finance, and Sándor Erdélyi [1839–1922], the Minister of Justice, both originally from Transylvania) by means of an open letter that he published in the newspaper *Fővárosi Lapok* (Téglás 1896). This was the year of the widespread millenary celebrations of the Hungarian conquest of the Carpathian Basin. At the start of the letter, Téglás stressed that

> Most a millenniumi esztendőben (...) a mikor a törvényhozás a hazai történelem kimagasló eseményeinek szinhelyeit országos alapból siet emlékekkel kitüntetni, kétszeresen lehangoló volna reánk nézve, ha nemzeti nagyságunk emlékeiben amugy is szükölködő (...) vidékünkön (...) a nagy Bethlen Gábor egyik sokat mondó emlékét a modern építkezés áldozatául ejtenék. Nem kevesebbről van ungyanis szó, mint arról: hogy Déván a fejedelmi kastélynak bejáró kapuját a királyi törvényszék kedvéért lerombolják.
> (It would be sad and hard to understand that in the millennium year, (...) when monuments marking important events in Hungarian history are being built throughout the country from national funds, and in a county (...) with so few such monuments cherishing the memory of our national greatness, (...) a monument that speaks to the great Gábor Bethlen should fall victim to modern construction, specifically, to demolishing the entrance gate to the princely castle for the sake of the Royal Tribunal.) (Téglás 1896: 2)

Téglás offered arguments for preserving the gate, including its special character, its historical and artistic value as a singular piece as well as part of the ensemble represented by the *Magna Curia* (Great Court). He underlined the importance of the keystone of the gateway arch, which was emblazoned

[19]Lajos Réthi was a school inspector in Hunedoara County and a member of the Society's Executive Committee (Szinnyei 1906: 821–24).

[20]Róbert Kun was a history and geography teacher at the State Central Secondary School for Technology and Science of Deva starting in July 1878 and was a founding member of the History and Archaeology Society of Hunedoara County, serving as its secretary and editor of its first yearbooks. He collected, copied, studied, and published many documents regarding the county's history (Téglás 1898: 139–47).

[21]*Hunyad*, issues no. 24, 25, 27 and 28 from 1897.

[22]The gate was named after the Prince of Transylvania, Gábor Bethlen (1613–1629), in whose time the castle underwent important changes, including construction of the gate.

with the coat of arms of Prince Bethlen, and of the entablature with its inscription, maintaining that the gate should be restored, or in any event not demolished. The only people who could stop its destruction were the two ministers, from whom Téglás requested a decision as quickly as possible, as construction work was to begin soon (Téglás 1896: 2–3).

Unfortunately, Bethlen Gate's fate was ill-omened. The undertaking to save it did not get the necessary backing from the authorities and it was demolished. The restoration project stalled as well. Soon after this, Szinte left Deva. The society's members subsequently focused on salvaging the gate's fragments. After the demolition, Téglás succeeded in persuading Zsigmond Blum,[23] who had obtained all the material resulting from the demolition at auction, to donate the remaining stone pieces of the gate to the Society (Téglás 1913).

In the last years of the nineteenth century, Society members strove to gather the necessary funds to restore the Bethlen Gate. At first, they wanted to reconstruct the gate in its original location. Shortly afterward, when the *Magna Curia* Castle park was turned over to the Deva mayor's office and became a City Park, a decision was made to erect the Gate at the park's entrance facing the newly built Administrative Palace (Executive Committee 1898: 310v). Moreover, at the prompting of Lajos Réthi, the Executive Committee of the Society decided to attach an informative plaque to the gate's foundation that would explain for posterity how the Bethlen Gate ended up at the park entrance (Executive Committee 1898: 311). Although the gate's reconstruction seemed imminent, it failed to happen, and discussions continued fruitlessly into the next century, while the stones were exposed to further damage and decay. Work on the gate's reconstruction began again in 1904. As the photograph and the reconstruction design of the gate had been mislaid, Szinte offered to redo the design (Executive Committee 1904b: 402). Despite all these good intentions, the gate was never rebuilt, and the last stone fragments were transported to the Society's Museum at the start of 1912, where they remained in the courtyard of the State Central Secondary School in Deva (Executive Committee 1912: 662).

At the same March 1897 meeting at which he presented the photograph and the design for the restoration of the Bethlen Gate, Szinte announced his transfer to the Hungarian Royal Central Secondary School in the 8th District of Budapest. He thanked the assembled members for the trust he had received from the Society over the years and expressed his wish to remain a member after his departure (Executive Committee 1897b: 300v). In Budapest, his research focused on ethnographic studies[24] and he began collaborating with the Museum of Ethnography (then the Ethnography Department of the Hungarian National Museum) (see Bata 2013, 2021; Bata and Tasnádi 2015).

Szinte's place in the Art Department of the State Central Secondary School for Technology and Science in Deva was briefly occupied by István Mérő (1873–1938) until 1900 and then by Henrik Szirtes (1876–?). Both were members of the History and Archaeology Society of Hunedoara County, the latter handling photographic documentation (e.g., field investigations, various events) within the Society (Executive Committee 1902: 374, 1904a: 399v). It seems that they did not collaborate with the National Monuments Commission. Nevertheless, the Commission remained highly interested in Hunedoara County's artistic and historic monuments. In the last years of the nineteenth century and the first years of the twentieth, István Möller (1860–1934), Ottó Sztehlo (1851–1923), and István Gróh (1867–1936) became particularly active in preserving the artistic and historical heritage of Hunedoara County. The post-1900 effort to copy medieval monument mural paintings in the County was taken up especially by Gróh. The archives of the National Monuments Commission contain drawings and watercolors of the church paintings in Densuş (Hungarian: Demsus), Gurasada, Râu de Mori (Hungarian: Malomvíz), Sântămăria Orlea (Hungarian: Őraljaboldogfalva), and Strei (Hungarian: Zeykfalva), all of which were produced on commission by Gróh (Gerecze 1905: 267–80).

7.4 Conclusions

The work of Gábor Szinte, during the 14 years he spent in Deva, must be understood through the two networks that he was part of. As a drawing teacher he was involved in the network coordinated by the National Monuments Commission whose main purpose was to investigate and protect historical monuments in Hungary. In addition, he had the opportunity to be a member of the History and Archaeology Society of Hunedoara County, where he carried out archaeological field research, especially on medieval and Roman monuments, and published parts of his results (Szinte 1897a, b, c, d, 1898).

[23]Zsigmond Blum was a merchant from Deva, and a member of the History and Archaeology Society of Hunedoara County starting with 1901 ("Társulati névkönyv," 1907: 182).

[24]We have little information about Szinte's ethnographic research during his stay in Deva. The 1896 annual report of the State Central Secondary School for Technology and Science contains notes about his research for the Millennial Exhibition related to agricultural exhibits from Hunedoara County. For this display Szinte created cottage industry models and took pictures of ethnographic landscapes and scenes of animal husbandry ("Adatok," 1896: 62). He also collaborated with Samu Kolumbán (1861–1908) and Oszkár Mailand, Society members whose efforts were mostly in this discipline.

Even if Szinte's historic interpretations of sites, monuments, and artifacts were not always accurate, the work he carried out during his time as an educator in Deva and his scholarly output more generally remain of utmost importance. This is because much of the information he had noted and drawings he produced include numerous elements that are lost to us today. His tireless work to preserve the county's historical heritage was undertaken with scholarly rigor and he was committed to promoting the heritage of Hunedoara County, not only locally, but also at the regional and national level.

Throughout his time in Deva (1883–1897), apart from his teaching activity at the State Central Secondary School for Technology and Science, Szinte was an active participant in the History and Archaeology Society of Hunedoara County. In 1884, he became a member of the Society and of its Executive Committee, and in 1893 he was listed as a member of the Society's Archaeology Department[25] (Mailand 1891: 121). The survey drawings that Szinte created for the National Monuments Commission were not produced in his capacity as a member of the Hunedoara Society. He became part of the Historical Monument Commission's network of collaborators due exclusively to his artistic abilities, which had been recognized by Gábor Téglás as soon as he arrived in Deva. Still, it is likely that the surveys of Bârsău, Streisângeorgiu, and Deva churches fueled his interest in investigating the county's historical monuments. The Society's documents reference a series of investigations that Szinte carried out on the medieval monuments in the county, but unfortunately only some of the resulting reports were preserved.

Szinte broadened his area of research over time with the Society's encouragement and financial support. The field research that Szinte conducted in the territory of Hunedoara County together with Téglás, Pál Király, Sándor Szőts, Ferencz Sólyom-Fekete, and Oszkár Mailand covered not only the field of art history, but also archaeology. In addition to his active participation in archaeological excavations (sometimes even as coordinator), Szinte was responsible for drafting archaeological surveys and drawings of the artifacts that illustrated his own, and his colleagues' publications.

[25] In 1881, the Society adopted a regulation that provided the six departments, including the Department for Archaeological and Historical Monuments, but the departments were not established before February 1883. Even so, the departments did not function because the members of the Society followed their own diverse scientific interests. In January 1893, it was decided again to re-establish the six departments: the Geography Department, the Natural Sciences Department, the Archaeology Department (with Gábor Téglás as president), the History Department, the Statistics Department, and the Commission for drafting the Society's yearbook. After the re-establishment of the Department, Szinte also became a member as a rapporteur (Bodó 2021: 34–37).

Szinte's departure from Deva in 1897 left a major gap in the Society's historical research program. Téglás bemoaned the fate of medieval monuments in Hunedoara County, which, in his view, "historicusukat és rokonszenves magyarázojukat veszítik el" (had lost their historian and sympathetic storyteller) when Szinte transferred to Budapest (Veress 1898: 159). With the departure of Szinte, no one from the Society has researched the medieval monuments from the county. The problems faced by the Society after 1900 and the start of the First World War in 1914 meant that research on medieval monuments had to wait a long time before being resumed.

References

Primary Sources

Archival Material

Executive Committee. (1882a, June 7). *Minutes of the Executive Committee meeting of the History and Archaeology Society of Hunedoara County*. Fond Societatea de Istorie și Arheologie a comitatului Hunedoara, Fond propriu (2/1880–1914), Serviciul Județean Hunedoara al Arhivelor Naționale (Fonds of the History and Archaeology Society of Hunedoara County, Society Fonds [2/1880–1914], Hunedoara County Service of the National Archives), Deva, Hunedoara County, Romania.

Executive Committee. (1882b, July 5). *Minutes of the Executive Committee meeting of the History and Archaeology Society of Hunedoara County*. Fond Societatea de Istorie și Arheologie a comitatului Hunedoara, Fond propriu (2/1880–1914), Serviciul Județean Hunedoara al Arhivelor Naționale (Fonds of the History and Archaeology Society of Hunedoara County, Society Fonds [2/1880–1914], Hunedoara County Service of the National Archives), Deva, Hunedoara County, Romania.

Executive Committee. (1885a, October 14). *Minutes of the Executive Committee meeting of the History and Archaeology Society of Hunedoara County*. Fond Societatea de Istorie și Arheologie a comitatului Hunedoara, Fond propriu (2/1880–1914), Serviciul Județean Hunedoara al Arhivelor Naționale (Fonds of the History and Archaeology Society of Hunedoara County, Society Fonds [2/1880–1914], Hunedoara County Service of the National Archives), Deva, Hunedoara County, Romania.

Executive Committee. (1885b, November 4). *Minutes of the Executive Committee meeting of the History and Archaeology Society of Hunedoara County*. Fond Societatea de Istorie și Arheologie a comitatului Hunedoara, Fond propriu (2/1880–1914), Serviciul Județean Hunedoara al Arhivelor Naționale (Fonds of the History and Archaeology Society of Hunedoara County, Society Fonds [2/1880–1914], Hunedoara County Service of the National Archives), Deva, Hunedoara County, Romania.

Executive Committee. (1886, May 5). *Minutes of the Executive Committee meeting of the History and Archaeology Society of Hunedoara County*. Fond Societatea de Istorie și Arheologie a comitatului Hunedoara, Fond propriu (2/1880–1914), Serviciul Județean Hunedoara al Arhivelor Naționale (Fonds of the History and Archaeology Society of Hunedoara County, Society Fonds

[2/1880–1914], Hunedoara County Service of the National Archives), Deva, Hunedoara County, Romania.
Executive Committee. (1894, March 4). *Minutes of the Executive Committee meeting of the History and Archaeology Society of Hunedoara County*. Fond Societatea de Istorie şi Arheologie a comitatului Hunedoara, Fond propriu (2/1880–1914), Serviciul Judeţean Hunedoara al Arhivelor Naţionale (Fonds of the History and Archaeology Society of Hunedoara County, Society Fonds [2/1880–1914], Hunedoara County Service of the National Archives), Deva, Hunedoara County, Romania.
Executive Committee. (1897a, February 3). *Minutes of the Executive Committee meeting of the History and Archaeology Society of Hunedoara County*. Fond Societatea de Istorie şi Arheologie a comitatului Hunedoara, Fond propriu (2/1880–1914), Serviciul Judeţean Hunedoara al Arhivelor Naţionale (Fonds of the History and Archaeology Society of Hunedoara County, Society Fonds [2/1880–1914], Hunedoara County Service of the National Archives), Deva, Hunedoara County, Romania.
Executive Committee. (1897b, March 10). *Minutes of the Executive Committee meeting of the History and Archaeology Society of Hunedoara County*. Fond Societatea de Istorie şi Arheologie a comitatului Hunedoara, Fond propriu (2/1880–1914), Serviciul Judeţean Hunedoara al Arhivelor Naţionale (Fonds of the History and Archaeology Society of Hunedoara County, Society Fonds [2/1880–1914], Hunedoara County Service of the National Archives), Deva, Hunedoara County, Romania.
Executive Committee. (1897c, September 8). *Minutes of the Executive Committee meeting of the History and Archaeology Society of Hunedoara County*. Fond Societatea de Istorie şi Arheologie a comitatului Hunedoara, Fond propriu (2/1880–1914), Serviciul Judeţean Hunedoara al Arhivelor Naţionale (Fonds of the History and Archaeology Society of Hunedoara County, Society Fonds [2/1880–1914], Hunedoara County Service of the National Archives), Deva, Hunedoara County, Romania.
Executive Committee. (1898, March 8). *Minutes of the Executive Committee meeting of the History and Archaeology Society of Hunedoara County*. Fond Societatea de Istorie şi Arheologie a comitatului Hunedoara, Fond propriu (2/1880–1914), Serviciul Judeţean Hunedoara al Arhivelor Naţionale (Fonds of the History and Archaeology Society of Hunedoara County, Society Fonds [2/1880–1914], Hunedoara County Service of the National Archives), Deva, Hunedoara County, Romania.
Executive Committee. (1902, June 22). *Minutes of the Executive Committee meeting of the History and Archaeology Society of Hunedoara County*. Fond Societatea de Istorie şi Arheologie a comitatului Hunedoara, Fond propriu (2/1880–1914), Serviciul Judeţean Hunedoara al Arhivelor Naţionale (Fonds of the History and Archaeology Society of Hunedoara County, Society Fonds [2/1880–1914], Hunedoara County Service of the National Archives), Deva, Hunedoara County, Romania.
Executive Committee. (1904a, April 3). *Minutes of the Executive Committee meeting of the History and Archaeology Society of Hunedoara County*. Fond Societatea de Istorie şi Arheologie a comitatului Hunedoara, Fond propriu (2/1880–1914), Serviciul Judeţean Hunedoara al Arhivelor Naţionale (Fonds of the History and Archaeology Society of Hunedoara County, Society Fonds [2/1880–1914], Hunedoara County Service of the National Archives), Deva, Hunedoara County, Romania.
Executive Committee. (1904b, June 2). *Minutes of the Executive Committee meeting of the History and Archaeology Society of Hunedoara County*. Fond Societatea de Istorie şi Arheologie a comitatului Hunedoara, Fond propriu (2/1880–1914), Serviciul Judeţean Hunedoara al Arhivelor Naţionale (Fonds of the History and Archaeology Society of Hunedoara County, Society Fonds [2/1880–1914], Hunedoara County Service of the National Archives), Deva, Hunedoara County, Romania.
Executive Committee. (1907, September 2). *Minutes of the Executive Committee meeting of the History and Archaeology Society of Hunedoara County*. Fond Societatea de Istorie şi Arheologie a comitatului Hunedoara, Fond propriu (2/1880–1914), Serviciul Judeţean Hunedoara al Arhivelor Naţionale (Fonds of the History and Archaeology Society of Hunedoara County Fund, Society Fonds [2/1880–1914], Hunedoara County Service of the National Archives), Deva, Hunedoara County, Romania.
Executive Committee. (1912, April 30). *Minutes of the Executive Committee meeting of the History and Archaeology Society of Hunedoara County*. Fond Societatea de Istorie şi Arheologie a comitatului Hunedoara, Fond propriu (2/1880–1914), Serviciul Judeţean Hunedoara al Arhivelor Naţionale (Fonds of the History and Archaeology Society of Hunedoara County, Society Fonds [2/1880–1914], Hunedoara County Service of the National Archives), Deva, Hunedoara County, Romania.
General Assembly. (1880, May 13). *Meeting of the Constituent Assembly of the History and Archaeology Society of Hunedoara County*. Fond Societatea de Istorie şi Arheologie a comitatului Hunedoara, Fond propriu (2/1880–1914), Serviciul Judeţean Hunedoara al Arhivelor Naţionale (Fonds of the History and Archaeology Society of Hunedoara County, Society Fonds [2/1880–1914], Hunedoara County Service of the National Archives), Deva, Hunedoara County, Romania.
Henszlmann, I. (1884, October 26). [Letter to Gábor Szinte]. Műemlékek Országos Bizottsága (1886/5), Tudományos Irattár, Magyar Építészeti Múzeum és Műemlékvédelmi Dokumentációs Központ (National Monuments Commission [1886/5], Scientific Archives, Hungarian Museum of Architecture and Monument Protection Documentation Centre), Budapest, Hungary.
Henszlmann, I. (1886, February 15). [Letter to Ágoston Trefort]. Műemlékek Országos Bizottsága (1886/5), Tudományos Irattár, Magyar Építészeti Múzeum és Műemlékvédelmi Dokumentációs Központ (National Monuments Commission [1886/5], Scientific Archives, Hungarian Museum of Architecture and Monument Protection Documentation Centre), Budapest, Hungary.
Szinte, G. (1884, June 30). [Letter to the National Monuments Commission]. Műemlékek Országos Bizottsága (1886/5), Tudományos Irattár, Magyar Építészeti Múzeum és Műemlékvédelmi Dokumentációs Központ (National Monuments Commission [1886/5], Scientific Archives, Hungarian Museum of Architecture and Monument Protection Documentation Centre), Budapest, Hungary.
Szinte, G. (1886a, January 23). [Letter to the National Monuments Commission]. Műemlékek Országos Bizottsága (1886/5), Tudományos Irattár, Magyar Építészeti Múzeum és Műemlékvédelmi Dokumentációs Központ (National Monuments Commission [1886/5], Scientific Archives, Hungarian Museum of Architecture and Monument Protection Documentation Centre), Budapest, Hungary.
Szinte, G. (1886b). [Drawings by Gábor Szinte of Bârsău church murals]. Tervtár (FM 42, FM 43, FM 44, FM 45, FM 46, FM, 47, FM 48, FM 49, FM 50, FM 51, FM 56, FM 57, FM 58), Magyar Építészeti Múzeum és Műemlékvédelmi Dokumentációs Központ (Plan Repository [FM 42, FM 43, FM 44, FM 45, FM 46, FM, 47, FM 48, FM 49, FM 50, FM 51, FM 56, FM 57, FM 58], Hungarian Museum of Architecture and Monument Protection Documentation Centre), Budapest, Hungary.
Szinte, G. (1886c). *The Annunciation scene*. Tervtár (FM 574), Magyar Építészeti Múzeum és Műemlékvédelmi Dokumentációs Központ (Plan Repository [574], Hungarian Museum of Architecture and Monument Protection Documentation Centre), Budapest, Hungary.
Téglás, G. (1884, September 24). *Field report*. Fond Societatea de Istorie şi Arheologie a comitatului Hunedoara, Fond propriu (38/1880–1918), Serviciul Judeţean Hunedoara al Arhivelor

Naționale (Fonds of the History and Archaeology Society of Hunedoara County, Society Fonds [38/1880–1918], Hunedoara County Service of the National Archives), Deva, Hunedoara County, Romania.

Téglás, G. (1891, June 7). *Annual report of the director of the museum.* Fond Societatea de Istorie și Arheologie a comitatului Hunedoara, Fond propriu (38/1880–1918), Serviciul Județean Hunedoara al Arhivelor Naționale (Fonds of the History and Archaeology Society of Hunedoara County, Society Fonds [38/1880–1918], Hunedoara County Service of the National Archives), Deva, Hunedoara County, Romania.

Téglás, G., & König, P. (1886, April 7). *Field report.* Fond Societatea de Istorie și Arheologie a comitatului Hunedoara, Fond propriu (38/1880–1918), Serviciul Județean Hunedoara al Arhivelor Naționale (Fonds of the History and Archaeology Society of Hunedoara County, Society Fonds [38/1880–1918], Hunedoara County Service of the National Archives), Deva, Hunedoara County, Romania.

[The teaching staff of the State Central Secondary School for Technology and Science in Deva]. (before 1898). Magyar Nemzeti Múzeum (561.1956) (Hungarian National Museum [561.1956]), Budapest, Hungary.

Newspaper and Magazine Articles

Dévai, S. (1880, October 5). Hunyadmegye és a kormány. *Ellenzék.*

Fodor, A. (1844a, September 24). Római régiségek Hunyad vármegyében. *Hon és Külföld*, 305–307.

Moldovan, Șt. (1853, September 23). Annotatiuni despre tiéra Hatiegului. *Foaie pentru minte, inimă și literatură*, 281–284.

Nemes, A. (1867, January 20). A Kolcvár alatti templom, Erdélyben. *Hazánk s a Külföld*, 48.

Régi oláh (ó-szláv) könyvek (1886, June 21). *Ellenzék.*

Sólyom-Fekete, F. (1879, November 1). Derítsük föl a multunkat és mentsük meg emlékeit. *Hunyad.*

Szinte, G. (1897a, June 12). Déva ékessége. *Hunyad.*

Szinte, G. (1897b, June 19). Déva ékessége. *Hunyad.*

Szinte, G. (1897c, July 3). Déva ékessége. *Hunyad.*

Szinte, G. (1897d, July 10). Déva ékessége. *Hunyad.*

Szinte, G. (1910, August). Déva vára. *Unitárius Közlöny*, 154–262.

Szőts, S. (1893, September 3). A dévai református templom és torony jelenlegi és leendő alakjában. *Vasárnapi Ujság*, 603.

Téglás, G. (1896, August 10). Kegyelem a dévai Bethlen-kapunak. *Fővárosi Lapok*, 2–3.

Téglás, G. (1913, March). Bethlen Gábor kastélya és kapuja Déván. *Uránia*, 117–119.

Uj lap a vidéken. (1877, April 8). *Vasárnapi Ujság*, 219.

Secondary Sources

A társulat igazgatása. (1886). *Az Országos Régészeti és Embertani Társulat Évkönyve*, 1879–1885, 81–83.

Adatok az 1895–96. év történetéhez. (1896). *A dévai M. Kir. Áll. Főreáliskola XXV–ik évi Értesítője az 1895–96. tanévről*, *24*, 56–62.

Az Országos Régészeti és Embertani Társulat tagjai. 1886 február 24-étől 1889 február 24. napjáig. (1889). *Az Országos Régészeti és Embertani Társulat Évkönyve*, 1886–1888, 67–76.

Bata, T. (2007). Huszka József és a fényképek. *Néprajzi Értesítő*, *LXXXIX*, 127–154.

Bata, T. (2013). Székelykapuk, fatemplomok nyomában. Szinte Gábor (1855–1914) rajztanár fotográfiái. *Néprajzi Értesítő*, *XCV*, 41–63.

Bata, T. (2021). On the trail: Székely gates and wooden churches. The photographic work of teacher Gábor Szinte (1855–1914). *Néprajzi Értesítő*, *CI*, 85–106.

Bata, T., & Tasnádi, Zs. (2015). *A székelykaputól a törülközőig. Szinte Gábor gyűjtései/From the Székely gate to the last towel. The Gábor Szinte collection.* Budapest: Néprajzi Múzeum.

Bodó, C. (2012). Începuturile Societății de Istorie și Arheologie a comitatului Hunedoara. *Sargetia* (Serie Nouă), *III*, 371–385.

Bodó, C. (2017). Archaeological research undertaken by the members of the History and Archaeology Society of Hunedoara County. In Cs. Szabó, V. Rusu-Bolindeț, G. T. Rustoiu, & M. Gligor (Eds.), *Adalbert Cserni and his contemporaries. The pioneers of archaeology in Alba Iulia and beyond* (pp. 235–249). Cluj-Napoca: Editura Mega.

Bodó, C. (2020). Museum of History and Archaeology Society of Hunedoara County: A brief review. *Sargetia* (Serie Nouă), *XI*, 179–192.

Bodó, C. (2021). *Din activitatea Societății de Istorie și Arheologie a comitatului Hunedoara.* Cluj-Napoca: Argonaut.

Burnichioiu, I. (1999). Despre tipologizare în arhitectură: Bisericile gotice din Transilvania cu închideri de altar în ax. In I. Kiss, & P. L. Szőcs (Eds.), *Arhitectura religioasă medievală din Transilvania. Középkori egyházi építészet Erdélyben. Medieval ecclesiastical architecture in Transylvania, I*, (pp. 107–120). Satu Mare: Editura Muzeului Sătmărean.

Burnichioiu, I. (2003). Revenirea la un subiect fără surse: Biserica din Bârsău în secolele XV–XVI. *Mediaevalia Transilvanica, V–VI*(1–2/2001–2002), 97–112.

Codrea, I.-C. (2009). Poarta principală a castelului Magna Curia din Deva. Noi considerații. *Ars Transilvaniae*, *XIX*, 37–40.

Codrea, I.-C. (2010). *Vechea biserică parohială a Devei.* Deva: Cetate Deva.

Coltofean, L. (2017). Zsófia Torma: A pioneer of prehistoric archaeology in nineteenth-century Transylvania. In Cs. Szabó, V. Rusu-Bolindeț, G. T. Rustoiu, & M. Gligor (Eds.), *Adalbert Cserni and his contemporaries. The pioneers of archaeology in Alba Iulia and beyond* (pp. 327–354). Cluj-Napoca: Editura Mega.

Deju, G., & Bodó, C. (2006). Bibliotecarii „Societății de Istorie și Arheologie a comitatului Hunedoara." *Sargetia*, *XXXIV*, 633–643.

Dumitriu, L. (2000). „Sfântul Nicolae" Church from Bârsău, Hunedoara County. *Cercetări arheologice*, *11*, 189–196. https://doi.org/10.46535/ca.11.05

Értesítő a Budapesti VIII. Ker. *Magyar Királyi Állami Főgymnasyumról 1913/1914 tanévben.* (1914). Budapest.

Fodor, A. (1844b). *Méhádia vagy a Hercules-fürdők és utazás Hunyadmegyén keresztül a mehádiai fürdőkre, onnan Drenkovára.* Kolozsvár.

Gáll, E. (2014). Precursorii arheologiei profesioniste din Transilvania. Gábor Téglás. *Studii și Cercetări de Istorie și Arheologie, 65*(3–4), 219–293.

Gerecze, P. (1905). *Magyarország műemlékei* (Vol. 1 – *A Műemlékek Országos Bizottsága rajztárának jegyzéke*). Budapest: A Műemlékek Országos Bizottsága Kiadványai.

Greceanu, E. (1971). Influența gotică în arhitectura bisericilor românești de zid. *Studii și Cercetări de Istoria Artei. Seria Artă Plastică*, *18*(1), 33–59.

Harsányi, P. (1915). Dr. Réthy László emlékezete. *Numizmatikai Közlöny*, *14*(I), 22–28. (1898)

Henszlmann, I. (1883). Évi jelentés. A műemlékek orsz. Bizottságának működéséről 1883 –ban. *Archaeologiai Értesítő* (Új folyam), *3*, IX–XI.

Horváth, A. (1990). Imre Henszlmann, and the origins of monuments preservation in Hungary. *Periodica Polytechnica Architecture*, *34*(3–4), 139–160.

Kun, R. (1884). A társulat 1881. május hó 30-tól 1883. május hó 30-ig kifejtett működésének vázlata. *A Hunyadmegyei Történelmi és Régészeti Társulat Évkönyve, II*(2), 83–110.
Kuun, G., Torma, Zs., & Téglás G. (1902). *Hunyadvármegye története. I. Hunyadvármegye földjének története az őskortól a honfoglalásig*. Budapest: Az Athenaeum Irodalmi és Nyomdai Részvény Társulat Könyvnyomdája.
Lővei, P. (2016). Monument conservation. In J. Sisa (Ed.), *Motherland and progress: Hungarian architecture and design 1800–1900* (pp. 555–573). Birkhäuser: Basel.
Mailand, O. (1891). A társulat életéből. 1891–1892. *A Hunyadmegyei Történelmi és Régészeti Társulat Évkönyve, VI* (1889–1890), 121–139.
Marosi, E., & Cennerné Wilhelmb, G. (1983). *Die ungarische Kunstgeschichte und die Wiener Schule 1846–1930: Ausstellung, Collegium Hungaricum, Wien, September 1983*. Budapest: Statistischer Verlag.
Molnár, M. (2001). *A concise history of Hungary*. Cambridge: Cambridge University Press.
Naghi, Gh. (1979). Memoriile lui Ştefan Moldovan din 1852 şi 1854 cu privire la şcolile româneşti din Haţeg. *Sargetia, XIV*, 421–430.
Rusu, A. A. (1993). Mănăstirea franciscană din Haţeg. *Ars Transilvaniae, 3*, 137–144.
Sarkadi Nagy, E. (2010). Adatok az eltűnt dévai templom történetéhez. In T. Kollár (Ed.), *Építészet a középkori Dél-Magyarországon. Tanulmányok* (pp. 935–966). Budapest: Teleki László Alapítvány.
Sisa, J. (2002). The beginnings of art history and museology in Hungary: Some semper connections. *Centropa: A journal of central European architecture and related arts, 2*(2), 129–135.
Sisa, J. (2020). From the Central Commission to the National Commission of Monuments: Changing institutional framework and principal projects in the Kingdom of Hungary. *Monumentorum Tutela. Ochrana pamiatok, 30*, 191–207.
Szádeczky, L. (1905). Dr. Gróf Kuun Géza emlékezete. *Erdélyi Muzeum, XXII*(8), 401–431.
Szinnyei, J. (1894). *Magyar írók élete és munkái* (Vol. III). Budapest: Hornyánszky Viktor Könyvkereskedése.
Szinnyei, J. (1903). *Magyar írók élete és munkái* (Vol. IX). Budapest: Hornyánszky Viktor Könyvkereskedése.
Szinnyei, J. (1906). *Magyar írók élete és munkái* (Vol. XI). Budapest: Hornyánszky Viktor Könyvkereskedése.
Szinte, G. (1893). Kolczvár. *A Hunyadmegyei Történelmi és Régészeti Társulat Évkönyve, VII*(1891–1893), 69–79.
Szinte, G. (1894). Kerek erőd Krivádia fölött. *Archaeologiai Értesítő* (Új folyam), *XIV*(2), 110–114.
Szinte, G. (1897e). Ulpia Trajana. Ezelőtt Sarmizegethusa, ma Várhely amphitheatruma. *A Hunyadmegyei Történelmi és Régészeti Társulat Évkönyve, VIII*(1893–1896), 35–37.
Szinte, G. (1898). A Poklisa és Totesd közti sírmező. *A Hunyadmegyei Történelmi és Régészeti Társulat Évkönyve, IX*(1896–1898), 94–96.
Tasnádi, Zs. (2006a). Vidéki rajztanárok a műemlékvédelem és a néprajz korai történetében. In Z. Fejős (Ed.), *Huszka József, a rajzoló gyűjtő* (pp. 76–80). Budapest: Néprajzi Múzeum.
Tasnádi, Zs. (2006b). Falképek és templomi berendezések rajzai. In Z. Fejős (Ed.), *Huszka József, a rajzoló gyűjtő* (pp. 34–75). Budapest: Néprajzi Múzeum.
Tasnádi, Zs. (2016). Néprajzi elemek a Képzőművészeti Főiskola rajztanárképzésében. *Néprajzi Értesítő, XCVIII*, 117–122.
Tárca. Nagyváradi dr. Sólyom-Fekete Ferenc. (1906). *A Hunyadmegyei Történelmi és Régészeti Társulat Évkönyve, XVI*(2), 127–128.
Társulati névkönyv. (1907). *A Hunyadmegyei Történelmi és Régészeti Társulat Évkönyve, XVII*(4), 158–168.
Téglás, G. (1898). Emlékezés Kun Róbertre. *A Hunyadmegyei Történelmi és Régészeti Társulat Évkönyve, IX*(1896–1898), 139–147.
Torma, K. (1879). A berekszói régi templom Hunyadmegyében. *Archaeologiai Közlemények, XIII*(1), 53–61.
Veress, E. (1898). Hivatalos értesítő. A társulati ülések krónikája. II. 1897 szept. 30-ikától 1898 okt. 26-ig. *A Hunyadmegyei Történelmi és Régészeti Társulat Évkönyve, IX*(1896–1898), 153–175.
Zichy, J., & Henszlmann, I. (1883). A magyarországi műemlékek országos bizottsága 1883–iki augusztus 16-án tartott rendes üléséének jegyzőkönyve. *Archaeologiai Értesítő* (Új folyam), *3*, VI–VIII.

Archaeological Sociability: Building Patriotic and Scientific Networks in Italy (1861–1915)

8

Fedra Alessandra Pizzato

Abstract

During the long nineteenth century, European archaeologists played an active role in nation building. In Italy, their contributions covered three main areas: (1) they developed theories on the origin, settlement, and identity of the Italian nation; (2) they pursued a thorough musealization of prehistoric finds and monumentalizing of archaeological sites; (3) they undertook a zealous popularization of the archaeological past using the scientific methods available at the time. Because these activities were embedded in a broader environment permeated by a value system that emphasized patriotism and nationalism, they were integrated into a system of social and political practices. More precisely, they reinforced sociability practices influenced by patriotic organizations dedicated to sustaining a national movement and construct a new Italian identity. These organizations were made up of a network of a patriotic bourgeoisie, humanists, natural scientists, young sportsmen, and even women interested in the national values supported by discovering ancient Italy's origins. This chapter deconstructs the role of prehistoric archeology in supporting national narratives within the context of a complex interplay of power relations, political agendas, and scientific communication strategies in late nineteenth-century Italy. The complex network established between the Italian Alpine Club (CAI), one of the most important patriotic associations in Italy (1861–1915), and prehistoric archaeologists is the basis of this historical analysis.

Keywords

Scientific networks · History of archaeology · Alpinism · Nationalism · Sociability practices

8.1 Introduction

In her book on the networks created by Spanish and British archaeologists in the twentieth century, historian of archaeology Margarita Díaz-Andreu underlined how geographical space, understood as both enabling and limiting mobility and communication, is an essential element for accessing both personal relationships and knowledge itself (Díaz-Andreu 2012: 3–8). Institutional networks are articulated by hierarchical scales and complex center-periphery relations that are often crucial for circulating scientific knowledge (see Nieto-Galan 2004; Papanelopoulou et al. 2009). Alongside institutional relations, non-disciplinary and transdisciplinary networks also emerge from the analysis of archaeologists' correspondence. Their epistolary exchanges reveal the existence of personal relationships that are not always purely scientific or academic. This makes it possible to examine the circulation of knowledge outside the disciplinary sphere throughout the history of archaeology, including the history of prehistoric archaeology.

The popularization and circulation of knowledge have aroused increased interest in the field of the history of science in recent decades. Starting with the pioneering work of Cooter and Pumfrey (1994), historians of science have begun to reflect on the extent to which the popularization of science influenced the communication between scientists (Secord 2004; Topham 2009a, b). This debate has revealed the impor-

F. A. Pizzato (✉)
Department of Cultures and Civilization, University of Verona, Verona, Italy
e-mail: fedraalessandra.pizzato@univr.it

L. Coltofean et al. (eds.), *Connecting People and Ideas*, Themes in Contemporary Archaeology,
https://doi.org/10.1007/978-3-031-81006-0_8

tance of transdisciplinarity[1] in the institutionalization of individual fields of science as well as in the internal development of various disciplines that make use of the scientific method. This chapter draws inspiration from this premise and the strand of cultural studies concerned with networks as viewed from the perspective of social theory. The latter approach is based on two studies that, albeit several decades old at this point, are still relevant for research in the history of science and archaeology. The first study, authored by the social philosopher Jürgen Habermas, dates to the1960s. Through the concept of *Öffentlichkeit*, Habermas defined and illustrated the processes characterizing communicating and enabling the construction of public opinion through the circulation of practices and ideas linked to civil life (Habermas 1962/1989). The second study, written by French social historian Maurice Agulhon (1977), analyzed the dynamics inherent in bourgeois sociability. Agulhon describes sociability as an organic system of various forms of interaction that may occur within a social group, such as rituals, folklore, and associationism. According to him, these practices are by no means negligible in the study of the circulation of ideas since the identities of social groups and their members are structured on this basis (Agulhon 1970, 1977). How members of the same group and different groups interact, as well as the power relations that structure society, depend on the forms of sociability. In other words, sociability is one of the ways in which the habitus of a group or social class manifests itself. Agulhon's ideas also include an interesting methodological reflection. He writes that the sources of the explosion of associationism coincide with the existence of a bourgeois model, with the presence of a condition of sociability of imitation and finally with an external contribution of a national kind (Agulhon 1977: 219–20). The bourgeoisie, the imitation of social behavior models and national contexts: these, according to the French historian, are the cornerstones on which the study of social circles and networks should be structured (Agulhon 1977: 220).

This chapter integrates cultural studies on bourgeois associationism, stemming from Agulhon's studies on sociability, with the study of sources related to the history of prehistoric archaeology and recent developments in the historiography of science. The aim is to analyze how informal networks have contributed to the circulation of knowledge and the socio-political legitimation of prehistory as a discipline. The idea of sociability as a vehicle of knowledge and practices enables the understanding of mechanisms that are at the root of two significant theses in the recent history of prehistoric archaeology: (1) the substantial link between the process of nation-building and the development, legitimization, and institutionalization of the discipline (Díaz-Andreu 2007); (2) and the ideological connection between the ideals of the nationalist and positivist bourgeoisie and the support given to the development of prehistoric studies in Italy in the nineteenth century (Guidi 1996).

The case study examined in this chapter concerns a typical bourgeois and national-patriotic sociability practice in Italy between 1861 and 1918: the associationism linked to hiking and mountaineering. This will be discussed using the example of the Club Alpino Italiano (CAI: Italian Alpine Club), which included among its members numerous people engaged in various ways in prehistoric research in the Italian Peninsula. The influence exerted by the CAI in Italy in supporting, both politically and socially, the ideals of the modernizing bourgeoisie (Mizrahi 1975) and the national-patriotic ideology was significant (Pastore 2003; Morosini 2012; Gai 2015; see also Hoibian 2000). The CAI also supported the positivist sciences, including prehistoric archaeology, because of their promotion of national ideals (Morosini 2009: 42–45), as discussed below. By analyzing the correspondence and publications of CAI members and the foremost Italian prehistoric scholars, this chapter reconstructs the practices, the value systems, and the network of transdisciplinary links that made it possible for the CAI to become a sounding board for theories and studies of prehistoric archaeology in Italy.

The chapter will first analyze the origins of the connection between prehistoric archaeology and mountaineering in Italy, highlighting how the encounters between actors of various backgrounds transformed into a structural link between the two fields. It then presents both the patriotic and scientific practices that characterized the prehistoric-mountaineering networks by means of a case study drawn from the practice of patriotic pilgrimage, followed by a discussion of scientific divulgation, amateur research, and local tourism by the middle classes. Finally, the chapter analyses a peripheral site of the Kingdom of Italy, highlighting how a local network of scholars and patriots, once again linked to the world of CAI, promoted archaeological excavations related to prehistoric periods and disseminated their main scientific theories (i.e., Pigorini's theory on Italian prehistory) in the local press.

[1]Although the phrase "interdisciplinary circulation of knowledge" has a long pedigree (see Darbellay 2012; for a discussion on terminology in history of archaeology see Coltofean-Arizancu and Díaz-Andreu 2021: 2–5), in this chapter I refer to the social-pedagogical theory term "transdisciplinarity". As a technical concept, it was purportedly adopted to describe the dynamics between academia (i.e., expert knowledge) and social society (Wickson et al. 2006). Hence transdisciplinarity shifts the emphasis to the circulation of knowledge among *social*, rather than *disciplinary* ambits. Furthermore, from the standpoint of pedagogical theory, transdisciplinarity includes the agency of social actors in the process of knowledge production (Soublis 2017: 65–66).

8.2 The Birth of Mountaineering and Prehistoric Archaeology in Italy

The birth of patriotic mountaineering in Italy can be dated between August 11 and 13, 1863, when the minister Quintino Sella (1827–1884), together with three other companions, led the first Italian ascent of Mount Monviso in the Alps. On his return, Sella wrote a famous letter that is considered to be the founding text of this new 'sporting' practice. The letter was later made public by the pro-government and moderate liberal newspaper *L'Opinione* (Sella 1863a). Sella (1863b: 62) defined mountaineering as a helpful way to improve the "Italian race" in physical and intellectual terms. Interestingly, he also said that thanks to the spread of mountaineering, "l'amore per le scienze naturali" (the love for the natural sciences) would grow. The need to study the natural sciences was in turn justified by the need not to let foreign scholars study "le cose nostre" (our things) (Sella 1863b: 62).

The Italian mountaineering movement emerged, as in other European countries, in the wake of two phenomena. The first of these was the emergence of a positivist bourgeoisie, interested above all in technical and scientific knowledge. The second factor was genuinely political and consisted of the need to bring the reborn Italian nation (1861) onto the same level as the other European nations, both from the point of view of the *power of the race* and of scientific culture and achievements (Pastore 2003; Morosini 2009: 30–52). To this end, the CAI aimed to become a movement with a solid vocation to develop a truly national pedagogy: "Gli italiani, educati alla scuola del CLUB ALPINO diventeranno forti e l'Italia diverrà un popolo di forti" (Italians, educated at the school of the ALPINE CLUB will become strong, and Italy will become a country of strong people). This sentiment was reiterated by one of its most famous members, the priest, geologist, and paleontologist Antonio Stoppani (1824–1891), in the volume *Il Bel Paese*, first published in 1876 (Stoppani 1908: 31). For many decades, this volume was one of the key texts used in Italian primary schools, testifying to the importance of the CAI as a model of education for Italians (Zanoni 2014; Gai 2015).

The importance of Sella's letter for the history of Italian archaeology can be easily demonstrated. The addressee of the letter was, in fact, none other than Bartolomeo Gastaldi (1818–1879), considered the first pioneer of prehistoric archaeology in Italy. Gastaldi had met Sella while they both attended the École des Mines in Paris and the courses of the famous French geologist Jean-Baptiste-Armand-Louis-Leoncé Elie de Beaumont (1798–1874), as well as the Jardin des Plantes, the Collège de France and the Sorbonne between 1846 and 1849 (Balan 2014). These institutions attracted large numbers of young people from various European countries interested in studying the natural sciences at the time. While Sella was interested in crystallography, Gastaldi broadened his studies to include geology, glaciology, and paleontology. The latter also met the great names in the discipline during his stay in Paris, including the physician, botanist, and geologist Charles Frédéric Martins (1806–1889) and archaeologist Gabriel de Mortillet (1821–1898), with whom he began to regularly collaborate.

The friendship between Sella and Gastaldi continued after their return to Turin in the early 1850s. Sella called his friend to the Istituto Tecnico (Technical Institute, then Polytechnic), a key institution in the formation of the patriotic middle class, catering to those with an industrial, mining, and engineering vocation. In 1861 Gastaldi became assistant to the chair of mineralogy and geology previously held by Sella, and, when Sella joined politics, he became a full professor (see Corsi 2007b). Gastaldi donated his entire prehistoric collection to the Institute, which served as a training tool for the first generations of geologists in the Kingdom of Italy. He also took part in the scientific and strategic enterprise of building a geological map of the Kingdom based on similar projects initiated by other European nations (Corsi 2007a).

After the discovery of lake-dwelling settlements in Switzerland (1854), Gastaldi became increasingly interested in prehistoric archaeology. In the early 1860s, he led several research trips to Arona, Imola, and the regions of Emilia and Lombardy in Italy. Although for Gastaldi prehistoric research was primarily aimed at tackling one of the major questions of the geological and theological debates of his time, i.e., the co-occurrence of humans and extinct Pleistocene megafauna, his trip to Emilia and the publication of his first work, *Nuovi cenni sugli oggetti di alta antichità trovati nelle torbiere e nelle marniere d'Italia* (1862), played a significant role in the birth of prehistoric archaeology as a science in Italy (Tarantini 2012: 23–33). During his trip, Gastaldi convinced Luigi Pigorini (1842–1925), a brilliant young numismatic enthusiast who would later become the mastermind of prehistoric research in Italy (see also Gustavsson, Chap. 2, this volume), of the importance of prehistoric research. Moreover, Gastaldi acted as a mediator of the meeting, decisive for developing prehistoric research in Italy, between Pigorini and the naturalist Pellegrino Strobel (1821–1895). Strobel and Pigorini contributed to a text (Pigorini and Strobel 1862) that appeared as an appendix to Gastaldi's above mentioned book, which marked the beginning of prehistoric archaeological studies in Italy. Gastaldi's support was also instrumental in Pigorini's first scientific trips in Italy and Switzerland (Fugazzola Delpino and Pellegrini 1994: 95).

Gastaldi maintained correspondence with the three founders of Italian prehistoric research, Pigorini, Gaetano Chierici (1819–1886), and Strobel, throughout his life. His correspondence with Pigorini, for example, although not extensive, is linked to both Gastaldi's activity as a geologist and as a museologist. In 1875, Gastaldi became director of the Museo Civico (Civic Museum) in Turin, the first capital of the Kingdom of Italy. The correspondence with Pigorini provides information mainly on the exchange of materials between the Museo Civico and the Regio Museo Nazionale Preistorico ed. Etnografico di Roma (Royal National Museum of Prehistory and Ethnography of Rome), which Pigorini directed, on the occasion of exhibitions, conferences and the setting up of the museum. In addition, it demonstrates how the young Pigorini was directly involved in the debate regarding the applicability of scientific theories to the origins and development of human races and on the flaws of the biblical explanations for the antiquity of man; these were two *foci* around which the scientific discussions of the time revolved on an international level. Gastaldi also shared with Pigorini, Strobel, and Chierici his adhesion to the ideals of the patriotic bourgeoisie and his support for the CAI (see Gastaldi 1864).

8.3 Patriotic Pilgrimages: From the Primacy to the CAI of Emilia

The previous section has shown that one of the founders of mountaineering in Italy was also involved in a group of individuals who contributed significantly to early prehistoric research in Italy. However, the link between mountaineering and prehistoric research is not limited to a fortunate prosopographic coincidence but is firmly grounded in ideology. At the roots of Italian nationalism lies a book written by the philosopher and patriot Vincenzo Gioberti (1801–1852), entitled *Del Primato morale e civile degli italiani* (Gioberti 1843). Viewed as the major theoretical outcome of liberal and moderate Italian thought in the early nineteenth century, the book explicitly addresses the relationship between the ancient and modern worlds, framing the practice of patriotic pilgrimage within the overall discussion of Italian identity and the national pedagogy. According to Gioberti (1843: 524–26), the ideas of modern young patriots devoted to the cause of Italian unity should be guided by two examples: the martyrs of the fatherland during the *Risorgimento* (Resurgence),[2] and the ancients.

In Gioberti's view, the physical return, in the form of a journey or patriotic pilgrimage, to places of ancient memory, i.e., the archaeological sites scattered throughout the peninsula, was an essential educational and political practice in the formation of the Italian patriot (Gioberti 1843: 524–26). It also established a ritual custom of transforming significant places from antiquity into veritable national monuments to directly connect the nation's past and present, represented as a continuous progression. The same was happening in other European nations, including France under Napoleon III, where pilgrimages to archaeological sites became an expedient means of consolidating national ethnic identity. In other words, the use of "rituals of commemoration to transform archaeological sites into monuments invoking historical continuity between past and present" (Dietler 1998: 74) spread across Europe during this time.

The practice of hiking, typical of mountaineering and sports clubs in the nineteenth century, lent itself well to being integrated into a system of actions promoting continuity between past and present when framed as a patriotic pilgrimage to ancient sites. In addition, the scientific purpose was expressly one of the foundational pillars of the CAI (Morosini 2009: 42–43). Consequently, at the same time that prehistoric archaeology was being granted the status of a science, the possibility of integrating excavations and patriotic trekking became an incentive for CAI members to participate in the great enterprise of discovering their national origins—an enterprise that had an invaluable political meaning.

It is therefore unsurprising to find Abbot Gaetano Chierici, a patriot and liberal priest as well as the co-founder of prehistoric studies in Emilia-Romagna (see Macellari 1994), actively supporting CAI initiatives. A naturalist by training, Chierici soon became major promoter of the CAI in Emilia. In 1875, in fact, when the Enza Section of the Italian Alpine Club was established, Chierici was among the founding members and later also became its president. Under his guidance, the mountaineers from Reggio Emilia, most of whom were middle-class members of the local political and cultural élite (e.g., doctors, lawyers, town councilors), often took on the role of archaeologists (Chierici 2007).

The most famous project carried out by the mountaineers of the Enza Section under Chierici's lead took place in 1877, when they were involved in the first excavations of the remains of the medieval castle of Canossa. Their activities also laid the foundations for today's Museo Narborre Campanile (Naborre Companini National Museum, named after the man who continued the excavations in 1893), now located in the historic headquarters of the Reggio Emilia CAI section. The site was acquired by the State in 1878, a year after the excavation, and was immediately declared a national monument. The archaeological site complex of Canossa lies on a clifftop and includes the remnants of a medieval castle on the summit and evidence of settlement ranging from prehistoric times to the Iron Age (tenth-ninth to first centuries BC). By the end of the nineteenth century, it had become a regular destination for mountaineers (Cervi 2020). This demonstrates the interest of Reggio Emilia CAI members in the country's antiquities and their active contri-

[2] *Risorgimento* refers to the set of protracted ideological, cultural, political, military, and socio-economic developments that led to the emergence of the united and independent Italian state (1815–1871).

bution to scientific practices and fieldwork ("Relazione," 1877; see also Desittere 1988; Cervi 2020).

Despite the presence of settlement traces since prehistoric times, the most important antiquities at the site of Canossa were those related to the Roman period and the Middle Ages. The mountaineers' interest in the country's history included the most distant past in which the Italian nation was supposedly rooted. Another example of the CAI's temporal focus is the excursion held on October 5, 1875, to Bismantova, near Castelnovo ne' Monti (Reggio Emilia). The trip, also led by Chierici, was then recounted for the general public by the Emilian paleontologist in a memoir published in the book *La montagna tra il Secchia e l'Enza. Memorie e studi di alpinisti reggiani* (Spagni 1876). Chierici's narration dwells on the epic aspects of the adventure, highlighting the horrid-sublime in perfect continuity with Sella's description in the letter to Gastaldi quoted above. In his narration, Chierici underlines how fear of the danger of the place was fed by continuous landslides,

> inutile avviso del pericolo d'erosioni celate, onde già staccaronsi gli enormi massi (…) coi dossi e le punte biancheggianti fra il verde delle querce e dei castagni, fanno spaventosa corona al piccolo santuario (…) ed. ogni sassolino che venga saltellando per quelle balze fa tremar di paura.
> (useless to warn of the danger of hidden erosion, from which the enormous boulders [...] with their bumps and white points among the green of the oaks and chestnut trees, make a frightening crown to the small sanctuary [...] and every pebble that comes jumping through those crags makes one tremble with fear.) (2007: 211)

The dreadful and the sublime represent opposing polarities of the perception of the Alpine landscape. Before the Enlightenment, the lack of knowledge about the Alps elicited horror and fear. By contrast, during the Romantic period and in connection with the process of Italian nation building, the Alps became part of the national cultural landscape. Alpinism played a key role in this process (Garimoldi 2002). The alpinist ideology combined the love for the Italian landscape with the view of the Alps as a way to develop strength and values. The potential danger made this trip a typically virile enterprise for the patriotic-mountaineering ethic (see Gai 2015). Chierici's report is also characterized by strongly spiritualistic accents; the itinerary followed by the mountaineers was therefore charged with a religious quality. Thus, by combining sporting practice with a sense of the sacred, the journey served as training for the mountaineers on more than one level.

What made the enterprise truly remarkable was the possibility of taking part in a scientific mission. Chierici's text recounts how, on hearing the news that new finds had emerged from the partially explored protohistoric site of Pietra di Bismantova, the mountaineers of the Enza Section not only wanted to ascend this mountain immediately, but "vollero anche sostener in comune la spesa dello scavo" (also wanted to share the cost of the excavation) (Chierici 2007: 211). In this way, they not only justified the practice of combining sport with bourgeois values related to concepts of utility and progress, but also served as patrons of science by contributing to the discovery of national roots. In this regard, it is worth noting that the Pietra di Bismantova was also linked to the national epic of the Divine Comedy, where it is mentioned by Dante in the fourth *canto* of Purgatory, which is also cited in Chierici's report (Chierici 2007: 212). In this context, the Enza Sections' subsidy of the excavations of the pre-Roman necropolis of Bismantova acquires an even stronger patriotic meaning, comparable to remembering ancient ancestors. The site of Pietra di Bismantova was particularly suitable for such patriotic actions and feelings:

> ivi è il sepolcreto, e la solitudine del luogo, in prospetto all'opera rovinosa del tempo che avvalla le montagne, fa mesto invito a meditare le genti che popolarono que' dirupj, e sol vi lasciarono a ricordarle entro vasi di creta poche ossa bruciate.
> (there is the burial ground, and the solitude of the place, in full view of the ruinous work of time that hollows out the mountains, a sad invitation to meditate extended by the people who populated those ravines, with nothing remaining but a few burnt bones in clay pots to remember them by.) (Chierici 2007: 213)

It was only with the discovery of a burial, a tomb of the nation's ancestors, that the patriotic mission could be declared a success. The attribution of the burial to the Ligurian people meant that the CAI mountaineering group of the Enza Section "può rallegrarsi dell'ajuto che le prime esplorazioni de' suoi soci recano alla scienza" (can rejoice in the contribution that the first explorations of its members have made to science) (Chierici 2007: 216). The bourgeois ethics of the archaeologist-mountaineers, centered on the idea of the "passatempo utile e modernizzatore" (useful and modernizing pastimes) of archaeology and mountaineering (Chierici 2007: 216) combined with nationalist intentions, therefore appeared to be fully satisfied by the mission on the Bismantova mountain.

Numerous other mountaineering enterprises of the Enza Section followed throughout Italy, and members of the city's elite continued to participate in them. By means of these activities, they were also involved in disseminating knowledge about the country's prehistory and promoting the usefulness of the new science. In this way, thanks to their "imprese preistoriche" (prehistoric enterprises), CAI members and archaeologists pioneered what was to become a well-known process in the structuring of national identities. This process consisted of "managing inherited images, creating new tables of value, transmitting the message amid the urban middle class, shaping public opinion in favor of Italian unity" (Gai 2015: 353; for more details, see Ascoli and Von Henneberg 2001; Banti and Ginsborg 2007a, b: XXIII–XLI; Banti 2020). The following section will discuss how these mechanisms for building national identities functioned also in less important centers of Italian prehistoric research, making scientific knowledge accessible to the Italian peripheries.

8.4 Paolo Lioy: "Minor Pompeii Buried Under the Mud of Lakes.."

Gaetano Chierici is not the only example of an archaeologist-mountaineer in the history of Italian archaeology. Often the most prominent members of the local sections of the CAI, although not originally prehistory scholars, became promoters of archaeological enterprises, if not archaeologists themselves. This was quite common in northern Italy, both at the provincial and the local level. This section will present a case study of the province of Vicenza in the Veneto region at the foot of a geological formation of volcanic origin known as the Berici Hills. After an initial focus on the provincial capital of Veneto and Paolo Lioy, the most enthusiastic proponent of the Berici CAI at that time, this section will discuss a more marginal area of the Vicenza province on the border with Austrian Trentino.

Paolo Lioy (1834–1911) was born into a noble family in Vicenza, in the Austrian-dominated Veneto. From adolescence onward, he was imbued with patriotism. He had as teachers some of the most well-known promoters of the anti-Austrian uprisings of 1848, one of the main features of the Italian *Risorgimento*. He began a legal career, but at the University of Padua he was almost exclusively involved with natural sciences. He had varied interests: in addition to his work as a naturalist, which led him to reorganize the collections in the civic museum of his hometown based on more modern methods, he was also interested in literature and popular traditions (Zavalloni 2005; see also Morosini 2012; Sari 2016: 11–34).

In the ensuing years, Lioy held important scientific and political positions both locally and nationally. He was secretary of the Accademia Olimpica (Olympic Academy) in Vicenza, the province's main cultural institution. During his term of office, he was responsible for the education of the city's workers and carried out a large scientific popularization campaign. Following the example of other scientists of the time, the most famous of whom in Italy was Abbot Stoppani (see Zanoni 2014; Gai 2015), Lioy also devoted himself to writing popular scientific works with the aim of developing a national pedagogy advocating modernization and patriotism. His commitment to national education also led him to a successful political career: he was a municipal and provincial councilor and then a parliamentary deputy for six terms, until 1888. In 1905 he was appointed senator of the Kingdom of Italy (Zavalloni 2005).

Lioy also had a remarkable cursus honorum in the field of Italian alpinism. Besides being one of the founders of the CAI section in Vicenza (Engel 1965: 188), which he eventually chaired, Lioy also became CAI national president, a position he held for five years, from 1885 to 1890. This position gave him both scientific and political prestige because it had previously been held by one of the most important Italian naturalists and politicians of the time, Quintino Sella.

Lioy's research in the field of prehistory began in 1864, when he undertook a series of archaeological surveys in the Berici Hills and at the nearby Fimon Lake (Sari 2016: 49). As he himself reports in one of his works, Lioy was inclined towards archaeology by the

> (...) scoperte di Ferdinando Keller a Zurigo, di Pigorini e di Strobel nelle terremare dell'Emilia, di Desor e di Stoppani nei laghi lombardi che m' invogliarono a tentare gli scavi a Fimon sulle rive ora asciutte. Gli scavi ebbero fortuna; ne venne alla luce appunto una delle umili Pompei.
> ([...] discoveries of Ferdinand Keller in Zurich, of Pigorini and Strobel in the terramare[3] of Emilia, of Desor and Stoppani in the lakes of Lombardy, which encouraged me to attempt excavations in Fimon on the now dry shores. The excavations were successful; a minor Pompeii came to light.) (Lioy 1901: 323)

Lioy learned of the Swiss pile-dwelling discoveries through his interactions with the scientific societies with which he was in constant contact. In particular, the Società Italiana di Scienze Naturali (Italian Society of Natural Sciences) played a decisive role in the dissemination of information about the pile-dwelling civilization and the subsequent discoveries of comparable sites along the shores of the Italian alpine lakes. The Society had been the driving force behind Italian research on this cultural phenomenon as early as 1863 when, at a meeting held in Milan on April 26, Édouard Desor (1811–1882) and Gabriel de Mortillet had announced the discovery of lake dwelling sites beyond the Alps.[4] Lioy's reaction to the news arriving from Milan was swift: in 1864 he began similar investigations in the Vicenza area. In 1863–1864 Vicenza was still Austrian and Lioy's contacts with patriotic societies such as the Italian Society of Natural Sciences were fueled as much by scientific news as by anti-Austrian nationalist motives. The search for the Italian roots of the Venetian lands thus served the dual purpose of contributing to a scientific resurgence and satisfying political demand. The zeal with which Lioy set about rediscovering the prehistoric civilizations of his country is therefore unsurprising.

Lioy first contacted various local scholars, beginning with the village doctor in Arcugnano near Lake Fimon, Dr. Lodovico De Favero, requesting information on possible

[3]The word "Terramare" (also known as *Terremarne, Terremare, Marniere*) defines an ancient culture mainly of the Po Valley (Emilia-Romagna, Italy) and dating to the Middle and Late Bronze Age (1700–1150 BC). The sites were excavated exhaustively between 1860s and 1910s and, in the same period, people from Terramare were identified as Italian ancestors.

[4]See "Seduta," 1863: 117; see also Zocchi 2011; Tarantini 2012: 28.

finds or indications of the presence of piles or pots emerging from the lake. In a letter dated August 16, 1864, De Favero reported that local farmers had indeed confirmed the presence of poles and strange findings during the construction of an irrigation canal (De Favero 1864). Lioy rushed to the lake to verify De Favero's account and gave a report on the discovery of a lake dwelling "station" at the meeting of the Italian Natural Science Society held in Biella on September 4. Ten days later, the news had already made the rounds in Veneto and crossed the regional borders: the antiquarian and numismatist Carlo Kunz (1815–1888), originally from Trieste but Venetian by adoption (Schingo 2004), wrote to Luigi Pigorini about Lioy's discoveries in Fimon and put forward the hypothesis that an ancient pile-dwelling civilization had existed in northern Italy (Kunz 1864). From that moment on, Lioy also established contact with the group of scholars from Emilia and, during the first years of excavation, he kept up a brief correspondence with Pigorini, which was followed by an exchange of materials between the two scholars.[5] Pigorini's relationship with Lioy proved exceptionally fruitful for the future dominus of Italian archaeology. It was Lioy who put Pigorini in contact with Pierpaolo Martinati (1812–1878), a renowned scientist in nineteenth-century Italy, who would become Pigorini's mentor and father-in-law, as evidenced by Lioy's letters preserved in the Pigorini Fund at the University of Padua.

> Lioy's work was well known internationally, to the extent that Isacco Tyler considerò la stazione di Fimon come una delle più istruttive d'Italia, e le relazioni che ne ho pubblicate e poi riassunte (…) furono riprodotte, insieme con le tavole, nelle opere magistrali di Keller a Zurigo, di John Edwards Lee a Londra, di Roberto Munro a Edimburgo, di Oscar Montelius a Stoccolma.
> (Isaac Tyler considered the Fimon station to be one of the most instructive in Italy, and the reports I published and then summarized [...] were reproduced together with the plates, in the masterly works of Keller in Zürich, John Edwards Lee in London, Robert Munro in Edinburgh, and Oscar Montelius in Stockholm.) (Lioy 1901: 326).[6]

Above all, however, Lioy's work as a popularizer, often supported by the CAI itself, led on the one hand to an interesting monograph (Lioy 1876), and on the other to a series of public conferences held at the Accademia Olimpica that kindled the enthusiasm of the Veneto elites (see Sari 2016: 34–56).

In addition to Lioy's Vicenza conferences, a sort of patriotic tourism developed in Fimon as a result of his public outreach efforts. In one of his popular works, Lioy presented the pilgrimage to Fimon as a tourist excursion comparable to those of the Grand Tour (see Díaz-Andreu 2019: 23–27):

> Vogliamo affacciarci un momento dinanzi ad una delle umili Pompei sepolte non sotto a lave, ma sotto a melme di laghi? I l viaggio è breve: basta recarsi nei dintorni del lago di Fimon vicentino. Chi sapeva che esistesse questa piccola conca di acqua, nascosta fra letti di torba, in solitudini agresti? Sino a pochi anni or sono era nota soltanto ai cacciatori, che appostati in capannucci di fascine aspettavano stormi di anitre, folaghe e frullini, fra i canneti (…).
> (Shall we look for a moment at this humble Pompeii buried not under lava, but under the mud of lakes? The journey is short: just go to the area around Lake Fimon in Vicenza. Who knew that this small basin of water existed, hidden among beds of peat, in rural solitude? Until a few years ago, it was only known to hunters, who waited for flocks of ducks, coots, and black-winged godwits amidst the reeds.) (Lioy 1901: 322)

The "tourists" did not fail to arrive, also thanks to the social network of the CAI, which acted as a promoter and organizer for the trips that were often conducted by Lioy himself, as had happened in Bismantova with Chierici. This was documented, for example, by Jacopo Mattielli (1819–1899), a learned Paduan, author of medical treatises and an acquaintance of Lioy's, who wrote:

> Lunedì 28 corre(nte), con la prima corsa che parte da Padova, capitanando un eletto drappello di miei amici, mi reco alla stazione di Vicenza, per salire immediatamente i Berici colli, e visitare il lago Fimon divenuto celebre per la dotta di, Lei penna, e per quelle escavazioni che vi si ammirano! Ci sarebbe possibile d'averla per quel giorno nella nostra compagnia, duce e maestro dell'importante escursione?
> (On Monday, June 28, with the first train from Padua, leading a select group of my friends, I went to the station in Vicenza, to climb immediately up the Berici Hills, and visit Lake Fimon, which your pen and the excavations that can be admired there have made famous! Would it be possible for us to have you [Lioy] in our company for that day, as leader and master of this important excursion?) (Mattielli 1865)

Once again, hiking, tourism, and patriotism were promoted in the scientific, popular, and hiking work of one of the members of the CAI, favoring the circulation of knowledge and the creation of social and political practices typical of bourgeois sociability.

8.5 Capillary Networks: On the Border of the Kingdom

As previously discussed, the CAI network not only covered the Italian provinces, but also the territory of small municipalities. Even at the local level, at least in the smaller cities, the bourgeois interest in prehistory and protohistory emerged. One example of this phenomenon can be observed at the end of the century in the municipality of Bassano Veneto (now Bassano del Grappa) where ceramic materials came to light in the locality of Angarano on land owned by a noble family

[5] The Fondo Pigorini at the University of Padua has six letters by Lioy; four letters are kept by the Biblioteca Bertoliana (Bertoliana Library) and the Biblioteca del Museo Civico di Vicenza (Civic Museum of Vicenza Library). See also Dal Lago and Bernardelli 2015.

[6] For more details about the personal contacts between Montelius and Lioy, see e.g., Gustavsson, Chap. 2, this volume.

but cultivated by sharecroppers. Once the owners and sharecroppers had overcome their opposition to the excavation, an entire pre-Roman necropolis of cremated burials was revealed (Roberti 1896). The local superintendent of antiquities who took charge of the excavations was a local nobleman and scholar, committed to the Italian national cause, Count Tiberio Roberti. However, the Roberti family, although of noble origin, was also an exponent of that patriotic and positivist bourgeoisie committed to the modernization of the economic and social fabric of the town (Pizzato 2015: 14–18).

Three people connected in different ways to the bourgeois associations and particularly to the mountaineering societies took an interest in the Angarano discovery and were instrumental in spreading the news about this discovery. On a national scale, the news was made known by the archaeologist Paolo Orsi (1859–1935), a member of the Accademia degli Agiati in Rovereto (Academy of Agiati), his hometown, and very active within the Società degli Alpinisti Tridentini (SAT: Society of Tridentine Alpinists) (Maurina and Sorge 2010). The SAT was the Trentino branch of CAI, known by a different name because of the dictates of the German Confederation which, together with the Austro-Hungarian Empire, directly controlled the provinces of Trento and Bolzano. The Confederation was less willing than the Austro-Hungarian Empire to allow its citizens to join Italian patriotic associations. Thus, the local mountaineering section took the name SAT although it maintained the characteristics of other mountaineering societies in the CAI. Above all, it was devoted to supporting the Italian nationalist cause (Morosini 2009: 77). It was probably through the networks of the bourgeois patriotic associations that Orsi came into contact with Count Roberti, who invited Orsi to publish the results of the excavation campaign at Angarano in the prestigious Italian scientific journal *Notizie dagli scavi* (Orsi 1894).

Another influential member of the SAT was Ottone Brentari (1852–1921), a historian and scholar of Trentino origin, who later became a literature professor at Bassano and authored the monumental and well-known *Storia di Bassano* (1884), which continued to be reprinted until the end of the twentieth century. It is not unlikely that Brentari himself acted as an intermediary between Orsi and the Roberti family. Brentari also wrote some articles in the local newspaper *La Provincia di Vicenza* (Brentari 1893a, b) about the discoveries in Bassano, which came out a year before being reported in the *Notizie dagli Scavi* (Orsi 1894). In these articles, Brentari dealt with the paleoethnological aspect of the discovery, emphasizing the pre-Roman antiquity of the necropolis. He had in fact previously written to Pigorini for advice on the findings at Angarano. Upon receiving a reply, he intended to draw attention to local antiquities with articles in the provincial newspaper celebrating the origins of the population in the Bassano area. He attributed the archaeological remains discovered at Angarano to the pile-dwelling civilization that Pigorini had identified as the point of origin of the Italian nation and had been rediscovered at Lake Fimon by Lioy. Brentari therefore provided a strongly nationalist and patriotic description of the discovery in his media reports (Brentari 1893a, b).

At a municipal level, the discovery at Angarano was promoted by the son of the Count, Superintendent Giuseppe Roberti. He published an article in the city periodical *Il Brenta* on the foundation of the Angarano settlement, considered the oldest part of the city, where the pre-Roman necropolis had come to light. He reiterated the hypothesis already put forward by Pigorini and Brentari that "sugli ameni poggi di San Giorgio [località in Angarano] avea posto dimora una gente uscita dalle palafitte e da stazioni analoghe alle terramare" (on the pleasant hillocks of San Giorgio [locality in Angarano] people had settled from pile dwellings and stations similar to those of the terramare) (Roberti 1895: 10). A passionate chronicler of the town's origins, Roberti produced numerous accounts of the natural beauties of his "piccola patria" (little homeland), to which he dedicated an article published in the pages of the *Bullettino Annuale del Club Alpino Bassanese* in which he extolled the beauties of Angarano (Roberti 1896).

Through contacts between members of mountaineering associations, Pigorini's theory about the discovery at Angarano eventually made its way to Bassano, a small Italian town on the border between Austrian-German Trentino and the Kingdom of Italy. This suggests that the discovery of the necropolis could potentially have played a strategic role in the nationalist discourse of the time, providing material proof of the ideological link between the Trentinean people and the terramare-dwelling ancestors of the Romans, and by extension, the Italians of the nineteenth century. In this sense, the fact that the discovery was made and promoted by two Trentinean scholars, Orsi and Brentari, and the local nobleman Roberti, all three linked by the practice of mountaineering sociability, shows the importance of the mountaineering network in spreading a shared sense of the origins of the Italian nation. It also reveals the extent to which mountaineering patriotism played a role on the one hand as a catalyst of interest and on the other as a disseminator of interpretations of Italian prehistory (see e.g., Guidi 1996).

The final chapter of the Bassano case is once again linked to a patriotic archaeologist who was a member of the CAI. The career of Giuseppe Gerola (1877–1938), also a Trentinean, took off when he was appointed director of the Museo Civico di Bassano (Civic Museum of Bassano) (Varanini 2000). In 1910, almost 20 years after the discovery of the Angarano necropolis, Gerola published a paper in which he recounted how the Bassano area "avesse palpitato già nella dimenticata preistoria di un remoto passato (…) nell'ansia della suscitata civiltà" (had already pulsed in the forgotten prehistory of a remote past [...]) when "schiere compatte di terramaricoli avevano abbandonato le lacustri loro palafitte" (groups of people from the terramare had abandoned their lake dwellings) to reach Angarano (Gerola 1910: 14–15). Once again, the topics of national origins, the piety of the ancients, the familial link between the terramare culture and modern Italians return in a tone similar to that previously encountered in Vincenzo Gioberti's *Primato* and in the writings of Gaetano Chierici.

8.6 Conclusions

Habermas (1962/1989) and Agulhon (1977) both stressed the role of sociability practices in structuring personal and collective identities. Building on this idea, this chapter has discussed the link between two different social groups, mountaineers and archaeologists, in Italy. It did so from the point of view of the researchers who actively participated in the mountaineering associations and of the mountaineers' contribution to the recovery and circulation of prehistoric knowledge. The connection between mountaineering and prehistoric research emerged organically and predictably for at least two reasons. Firstly, the practice of "pellegrinaggio archeologico" (archaeological pilgrimage) meshed well with ideas about exploring uncharted territory in the larger context of a nationalist mindset. The latter was the founding ideal of the CAI as articulated in the 'bible' of Italian patriotism, i.e., Gioberti's *Primato* (1845). Secondly, scientific practice, particularly geology, had served as a structural element of the ideology of CAI members since its inception, as laid out in Stoppani's *Il Bel Paese*. Because prehistoric archaeology was originally linked to geology and the natural sciences, the combination of scientific-archaeological practice and mountaineering spread rapidly and aroused extraordinary interest among the members of the national-patriotic and modernizing bourgeoisie.

The chapter has also revealed some of the consequences of the relationship between CAI and scholars of prehistory. These mainly concern the circulation of knowledge through the consolidation of broad, transdisciplinary networks based on bourgeois sociability (Agulhon 1977). Firstly, the broad participation and interest of CAI members in excavations and discussions of archaeological discoveries facilitated the dissemination of scientific theories on the origins of the nation as well as scientific practices outside the narrow circle of specialists. This happened at a very early stage of archaeology's development as a discipline compared to its institutionalization, which started in the 1870s. Good examples in this regard are the case of Chierici, who served as a CAI guide, and Lioy. Secondly, the overlap between nationalism, bourgeois science, and the ideals of modernization enabled the development of innovative social practices involving prehistoric research that were distinct from those of classical archaeology. In this way, scientific practices such as archaeological excavation, museum visits, excursions to prehistoric sites, and the dissemination of scientific discoveries were integrated into the political framework of nationalism. Thirdly, the sociability practices inherent in prehistoric archaeological research and the resulting discoveries led to an early and quick legitimization of the discipline among urban elites, widening its audience and defining it as a national and nationalist science par excellence. This was a process that, within a few decades, also involved the more peripheral centers in the north of the country, including the small town of Bassano del Grappa. In other words, the formation of an ideological horizon, which included the discourse on national origins that the CAI helped to spread among Italian elites, laid the foundations for a social legitimation of prehistoric studies. As highlighted by Coltofean, Arnold, and Bartosiewicz (Chap. 1, this volume), this process of heightened exposure for the discipline frequently entails bolstering individual prestige, garnering public endorsement for archaeology as a scholarly field, securing funding for excavation endeavors, and earning acknowledgement within archaeological communities.

As discussed in this chapter, personal connections were at the very heart of mountaineering organizations as well as prehistoric research. If one follows a prosopographic approach to the study of the major figures of CAI and prehistoric research, one discovers a significant number of actors who operated transversally between both groups from the 1860s to the First World War. These nodes were scholars, scientific communicators, and members of the local elite such as Pigorini, Strobel, Chierici, Lioy, Roberti, Brentari, Orsi and Gerola. This overlap took place nationally as well as provincially and locally, leading to a rapid and effective transfer of knowledge by means of a networking system that involved people, objects, and ideas connecting the peripheral sphere to the center. In this sense, the CAI acted as a "non-human hub," meaning a non-human element of the network that possesses an exceptionally high degree of connections, as described by Coltofean, Arnold, and Bartosiewicz (Chap. 1, this volume). This network functioned based on the ideological factors shared by the two groups, factors that characterized not only Italian society, but the entirety of European societies during this nation building process.

Scientific endeavor, patriotism, the search for origins, sport, and recreational activities—these all contributed to guiding the interests, social practices, and forms of sociability typical of Italian mountaineering societies at the end of the nineteenth century. Much more than classical archaeology, which also boasted a long tradition, prehistoric archaeology adopted a positivist guise consisting of a Darwinian outlook, positivist philosophy and a close connection with geology and other natural sciences, all of which made it particularly suited to the bourgeois ethos. The bourgeois aesthetic sentiment was reflected in the Romantic sentiment of the sublime embodied in Alpine nature, as the Chierici example demonstrates. Moreover, prehistoric archaeology, born as a positivist science, appeared ideally suited to respond to the demands of modernization that the Italian bourgeoisie was advocating. Modernity and scientific status made these disciplines excellently aligned with the ethics and goals of the bourgeoisie. Furthermore, bourgeois society aspired to develop a national education system that would model how to behave like Italians and the search for the origins of the nation was in sync with this goal. For all these reasons, the CAI network acted as a sounding board and a privileged channel for the diffusion of scientific theories at various levels.

Acknowledgements The research for this chapter started many years ago, in 2017, triggered by the discussions held in two work-in-progress seminars at the Centre d'Història de la Ciència (Universitat Autònoma de Barcelona, Spain) and at the Scuola di Dottorato in Scienze Storiche, Geografiche e dell'Antichità at Ca' Foscari University (Venice, Italy). In particular, I am indebted to the following scholars for the support they provided during the production of this chapter: Massimiliano Badino, Francesco Cassata, Luca Ciancio, Margarita Díaz-Andreu, Marco Fincardi, Oliver Hochadel, Agustí Nieto-Galan, Simon Levi Sullam, and Gian Maria Varanini. I would like to also thank Luigina Mortari, Roberta Silva, and the Teaching and Learning Center of the University of Verona for their constant support of my academic work. Finally, I thank Michele Cupitò and the staff of the Fondo Pigorini at the University of Padua for supporting my archival research, as well as Bettina Arnold, László Bartosiewicz, Laura Coltofean, and the two anonymous reviewers for the feedback on my chapter.

References

Primary Sources

Archival Material

De Favero, L. (1864). [Letter to Paolo Lioy]. Fondo Paolo Lioy (not yet catalogued), Biblioteca Musei Civici di Vicenza, Vicenza, Italy.

Gastaldi, B. (1864, March 23). [Letter to Luigi Pigorini]. Fondo Pigorini (Busta 5, E-F-G; Fasc 4; Auth 6; Letter 1), Padua, Italy.

Kunz, C. (1864, September 15). [Letter to Luigi Pigorini]. Fondo Pigorini (Busta 6, H; Fasc 3; Auth 7; Letter 1), Padua, Italy.

Mattielli, J. (1865, August 24) [Letter to Paolo Lioy]. Fondo Paolo Lioy (not yet catalogued), Biblioteca Musei Civici di Vicenza, Vicenza, Italy.

Articles, Books and Chapters

Relazione delle gite fatte a Canossa dai soci del Club Alpino sezione dell'Enza nei giorni 31 maggio e 10 giugno 1877 e proposta di scavi edite per cura del Club medesimo. (1877). Reggio Emilia: Tipografia di Stefano Calderini.

Gastaldi, B. (1862). *Nuovi cenni sugli oggetti di alta antichità trovati nelle torbiere e nelle marniere d'Italia*. Torino: Marzorati.

Gerola, G. (1910). *Bassano*. Bergamo: Istituto italiano di arti grafiche.

Gioberti, V. (1843). *Del primato morale e civile degli italiani*. Bruxelle: Meline Cans e Compagnia.

Gioberti, V. (1845). *Del primato morale e civile degli italiani*. Losanna: S. Bonamici e compagnia.

Lioy, P. (1876). *Le abitazioni lacustri di Fimon*. Venice: Presso la Segreteria del R. Istituto nel Palazzo Ducale.

Lioy, P. (1901). *Storia naturale in Campagna*. Milano: Fratelli Treves.

Pigorini, L., & Strobel, P. (1862). Le terramare dell'Emilia. Prima relazione di Luigi Pigorini alunno del Museo di antichità di Parma e Pellegrino Strobel professore nell'università parmense diretta al signor Bartolomeo Gastaldi in Torino. In B. Gastaldi (Ed.), *Nuovi cenni sugli oggetti di alta antichità trovati nelle torbiere e nelle marniere d'Italia*. Torino: Marzorati.

Sella, Q. (1863b). *Una salita al Monviso. Lettera di Quintino Sella a B. Gastaldi*. Torino: Tipografia dell'Opinione.

Stoppani, A. (1908). *Il Bel Paese*. Milano: Cogliati.

Newspaper and Magazine Articles

Brentari, O. (1893a, February 8). Avanzi romani a Bassano. *La Provincia di Vicenza*.

Brentari, O. (1893b, February 18). Scoperta archeologica. *La Provincia di Vicenza*.

Orsi, P. (1894). Di una antichissima necropoli e di altri avanzi romani scoperti presso Bassano Veneto. *Notizie dagli Scavi*.

Roberti, G. (1895). Ancharianus. *Il Brenta*, 10–11.

Sella, Q. (1863a, September 3–10). Lettera a Bartolomeo Gastaldi. *L'Opinione*.

Secondary Sources

Agulhon, M. (1970). *La République au villages: Les populations du Var de la Révolution à la Seconde République*. Paris: Plon.

Agulhon, M. (1977). *Le cercle dans la France bourgeoise (1810–1848): Etude d'une mutation de sociabilité*. Paris: A. Colin.

Ascoli, A. R., & Von Henneberg, K. (Eds.). (2001). *Making and remaking Italy: The cultivation of national identity around the Risorgimento*. Oxford and New York: Berg.

Balan, B. (2014). La naissance de la tectonique moderne en Pennsylvanie au XIXe siècle. *Revue D'histoire Des Sciences, 67*(1), 71–92. https://doi.org/10.3917/rhs.671.0071

Banti, A. M. (2020). *The nation of the Risorgimento: Kinship, sanctity, and honour in the origins of unified Italy*. London: Routledge.

Banti, A. M., & Ginsborg, P. (Eds). (2007a). *Storia d'Italia* (Annale, 22—*Il Risorgimento*). Torino: Einaudi.

Banti, A. M., & Ginsborg, P. (2007b). Per una nuova storia Risorgimento. In A. M. Banti, & P. Ginsborg (Eds.), *Storia d'Italia* (Annale, 22—*Il Risorgimento*) (pp. XXIII–XLI). Torino: Einaudi.

Cervi, G. (2020). Gaetano Chierici e il Club Alpino Italiano. *Bullettino di Paletnologia Italiana, 100*(III), 221–230.

Chierici, G. (2007). *Tutti gli scritti di archeologia (a cura di Paolo Magnani)*. Reggio Emilia: Diabasis.

Coltofean-Arizancu, L., & Díaz-Andreu, M. (Eds). (2021). *Interdisciplinarity and archaeology: Scientific interactions in nineteenth- and twentieth-century archaeology*. Oxford and Philadelphia: Oxbow Books.

Cooter, R., & Pumfrey, S. (1994). Separate spheres and public places: Reflections on the history of science popularization and science in popular culture. *History of Science, 32*, 237–267.

Corsi, P. (2007a). Introduction to thematic set of papers on geological surveys. *Earth Sciences History*, *26*(1), 5–12.

Corsi, P. (2007b). Much ado about nothing: The Italian Geological Survey, 1861–2006. *Earth Sciences History*, *26*(1), 97–125.

Dal Lago, A., & Bernardelli, A. (2015). Paolo Lioy. Gli scavi nelle valli di Fimon e i rapporti con L. Pigorini. In A. Guidi (Ed.), *150 anni di Preistoria e Protostoria in Italia* (Studi di Preistoria e Protostoria, 1) (pp. 628–630). Firenze: Istituto Italiano di Preistoria e Protostoria Firenze.

Darbellay, F. (2012). The circulation of knowledge as an interdisciplinary process: Travelling Concepts, analogies and metaphors. *Issues in Integrative Studies*, *30* 1–18.

Desittere, M. (1988). *Paletnologia e studi preistorici nel'Emilia-Romagna dell'Ottocento*. Reggio Emilia: Comune di Reggio nell'Emilia, Civici musei.

Dietler, M. (1998). A tale of three sites: The monumentalization of Celtic oppida and the politics of collective memory and identity. *World Archaeology*, *30*(1), 72–89.

Díaz-Andreu, M. (2007). *A world history of nineteenth-century archaeology: Nationalism, colonialism, and the past*. Oxford and New York: Oxford University Press.

Díaz-Andreu, M. (2012). *Archaeological encounters: Building networks of Spanish and British archaeologists in the 20th century*. Newcastle upon Tyne: Cambridge Scholars Publishing.

Díaz-Andreu, M. (2019). *A history of archaeological tourism: Pursuing leisure and knowledge from the eighteenth century to World War II* (Springer Briefs in Archaeology). Cham: Springer.

Engel, C. (1965). *Storia dell alpinismo*. Torino: Einaudi.

Fugazzola Delpino, M. A., & Pellegrini, E. (1994). Pigorini: cenni biografici. In M. Brea Bernabò, & A. Mutti (Eds.), *"... Le terremare si scavano per concimare i prati ...". La nascita della paletnologia a Parma nel dibattito culturale della seconda metà dell'Ottocento* (pp. 95–96). Parma: Silva Editore.

Gai, A. (2015). Guilty, sexual and local: Nature for the nation in the Italian narrative of the 'Bel Paese" (19th century). *Environment and History*, *21*(3), 351–380. https://doi.org/10.3197/096734015X14345369355782

Garimoldi, G. (Ed.). (2002). *Dall'orrido al sublime. La visione delle Alpi*. Milano: Biblioteca di via Senato.

Guidi, A. (1996). Nationalism without a nation: The Italian case. In M. Díaz-Andreu, & T. Champion (Eds.), *Nationalism and archaeology in Europe* (pp. 108–118). London: University College London Press.

Habermas, J. (1989). *The structural transformation of the public sphere: An inquiry into a category of bourgeois society* (Th. Burger & F. Lawrence, Trans.). Cambridge, MA: Massachusetts Institute of Technology Press. (Original work published 1962)

Hoibian, O. (2000). *Les alpinistes en France, 1870–1950. Une histoire culturelle*. Paris: L'Harmattan.

Macellari, R. (1994). Gaetano Chierici, prete e preistorico. In M. Brea Bernabò, & A. Mutti (Eds.), "... Le terremare si scavano per concimare i prati ...". La nascita della paletnologia a Parma nel dibattito culturale della seconda metà dell'Ottocento. Parma: Silva Editore.

Maurina, B., & Sorge, E. (Eds). (2010). *Orsi, Halbherr, Gerola. L'archeologia Italiana nel mediterraneo*. Rovereto: Osiride Edizioni.

Mizrahi, R. (1975). *Genèse des représentations urbaines de l'espace alpin. L'espace alpin comme matière à pratiques urbaines et à rapports de pouvoir*. Paris: Cordes.

Morosini, S. (2009). *Sulle vette della patria. Politica, guerra, e nazione nel Club alpino italiano (1863–1922)*. Milano: Franco Angeli.

Morosini, S. (2012). *Le Alpi e l'Unità d'Italia. Lo spazio alpino e il processo di Unità nazionale (1861–1900)*. Bergamo: Club Alpino Italiano, Sezione di Bergamo.

Nieto-Galan, A. (2004). Free radicals in the European periphery: 'Translating' organic chemistry from Zurich to Barcelona in the early twentieth century. *The British Journal for the History of Science*, *37*(2), 167–191.

Papanelopoulou, F., Nieto-Galan, A., & Perdiguero, E. (Eds.). (2009). *Popularizing science and technology in the European periphery, 1800–2000*. Aldershot: Ashgate.

Pastore, A. (2003). *Alpinismo e storia d'Italia. Dall'Unità alla Resistenza*. Bologna: Il Mulino.

Pizzato, F. A. (2015). Archeologia locale, racconto nazionale. La collezione preromana del museo civico di Bassano del Grappa e la costruzione di identità nel periodo post-unitario. *Ateneo Veneto* (3rd series), *13*(2), 11–30.

Roberti, G. (1896). Il Castellaro d'Angarano. *Bullettino Annuale del Club Alpino Bassanese, II*.

Sari, C. (2016). *A colloquio con Paolo* Lioy. *Letteratura, scienza, politica (1851–1905)*. Milano: Franco Angeli.

Schingo, G. (2004). Carlo Kunz. *Dizionario biografico degli Italiani* (Vol. 62). Retrieved May 30, 2021, from https://www.treccani.it/enciclopedia/carlo-kunz_(Dizionario-Biografico)

Secord, J. (2004). Knowledge in transit. *Isis*, *95*, 654–672.

Seduta del 24 aprile 1863. (1863). *Atti della Società Italiana di Scienze Naturali, V*, 17–27.

Soublis, T. (2017). Transdisciplinary pedagogy: A competency-based approach for teachers and students to promote global sustainability. *Journal of Interdisciplinary Studies in Education*, *5*(2), 64–72. https://www.ojed.org/index.php/jise/article/view/1561

Spagni, E. (1876). *La montagna tra il Secchia e l'Enza. Memorie e studi di alpinisti reggiani*. Reggio Emilia: Tipografia di Stefano Calderini.

Tarantini, M. (2012). *La nascita della paletnologia in Italia (1860–1877)*. Firenze: Insegna del Giglio.

Topham, J. R. (2009a). Introduction. In J. R. Topham (Ed.), Focus: Historicizing "popular science" [Special section]. *Isis*, *100*(2), 310–318. https://doi.org/10.1086/599551

Topham, J. R. (Eds.). (2009b). Focus: Historicizing "popular science" [Special section]. *Isis*, *100*(2), 310–368.

Varanini, G. M. (2000). Giuseppe Gerola. In *Dizionario biografico degli Italiani* (Vol. 53). Retrieved July 30, 2021, from https://www.treccani.it/enciclopedia/giuseppe-gerola_(Dizionario-Biografico)

Wickson F., Carew, A. L., & Russell, A. W. (2006). Transdisciplinary research: characteristics, quandaries and quality. *Futures*, *38*(9), 1046–1059. https://doi.org/10.1016/j.futures.2006.02.011

Zanoni, E. (2014). *Scienza, patria e religione. Antonio Stoppani e la cultura italiana dell'Ottocento*. Milano: Franco Angeli.

Zavalloni, F. (2005). Paolo Lioy. *Dizionario biografico degli Italiani* (Vol. 65). Retrieved July 30, 2021, from https://www.treccani.it/enciclopedia/paolo-lioy_(Dizionario-Biografico)

Zocchi, P. (2011). Natura e patria. I congressi della Società Italiana di Scienze Naturali nel processo di costruzione dell'identità nazionale. *Atti della società italiana di scienze naturali e del museo civico di Milano*, *152*(III), 123–156.

Celticism and the *Volk*: Tracing the Ideas and Networks that Shaped Irish Archaeology

9

Andrew Whitefield

Abstract

The scholarly networks that shaped the discipline of archaeology were often highly politicized. Antiquarianism in Ireland was largely a colonial endeavor. The majority Gaelic-Catholic population were excluded from the social and professional circles from which antiquaries were drawn. By denigrating Ireland's past, colonial antiquaries made the case for the imposition of Britain's cultural values in Ireland. During the later-eighteenth century, romantic nationalism and the *Ossian* vogue generated international scholarly interest in Ireland's 'Celtic' past. The Celtic Revival movement was initiated in the early nineteenth century by Anglo-Irish intellectuals. As the movement developed it took root among the expanding Gaelic Irish middle class. 'Celtic' antiquities, art, and literature became the iconography of a new ethnic nationalism. During the interwar years, Nazi Germany 'weaponized' ethnicity. The 'Celtic' regions of northern Europe were seen to preserve the lifeways, traditions, and values of the imagined golden age of Nordic Aryan supremacy in Europe. This was the past the Nazis wanted to recreate in the present. Archaeologists, folklorists, and linguists were deployed by the Nazis to infiltrate Celtic movements. Ireland's leading archaeologist during the interwar years was an Austrian Nazi. The legacy of this period remains deeply embedded in archaeological practice in Ireland to this day.

Keywords

Nationalism · Celtic · Irish · Folk · Volk · Nazi · Networks

A. Whitefield (✉)
School of Geography, Archaeology and Irish Studies, University of Galway, Galway, Ireland
e-mail: andrew.whitefield@universityofgalway.ie

9.1 Introduction

From its tentative beginnings in the antiquarianism of the eighteenth century, archaeology in Ireland has been shaped by the politics, religion, and the social status of its practitioners. Eighteenth-century antiquarianism was largely the preserve of a (male) professional class dominated by British colonists and their descendants. Many antiquaries of the Anglo-Irish Protestant 'Ascendancy', whose political, economic, and social hegemony had been enshrined in law, sought to justify the often-brutal repression of the largely Catholic Gaelic Irish population by denigrating Ireland's past. The few prominent Catholic antiquaries, meanwhile, pointed to the richness of Ireland's antiquities and medieval manuscripts as evidence of the country's ancient cultural sophistication.

Ireland began to build a new 'Celtic' national identity in the late-eighteenth century, born out of romantic nationalism and the development of comparative philology. The romantic conception of the northern Celt began with the *Ossian* vogue. The Scottish author James MacPherson (1736–1796) claimed to have reproduced from oral and written sources the authentic testimony of a third-century bard, whose heroic deeds were intimately bound-up with the rugged landscapes of the Scottish Highlands. MacPherson's main source was his own imagination, but he deftly captured the European literati's appetite for a hero of the north to rival Homer. *Ossian* was quickly exposed as a literary fraud, but in terms of its cultural legacy, this was almost beside the point. MacPherson created the epitome of the heroic yet melancholic northern Celt. Many of MacPherson's principal characters had been appropriated from Irish myth. This elevated Ireland's standing among the European intellectuals as a surviving reservoir of ancient European culture and tradition that remained largely uncorrupted by the modern world. Thus, Ireland's ancient 'Celtic' artifacts, monuments, and manuscripts, along with its language, assumed their place in a venerable European lineage. Ireland moved out from under

L. Coltofean et al. (eds.), *Connecting People and Ideas*, Themes in Contemporary Archaeology,
https://doi.org/10.1007/978-3-031-81006-0_9

Britain's shadow. Archaeologists, folklorists, and philologists played an important role in the Celtic Revival movement, helping shape a positive new vision of Ireland's past.

The romantic Celtic trope owed as much to eighteenth- and nineteenth-century political concerns and ambitions as it did to the ancient past. The nebulous mythical narratives of folk traditions were easily manipulated to support the interests of the interpreter. The 'national essence' of an ethnically pure people rooted in a timeless landscape, bound by the simple pleasures and traditions of rural life, evolved in the imaginations of the urban middle classes. This bucolic nationalism was a by-product of modernity—a nostalgic response to the uncertainties of emerging industrialization. Among small-scale rural communities, the descent into the Great Famine (1845–1852) and its devastating aftermath of poverty and emigration scarred the collective memory—its tragedies written into the landscape (Leerssen 2004: 34–35).

Such concerns did not weigh heavily on the networks of European philologists, folklorists, and archaeologists that fetishized rural Ireland. For them, 'Celtic' Ireland, along with the 'Celtic' regions of Scotland, Wales, Brittany, and Cornwall, were akin to open-air laboratories, where the last survivors of an ancient racial group could be studied in the flesh (McMahon 2016: 251). Although peripheral to the putative epicenter of ancient European civilization in the Nordic regions of Scandinavia and Germany, the systematic study of the language, material culture and traditions of the surviving northern Celts provided an evidence base for the study of the first Golden Age of Nordic Aryan civilization (Leerssen 2004: 29–34). Ethnicity increasingly became a dominant concern in such research, and archaeology was no exception. Many practitioners espoused or accepted racist ideologies. Ireland's leading archaeologist during the inter-war years was an Austrian Nazi, Adolf Mahr.

This chapter examines how networks of European archaeologists, along with colleagues in the allied disciplines of folklife studies and philology, were instrumental in transforming Irish archaeology from a neglected offshoot of British imperial myth to the enduring exemplar of the 'Celtic' tradition. It finds that the entwining of the archaeological imagination and political ideology remain deeply embedded in the discipline today.

9.2 Antiquarianism in Early Modern Ireland

British colonial settlers overwhelmingly dominated the social and professional elite in Early Modern Ireland. The majority Gaelic-Catholic population suffered systemic discrimination under the notorious Penal Laws. Enacted at the turn of the seventeenth century, this legislation was designed to suppress Irish Catholic support for the Stuart royal dynasty and underpin the Protestant hegemony (Boyce 1995: 94; O'Halloran 2012: 195). Catholics were prevented from holding political office and barred from the professions, all but excluding them from the social networks from which antiquaries were drawn (O'Halloran 1995, 2004, 2012).

The dominance of the Protestant Ascendancy in Ireland was justified by the supposed cultural superiority of the British colonists. Any suggestion that ancient Gaelic culture could have independently advanced beyond barbarism might represent a challenge to the foundations of the precarious status quo. The networks of Ascendancy antiquaries formed clubs and learned societies which generally worked within the framework of the conventional British historical narrative. Since archdeacon and historian Giraldus Cambrensis (1146–1223) in the twelfth century, Ireland had been characterized as beholden to the civilizing influence of the British (Foster 1983: 170–71; O'Halloran 2004: 98). Catholic antiquaries, meanwhile, directed their research towards undermining this perspective by foregrounding the richness of the manuscripts and artifacts of elite pre-Colonial Gaelic culture (MacDonagh 1983: 2; O'Halloran 2012: 196). They argued that ancient pagan Ireland had developed a sophisticated and literate society, laying the foundations for Early Christian Ireland to become one of Europe's outstanding centers of monastic and secular learning: "the island of saints and scholars" (O'Halloran 1990).

The principal source of Gaelic Irish historiography was the late-eleventh century *Leabhar Gabhála Érenn* (*LGÉ*)—the *Book of the Taking of Ireland.* Compiled by monastic authors, the *LGÉ* charts the history of Ireland through a series of invasions (Waddell 2005: 19ff.; Ní Lionáin 2012: 36ff.). Although often portrayed as the outcome of neutral encounters between the illiterate bards (the custodians of Gaelic oral histories) and monastic scribes, the *LGÉ* is a Christian historiography which follows many of the conventions of European national histories in the early Middle Ages (Scowcroft 1988: 63; Waddell 1991: 350). Gaelic Irish myth-histories were assimilated into a biblical framework extending back to Noah and the Old Testament Patriarchs. The *LGÉ* thereby subsumed the competing traditions of regional dynasties under the central Tara kingship (Scowcroft 1988: 42). The outcome was a national past designed to create a national sense of identity, unified by Christian orthodoxy (Kidd 1999: 157; O'Halloran 2004: 14–16).

The *LGÉ* identifies a mythical patriarch, Míl, as the common ancestor of the Irish, to whom all Ireland's rulers (prior to the Anglo-Norman invasion) could trace their bloodline. Míl's sons—the Milesians—journeyed to Ireland from Spain (where Míl ruled). Míl's genealogy is traced via Egypt and Scythia to Japheth, son of Noah (O'Halloran 2004: 15). Alongside the biblical framework, and Phoenician (via Scythia) and Egyptian ancestry, the *LGÉ* draws on stories of supernatural beings and giants to explain the origins of the

Irish. Ancient monuments and auspicious places in the Irish landscape are associated with mythical figures and events. The narrative moves imperceptibly between verifiable sources and fictitious accounts, without distinguishing their relative credibility (Scowcroft 1987: 81). Much of the source material is in any case reworked beyond recognition (Waddell 2005: 20).

Interest in the *LGÉ* was heightened during the seventeenth century, most notably through the work of Geoffrey Keating (c. 1517–1644) Keating was a Catholic priest of Old English descent, his ancestors having arrived in Ireland among the first waves of settlers after the Anglo-Norman conquest. Later generations of the family remained Catholic after the Anglican Reformation (Cunningham 1986). Keating was also a Gaelic-speaker. He re-worked the *LGÉ* and other medieval manuscripts into an accessible narrative history written in modern Gaelic. Keating's *Foras Feasa ar Éirinn* (Compendium of Knowledge about Ireland) was published in 1634 against the backdrop of the arrival of new waves of British Protestant settlers, to whom most Catholic estates were transferred under the Penal Laws (O'Halloran 2004: 15ff). *Foras Feasa* provided a counterpoint to disparaging Anglophile narratives and was widely circulated (O'Halloran 1995: 139; Kidd 1999: 155). Keating described a fabulous pre-Christian Milesian civilization in Ireland, primed for the immediate acceptance of Christianity by virtue of their moral and intellectual sophistication (O'Halloran 1990: 9). Keating sought to bridge the divide between the Old English Catholics and the Gaelic Irish by characterizing the arrival of the Anglo-Normans as just the latest in the series of invasions described in the *LGÉ* (Cunningham 1986: 127). This helped foster a new sense of national identity among the Old English, interweaving their heritage with that of the Gaelic Irish (Kidd 1994: 1198–99).

Foras Feasa was published in English as *The General History of Ireland* in 1723. The "Catholic tone and perspective" of Keating's original was carefully diminished in translation, increasing its popularity among influential Protestants (O'Halloran 1990: 12). By the early-eighteenth century, Ireland's leading Catholic antiquaries were in any case largely unconcerned with the characterization of Milesian Ireland as proto-Christian. Instead, their focus was on the cultural sophistication of the pre-Christian Gaelic Irish in their own right. Sylvester O'Halloran (1728–1807), a Limerick surgeon, went as far as claiming the arrival of Christianity in Ireland had a negative impact, replacing the martial prowess and independence of spirit of the Milesians with "patience, humility and meekness" (S. O'Halloran 1772 quoted in O'Halloran 1990: 9).

The eighteenth century's most influential Catholic antiquary was Charles O'Conor (1710–1791), a Gaelic Catholic of noble ancestry, whose family had managed to retain some of their ancestral lands (O'Halloran 1990, 2012). Like Sylvester O'Halloran, O'Conor's antiquarianism focused on extolling the greatness of Milesian civilization, and so the pedigree of the Milesian's Gaelic descendants. Unlike Sylvester O'Halloran, O'Conor was not satisfied that the Gaelic sources alone were sufficient proof of the veracity of the Milesian legends and sought corroborative evidence from international scholarship (O'Halloran 2004: 25). O'Conor believed he had found this evidence in a late work of English scientist Isaac Newton (1643–1727)—*Chronology of Ancient Kingdoms Amended* (1728)—which used astronomy and genealogy to calculate the dates of events described in classical literature and the Old Testament (O'Halloran 2004: 26). The similarities between Newton's account and the stories in the *LGÉ* and *Foras Feasa* were striking. O'Conor saw uncanny parallels between Míl and Newton's description of the classical hero Hercules (O'Conor 1753: 19 quoted in O'Halloran 2004: 26). Neither O'Conor nor his detractors were aware that these coincidences were the result of a common reliance on medieval Christian chroniclers. But British antiquaries had already incorporated aspects of Newton's chronology, including Hercules, into their own origin myths. O'Conor failed to gain the acceptance for Milesian Ireland he had hoped for among British scholars (O'Halloran 2004: 27).

O'Conor was pragmatic in the face of the Ascendancy's repression of Catholics, and this enabled him to work at the intersection between Catholic and Protestant antiquaries over the course of his long career. He was unconcerned by Ascendancy representations of the Irish peasantry as uncivilized: this was a perspective he shared. Instead, it was the conflation of the Catholic gentry with the unlettered lower social orders to which O'Conor (like Keating before him) objected (Kidd 1994: 1211; O'Halloran 2012: 196). For O'Conor, it was social rank rather than ethnicity or religion that distinguished the civilized from the barbaric. It followed that the higher castes—be they Protestant or Catholic—should rule over the peasantry together (Kidd 1994: 1212).

9.3 British Antiquarianism, the Ascendancy, and the 'True Religion'

> (…) the true religion has chiefly since the repeopling [of] mankind after the Flood, subsisted in [Britain]: and here we made the best reformation from the universal pollution of Christianity, Popery. William Stukeley (1743: iv)

Religion and national identity were central to the shaping of antiquarianism in Britain at the turn of the eighteenth century. During the seventeenth century, John Aubrey (1626–1697), a prominent Royalist of Welsh ancestry, had recognized the potential of antiquities in the promotion of a shared sense of British national identity over the separate national pasts of England, Scotland, and Wales (Harvey 2003: 478). Aubrey challenged the conventional wisdom that Britain's most famous ancient monument, Stonehenge, was a Roman temple. He noted that similar monuments were widespread in parts of Britain, such as in Scotland, as well as in Ireland, where the Romans had little or no presence (Piggott 1989: 114–15). Aubrey concluded that Stonehenge was a temple constructed by ancient Britons before the Roman conquest. But he reverted to speculative literary accounts to identify Stonehenge's pre-Roman architects. The Druids were the only religious order that historical scholars had placed in Britain before the Romans. Britain and Ireland's megalithic monuments thus became Aubrey's *Templa Druidum* (Piggott 1986).

It was William Stukeley (1687–1765)—a medical doctor who became an Anglican priest—who went on during the eighteenth century to develop Aubrey's "modest suggestion" (Piggott 1986: 116) of a pre-Roman Druidic priesthood into an enduring origin myth for Britain and the Anglican Church. A friend (and biographer) of Isaac Newton, under whose tenure as president he was appointed Fellow of the Royal Society, Stukeley was keen to extend antiquarianism's methods beyond literary sources (Haycock 2002: 1–12). But like Newton's, Stukeley's empirical reasoning was framed within the orthodox Anglican doctrine that accepted the literal truth of the bible (Cook 2004: 179).

Stukeley argued that Anglicanism predated Roman Catholicism. He sought to prove his case through the examination of the pre-Roman 'Druidic' temples identified by Aubrey, of which Stonehenge and Avebury were the outstanding examples. At Stonehenge, Stukeley concluded that the Roman foot had not been the unit of measurement used in the monument's construction. Instead, he claimed to have rediscovered a more ancient architectural measurement, which he named the "Druid's cubit" (Hill 2008: 42). "[P]robably deriv'd from Noah or Adam", Stukeley argued that the Druid's cubit had similarly determined the dimensions of the earliest biblical temples, such as Solomon's Temple, as well as the great temples of Egypt (Stukeley 1740: 6 quoted in Haycock 2002: 168). It followed that the Druids were proto-Christian antecedents to Anglicans (Haycock 2002).

To explain how the Druids came to Britain, Stukeley reverted to the narrative history compiled by the seventeenth-century lawyer Aylett Sammes (1636–1679) (see Hutton 2016: 129). Sammes had adapted the work of the French scholar Samuel Bochart (1591–1667), determining that the first colonists of Britain were Phoenician seafarers who reached southwest Britain from Tyre in ancient Egypt (Haycock 2002: 166–69). In Stukeley's (1740: 6–8) account, the expedition was led by Hercules, himself identified as a Tyrian Phoenician. The voyagers had set out in search of tin. Stukeley identifies the Druids as the Phoenicians' priestly caste. Having preserved the teachings of Abraham, the Druids had constructed the great temples of the ancient world (Haycock 2002). According to Stukeley (1740: 7), the Druids who voyaged with Hercules were responsible for the construction of Stonehenge and Avebury. The British Druids were eventually driven to Ireland by the Romans, leading to the construction of temples such as Newgrange, and the establishment of "the most flourishing seat of learning" by the British colonists (Stukeley 1763: 11).

Eighteenth-century Ireland's most prominent antiquary was a British military engineer (Leerssen 1986: 99; O'Halloran 1989: 1–2). Major (later General Sir) Charles Vallancey (1721–1812) was instrumental in establishing the (Royal) Dublin Society's Committee of Antiquities in 1772, and a founder member of the Royal Irish Academy (RIA) in 1785. He was also a Fellow of the Royal Society of London and the Antiquaries Committee in London (Pittock 2008: 102). Like Stukeley, Vallancey was a passionate Orientalist. Vallancey became enchanted with rural Ireland through his travels as an engineer. The lives of the small-scale Gaelic speaking communities he encountered were so far removed from his own life experience that he believed they preserved the traditions of the first Phoenician settlers (Vallancey 1783/1786). What is more, he believed that the Phoenician language had been substantially preserved as Gaelic Irish. He cited a monologue in a comedy by the classical Roman playwright Plautus as evidence, contradicting contemporary language experts (Vance 1980: 226; O'Halloran 2004: 42). Philology was an emerging field in European scholarship, and Vallancey did not allow his lack of any notable linguistic skill or knowledge to stand in the way of his linguistic theorizing (Vance 1980: 226–27). He traced Gaelic back to the language spoken in the Garden of Eden.

A man of great self-confidence and enthusiasm, Vallancey was not unduly constrained by the tension between his position within the colonial Establishment and his valorization of ancient Gaelic civilization (Vance 1980: 226). He argued that both Britain and Ireland were ancient Phoenician colonies, but that they had been colonized by separate groups (Leerssen 1996: 100–1). This accounted for the similarities in the medieval historiographies and the establishment of an independent high culture in pre-Christian Ireland. Vallancey's limitations as a linguist meant that his main source of Irish historiography was the English translation of *Foras Feasa*, which he treated as a primary source representing the entire corpus of Irish medieval historiography (O'Halloran 2004: 42). Vallancey believed that previous attempts by Irish scholars to historicize ancient myths had failed because of the ineptitude of the translators. He argued, for example, that "Iran" (Persia) had been mistakenly interpreted as "Eirin" (Ireland) (O'Halloran 2004: 48). His understanding of archaeological monuments was similarly eccentric. He interpreted the round belltowers of Irish early Christian monastic sites as Phoenician astral observatories (Roling 2018: 763).

Despite his hubris, Vallancey's arguments in support of the Phoenician-Milesian origins of the Irish were welcomed by Gaelic-Irish antiquaries, as well as by politicians seeking greater freedom for Catholics (Leerssen 1986: 100–1; Kidd 1994: 1204). Charles O'Conor cultivated his friendship, flattering his scholarship and assisting him in the translation of Gaelic manuscripts (Vance 1980: 227; O'Halloran 1995: 51–52). Not having suffered the indignity of being equated with the Catholic peasantry, Vallancey took a greater interest than O'Conor in the "pagan customs of the common people", whom he described as "the most valuable monuments of antiquity" (Vallancey 1772/1818: 49). He argued that the Gaelic language should be used for the purposes of Anglican proselytization, that support for Catholic education should be withdrawn, and that there should be no further relaxation of anti-Catholic legislation (O'Halloran 2004: 45).

Political unrest in Ireland during the 1790s culminated in the Irish Rebellion of 1798 and the Act of Union of 1800. Vallancey's detractors, such as the Irish Protestant clergyman Edward Ledwich (1738–1823), pointed to the Rebellion as proof that the Gaelic Irish were irredeemably barbarous (Leerssen 1986: 102). Vallancey's ideas went out of fashion among the Anglo-Irish. But the attention his enthusiasm and industry had brought to the Gaelic past was not so easily extinguished among the Gaelic Irish. Cooney (1996: 160) argued that aspects of Vallancey's romantic antiquarianism "are still central to the public view of Irish prehistory today". Despite his faulty scholarship, Vallancey helped ensure the enduring centrality of Irish mythology in Gaelic identity and European romanticism (Leerssen 1996: 161). There can be no doubt that the collection and preservation of Irish medieval manuscripts that Vallancey's enthusiasm inspired secured a wealth of material for future scholarship (McDowell 1985: 6–7).

9.4 The Literary Invention of the Northern Celt

Despite the apparent common ground between the marginalized Gaelic-Irish and their Gaelic-speaking counterparts in Scotland, Wales, Cornwall, and the Isle of Man, there was little sense of any pan-Gaelic unity in the presentation of their competing and often irreconcilable national historiographies (Kidd 1999: 185). From the 1750s, Scottish Gaelic identity had been undermined by the development of new social theories by Enlightenment thinkers. Stadialism proposed that the history of human development could be divided into progressive stages, from hunter-gatherer, through agrarian, to the commercial and cultural sophistication of the modern age (Kidd 2004: 508). From this perspective, the traditional way of life in the Gaelic Highlands was an embarrassing anachronism. The Highlanders were seen to have more in common with the "primitive" small-scale societies encountered by travelers to places such as North America, than with their Lowland Scottish neighbors (Kidd 1994: 1206).

The response to this denigration of Highland life by the Scottish writer James MacPherson—himself a native Highlander—transformed the nascent romantic movement in Europe. During the early 1760s, MacPherson presented what he claimed were translations of traditional Gaelic verse he had recorded from oral recitals while touring the Scottish Highlands and Islands. He also drew on a small number of manuscripts collected during his travels (O'Halloran 1989: 69; Porter 2001: 414–15). MacPherson's *Fragments of Ancient Poetry collected in the Highlands of Scotland and translated from the Galic or Erse Language* was published in 1760 to great acclaim. It was quickly followed by two epic ballads, *Fingal* (1761) and *Temora* (1763). Collectively, these works are referred to as *Ossian* (e.g., Gaskill 2004).

MacPherson claimed his works faithfully recounted the wistful poetry of a blind and aged third-century Scottish bard named Ossian, whose verse recalling his heroic youth had been passed down unaltered in Gaelic oral tradition (Gaskill 1994, 2004). But while plots and passages from traditional ballads did feature in his long poems, MacPherson's imagination was his principal source (Stafford 1988: 152; see Mackenzie 1805). No Highland crofter could be identified who was able to recite the *Ossian* ballads in the original Gaelic, and corroborating ancient manuscripts were not forthcoming (Hall 2007: 4). Nevertheless, in terms of *Ossian's* influence on European literature, art, and music, this is almost beside the point. MacPherson expertly captured the *Zeitgeist* of the late-eighteenth century vogue for Primitivism, with its sentimental hankerings for the lost innocence of earlier times. He connected the nostalgic literati with the past they wanted to hear about, in a style that was pleasing to their tastes and sensibilities (O'Halloran 1989: 71). MacPherson did not invent the Celtic aesthetic in European literature, but through *Ossian* he exemplified the nascent concept (Watson 2009: 150–54).

Earlier in the eighteenth century, the Breton scholar Abbé Paul-Yves Pezron (1639–1706) had proposed that the Breton language was the modern survival of the language spoken by Caesar's *Celtae* (James 1999: 44–45; Collis 2003: 48–49). The *Celtae* (like the Greek *Keltoi*) referred to by Classical writers in the final centuries BC were vaguely defined, the name having been applied variously to peoples in neighboring parts of Anatolia and continental Europe (Chapman 1992: 33–34; Collis 1997: 103). Pezron's motivation was his patriotic desire to elevate the historical pedigree of the Breton people, but he also linked the Breton language to Welsh, capturing the interest of the Welsh philologist Edward Lhuyd (1660–1709) (Morgan 1983: 67). Lhuyd—who had spent many years establishing similarities between the languages of Wales, Ireland, Scotland, Cornwall, and Brittany—arranged for Welsh and English translations of Pezron's (1703/1706) *L'Antiquité de la nation et de la langue des Celtes, autrement appellez Gaulois* (James 1999: 44–46).

The (1706) English translation was given the extended title *The Antiquities of Nations, More particularly of the Celtae or Gauls, Taken to be Originally the same People as our Ancient Britains* [sic]. Lhuyd's *Archaeologia Britannica* (1707), attached the name 'Celtic' to the broader language grouping for the first time (Collis 2003: 49–50). Pezron and Lluyd never met, but together they gave rise to the fallacy that not only had the hypothesized language of the ancient continental *Celtae* been preserved on the islands of the British-Irish archipelago, but that the newly identified modern Celtic speakers must be *ethnically* Celtic (e.g., Collis 2017: 60–61). *Ossian* was the imagined personification of the imagined northern Celt (Kidd 1999: 185). Just as Pezron had sought to improve the standing of the marginalized Bretons among the wider population of France, MacPherson sought respectability for the Highland people in the eyes of the Lowland Scots and the English (Hall 2007: 9–14).

The rehabilitation of the Highlands that MacPherson contrived came in part at Ireland's expense. As critics in both Britain and Ireland were quick to point out, MacPherson extensively appropriated Irish mythology in his *Ossian* 'translations'. The main protagonist was adapted from Oisín, son of Fionn Mac Cumhaill (MacPherson's *Fingal*), both heroes of the Irish mythic Fenian Cycle (Sheehy 1980: 14; O'Halloran 1989: 74). Nevertheless, interest among the literate public in Britain quickly began to extend to the wider 'Celtic Fringe' of Wales, the Isle of Man, Cornwall, and Ireland. The spirit of MacPherson's *Ossian*—the essence of his Highland Celt—was infused with the brooding majesty of the Highland landscapes, replete with mossy stones, ancient ruins, mournful winds, and spectral mists (e.g., Geikie 1905: 115–17 quoted in Gaskill 2004: 5). Landscapes and literature that invoked Gaelic, Celtic or Druidic traditions fueled the vogue. For William Stukeley—whose books on Avebury and Stonehenge had become much sought after—the authenticity of *Ossian* was evident from the extent to which MacPherson captured his own sense of the Celtic (Stafford 1988: 172; see Stukeley 1763).

The reception of *Ossian* in Ireland, however, was more complicated. MacPherson's appropriation of Irish folk-poems and their mythic heroes was compounded by his disparaging portrayal of Irish Gaelic culture. MacPherson ridiculed the Milesian origin myths, claiming instead that Ireland's original settlers were Scots, and that Irish Gaelicism was merely a degenerate derivative of the Scottish epitome (Curley 2009: 145). Gaelic scholarship in Ireland (and Scotland) was "generally limited and superficial" (O'Halloran 1989: 82). Charles O'Conor was among the few qualified to offer anything approaching a scholarly critique. But O'Conor's interest in upholding the honor of Irish mythical poetry was limited. He doubted the historical significance of the bardic tradition, with its "crude tales", which he considered "mere amusements for the vulgar" (cited in O'Halloran 2004: 103). The thrust of his critique was not therefore MacPherson's faulty historiography or appropriation of Irish sources, but rather the characterization of Irish Gaelicism as an invariably unlettered, oral culture of savage simplicity (Kidd 1994: 1211). Vallancey, while broadly supportive of O'Conor's stance, brought little to the debate. Like O'Conor, he saw no literary merit in the Gaelic stories. He believed in the primacy of Irish Gaelicism over Scottish but believed the ultimate origins of the stories to be Persian (O'Halloran 1989: 82). For O'Conor and Vallancey, MacPherson's contrived style and Anglicizations were *Ossian*'s redeeming features. For contemporary Gaelic Highlanders (and perhaps many of their Irish counterparts, had they been surveyed), these were the least appealing aspects of the poems (Hall 2007: 24).

For better or worse, *Ossian* proved "une révélation pour l'Europe" (a revelation for Europe) (Van Tieghem 1917: 47), exemplifying the moral struggle between traditional rural life and the 'civilized', yet hollow and unprincipled, urban existence (Gaskill 1994). Although the wider European literati may have understood little of the plight of contemporary Scottish Highlanders or the Gaelic Irish, Ossian's wistful reflections on a simpler yet heroic past struck a chord in many countries (Kidd 1999: 185; Barnaby 2004). Reaction against suppressed nationality and the generalizing tendencies of the Enlightenment were strongest in Germany and Scandinavia (Kearney 2007: 131). Ethnic distinctions between Celts, Germans, and Scandinavians were yet to be constructed (Wilson 1973: 828). *Ossian* could therefore be adopted as part of German and Scandinavian national heritages—a counter to slurs of Nordic barbarity (Okun 1967: 330; O'Halloran 1989: 72). For the first time there was evidence of ancient Epic poetry—and thus ancient high civilizations—north of the Alps. The personification of a long-awaited alternative to the cultural supremacy of Mediterranean-Classical civilization, *Ossian* was celebrated as the "Homer of the North" (e.g., Oz-Salzberger 2006: 386).

9.5 Celticism and the *Volk*

> A nation (…) has nothing more valuable than the language of its fathers. In it lives its entire spiritual treasury of tradition, history, religion, and principles of life, all its heart and soul. To deprive such a nation of its language, or to demean it, is to deprive it of its sole immortal possession transmitted from parents to children. Johann Gottfried Herder (quoted in Wilson 1973: 837)

In Germany and Scandinavia, *Ossian* helped bring about the elevation of folk traditions from the preserve of provincial low culture to the focus of Nordic high culture. This transformation owes much to the German philosopher Johann Gottfried Herder's (1744–1803) immediate embrace of MacPherson's work toward the end of the eighteenth century. Herder was the impetus behind romantic nationalism and its celebration of passion and instinct over reason (Wilson 1973: 820). This is exemplified by his pronouncements on the authenticity and antiquity of *Ossian* before having seen either MacPherson's English originals or the later German translations (Reed 1965: 556; O'Halloran 1989: 71). The poems inspired Herder's redefinition of the popular understanding of the word *Volk*, shifting the inference from (lower) social class to regional ethno-cultural association (Shellenberg 2012: 11). Beguiled by MacPherson's sublime narrative, Herder described "an ancient Celtic hut on a rugged mountain, amidst cold and storm and mist" as his vision of Eden (Porter 2001: 419). For Herder, folk poems were "the archives of nationality" (Wilson 1973: 825–26)—the antithesis of the mores of French formalism and the transnational aristocracy that ruled the German princely states (Ó Giolláin 2012: 81–82).

MacPherson's allegorical use of landscape and weather to evoke emotions—particularly brooding melancholy and wistfulness—was eminently translatable into visual art. The tombs of fallen heroes feature prominently in Ossian's verse: ever-present reminders of the ancestors, whose heroic deeds and sufferings he recounted (Hall 2007: 18–21). Megalithic tombs and burial cairns provided tangible human-wrought links to Ossian's world. Windswept North European landscapes—often replete with archaeological remains—became the subject of new regional iconographies in Germany and Scandinavia (Okun 1967; see Fig. 9.1).

Fig. 9.1 Christian Faedder Høyer (1775–1855), *Ossian is resting by a dolmen*. Oil on paper laid on panel. Megalithic monuments, including dolmen, feature prominently in the sublime *Ossianic* landscapes of northern Europe. (Photo © Bruun Rasmussen Kunstauktioner; reproduced with the permission of Bruun Rasmussen Kunstauktioner, Denmark)

The radical changes brought about by the French Revolution (1789–1799) and the Napoleonic Wars (1803–1815) provided a further impetus for change in the way nations understood their origins and identities at the turn of the nineteenth century (Smith 1986: 155). National pride was severely dented across Scandinavia and Germany: Denmark was declared bankrupt in 1813, Norway was ceded to Sweden, and the German states had succumbed to Napoleonic control (Lund 2014: 23). These events led to an upsurge of interest in Herder's vison of national origins as preserved in folk ballads and fairy tales: his *Volkstum* (Veit 1989: 50). Herder equated language with ethnicity, and this gave rise to the discipline of comparative philology (Tristram 1996: 43). French was the predominant language of the educated in eighteenth-century Germany. The German of the common people was considered vulgar. The reassertion of the German language was central to Herder's *völkisch* philosophy (Wilson 1973: 826). Previously marginalized north European languages such as Gaelic—so long denigrated as barbaric—became the subject of intense philological interest (Tristram 1996).

Among the first to develop Herder's philology were the brothers Jakob and Wilhelm Grimm (1785–1863; 1786–1869), who began publishing fairy stories they had collected in 1812 (Wilson 1973: 828). Jakob Grimm worked with linguist Franz Bopp (1791–1867) in Berlin on a formal method of etymological analysis based on grammar and structure. Together they created a system of comparative linguistics that raised philology status to that of a prestigious science in Germany (Leerssen 2006: 260). The complex of Indo-German/Indo-European languages had gained wide acceptance by the 1820s, although a connection between these and the Gaelic/'Celtic' languages was not obvious. Thus, although already the subject of ridicule among many scholars, the trope of Phoenician origins of Gaelic language persisted among some antiquaries (Leerssen 1996: 73–75). In Ireland, antiquarian Sir William Betham (1779–1853) sought to honor Vallancey's legacy at the RIA by continuing to study the relationship between the Irish language (which he did not understand) and the language of the ancient Phoenicians (which nobody understood) (Waddell 2005: 109). Betham was an influential figure at the RIA from 1826 through to the middle of the nineteenth century. It was not until 1854—the year after Betham's death—that Johann Kaspar Zeuss (1806–1856) of Munich (another of Jakob Grimm's collaborators) was able to make a sufficiently persuasive case for the inclusion of Gaelic/'Celtic' within the Indo-German/Indo-European language group (Leerssen 1996: 88; see Fig. 9.2).

Fig. 9.2 The German philologist Johann Kaspar Zeuss (1806–1856) is commemorated on this 2006 Irish stamp. (Photo © The Royal Irish Academy; reproduced with the permission of *An Post*)

A new, more methodical, approach to the study of Gaelic antiquity had begun to develop among some of Betham's RIA colleagues during the 1830s. The leading figure was George Petrie (1790–1866), a Protestant of Scottish parentage (Hutchinson 2001: 508). He represented a revival of interest in Irish antiquity and identity among the Protestant middle class. While the rural Irish peasantry continued to suffer derision and oppression, the passing of the Catholic Emancipation Act of 1829 meant the Catholic middle class also represented an expanding audience for Irish history (Sheehy 1980: 7). Petrie supported political union with Britain, but not the immoral treatment of Catholics or the denigration of Gaelic culture (Hutchinson 2001: 508). A talented writer and musician, Petrie trained as an artist. He was able to bring his enthusiasm for ancient Irish antiquities and culture to a wide audience through his journalism and artwork, initially making a living as a topographical artist and illustrator of guidebooks for the developing tourist market (Sheehy 1980: 17). As editor of *Dublin Penny Journal* and the *Irish Penny Journal*, he was a popularizer, propogandist and moralizer—encouraging discussion of Irish history and society (Hutchinson 2001: 509).

Petrie was elected to the RIA in 1828, where his organization and expansion of the museum collection of artifacts, including 'Celtic' art and manuscripts, became a leading visitor attraction (Hutchinson 2001: 518). Archaeological excavation in nineteenth-century Ireland was a "haphazard affair" (Harbison 1988: 13). It would be more than a century before sustained programs of systematic excavation began in Ireland (e.g., Waddell 2000: 10). The RIA's collection of portable artifacts was overwhelmingly the product of chance finds discovered during the work of agricultural improvement or cutting turf for fuel (Herity and Eogan 1977: 12). Petrie was instrumental in amassing some 1500 artifacts for the RIA, through appeals to landowners, agricultural workers, and private collectors (Harbison 1988: 12). Acquisitions of antiquities from Dublin jewelers stemmed the flow of Irish artifacts into international private collections, or worse, being melted down for bullion (Herity and Eogan 1977: 10).

Petrie's prominence in the RIA—Ireland's leading scholarly network—led to his appointment as Superintendent of the Topographical Section of the Ordnance Survey of Ireland (OSI) from 1835 to 1842 (Waddell 2005: 107–8). Petrie engaged John O'Donovan (1806–1861) and Eugene O'Curry (1794–1862) (both Gaelic-speaking Catholics) in the development of the OSI 'memoirs': written commentaries that would supplement county maps with information incorporating (among other things) local history and antiquities (Sheehy 1980: 20). Petrie dispatched O'Donovan to carry out research in the field, while O'Curry was tasked with extracting information from medieval manuscripts, such as the *LGÉ* (Waddell 2005: 99). In the end, the enthusiasm and zeal with which the three approached their work, was their undoing. The British Prime Minister, Robert Peel (1788–1850)—horrified by the precision with which local history and traditions were being recorded and recognizing the potential of the material to advance political nationalism—cut off the government funding (Leerssen 1996: 104).

The more organized and scholarly approach of Petrie, O'Donovan, and O'Curry did not lead to their abandonment of Gaelic origin myths. The three remained convinced of the literal truth of the medieval manuscripts and sought to correlate the mythical events described in the texts with monuments, artifacts, and places (O'Halloran 2004: 184). Major archaeological sites such as Newgrange were described as the burial grounds of mythical warriors or the sites of mythical battles (Waddell 2005: 113–16). Alongside artists such as Henry O'Neill (1798–1880) and Andrew Nichol (1804–1886), Petrie's depictions of archaeological landscapes, structures and artifacts helped establish ('Celtic') antiquities such as monastic high crosses as symbols of Irish nationality in the developing cultural revival (Sheehy 1980: 20–23; Hutchinson 2001: 509–10; see Fig. 9.3).

In 1856, the distinguished eye surgeon Sir William Wilde (1815–1876)—whose famous son was named after Ossian's son, Oscar—undertook the task of compiling a catalogue of the RIA antiquities collection (Vance 1980: 221). Wilde's romanticism extended to his organization of the collection. He rejected the Three Age System devised by the archaeologist Christian Jürgensen Thomsen (1788–1865), which established a relative chronology based on the observation that artifacts had been made successively of stone, bronze, and iron. Instead, Wilde attributed developments in artifacts

Fig. 9.3 George Petrie (1790–1866), *The Last Circuit of Pilgrims at Clonmacnoise County Offaly*. Graphite and watercolor on paper. The 'Celtic' high cross and round tower. Petrie helped establish monastic high crosses and round towers as symbols of Irish identity. (Photo © National Gallery of Ireland; reproduced with the permission of the National Gallery of Ireland)

and monuments to the legendary invaders of Ireland identified in the *LGÉ* (Waddell 2005: 134–36). Drawing on his medical background, Wilde (1850) also devised a scheme of ethnological classification, which he claimed identified "two separate races" in ancient Ireland. These observations ran parallel to work across Europe on the classification of race and intelligence through craniological characteristics (Díaz-Andreu 2007: 345–49). In Sweden, anatomist and anthropologist Anders Retzius (1796–1860) developed the cephalic index in 1845. This influential schema distinguished between *dolichocephalic* (long-headed) and *brachycephalic* (short-headed) types. *Dolichocephalic* people—including Scandinavians, Germans, English, and French—were considered more intelligent, while Lapps, Finns and Bretons were among the less-favored *brachycephalic* types (Díaz-Andreu 2007: 347).

Zeuss's major philological work, *Grammatica Celtica* (1853), equated the history of the 'Celtic' language with the ethnicity of its speakers. Comparative philology's expression of the utility of language as a determiner of historical ethnicity enabled the discipline to play an important role in increasing ethnic self-awareness and the development of national identities (Tristram 1996). As philology developed in the second half of the nineteenth century, the discipline borrowed increasingly from scientific methods—particularly botany—in its classification of linguistic traits. Laws were established that were seen to control the variability of languages over time, enabling language to be presented as a stable signifier of race (Tristram 1996: 56–57). In contrast to biological evolution and the principle of improvement through natural selection, changes in language were seen as a denigration or corruption of the perfect original. When the 'Celtic' languages were admitted to the Indo-European family, Irish was seen as the least corrupt, and therefore foremost exemplar of them (Tristram 1996: 44).

In the second half of the nineteenth century, German philologist and Celticist Ernst Windisch (1844–1918) and his long-term collaborator Whitley Stokes (1830–1909) "caught the torch from the hands of Zeuss, O'Curry and O'Donovan" (Knott 1919: 264–66). Windisch had been introduced to Stokes—a Dublin lawyer and Celtic scholar—in the early 1870s by Standish O'Grady, himself an eminent antiquarian, Celticist, and founder member of the Irish Ossianic Society (Rockel 1990: 32). Whitley's father, William (1804–1878), combined his career as an eminent Dublin heart surgeon with a passion for antiquarianism (Andrews 2009). Petrie, O'Curry, and O'Donovan were regular visitors to the Stokes' household, and having worked closely with Petrie over many years, William published Petrie's biography in 1868. Being raised at the heart of this scholarly network had a tremendous influence on the young Whitley, as well as his sister, Margaret McNair Stokes (1832–1900), who went on to become a noted antiquarian and illustrator (Clinton and Sturgeon 2009). Windisch's friendship and scholarly collaboration with Whitley Stokes lasted until Stokes' death in 1909, interlinking the pre-eminent networks of German and Irish Celticists (Knott 1919: 266; Rockel 1990: 33).

Windisch's great contribution to Celtic philology was the simplification of Zeuss's *Grammatica Celtica,* rendering it more accessible to a new generation of scholars, whose enthusiasm he captured with his translations of the great Irish sagas such as the *Táin Bó Cualnge* (Ó Lúing 1994:

257). The standardized Gaelic that formed the basis of the language revival at the turn of the twentieth century owes much to the network of German Celticists that Windisch inspired through his teaching at the University of Leipzig (Knott 1919: 265–67; O'Toole 2008: B7). Windisch's greatest prodigies—Heinrich Zimmer (1890–1943) and Kuno Meyer (1858–1919)—went on to become dominant figures in Celtic philology and folklife. Windisch introduced Zimmer and Meyer to his network of Irish Celticists and instilled in them a deep affection for Ireland and its people, along with methodological rigor and a formidable work-ethic (Ó Lúing 1994: 259–63). Both combined their academic interests with political activism. Zimmer—dubbed the "Tiger of Greifswald" for his brusque manner—spoke up for Irish tenant farmers during the Land War in the early 1880s (Ó Lúing 1991: 15). Yet he also approved of the suppression of the speaking of Irish and Welsh languages in schools by the British authorities. Ritual humiliation and beatings were common punishments for Irish and Welsh schoolchildren slipping into the mother-tongues. Zimmer endorsed a similarly robust approach in Polish-speaking areas and sought to encourage the Catholic church to withdraw its support for the Polish language (Gmerek 2003: 33–34).

Zimmer argued on linguistic grounds that Britain played no role in the Celticisation of Ireland (Waddell 1991: 351). Such validation of Ireland's prehistoric independence from Britain was a great boost to the Celtic Revival movement in Ireland (Tierney 1963: 340). The academic standing and resources of the German Celticists advanced the cause of Irish nationalists of all hues. Home Rule—the campaign for a national parliament for Ireland, albeit subservient to the imperial government in London—had considerable support among middle-class nationalists, as well as many British Liberals—including prime minister William Gladstone (1809–1898) who tabled unsuccessful Home Rule Bills in 1886 and 1893. While the third Home Rule Bill was enacted by the Liberal government in 1914, it lacked the support of Ulster Unionists, and implementation was interrupted by the First World War. The partition of Ireland following the Anglo-Irish Treaty in 1921 superseded the Home Rule movement (MacDonagh 1983: 59–65). One of the most influential figures in both the Celtic Revival and Home Rule movements was George Coffey (1857–1916), who qualified as a barrister, but is best known as a political activist and archaeologist (Crooke 2000: 137; O'Halloran 2007: 190). Though not an archaeologist by training, Coffey became curator of antiquities at the Dublin Museum of Science and Art (the future National Museum of Ireland). He went on to become the Museum's first Keeper of Antiquities. Based on his stylistic interpretation, Coffey argued that there was no British influence on Irish 'Celtic' antiquities. He therefore came independently to the same conclusion as Zimmer, arguing that Ireland was invaded and settled by Continental European Celts (Coffey 1910). The idea of the direct colonization of Ireland by European Celts was thus firmly established as one of the great themes in Irish archaeology (Waddell 2005: 184).

Meyer co-founded the School of Irish Learning in Dublin in 1903, with the aim of bringing the foremost Celticists from continental Europe to teach a new generation of Irish Celtic scholars (Mac Cana 2004: 1). "Conservative and aristocratic" (Ó Drisceoil 1991: A8), Meyer was nevertheless a "militant propagandist", who approached Celtic Studies with missionary zeal (Best 1923: 181–82). The successful promotion of the Irish Language by the Gaelic League—chaired by Meyer's friend, the linguist and future first President of Ireland, Douglas Hyde (1860–1949)—created the social and political milieu in which the School of Irish Learning thrived (Ó Lúing 1991: 16–18). Such was the influence of Meyer and his fellow German Celticists on the development of Celtic studies at the School of Irish Learning, and within the field of Celtic scholarship in the colleges of the National University of Ireland, that the inability to speak German was a significant drawback to aspiring Celtic scholars (Ó Lúing 1994: 265). Meyer, however, conflated Celtic Studies and German Imperial interests to try to gain support for Germany during a lecture tour of America during the First World War (Ó Drisceoil 1991: A8). For this, his status as freeman of the cities of Cork and Dublin was rescinded.

9.6 Irish Archaeology and the *Volk*

In 1920, Julius Pokorny (1887–1970) succeeded Meyer as Professor of Celtic Philology at the University of Berlin (Murphy 2016: 200). Pokorny was a leading authority on Old Irish and a friend of Douglas Hyde. He was also a German nationalist and Nazi sympathizer (Ó Dochartaigh 2004: 88; Murphy 2016: 201). Pokorny's research included *Rasse und Volk in Irland* (Pokorny 1917), which racially profiled the people of Ireland. He believed that 'Celtic' people were Aryans, and that nearly half of the Irish could be regarded as "pure Aryan" on account of them apparently having blue eyes (Ó Dochartaigh 2004: 45). Pokorny's research was assisted by Gustaf Kossinna (1858–1931), who had trained as a philologist before becoming an archaeologist (Veit 1989: 37; Heinz 2002: 294). Appointed to the first Chair of prehistoric archaeology in Berlin, Kossinna's *siedlungsarchäologische Methode* (settlement archaeology method) defined ethnocentric German prehistory (Veit 1989: 39; Cornell et al. 2008: 41–42). Kossinna sought to identify the historic 'homelands' of ethnic groups based on the interpretation of material culture, thereby adding further support to ethno-linguistic models (Arnold 1990: 464–65; Wiwjorra 1996: 175–77, Veit 2002). *Siedlungsarchäologie* gained huge political significance as a 'scientific' justification for the Nazi expansion into areas claimed as historic 'homelands' of the *Germani* (Sklenář 1983: 159–61; Arnold 1990: 464).

Kossinna believed that the *Germani*—his ideal Aryan, Nordic race—were the true pioneers of agricultural civilization (Childe 1925: 14–17). This brought together archaeological evidence, language and territory, projecting these back onto the (apparently *dolichocephalic*) Ertebølle 'folk'—the Scandinavian ancestors of the (Indo-) German people (Childe 1925: 15). As archaeologist Gordon Childe (1892–1957) pointed out, this immediately raised the issue as to the genetic origins of the grains and animals that it was claimed were domesticated in the northern European homelands (Childe 1925: 16). A further problem was that the earliest dolmens had been identified in Iberia. Also, the pottery in northern European dolmens bore no relationship to Ertebølle pottery. Kossinna's solution invoked ancient trade links between Scandinavia and Iberia, with Ireland as the staging post (Childe 1925: 16). Kossinna's scheme found favor in Sweden, where archaeologist Oscar Montelius (1843–1921) was instrumental in founding the Svensk-Tyska föreningen (Swedish-German Association) in 1913 (Cornell et al. 2008: 46).

'Aryan', 'Indo-German', and 'Nordic' became interchangeable ethnic identifiers (Arvidsson 2006: 143). By the 1920s, the dominant view of archaeologists in Southern Scandinavia and Germany assumed ethnic continuity extending back through the Iron Age, Bronze Age, and into the Neolithic. Folkloric research was used to project a nationalistic view of contemporary rural life back into prehistory (Trigger 1984: 358). A suite of cultural-historical disciplines encompassing philology, archaeology, folklore, and ethnology evolved, all engaged in creating a perception of a common, stable, racial identity for the Indo-German people (Prescott 2016: 20–21). Swedish *folklivsforskning* (in the sense of regional ethnology) was at the forefront of European folklife research (Ó Giolláin 2017: 9). Eminent Scandinavian folklorists carried out research in Ireland during the interwar years. The Norwegian Reidar Th. Christiansen (1886–1971) and the Swedish Carl Wilhelm von Sydow (1878–1952)—both Irish speakers—were influential in the establishment of the *An Cumann le Bealoideas Eireann* (Folklore of Ireland Society) in 1927, with Douglas Hyde as Treasurer (Ó Giolláin 2017: 7). The Swedish ethnologists Åke Campbell (1891–1957) and Albert Eskeröd (1904–1957; previously known as Albert Nilsson) carried out extensive fieldwork in Ireland during the 1930s (Lysaght 1993: 22–29). Leading Irish folklorists established enduring networks with colleagues in Scandinavia and Germany, engaging in extended study trips and lecture tours (Ó Giolláin 2012: 84–85, 2017: 8–9). It was a Swedish folklorist, Bo Almqvist (1931–2013) from Uppsala University, who established folklore studies at University College Dublin (UCD), where he was appointed Professor of Folklore and Director of the National Folklife Archive in 1972 (Lunney 2009; Ó Giolláin 2017: 9).

In addition to the founding of the Folklore of Ireland Society, 1927 saw the establishment of a committee to report on the purpose of the National Museum of Ireland (NMI). The NMI had been founded as the Dublin Museum of Science and Art (DMSA) by an act of the Westminster parliament in 1877 as a subsidiary to the Department of Science and Art at South Kensington, London (Crooke 2000: 114). During the 1890s, both the RIA and Trinity College Dublin transferred their antiquities collections to the DMSA, and so in effect, to the imperial civil service (Ó Síocháin et al. 2012: 3). The DMSA was initially run along the lines of a provincial city museum in Britain. The most prominent exhibition space was given to material on loan from South Kensington which celebrated British industrialization and imperialism (Crooke 2000: 125). The RIA's collection was housed in a side-room on the first floor. But the subordination of the Irish antiquities collection did not survive the political upheavals of the early-twentieth century. George Coffey's tenure as Keeper of Antiquities began in 1900 (Mitchell 1985: 135). He had previously described the "South Kensington system" as "an attempt to administer art as a department of government, controlled from an Imperial centre" (quoted in Cooke 2012: 73). Coffey set about reorganizing the museum as celebration of Irish culture (Mitchell 1985: 135). The shift towards a more nationalistic culture at the museum gained momentum with the Count (George Noble) Plunkett (1851–1948) as director in 1907 (Cooke 2015: 207). Coffey retired in 1914, and Count Plunkett was removed from office in 1916 by the British authorities due to his association with the 1916 Rising (Raftery 1988: 22). The establishment of the Irish Free State in 1922 was the catalyst for the rebranding of the museum as the National Museum of Ireland (Crooke 2000: 141). It was the Free State government's education department that convened the committee to report on the purpose of the NMI. The committee was chaired by Nils Lithberg (1883–1934), Professor of Nordic and Comparative Folklife Research, and Director of the Nordic Museum in Stockholm (Crooke 2000: 142). Ethnic continuity, folklife, and the rooting of peasant culture in the native soil, were central concepts at the Nordic Museum under Lithberg (Köstlin 1996: 170). These ideals were complimentary to the Free State government's vision of Ireland as a rural society, grounded in the values and traditions of small-scale farming communities—in opposition to urban, industrialized Britain (e.g., Brown 1981: 19; Duffy 1997: 70).

Lithberg argued that folk culture preserved "the principal characteristics of the native race" such that Irish peasants preserved "earlier forms of civilisation which the European peoples [had] abandoned perhaps thousands of years ago" (Lithberg 1927 quoted in Cooke 2012: 177). Archaeology and folklife existed on a continuum, and the purpose of the

NMI "should be to give consecutive representations of the native civilization of the country from the time when the human mind first showed its creative power until the present day" (Lithberg 1927 quoted in Crooke 2000: 143). Pointedly, while Lithberg recommended that the Museum acquire comparative material—both archaeological and ethnological—from Europe, he did not recommend the display of comparative British material (Carew 2008: 22). The colonial narrative that saw Britain as the civilizing influence in Ireland, and Irish archaeology as a derivative of the British discipline, was replaced with the model that emphasized direct European ancestry. The idea that the origins of the Irish lay in continental Europe was uncontroversial in the government of the Free State, which in many ways was defining itself in opposition to the former colonial power (e.g., Richards 1999: 100–1; Carew 2018: 10–11).

Responsibility for the implementation of Lithberg's recommendations would fall to the new Keeper of Irish Antiquities at the NMI. The RIA had favored the appointment of Cyril Fox (1882–1967), a Cambridge archaeologist who went on to become Director of the Museum of Wales. Fox's appointment was, however, vetoed by the Department of Education in favor of Walter Bremer (1887–1926), a prehistorian at the University of Marburg, where he was assistant to Paul Jacobsthal (1880–1957), Professor of Classical Archaeology (Scott-Fox 1988: 79; Ryan 2009: 545). Although Jacobsthal hoped Bremer's role at Marburg would expand such that he would become the University's first Chair in Prehistory, with no immediate prospect of promotion, Bremer took the opportunity to develop his understanding of Ireland's place in the story of German prehistory (Crawford and Ulmschneider 2011: 135). Bremer's archaeological philosophy was close to that of Lithberg's. He was convinced that the "aboriginal population of Ireland were of Iberian origin", and that Ireland was more than an "insignificant appendage to Great Britain' in prehistory" (Bremer 1928: 6–8). He believed lithics recovered from Islandmagee in County Antrim were of a type manufactured in Spain prior to 4000 BC, and that fragments of bronze found in a drawer at the NMI were part of an axe of Scandinavian type (Macalister 1928: 3). Here was apparent proof of Kossinna's theory that Ireland was a staging post between Iberia and Scandinavia in prehistory.

Less than a year after his appointment, Bremer contracted typhoid and died. In the search for Bremer's successor, the Department of Education again advertised in the German newspapers (Mullins 2007). The successful candidate was the Austrian archaeologist and museum curator Adolf Mahr (1887–1951). Having studied anthropology, ethnology, and prehistory at the University of Vienna, Mahr joined the NMI from the Anthropological and Ethnological division of Vienna's Natural History Museum in November 1927 (Stephan and Gosling 2004: 110). He arrived in Ireland having already established an extensive network of European archaeologists and ethnologists. Gordon Childe had stayed with Mahr while studying at the museum in Vienna in the 1920s (Green 1981: 43–44). Mahr was, however, dismissive of the leading Irish archaeologist of the period, Professor R. A. S. Macalister (1870–1950), who held the Chair at UCD from 1909 until 1943 (Carew 2018: 41–42). The implementation of the Lithberg report was Mahr's mandate for radical change:

> At present, there is a commission busy that is reorganising the museum according to modern ideas. I have been asked to give them advice, and together with Swedish expert, Dr. Lithberg, we have been invited to help them. Practically it means that Lithberg and I make out the whole plan, and the commission more or less just 'rubber stamp' it. (Mahr 1927 quoted in Mullins 2007: 30)

Mahr was an adherent of Kossinna's *Siedlungsarchäologie.* He argued that "(…) the present national and racial boundaries are a thing of very late origin and that the boundaries which separated mankind in the Stone Age are probably more important than the makeshifts of our present politicians" (quoted in Mullins 2007: 35). Mahr saw Ireland as the "only surviving Celtic state", its countryside a laboratory where customs and traditions with their origins deep in prehistory could be witnessed in the present. He considered the small-scale farming communities of Ireland's Atlantic seaboard to be the "backbone of European civilisation", the "most important continuum by which the archaeological past and the present society are connected in an unbroken succession" (quoted in Wallace 2007: 198). Rural Ireland was in effect an open-air folk museum, which instead of being curated by ethnologists—as in the case of the open-air folk museums of Scandinavia and Germany—was curated by politicians, keen to reinforce Ireland's identity as the last bastion of ancient Europe (Cooke 2012: 79–80). This conception of Irishness—ethnically 'pure', rooted in the soil, patriarchal, conservative—was very much in keeping with the *Blut und Boden* (blood and soil) movement of *völkisch* 'agrarian romanticism' espoused by the German National Socialists (e.g., Leerssen 2004: 35; McMahon 2009: 588).

Adolf Hitler (1889–1945) was appointed German Chancellor on 30 January 1933, and by 1 April 1933 Mahr's application to join the National Socialists had been processed (Ó Dochartaigh 2004: 94). A little over a year later, on the recommendation of the Department of Education, Éamon de Valera (1882–1975), President of the Executive Council of the Irish Free State, approved Mahr's promotion to Director of the NMI (O'Donoghue 1998: 5). 1934 was also the year that Mahr was appointed to the political rank of *Ortsgruppenleiter*, the most senior Nazi officer in the Irish *Auslandsorganisation*—the organization that oversaw Nazi party affairs in foreign territories (O'Driscoll 2004: 160). As such, Mahr outranked even the German ambassador in Ireland (Wallace 2007: 170). He chaired Nazi party meetings in Ireland, organized social events for German expats (including an Irish branch of the Hitler Youth), and was responsible for regular reporting on the local political situation to the central German intelligence authorities (O'Donoghue 1998: 19–28; Mullins 2007: 56–67). Hitler conferred on Mahr the title "Herr Professor" for his services to National Socialism (Keogh 1998: 106). Mahr's prominence in influential social networks in 1930s Dublin gave him personal access to leading figures, including Éamon de Valera (Wallace 2007: 205–6). One of Mahr's former assistants recalled his making an impromptu visit to de Valera's house to persuade him to write a personal cheque to fund an archaeological excavation, which would later be refunded by the Minister for Finance (Kilbride-Jones 1993: 30).

In 1935, Julius Pokorny was removed from his post as Professor of Celtic at the University of Berlin under the Nuremberg laws, having been identified as having Jewish ancestry (Keogh 1998). He was replaced by Ludwig Mühlhausen (1888–1956)—a former student of Windisch and an ardent National Socialist—who made multiple research visits to Ireland in the 1920s and 1930s (O'Donoghue 1998: 1–2; Hutton 1999: 127). Mühlhausen and Mahr were among the founders of the Deutsche Gesellschaft für Keltische Studien (DGKS: The German Society of Celtic Studies) in 1936 (Hutton 1999: 126). The DGKS had close links to the *Abwehr* (Nazi military intelligence), who recognized the potential strategic benefits of forming alliances with Celtic nationalist movements (Albrecht 1999: 98; Bowd 2013). Just as Mahr used his position to gather detailed maps of Ireland, some of which appear to have been passed on to a German military topographical unit (O'Donoghue 1998: 25), Mühlhausen used his travels in rural Ireland as a cover for espionage. He gathered maps and documents, and took thousands of photographs, some of which ended up in German military documents (O'Donoghue 1998: 12–16).

Both Mühlhausen and Mahr had a strong interest in folkloric research in Ireland. Mühlhausen contributed to the work of the Irish Folklore Commission (IFC), successor of the Folklore of Ireland Society, of which Mahr was a board member (O'Dowd 2012: 108; see Fig. 9.4). Mahr was a great admirer of Swedish folk museums and aspired to a "Scandinavian level" of efficiency at the NMI (Mahr 1937: 420). He had a strong philosophical commitment to folk studies, arguing that "folk research fills the gap between national archaeology and national history" (quoted in Wallace 2007: 198). Mahr (1936: 3) argued that the "National Irish style" of "handicrafts" had its origins in Neolithic art and artifacts. He believed that small-scale farming communities connected present society to the archaeological past "in unbroken succession" yet had "remained all too long the Cinderella of cultural research" (quoted in Wallace 2007: 198). In line with Lithberg's recommendations, Mahr began to assemble a folklore collection at the NMI. In 1937, he organized a major exhibition of Irish folklife at the NMI to celebrate the work of Scandinavian folklorists in Ireland. De Valera and his Minister for Education, Tom Derrig (1897–1956), gave speeches at the opening on the importance of folklife research in Ireland (O'Dowd 2012).

Mahr's extreme politics were, however, insufferable to some. Daniel Binchy (1899–1989), a prominent Irish scholar who had served as an Irish diplomat in Germany from 1929 to 1932 and had written critically on Hitler and the rise of the Nazis (see Binchy 1933), avoided board meetings of the IFC because of his profound objection to Mahr's views (Briody 2007: 62). Similarly, Mahr was appointed to the executive of the new Quaternary Research Committee in the 1930s, which was to oversee a program of palaeobotanical work in Irish peat bogs led by Danish expert Knud Jessen (1884–1971), Professor of Botany at the University of Copenhagen (Mitchell 1990: 11). Jessen's assistant, Hagbard Jonassen (1903–1977)—a committed pacifist—"found it very difficult not to explode when having to listen to Mahr's fascist tirades" (Mitchell 1990: 13–14). With the support of the previous incumbent, Gordon Childe, Mahr was elected president of the British Prehistoric Society in 1937 (see Cahill 2005: 48). Although not an active president, he gave a presidential address in London in October that year (Evans 1989: 440). After drinks at the reception that followed, Mahr was vocal in his support for the Nazi system and espoused antisemitic views (Phillips 1987: 65–66).

Ultimately, Mahr's regard for Ireland was "that of a scientist in a laboratory" (Wallace 2007: 179). Ireland's 'Celtic' language and culture, and the preserved archaeological landscapes inhabited by its traditional farming communities, constituted an ideal testbed for his *völkisch* theorizing. Ireland's strategic importance as Britain's neighbor meant it was also a place he could advance his National Socialist political ambitions (cf. Wallace 2007: 178–83). Mahr's political extremism was inseparable from his approach to archaeology (cf. Veit 1989: 50). Mahr viewed Ireland's people (and, it seems, women in general) with derision, as he demonstrated when an Irish newspaper had the temerity to send a female reporter to interview him:

> (…) in a little educated country, the press is almost the only source of information for a vast majority of people. I was annoyed that a newspaper dared to send me such a poor creature. It will not happen again as I told the head reporter what I thought earlier (…). (quoted in Wallace 2007)

Mahr left Ireland in the spring of 1939, having secured eight weeks' leave from the NMI to attend the International Congress of Archaeology in Berlin, and to take a holiday in his native Austria—recently annexed by Germany (Mullins 2007: 90–91). Mahr also arranged to attend meetings with prominent Nazis in Nuremberg (Keogh 1998: 150). He did not return to Ireland. Both Mahr and Mühlhausen worked for the German Foreign Office in wartime Berlin, establishing

Fig. 9.4 Adolf Mahr (1887–1951; fourth from right) is among the members and associates of the Irish Folklore Commission photographed here in 1937. Also in the photo (front-left) is the Swedish folklorist Åke Campbell (1891–1957). (Photo © National Folklore Collection; reproduced with the permission of the National Folklore Collection, UCD, Dublin, Ireland)

and directing nightly propaganda radio broadcasts to Ireland, *Irland-Redaktion* (O'Donoghue 1998: 64–74). Mahr was arrested for war crimes and imprisoned by British soldiers in Germany in 1946. His legacy to Irish archaeology was, however, secure. He established the NMI—*not* the universities—as the "academy of the future" (O'Sullivan 2009: 525).

The leading archaeologists of the next generation mentored by Mahr included Joseph Raftery (1913–1992), future Director of the NMI, whose son, Barry (1944–2010), later become Professor of Celtic Archaeology at UCD. Barry Raftery's predecessor at UCD was Rúaidhrí de Valera (1916–1978), son of Éamon. He was a Celtic Studies graduate who taught Old Irish and Welsh before studying archaeology and spent time at the NMI under Mahr as well (Mullins 2007: 66). Likewise, Michael Duignan (1907–1988), who became Professor of Archaeology at University College Galway (now University of Galway). Seán Ó Ríordáin (1905–1957), perhaps Mahr's most influential protégé, became Professor of Archaeology at University College Cork, before taking over the Chair at UCD (Wallace 2004). Ó Ríordáin's training under Mahr incorporated extensive study trips to major European universities and museums, with introductions to Mahr's extensive network of archaeologists and folklorists. Between 1932 and 1933, Ó Ríordáin was mentored by eminent scholars ranging from Childe to Mühlhausen to Gerhard Bersu, the then director of the Römisch-Germanische Kommission at Frankfurt (see Wallace 2004). Bersu's Jewish ancestry, along with his outspoken rejection of the appropriation of archaeology for nationalistic and racist purposes, led to his removal from office by the National Socialists in 1935 (Arnold 1990: 472). In a letter dated May 18 1933, Mahr warned Ó Ríordáin to "be careful with [Raymond] Lantier [Keeper at the French Musée des Antiquités Nationales] and do not speak about political developments in Germany because he is a great friend of Bersu [and Wilhelm] Unverzagt [Director of the Museum of Pre- and Early History, Berlin], and there is no use in your being mixed up with inter-continental quarrels" (Wallace 2004: 259). Having been interned as an "enemy alien" on the Isle of Man for a time during the War (Evans 1989: 437), Bersu was appointed to a professorship at the Royal Irish Academy in 1947 (McDowell 1985: 91). According to research carried out by Patrick Wallace (one of Mahr's successors as Director at the NMI), Mahr suspected Bersu of conspiring to ensure that the Irish authorities rejected his November 1947 petition to Éamon de Valera that he be permitted to return to Ireland (Wallace 2007: 176). Wallace (2007: 176) goes on to describe Mahr's "enmity towards Bersu" as longstanding. Wallace's research has made an important contribution in the bringing to light of Mahr's misuse of Irish archaeology in support of National Socialism. It is difficult to reconcile Wallace's findings with his conclusion that "Mahr was not an anti-Semite" (Wallace 2007: 173).

Wallace (2008: 168–69) has described the "first generation of Irish born and educated archaeologists in the postcolonial period" as "strongly nationalist", and the archaeology they practiced as "a justification, even a celebration, of indigenous Irish culture rather than an objective and dispassionate study of the island's past" (Wallace 2008: 168–69). German archaeology, though diminished, continued to exert a strong influence on the Irish discipline in the post-war period (Cooney 1995: 263). Scholarly networks between German and Irish archaeologists remained strong. Many Irish archaeologists continued to study at German universities (Ryan 2009: 545). There was a reluctance to re-engage in theoretical debates in post-war Germany, with the emphasis instead placed on the acquisition of data and its processing within a 'neutral' scientific framework (Härke 1995: 47; Gramsch and Sommer 2011: 22). Disciplinary norms encouraged consensus rather than critical scrutiny (Härke 1995: 49; Cornell et al. 2008: 55). Similar observations have been made about archaeology in Ireland in the post-war period (Cooney 1995: 269; Fontijn and Van Reybrouck 1999: 55, 62; Woodman 2000: 2). There has been no "fundamental change in approaching the interpretation of the [archaeological] data" since Mahr held office at the NMI (Cooney 1995: 268). The enduring paradigm in Irish archaeology—which envisages a timeless rural landscape, with folk traditions and practices scarcely changed since the arrival of the first Neolithic farmers—stems from the *völkisch* ideology of Lithberg and Mahr (Whitefield 2020: 168; cf. Cooney 1995: 269).

9.7 Conclusions

Prehistorian Ulrich Veit (1989: 50) observed that there can be no archaeological research outside of the socio-political circumstances in which that research takes place. Eighteenth-century antiquarianism in Ireland was dominated by the exclusive network of the colonial British elite. Their interest was in maintaining minority rule based upon Britain's supposed cultural superiority. This contemporary political aspiration drew on an imagined past that saw Ireland as forever having been beholden to Britain's civilizing influence. The marginalized Gaelic-Irish historiography of the period combined traditional Irish mythic narratives with conventional Christian historiography. Monastic scribes had embedded Christian orthodoxy into national origin myths across Europe in this way. Overlapping accounts of the Old Testament Patriarchs, Hercules, and the Phoenicians provide evidence of the shared Biblical origins of key aspects of both Irish and British origin myths.

The rise of primitivism and romantic nationalism in Europe, fueled by Herder's *völkisch* philosophy and (somewhat ironically) by the evocative prose of MacPherson's *Ossian*, thrust Ireland's 'Celtic' past to the forefront of the romantic response to the Enlightenment. Ireland's Celtic Revival movement was in the ascendant from the 1830s, attracting powerful support from across Europe, in particular the Nordic regions of Scandinavia and Germany, still reeling in the aftermath of the Napoleonic wars. Romantic nationalism, with its populist nostalgia for imagined golden ages—where language, ethnicity, material culture, rural traditions and rustic landscapes were claimed as clear and unequivocal national characteristics—was fueled by the intense nation-building of the later-nineteenth and early twentieth centuries.

During the interwar years, in the immediate aftermath of the partition of Ireland, archaeology in Ireland and across Europe became an institutionalized discipline, through the creation of national museums, university professorships, and legislation to protect antiquities (Díaz-Andreu and Champion 1996: 6). This was the period in which *völkisch* archaeology was at its height. Kossinna's *Siedlungsarchäologie* (as mediated through Childe's culture-historical archaeology), and the local influence of Lithberg and Mahr's conservative folk archaeology in Ireland, became deeply embedded within the discipline. The embrace of ideologically tainted scholars like Mahr by the Irish political leadership can also be seen as a reflection of Ireland's long struggle against British imperialism. Like other countries under the yoke of colonial Britain, notably India, the 'enemy of my enemy is my friend' phenomenon is almost certainly one of the reasons for the cordial relations between Ireland and Germany. A similar phenomenon has been documented by Leach (2008) in Brittany.

Stripped of its reinterpretation by the far right, Herder's utopian vision of nostalgic nationalism, rooted in the immemorial practices and customs of small-scale rural communities, sat comfortably with de Valera's aspirations for the Irish nation (e.g., Leerssen 2004: 35; Kearney 2007: 139). The continuing political celebration of traditional folkways may help explain the reluctance of many archaeologists in Ireland to challenge the culture historical paradigms on which the national discipline was founded (cf. Leerssen 2004: 35). The major theoretical developments of the later-twentieth century—under the broad headings of processualism and post-processualism—have had limited impact on disciplinary practice in Ireland, where cultural continuity and the maintenance of disciplinary consensus have remained important considerations (Cooney 1995: 264; Woodman 2000: 2–3). While the small professional elite of Irish archaeologists have worked closely with British colleagues, there remain certain fundamental differences of approach. These may in part stem from the historic ties with the German discipline and the continuing pervasiveness of the archaeology-folklife continuum in Ireland (cf. Cooney 1995: 264–68). Whitefield (2020) has observed that this can lead to ostensibly similar archaeological evidence being interpreted very differently on opposing sides of the Irish Sea.

Archaeology's political attractiveness lies in its ability to lend scientific credibility "while simultaneously providing enough ambiguity to support an almost infinite number of interpretations" (Arnold 2006: 174). Archaeologists working within a nationalistic framework (whether consciously or not) lack the critical distance to make disinterested scientific judgements (Prescott 2016: 20–21). By virtue of the history of the discipline, and the social conditions in which it is practiced, archaeology can never be entirely neutral or apolitical. Yet "archaeology is still perceived as telling "true" stories about the past" (Nilsson Stutz 2013: 179). The "national essence" of romantic nationalism is a political construct, not a historical truth traceable by archaeologists (Leerssen 2004: 34–35). But nationalist politics has been a pervasive force in shaping archaeology and the allied cultural historical disciplines of linguistics and folklife studies/ethnology since their inception (Prescott 2016 20). Can this relationship ever be unpicked (Brück and Nilsson Stutz 2016: 1)? At the very least, continuous self-reflection must be seen as an essential aspect of such research (Veit 1989: 50), alongside the enabling and encouragement of critical perspectives from outside established disciplinary networks.

Acknowledgements Thanks to Julie Arendse Voss at Bruun Rasmussen Kunstauktioner for supplying the image of the Høyer painting (Fig. 9.1). Thanks to Zac Johnson and Emma Timoney for reading the draft, and to Cathy, Evan, Libby, and Fionn for putting up with me while I was writing it. I am grateful to the College of Arts, Social Sciences and Celtic Studies at the University of Galway for research support funding which helped with expenses incurred during my research. The success of the conference session was a huge credit to its organizers. Their scholarship and editorial skills have been of tremendous benefit to this chapter and its author.

References

Primary Sources

Binchy, D. (1933). Adolf Hitler. *Studies: An Irish Quarterly Review, 22*(85), 29–47.

Bremer, W. (1928). *Ireland's place in prehistoric and early historic Europe*. Dublin: Hodges Figgis.

Macalister, R.A.S. (1928). *Ireland's place in prehistoric and early historic Europe: A translation of an essay by the late Walter Bremer, Keeper of Irish antiquities at the National Museum of Ireland*. Dublin: Hodges Figgis.

Mackenzie, H. (1805). *Report of the Committee of the Highland Society of Scotland, appointed to enquire into the nature and authenticity of the poems of Ossian*. Edinburgh: Constable.

Mahr, A. (1936). Irish Prehistoric Handicraft. *North Munster Antiquarian Journal, 1*(1), 2–9.

Mahr, A. (1937). New aspects and problems in Irish prehistory. *Proceedings of the Prehistoric Society, 3*, 261–436.

O'Conor, C. (1753). *Dissertations on the ancient history of Ireland: wherein an account is given of the origine, government, letters, sciences, religion, manners and customs, of the antient inhabitants*. Dublin: J. Hoey.

Pezron, P.-Y. (1703/1706). *The antiquities of nations; more particularly of the Celtae or Gauls, taken to be originally the same people as our ancient Britains* (D. Jones, Transl.). London: S. Ballard. (Original work published 1703)

Pokorny, J. (1917). Rasse und Volk in Irland. *Irische Blätter, 7*, 524–528.

Stukeley, W. (1740). *Stonehenge, a temple restored to the Druids*. London: W. Innys, R. Manby.

Stukeley, W. (1743). *Abury, a temple of the British Druids, with some others described*. London: W. Innys, R. Manby, B. Dod, & J. Brindley.

Stukeley, W. (1763). *A letter from Dr Stukeley to Mr MacPherson, on his publication of Fingal and Temora, with a print of Cathmor's Shield*. London: Richard Hett.

Vallancey, C. (1818). *An essay on the antiquity of the Irish language* (3rd ed.). London: Richard Ryan. (Original work published 1772)

Vallancey, C. (1786). Conclusion. Miscellaneous. *Collectanea de Rebus Hibernicis, 3*(12), 554–646. (Original work published 1783)

Van Tieghem, P. (1917). *Ossian en France*. Paris: Rieder.

Secondary Sources

Albrecht, B. (1999). Bretonische Lautaufnahmen und ihr Bearbeiter Rudolf Thurneysen. In S. Heinz (Ed.), *Die Deutsche Keltologie und ihre Berliner Gelehrten bis 1945* (pp. 91–100). Frankfurt am Main: Peter Lang.

Andrews, H. (2009). Stokes, William (1804–1878). In *Dictionary of national biography*. Retrieved November 23, 2021, from https://doi.org/10.3318/dib.008336.v1

Arnold, B. (1990). The past as propaganda: Totalitarian archaeology in Nazi Germany. *Antiquity, 64*(244), 464–478.

Arnold, B. (2006). Pseudoarchaeology and nationalism. In G. Fagan (Ed.), *Archaeological fantasies: How pseudoarchaeology misrepresents the past and misleads the public* (pp. 154–179). London: Routledge.

Arvidsson, S. (2006). *Aryan idols: Indo-European mythology as ideology and science* (S. Wichmann, Transl.). Chicago: University of Chicago Press.

Barnaby, P. (2004). Timeline: European reception of Ossian. In H. Gaskill (Ed.), *The reception of Ossian in Europe* (pp. xxi–lxviii). London: Thoemmes Continuum.

Best, R. (1923). Kuno Meyer. *Ériu, 9*, 181–186.

Bowd, G. (2013). *Fascist Scotland: Caledonia and the Far Right*. Edinburgh: Birlinn.

Boyce, D. G. (1995). *Nationalism in Ireland* (3rd ed.). London: Routledge.

Briody, M. (2007). *The Irish Folklore Commission 1935–1970: History, ideology, methodology*. Helsinki: Finnish Literature Society

Brown, T. (1981). *Ireland: A social and cultural history*. London: Fontana.

Brück, J., & Nilsson Stutz, L. (2016). Is archaeology still the project of nation states? An editorial comment. *Archaeological Dialogues, 23*(1), 1–3.

Cahill, M. (2005). The strange case of the Strangford Lough hoard. *Journal of the Royal Society of Antiquaries of Ireland, 135*, 5–118.

Carew, M. (2008). 'The glamour of ancient greatness': The importance of the Lithberg Report to Irish archaeology. *Archaeology Ireland, 22*(1), 20–22.

Carew, M. (2018). *The quest for the Irish Celt: The Harvard archaeological mission to Ireland, 1932–1936*. Newbridge: Irish Academic Press.

Chapman, M. (1992). *The Celts: The construction of a myth*. Basingstoke: Macmillan Press.

Childe, V. G. (1925). *The dawn of European civilization*. London: Kegan Paul, Trench, Trubner and Co.

Clinton, G., & Sturgeon, S. (2009). Stokes, Whitley (1830–1909). *Dictionary of national biography*. Retrieved November 23, 2021, from https://doi.org/10.3318/dib.008335.v1

Coffey, G. (1910). Archaeological evidence for the intercourse of Gaul with Ireland before the first century. *Proceedings of the Royal Irish Academy: Archaeology, culture, history, literature, 28*, 96–106.

Collis, J. (1997). The origin and spread of the Celts. *Studia Celtica, 30*, 17–34.

Collis, J. (2003). *The Celts: Origins, myths and inventions*. Stroud: Tempus Publishing.

Collis, J. (2017). Celts ancient and modern: Recent controversies in Celtic studies. *Studia Celtica Fennica, 14*, 58–71.

Cook, J. (2004). The discovery of British antiquity. In K. Sloan (Ed.), *Enlightenment: Discovering the world in the eighteenth century* (pp. 178–191). London: British Museum.

Cooke, P. (2012). The National Museum of Ireland: An ideological history. *Australasian Journal of Irish Studies, 12*, 70–85.

Cooke, P. (2015). History, materiality and the myth of 1916. In J. Brück, & L. Godson (Eds.), *Making 1916: Material and visual culture of the 1916 Rising* (pp. 203–216). Liverpool: Liverpool University Press.

Cooney, G. (1995). Theory and practice in Irish archaeology. In P. Ucko (Ed.), *Theory in archaeology: A world perspective* (pp. 263–277). London: Routledge.

Cooney, G. (1996). Building the future on the past: Archaeology and the construction of national identity in Ireland. In M. Díaz-Andreu, & T. Champion (Eds.), *Archaeology and nationalism in Europe* (pp. 146–163). London: University College London Press.

Cornell, P., Borelius, U., Kresa, D., & Backlund, T. (2008). Kosinna, the Nordische Gedanke and Swedish archaeology: Discourse and politics in German and Swedish archaeology 1900–1950. *Current Swedish Archaeology, 15–16*, 39–59.

Crawford, S., & Ulmschneider, K. (2011). Paul Jacobsthal's *Early Celtic Art*, his anonymous co-author, and National Socialism: New evidence from the archives. *Antiquity, 85*(327), 129–141.

Crooke, E. (2000). *Politics, archaeology and the creation of the National Museum of Ireland*. Dublin: Irish Academic Press.

Cunningham, C. (1986). Seventeenth-century interpretations of the past: The case of Geoffrey Keating. *Irish Historical Studies, 25*(98), 116–128.

Curley, T. (2009). *Samuel Johnson, the Ossian fraud, and the Celtic revival in Great Britain and Ireland*. Cambridge: Cambridge University Press.

Díaz-Andreu, M. (2007). *A world history of nineteenth-century archaeology: Nationalism, colonialism, and the past*. Oxford: Oxford University Press.

Díaz-Andreu, M., & Champion, T. (1996). Nationalism and archaeology in Europe: An introduction. In M. Díaz-Andreu, & T. Champion (Eds.), *Archaeology and nationalism in Europe* (pp. 1–23). London: University College London Press.

Duffy, P. (1997). Writing Ireland: Literature and art in the representation of Irish place. In B. Graham (Ed.), *In search of Ireland: A cultural geography* (pp. 64–83). London: Routledge.

Evans, C. (1989). Archaeology and modern times: Bersu's Woodbury 1938 and 1939. *Antiquity, 63*(240), 436–450.

Fontijn, D., & Van Reybrouck, D. (1999). The luxury of abundance: Syntheses of Irish prehistory. *Archaeological Dialogues, 6*(1), 55–73.

Foster, R. F. (1983). History and the Irish question. *Transactions of the Royal Historical Society, 5*(3), 169–192.

Gaskill, H. (1994). Ossian in Europe. *Canadian Review of Comparative Literature, 21*, 643–678.

Gaskill, P. (2004). Introduction: 'Genuine poetry...like gold'. In H. Gaskill (Ed.), *The reception of Ossian in Europe* (pp. 1–20). London: Thoemmes Continuum.

Geikie, A. (1905). *Landscape in history and other essays*. London: Macmillan.

Gmerek, K. (2003). Parczewski the bard. *Polish Anglo-Saxon Studies, 10–11*, 29–36.

Gramsch, A., & Sommer, U. (2011). *A history of Central European archaeology. Theory, methods, and politics*. Budapest: Archaeolingua.

Green, S. (1981). *Prehistorian: A biography of V. Gordon Childe*. Bradford-on-Avon: Moonraker Press.

Hall, S. T. (2007). James MacPherson's Ossian: Forging ancient Highland identity for Scotland. In A. Wawn, G. Johnson, & J. Walter (Eds.), *Constructing nations, reconstructing myth: Essays in honour of T. A. Shippey* (pp. 3–26). Turnhout: Brepols Publishers.

Harbison, P. (1988). *Pre-Christian Ireland. London:* Thames & Hudson.

Harvey, D. (2003). 'National' identities and the politics of ancient heritage: Continuity and change at ancient monuments in Britain and Ireland, c. 1675–1850. *Transactions of the Institute of British Geographers, 28*(4), 473–487.

Haycock, D. (2002). *William Stukeley: Science, archaeology and religion in eighteenth-century England*. Woodbridge: Boydell.

Härke, H. (1995). The Hun is a methodical chap': Reflections on the German tradition of pre- and protohistory. In P. Ucko (Ed.), *Theory in archaeology: A world perspective* (pp. 45–58). London: Routledge.

Heinz, S. (2002). Ur- und Frühgeschichtliche Erkenntnisse in den Arbeiten des Keltologen Julius Pokorny. In A. Leube (Ed.), *Prähistorie und Nationalsozialismus. Die mittel- und osteuropäische Ur- und Frühgeschichtsforschung in den Jahren 1933–1945* (pp. 293–304). Heidelberg: Synchron.

Herity, M., & Eogan, G. (1977). *Ireland in prehistory*. London: Routledge.

Hill, R. (2008). *Stonehenge*. London: Profile Books.

Hutchinson, J. (2001). Archaeology and the Irish Rediscovery of the Celtic Past. *Nations and Nationalism, 7*, 505–519.

Hutton, C. (1999). *Linguistics and the Third Reich: Mother-tongue fascism, race and the science of language*. London: Routledge.

Hutton, R. (2016). The rediscovery of the British Druids. In J. Parker (Ed.), *The Harp and the Constitution: Myths of Celtic and Gothic origin* (pp. 119–133). Leiden: Brill.

James, S. (1999). *The Atlantic Celts: Ancient people or modern invention?* London: British Museum Press.

Kearney, H. (2007). *Ireland: Contested ideas of nationalism and history*. New York: New York University Press.

Keogh, D. (1998). *Jews in twentieth-century Ireland: Refugees, antisemitism and the Holocaust*. Cork: Cork University Press.

Kidd, C. (1994). Gaelic antiquity and national identity in enlightenment Ireland and Scotland. *The English Historical Review, 109*, 1197–1214.

Kidd, C. (1999). *British identities before nationalism: Ethnicity and nationhood in the Atlantic world, 1600–1800*. Cambridge: Cambridge University Press.

Kidd, C. (2004). Subscription, the Scottish Enlightenment and the moderate interpretation of history. *Journal of Ecclesiastical History, 55*(3), 502–519.

Kilbride-Jones, H. (1993). Adolf Mahr. *Archaeology Ireland, 7*(3), 29–30.

Knott, E. (1919). Ernst Windisch, 1844–1918. *Studies: An Irish Quarterly Review, 8*(30), 264–267.

Köstlin, K. (1996). Perspectives of European ethnology. *Ethnologia Europaea, 26*(2), 169–180.

Leach, D. (2008). Bezen Perrot: The Breton nationalist unit of the SS, 1943–5. *e-Keltoi: Journal of Interdisciplinary Celtic Studies, 4*, 1–38.

Leerssen, J. (1986). On the edge of Europe: Ireland in search of oriental roots, 1650–1850. *Comparative Criticism, 8*, 91–112.

Leerssen, J. (1996). *Remembrance and imagination: Patterns in the historical and literary representation of Ireland in the nineteenth century*. Cork: Cork University Press.

Leerssen, J. (2004). Sublime landscape and national character. In G. Holfter, M. Krajenbrink, & E. Moxon-Browne (Eds.),

Connections and identities: Austria, Ireland and Switzerland (pp. 24–36). Bern: Peter Lang.
Leerssen, J. (2006). *National thought in Europe: A cultural history*. Amsterdam: Amsterdam University Press.
Lund, J. (2014). The relationship between archaeology and philology in Denmark in the first half of the 19th century seen through the lens of Friederich Münter, Peter Oluf Brøndsted and Christian Jürgensen Thomsen. In V. Mordvintseva, H. Härke, & T. Shevchenko (Eds.), *Archaeological and linguistic research: Materials of the Humboldt conference* (pp. 19–28). Kiev: Ribe Stift.
Lunney, L. (2009). Almqvist, Bo Gunnar (1931–2013). *Dictionary of national biography*. Retrieved July 5, 2022, from https://www.dib.ie/biography/almqvist-bo-gunnar-a10018
Lysaght, P. (1993). Swedish ethnological surveys in Ireland 1934–5 and their aftermath. In Hugh Cheape (Ed.), *Tools and traditions: Studies in European ethnology presented to Alexander Fenton* (pp. 22–31). Edinburgh: National Museums of Scotland.
Mac Cana, P. (2004). Ériu 1904–2004. *Ériu, 54*, 1–9.
MacDonagh, O. (1983). *States of mind: A study of Anglo-Irish conflict, 1780–1980*. London: George Allen and Unwin.
McDowell, R. (1985). The main narrative. In T. Ó Raifeartaigh (Ed.), *The Royal Irish Academy: A bicentennial history, 1785–1985* (pp. 1–92). Dublin: Royal Irish Academy.
McMahon, R. (2009). Anthropological race psychology 1820–1945: A common European system of ethnic identity narratives. *Nations and Nationalism, 15*(4), 575–596.
McMahon, R. (2016). *The races of Europe: Construction of national identities in the social sciences, 1839–1939*. London: Palgrave Macmillan.
Mitchell, F. (1985). Antiquities. In T. Ó Raifeartaigh (Ed.), *The Royal Irish Academy: A bicentennial history, 1785–1985* (pp. 93–163). Dublin: Royal Irish Academy.
Mitchell, F. (1990). *The way that I followed: A naturalist's journey around Ireland*. Dublin: Country House.
Morgan, P. (1983). From a death to a view: The hunt for the Welsh past in the Romantic period. In E. Hobsbawm, & T. Ranger (Eds.), *The invention of tradition* (pp. 43–100). Cambridge: Cambridge University Press.
Mullins, G. (2007). *Dublin Nazi No. 1: The life of Adolf Mahr*. Dublin: Liberties Press.
Murphy, B. (2016). *Forgotten patriot: Douglas Hyde and the foundation of the Irish presidency*. Dublin: Collins Press.
Ní Lionáin, C. (2012). *Lebor Gabala Erenn*: The use and appropriation of an Irish origin legend in identity construction at home and abroad. *Archaeological Review from Cambridge, 27*(2), 33–51.
Nilsson Stutz, L. (2013). Claims to the past: A critical view of the arguments driving repatriation of cultural heritage and their role in contemporary identity politics. *Journal of Intervention and Statebuilding, 7*(2), 170–195.
Ó Dochartaigh, P. (2004). *Julius Pokorny, 1887–1970: Germans, Celts and nationalism*. Dublin: Four Courts Press.
O'Donoghue, D. (1998). *Hitler's Irish voices: The story of German radio's wartime Irish service*. Belfast: Beyond the Pale Publications.
O'Dowd, A. (2012). Bygones and relics and museum acquisitions: Some thoughts on the Irish folklife collection in the National Museum of Ireland. In S. Ó Síocháin, P. Garvey, & A. Drazin (Eds.), *Exhibit Ireland: Ethnographic collections in Irish museums* (pp. 105–126). Dublin: Wordwell Books.
Ó Drisceoil, P. (1991, October 12). A German zealot. *The Irish Times*. https://nuigalway.idm.oclc.org/login?url=https://www-proquest-com.nuigalway.idm.oclc.org/historical-newspapers/our-cultural-nationalism-catholicism-are-utterly/docview/529130782/se-2?accountid=12899
O'Driscoll, M. (2004). *Ireland, Germany and the Nazis: Politics and diplomacy, 1919–1939*. Dublin: Four Courts Press.
Ó Giolláin, D. (2012). Archives and artefacts: Collecting, collections and ethnography in an Irish and comparative European context. In S. Ó Síocháin, P. Garvey, & A. Drazin (Eds.), *Exhibit Ireland: Ethnographic collections in Irish museums* (pp. 81–104). Dublin: Wordwell Books.
Ó Giolláin, D. (2017). *Irish Ethnologies*. Notre Dame: University of Notre Dame Press.
O'Halloran, C. (1989). Irish re-creations of the Gaelic past: The challenge of MacPherson's Ossian. *Past and Present, 124*, 69–95.
O'Halloran, C. (1990). 'The island of saints and scholars': Views of the early Church and sectarian politics in eighteenth-century Ireland. *Eighteenth-Century Ireland, 5*, 7–20.
O'Halloran, C. (1995). An English orientalist in Ireland: Charles Vallancey (1726–1812). In J. Leerssen, A. H. van der Weel, & B. Westerweel (Eds.), *Forging in the smithy: National identity and representation in Anglo-Irish literary history* (pp. 161–173). Amsterdam: Editions Rodopi B.V.
O'Halloran, C. (2004). *Golden Ages and barbarous nations: Antiquarian debate and cultural politics in Ireland,* c. *1750–1800*. Cork: Cork University Press.
O'Halloran, C. (2007). Review of 'Foundation myths: The beginnings of Irish archaeology by John Waddell'. *Journal of the Galway Archaeological and Historical Society, 59*, 188–190.
O'Halloran, C. (2012). Negotiating progress and degeneracy: Irish antiquaries and the discovery of the 'folk', 1770–1844. In T. Baycroft, & D. Hopkin (Eds.), *Folklore and nationalism in Europe during the long nineteenth century* (pp. 193–206). Leiden: Brill.
Ó Lúing, S. (1991). *Kuno Meyer 1858–1919: A biography*. Dublin: Geography Publications.
Ó Lúing, S. (1994). Celtic scholars of Germany: A brief survey. *Zeitschrift für celtische Philologie, 46*(1), 249–271.
Ó Síocháin, S., Garvey, P., & Drazin, A. (2012). Introduction: Ireland's new ethnographic horizons. In S. Ó Síocháin, P. Garvey, & A. Drazin (Eds.), *Exhibit Ireland: Ethnographic collections in Irish museums* (pp. 3–21). Dublin: Wordwell Books.
O'Sullivan, M. (2009). The life and legacy of R. A. S. Macalister: A century of archaeology at UCD. In G. Cooney, K. Becker, J. Coles, M. Ryan, & S. Sievers (Eds.), *Relics of old decency: Archaeological studies in later prehistory. Festschrift for Barry Raftery* (pp. 521–530.). Dublin: Wordwell Books.
O'Toole, F. (2008, July 5). Our cultural nationalism and Catholicism are utterly European. *The Irish Times*. https://www.irishtimes.com/news/our-cultural-nationalism-and-catholicism-are-utterly-european-1.943046
Okun, H. (1967). Ossian in painting. *Journal of the Warburg and Courtauld Institutes, 30*, 327–356.
Oz-Salzberger, F. (2006). The Enlightenment in translation: Regional and European aspects. *European Review of History*, 13(3), 385–410.
Phillips, C. (1987). *My life in archaeology*. Gloucester: Alan Sutton.
Piggott, S. (1986). William Stukeley: New facts and an old forgery. *Antiquity*, 60(229), 115–122.
Piggott, S. (1989). *Ancient Britons and the antiquarian imagination: Ideas from the Renaissance to the Regency*. London: Thames & Hudson.
Pittock, M. (2008). *Scottish and Irish romanticism*. Oxford: Oxford University Press.
Porter, J. (2001). 'Bring me the head of James MacPherson': The execution of Ossian and the wellsprings of folkloristic discourse. *Journal of American Folklore, 114*, 396–435.
Prescott, C. (2016). Is there an alternative to the nation state? *Archaeological Dialogues, 23*(1), 18–27.
Raftery, J. (1988). A backward look. *Archaeology Ireland, 2*(1), 22–24.
Reed, E. (1965). Herder, primitivism and the age of poetry. *The Modern Language Review, 60*(4), 553–567.
Richards, S. (1999). Breaking the 'cracked mirror': Binary oppositions in the culture of contemporary Ireland. In C. Graham, & R. Kirkland

(Eds.), *Ireland and cultural theory* (pp. 99–118). Baisingstoke: MacMillan.

Rockel, M. (1990). The growth of Celtic studies in Berlin and Leipzig. *Celtic Cultures Newsletter, 6*, 26–33.

Roling, B. (2018). Phoenician Ireland: Charles Vallancey (1725–1812) and the Oriental roots of Celtic culture. In K. Enenkel, & K. Ottenheym (Eds.), *The quest for an appropriate past in literature, art and architecture* (pp. 750–770). Leiden: Brill.

Ryan, M. (2009). A long 1923. In G. Cooney, K. Becker, J. Coles, M. Ryan, & S. Sievers (Eds.), *Relics of old decency: Archaeological studies in later prehistory. Festschrift for Barry Raftery* (pp. 543–548). Dublin: Wordwell Books.

Scott-Fox, C. (1988). A backward look. *Archaeology Ireland, 2*(2), 79.

Scowcroft, R. M. (1987) Leabhar Gabhála Part I: The growth of the text. *Ériu, 38*, 81–140.

Scowcroft, R. M. (1988). Leabhar Gabhála Part II: The growth of the tradition. *Ériu, 39*, 1–66.

Sheehy, J. (1980). *The rediscovery of Ireland's past: The Celtic revival 1830–1930. London:* Thames & Hudson.

Shellenberg, R. (2012). The impact of Ossian: Johann Gottfried Herder's literary legacy. In M. Campbell, & M. Perraudin (Eds.), *The voice of the people: Writing the European folk revival, 1760–1914* (pp. 9–20). London: Anthem.

Sklenář, K. (1983). *Archaeology in Central Europe: The first 500 years*. Leicester: Leicester University Press.

Smith, A. (1986). *The ethnic origins of nations*. Oxford: Blackwell.

Stafford, F. (1988). *The sublime savage: A study of James MacPherson and the poems of Ossian*. Edinburgh: Edinburgh University Press.

Stephan, A., & Gosling, P. (2004). Adolf Mahr (1887–1951): His contribution to archaeological research and practice in Austria and Ireland. In G. Holfter, M. Krajenbrink, & E. Moxon-Browne (Eds.), *Connections and identities: Austria, Ireland and Switzerland* (pp. 105–120). Bern: Peter Lang..

Tierney, M. (1963). What did the Gaelic League Accomplish? 1893–1963. *Studies, 52*, 337–345.

Trigger, B. (1984). Alternative archaeologies: Nationalist, colonialist, imperialist. *Man, 19*(3), 355–370.

Tristram, H. (1996). Celtic linguistic taxonomy in the nineteenth century. In T. Brown (Ed.), *Celticism* (pp. 35–60). Amsterdam: Rodopi.

Vance, N. (1980). Celts, Carthaginians and constitutions: Anglo-Irish literary relations, 1780–1820. *Irish Historical Studies, 22*(86), 216–238.

Veit, U. (1989). Ethnic concepts in German prehistory: A case study on the relationship between cultural identity and archaeological objectivity. In S. Shennan (Ed.), *Archaeological approaches to cultural identity* (pp. 35–56). London: Unwin Hyman.

Veit, U. (2002). Gustaf Kossinna and his concept of national archaeology. In H. Härke (Ed.), *Archaeology, ideology and society: The German experience* (pp. 41–66). Frankfurt am Main: Peter Lang.

Waddell, J. (1991). The Celticization of the West: An Irish perspective. In C. Chevillot, & A. Coffyn (Eds.), *L'Age du Bronze Atlantique. Ses faciès, de l'Écosse a l'Andalousie et leurs relations avec le bronze continental et la Méditerranée* (pp. 349–366). Beynat-et-Cazenac: L'Association des Musées du Sarladais.

Waddell, J. (2000). *The prehistoric archaeology of Ireland* (2nd ed.). Dublin: Wordwell Books.

Waddell, J. (2005). *Foundation myths: The beginnings of Irish archaeology*. Bray: Wordwell Books.

Wallace, P. (2004). Adolf Mahr and the making of Séan P. Ó Ríordáin. In H. Roche, E. Grogan, J. Bradley, J. Coles, & B. Raftery (Eds.), *From megaliths to metal: Essays in honour of George Eogan* (pp. 254–263). Oxford and Philadelphia: Oxbow Books.

Wallace, P. (2007). An error of judgment: Rescuing Adolf Mahr's contribution to the National Museum of Ireland. In M. Dungan (Ed.), *Speaking ill of the dead* (pp. 167–211). Dublin: New Island.

Wallace, P. (2008). Irish archaeology and the recognition of ethnic difference in Viking Dublin. In J. Habu, C. Fawcett, & J. Matsunga (Eds.), *Evaluating multiple narratives: Beyond nationalist, colonialist, imperialist archaeologies*. New York: Springer.

Watson, G. (2009). Celticism: Macpherson, Mathew Arnold and Ireland. In S. Ó Síocháin (Ed.), *Social thought on Ireland in the nineteenth century* (pp. 147–166). Dublin: University College Dublin Press.

Whitefield, A. (2020). Shaky foundations: Romantic nationalism and the development of the 'Irish model' of Neolithic settlement. In A. Barclay, D. Field, & J. Leary (Eds.), *Houses of the dead?* (Neolithic Studies Group Seminar Papers, Vol. 17) (pp. 159–175). Oxbow and Philadelphia: Oxbow Books.

Wilde, W. (1850). *The beauties of the Boyne and its tributary the Blackwater* (2nd ed.). Dublin: James McGlashan.

Wilson, W. (1973). Herder, folklore and romantic nationalism. *Journal of Popular Culture, 6*(4), 819–836.

Wiwjorra, I. (1996). German archaeology and its relation to nationalism and racism. In M. Díaz-Andreu, & T. Champion (Eds.), *Archaeology and nationalism in Europe* (pp. 164–188). London: University College London Press.

Woodman, P. (2000). Hammers and shoeboxes: New agendas for prehistory. In A. Desmond, G. Johnson, M. McCarthy, J. Sheehan, & E. Shee Twohig (Eds.), *New agendas in Irish prehistory: Papers in commemoration of Liz Anderson* (pp. 1–14). Bray: Wordwell Books.

The Young Bairrão Oleiro and the Building of an International Scientific Network

10

Ana Cristina Martins

O estudo exclusivamente teórico
traz sempre consigo grandes desvantagens,
e, por isso, procurei orientar as minhas investigações,
de forma a libertar-me, o mais possível delas.
(Exclusively theoretical studies are always
subject to major disadvantages. This is why I my research
was designed to minimize these disadvantages as far as
is possible.)

(Oleiro 1949a)

Abstract

In the late 1940s, João Manuel Bairrão Oleiro (1923–2000) left for Madrid with a scholarship from the Instituto para a Alta Cultura, a Portuguese state body that financed science and culture. The aim of the young graduate of the University of Coimbra was very clear: to specialize in classical art and archaeology in order to further develop this field in Portugal. This chapter examines the experience he gained during his first research stay in Spain, which allowed him to achieve his goal through the courses and the lectures he attended, the museums he visited, the collections he studied and the personal acquaintances he established with Spanish archaeologists and, through them, French and Italian ones. The chapter also discusses how the Spanish research stay equipped Bairrão Oleiro with knowledge that enabled him to build a wide international scientific network that would be expanded in later years, and on which he would build his own academic path.

Keywords

João Manuel Bairrão Oleiro · Classical archaeology · University of Coimbra · Internationalization · Scientific network · Portugal · Spain

A. C. Martins (✉)
IHC NOVA FCSH – University of Évora Pole and IN2PAST, Évora and Lisbon, Portugal
e-mail: acmartins@uevora.pt

10.1 Introduction

As the first university founded in Portugal (Lisbon, 1290), the Universidade de Coimbra (UC: University of Coimbra) has always maintained strong connections with the study of Greco-Latin sources. Some of the leading figures in this field of research and teaching have passed through the doors of its Faculty of Letters, including the German literary critic and lexicographer Carolina Michaëlis (1851–1925), the first woman to lecture at a Portuguese university (Real 2011: 51). Correspondingly, the UC also sustained a very close relationship with the Instituto de Coimbra (Institute of Coimbra, founded in 1852), an entity promoting scientific and cultural debate and dissemination of research through its journal *O Instituto, Jornal Científico e Literário* (1853), which drew on contributors including the archaeologist Francisco Martins Sarmento (1833–1899) (Leonardo et al. 2009; Ferreira 2011). Highlighting the importance of international exchanges, this Institute also engaged with corresponding members internationally and contributed towards the organization of the congresses of the Portuguese and Spanish Associations for the Progress of Science, initiated in 1921. Furthermore, a museum was founded in 1873 and an archaeology section was created within the Literature, Fine Arts and Arts class of the Institute. This opened the path for the archaeological excavations supervised by archaeologists Augusto Filipe Simões (1835–1884) and Vergílio Correia (1888–1944) among others, which focused especially on the Roman ruins of the site of Conimbriga (Ferreira 2012: 77–78). The Museu Nacional de Machado de Castro (Machado de Castro Museum, currently the Machado de

L. Coltofean et al. (eds.), *Connecting People and Ideas*, Themes in Contemporary Archaeology,
https://doi.org/10.1007/978-3-031-81006-0_10

Castro National Museum), which was inaugurated in 1913, incorporated Simões and Correira's collected finds (Ferreira 2012: 86) but by 1950 a decline began due to the need to internationalize science in a planned, centralized, and publicly financed way (Rollo et al. 2012). Nevertheless, such factors did not impact UC's dominance in the field of classical studies. This is reflected in the career profile of the philologist Maria Helena da Rocha Pereira (1925–2017), a grant holder from the Instituto para a Alta Cultura (IAC: Portuguese Institute for High Culture, active between 1936/1952–1976) at the University of Oxford, where she was a disciple of, among others, John D. Beazley (1885–1970), the art historian, archaeologist, and a renowned specialist in Classical ceramics. Her trajectory of academic success (she became a Full Professor in 1964) is especially notable given how few women were able to completely dedicate themselves to university life with such success at that time.

It is in the general context outlined above that João Manuel Bairrão Oleiro (1923–2000) took his first steps as a promising academic, deepening his passion for Roman archaeology and setting off abroad for the first time in search of knowledge and experience that would enable him to contribute to renewing archaeological activities in Portugal (Maciel 1996). This chapter examines how he achieved this goal due to an extensive network of personal, public as well as private, individual, and collective contacts.

10.2 A Desire Greater than Reality

João Manuel Bairrão Oleiro (Fig. 10.1) was born in Abrantes on January 15, 1923, the year that Portugal held its seventh presidential election. His father, Diogo Armando da Silva Oleiro (1887–1962), was an enthusiastic collector of antiquities, and was instrumental in the establishment of the Museu Regional Dom Lopo de Almeida (Dom Lopo de Almeida Regional Museum) in the same city by Ministerial Decree in 1921. He eventually became the director-conservator of this museum (Charneca 2005: 94, 96–103). Bairrão Oleiro thus grew up in a family which was particularly dedicated to the arts and letters, coupled with his own undeniable interest in studying the past across its most diverse dimensions: "desde muito novo dediquei especial atenção a questões de arqueologia e História da Arte. Ajudei meu Pai em trabalhos vários e fiz com ele diversas visitas de estudo, tomando apontamentos e fazendo fotografias" (From a very young age, I paid particular attention to questions about archaeology and the history of art. I helped my father with various jobs that he did and made various study visits with him, taking notes and photographs) (Oleiro n.d.). Following the death of his father, Bairrão Oleiro directed the Dom Lopo de Almeida Regional Museum; he did so *pro bono* for 16 consecutive years between 1963 and 1979.

Fig. 10.1 João Manuel Bairrão Oleiro (1923–2000). (Photo: Unknown author; copyright free, available through the website of the NOVA University of Lisbon, School of Social Sciences and Humanities [NOVA FCSH] and the Doação Bairrão Oleiro [Donation Bairrão Oleiro], Lisbon, Portugal, https://www.fcsh.unl.pt/faculdade/bibliotecas/bmsc/espolios-e-doacoes/doacao-bairrao-oleiro/)

It therefore comes as no surprise, due to his own experiences and inclinations as well as his family's social circle and its proximity to his place of birth, that the young Bairrão Oleiro enrolled in the Historical-Philosophical Sciences degree program in the Faculty of Letters at the UC in 1942. At the same time, he continued his studies in Spanish, French and English, which proved helpful to him in his education as well as his future career. This all took place against the backdrop of the totalitarian policies of the Estado Novo (1933–1974), already rooted in the country at that time, and in an Europe immersed in the convulsions of the Second World War. Bairrão Oleiro completed his degree in 1947 with the classification of "Good" (14 out of 20), an excellent average given the parameters of university standards prevailing at that time. However, it was the lectures in archaeology and art history by art historian Aarão de Lacerda (1890–1947)[1] that generated Bairrão Oleiro's greatest enthusiasm; in those subjects he received a final grade of 17. In addition, in late 1945, he published in the journal *Museu* some brief notes on a Neolithic find. Thus, Bairrão Oleiro started out as a researcher from the initial stages of his career, and above all as an archaeologist, as "o maior desejo é dedicar-me à arqueologia, ciência para q. sinto irresistível atracção" (the great-

[1]The same subjects were taught by the art historian, archaeologist, and museologist Vergílio Correia, the former conservator at the Ethnographic Museum of Belém in Lisbon, between 1912 and 1915, and during his term as director of the Machado de Castro Museum in Coimbra, between 1929 and 1944, the year of his death.

est desire is to dedicate myself to archaeology, a science for which I feel an irresistible attraction) (Oleiro n.d.).

Perhaps through the influence of his father, in 1946 Bairrão Oleiro enrolled in the five-week long 4th University Summer Course in Santiago de Compostela. This gave him the opportunity to attend classes on art, archaeology and literature while studying the Compostelan Romanesque under the guidance of priest, archaeologist, and writer Jesus Carro García (1884–1973). He achieved this by obtaining one of the two study grants awarded that year by the UC to attend its counterpart in Spain. These first achievements were facilitated by his undergraduate dissertation, for which he visited various archaeological sites, monuments and museums in the Algarve, the collections at the Museu Etnológico do Dr. Leite de Vasconcelos[2] (Dr. Leite de Vasconcelos Ethnological Museum, currently the National Museum of Archaeology), as well as by his participation in the excavations at the Roman ruins of Conimbriga. As a result of his positive experience at the summer course, which had been so important in helping him to reach his goals, he actively sought out other institutions to finance his academic specialization abroad.

Bairrão Oleiro's dedication and proficiency as a researcher led to an invitation to join the Instituto de Estudos Históricos Doutor António de Vasconcelos (Doctor António de Vasconcelos Institute of Historical Studies) at the UC, then run by Professor Torquato de Sousa Soares (1903–1988). Torquato Soares approached the IAC to request a grant for Bairrão Oleiro to comply with the integrated scientific research needs of the Institute of Historical Studies. The proposal described Bairrão Oleiro as a scholar "que se propõe estudar alguns aspetos da romanização na parte do território português compreendido entre o Cabo de S. Vicente e o Guadiana, tomando como base testemunhos arqueológicos" (who plans to study aspects of Romanization based on archaeological evidence in the area of Portuguese territory between Cabo de S. Vicente and the Guadiana) (Oleiro 1947). Furthermore, Torquato Soares saw in this grant an opportunity both for the Institute of Historical Studies and for science in Portugal,

> uma vez que não há em Portugal nenhum especialista de arqueologia romana, e que o candidato que propomos deu já as suas provas no trabalho apresentando como dissertação de licenciatura: "Alguns aspetos da vida económica do antigo Algarve ao tempo da ocupação romana, comprováveis pelos testemunhos arqueológicos".
> (because Portugal currently lacks an archaeologist specializing in Roman archaeology, and the candidate we propose has already demonstrated his suitability for this work through the submission of his undergraduate dissertation "Some aspects on the economic life of classical Algarve at the time of Roman occupation, as demonstrated by the archaeological evidence.") (Oleiro 1947)

Bairrão Oleiro's undergraduate dissertation served as the foundation and guiding framework for the study plan, he had set out for the grant he received at the IAC. In addition to correlating archaeological evidence and literary sources to identify the stages and singularities of the Roman presence in what would become Portuguese territory, Bairrão Oleiro sought to begin his study in the Algarve by analyzing,

> I – O meio geográfico. Exame de alguns dos textos dos antigos geógrafos. // II – Povos que habitaram a zona entre o "Sacrum" e o "Anas". // III – Cidades principais: "Ossonoba", "Balsa", "Baesuris", "Lacobriga". (O problema das localizações). // IV – "Villae". // V – "Aedificia". // VI – Banhos (Milreu, por exemplo) e Termas (Monchique). // VII – Estradas (seu traçado conjectural). // VIII – Vida económica (alguns aspectos principais, como tinturaria, mineração, etc., com especial referência à pesca e indústrias correlativas, como actividade dominante). // IX – Cultos e Santuários.
> (I – The geographic context of the classical documentary sources. // II – The peoples inhabiting the zone between "Sacrum" and "Anas". // III – The main cities: "Ossonoba", "Balsa", "Baesuris", "Lacobriga", particularly their locations. // IV – "Villae". // V – "Aedificia". // VI – Baths (Milreu, for example) and thermal spas (Monchique). // VII – Roads (their conjectural layout). // VIII – Economic life (some key aspects, such as textile dyeing, mining, etc., with a particular reference to fishing and its associated industries as the dominant activity). // IX – Religions and sanctuaries.) (Oleiro 1947)

Having obtained a grant in 1948 to study thanks to the IAC, Bairrão Oleiro was able to secure additional funding for a study trip to Spain, specifically to the museums in Barcelona, focusing on Roman archaeology (Oleiro 1949b). Furthermore, probably due to the appreciation shown of the work undertaken during his scholarship in Spain, Bairrão Oleiro "será considerado bolseiro de colaboração com o Conselho Superior de Investigações Científicas do País vizinho" (will be considered a grant holder supported by the Spanish Higher Council for Scientific Research) (Oleiro 1949b). This meant that he received a monthly stipend from the IAC that not only funded his work and provided his research plan with greater prestige but also placed greater responsibility on his subsequent performance (Martins 2022, 2023).

10.3 Madrid: The Beginning of an International Career

Bairrão Oleiro arrived in Madrid on May 7, 1949, the year Portugal joined NATO and the Spanish head of state, *caudillo* Francisco Franco (1892–1975), visited Lisbon. Bairrão Oleiro immediately sought out José Maria Albareda (1902–1966), one of the first members of Opus Dei (1928) and a close friend of its founder José María Escrivá de Balaguer (1902–1975). Albareda was of special significance

[2] José Leite de Vasconcelos (1858–1941) was founder and first director of the Portuguese Ethnology/Ethnographic Museum (Lisbon, 1893), as well as ethnographer, philologist, and university professor in Lisbon.

for the young grant holder, as he was the Secretary-General of the Consejo Superior de Investigaciones Científicas (CSIC: Higher Council for Scientific Research, founded in 1939) and the future rector of the Universidad de Navarra (University of Navarra) between 1960 and 1966. Bairrão Oleiro requested an audience but was unable to locate Albareda. Two days later he was finally able to meet with Albareda's secretary who delivered a letter addressed to the director of the Museu Arqueològic Provincial de Barcelona (Provincial Archaeological Museum of Barcelona), today the Museu d'Arqueologia de Catalunya (MAC: Archaeology Museum of Catalonia) and ordered the payment of Bairrão Oleiro's first month of the CSIC grant (Oleiro 1949b, i). As soon as the funds had been secured, Bairrão Oleiro attempted to purchase a train ticket to get to Catalonia. However, the difficulties encountered in this process forced him to delay the journey until the following week.

Having to remain in the Spanish capital for longer than foreseen, he approached Blas Taracena Aguirre (1895–1951), the director of the Museo Arqueológico Nacional (MAN: National Archaeological Museum) and secretary of the Instituto Diego Velásquez (IDV: Diego Velázquez Institute) of the CSIC between 1942 and 1951. Aguirre was also involved in the major international archaeological projects in Spain at the time (Gómez Barreda 2016).[3] Taracena welcomed his Portuguese visitor hospitably and granted him every facility for working at the MAN, including entry to its library and to the IDV, "com livre acesso aos livros, ficheiros fotográficos, etc." (with free access to books, photo archives, etc.) (Oleiro 1949a). Thus did Bairrão Oleiro begin "a colher elementos nestes centros de estudo, cuja organização me parece modelar" (to collect information in these study centers whose organization is of high standard) (Oleiro 1949c).

It was at the IDV that Bairrão Oleiro met Antonio García y Bellido (1903–1972), a Full Professor in classical archaeology at the Universidad Central (Central University, today Universidad Complutense/Complutense University) in Madrid (García y Bellido 1951; Olmos 1994; Bendala et al. 2005; Schattner 2005; Mederos 2011–2012, 2017). Bairrão Oleiro presented his work plans to García y Bellido who considered them of such interest and relevance that he made the young researcher's stay as productive as possible. In the meantime, Taracena and Bellido informed Bairrão Oleiro of an archaeology course planned by the IDV in Cartagena in southeastern Spain and the Balearic Islands between May 18 and June 28, 1949. This was the Curso de Arqueología en el Sudeste y Baleares (Archaeology Course in the Southeast and the Balearic Islands). Enthusiastic at this opportunity to attend and having secured the agreement of Taracena and Bellido, Bairrão Oleiro requested authorization from the IAC Secretary-General, António de Medeiros Gouveia (1900–1972) in order to enroll,

> não só porque nele teria oportunidade (talvez única) de assistir a trabalhos práticos, mas também porque ficaria conhecendo uma das regiões, arqueologicamente, mais ricas de Espanha. Neste curso tomarão parte professores de arqueologia, diretores de museus, especialistas estrangeiros, etc.
> (not only because this provides a possibly unique opportunity to assist with practical activities, but also because this would provide knowledge about one of the archaeologically richest regions of Spain. This course has attracted the attention of archaeology professors, museum directors, international specialists, etc.) (Oleiro 1949c)

Bairrão Oleiro also took the opportunity to introduce another subject into his letter addressed to Medeiros Gouveia from his personal residence, located on the fourth floor of number seven on Calle Serrano, i.e., next to MAN. He wrote that in the opinion of professors Tarcena and Bellido his internship should begin not in Barcelona but rather in Madrid. The justification seemed simple and understandable as in Madrid he would be working more on prehistoric than on Roman archaeology. Moreover, "aqui [Madrid] teria mais facilidades bibliográficas por poder utilizar as bibliotecas do Museu, do Instituto Velázquez e as particulares dos citados professores" (here [Madrid] he would have the opportunity to access more bibliographic resources in the libraries of the Museum and of the Velázquez Institute, and in the private libraries of the aforementioned professors) (Oleiro 1949c). In addition, being located in Madrid would give him easier access to the collections and museums in the rest of the country and make it possible to participate in archaeological digs. Hence, contrary to his original plans, Bairrão Oleiro was requesting permission to stay in the Spanish capital during September, "fazendo uma preparação teórica, finda a qual me deslocaria a Barcelona onde veria, nas ruínas de Ampúrias, cortes estratigráficos e excavações dirigidas pelo Professor Almagro, acompanhando os trabalhos de duas escolas diferentes" (undertaking a theoretical preparation, after which I would travel to Barcelona where I would see the stratigraphic cross-sections and excavations led by Professor Almagro in the ruins of Ampurias, following the work of two different schools [Madrid and Catalonia]) (Oleiro 1949c).

Five days later, Bairrão Oleiro received a positive response from Medeiros Gouveia (Oleiro 1949d). The answer reveals the esteem in which he was held by the IAC head office due to his serious dedication and the proficiency with which he complied with his tasks, coupled with the quality recommendations provided by his supervisors. The IAC letter arrived on the eve of his departure for Cartagena where the IDV

[3] One of these projects in Spain was *Corpus vasorum antiquorum*, initiated by the Union Académique Internationale (Brussels, founded in 1919).

course would take place, a journey on which he accompanied Taracena and Bellido. He nevertheless committed to returning to Madrid as soon as the course was finished on June 16 (Oleiro 1949f). In Madrid, he was to resume his studies at MAN in the morning and the IDV in the afternoon (Oleiro 1949a), at least until leaving for Barcelona (Oleiro 1949e).

In Bairrão Oleiro's own words, the course turned out to be "extraordinariamente interessante e proveitoso para todos q. nele tomaram parte" (extraordinarily interesting and beneficial for all those who took part") (Oleiro 1949f). There were many participants in the course, especially professors, representing various leading Spanish, French and Italian archaeological institutions.[4] In addition to Bairrão Oleiro, there was another Portuguese participant, Fernando Russel Cortez (1913–1994), from the Museu Soares dos Reis[5] (Soares dos Reis Museum, founded in Oporto in 1833). In Cartagena, on the morning of the first course day, May 28, the participants made a trip to the island of Escombreras in order to observe the collection of Roman amphoras rescued by Spanish navy divers. At midday, the official opening sessions took place, presided by Admiral Francisco Bastarreche[6] (1882–1962), head of the naval base in Cartagena and an archaeology enthusiast (Gracia Alonso 2021). The first lesson was then in the care of Taracena. In the afternoon, participants visited Cartagena's archaeological ruins and the museum where the prehistorian, epigraphist, and numismatist Antonio Beltrán (1916–2006) held a conference. In the evening, they listened to medieval music expert Agustín Isorna's (1914–1966) lecture on the polyphonic and popular dimensions of Spanish music.

On the following day, the course participants set off by car for the excavation sites of Villajoyosa, Benidorm, Peñon de Ifach and Jávea. On their return, they stopped off in Alicante for the official inauguration of the Museo Arqueológico Provincial de Alicante (Provincial Archaeological Museum of Alicante, today the Museo Arqueológico de Alicante/Archaeological Museum of Alicante), while also attending two lectures, one given by its director, Father José Belda Dominguez (1890–1969), and the other by Bellido. On the morning of June 30, they traveled to Múrcia, where they spent the afternoon engaged in individual study in the Museo de Cartagena (Museum of Cartagena, founded in 1943) followed by a brief speech given by the Portuguese archaeologist and museologist F. Russell Cortez (Oleiro 1949a). On the following day, they sailed to Ibiza on the ship Magallanes and, on July 1, visited the city and attended a conference given by the French specialist Fernand Benoit (1892–1969) on the "Greek Marseille". The Ibiza sojourn continued on July 2 with Bairrão Oleiro highlighting San José for the "festival típico que, para nós, tinha extraordinário interesse pelas sobrevivências que acusa e pelo muito que pode sugerir à arqueologia" (typical festival that, for us, held extraordinary relevance due to the surviving features that it registers and everything that these might suggest for archaeology) (Oleiro 1949a). This statement expresses the level of attention he paid to the realities around him and the awareness of the importance of ethnographic records not only to archaeological prospecting and excavations but also to understanding the lived experiences represented by such contexts.

In the morning of July 3, participants sailed from Ibiza to Mallorca where they remained before sailing on to Menorca in the same vessel Magellan. Having arrived at Menorca, they were welcomed into the Ateneo Científico, Literario y Artístico (Scientific, Literary and Artistic Athenaeum, founded in 1905) of Mahón. July 7 and 8 were dedicated to visiting megalithic monuments and attending the conference organized by Jole Bovio Marconi (1897–1986), the director of Palermo's Museo Archeologico Regionale 'Antonino Salinas' ('Antonino Salinas' Regional Archaeological Museum) and of the Soprintendenza alle Antichità della Sicilia Occidentale (Superintendence of Antiquities of Western Sicily) (for more about Bovio Marconi and her network, see Crisà, Chap. 12, this volume). This event was not significantly commented on by Bairrão Oleiro, even though

[4] "D. Blas Taracena, director of the National Archaeological Museum; D. António Garcia y Bellido, Full Professor, Central University; António Beltrán, director of the Museum of Cartagena; Raymond Lantier, director of the Museum of National Antiquities, Saint-Germain-en-Laye; Fernand Benoit, from the Museum of Borély, Marselle; Madame Jole Bovio Marconi0, from the Museum of the University of Palermo; Alejandro Ramos Folques, director of the Museum of Elche; Octavio Gil Farrés and Augusto Fernandez Avilés, keepers of the National Archaeological Museum; Paquita Rodríguez Pedroviejo, director of the Museum of Malaga; Samuel Ventura, director of the Museum of Tarragona; Concepcion Fernandez Chicarro, secretary of the Museum of Sevilla; Clarissa Millán, curator from the National Museum of Archaeology; Joaquin Sanchez Jiminez, director of the Museum of Albacete; Russell Cortez, from the Museum of OPorto; Maria Luisa Galván and Isabel Ceballos, curators of the National Museum of Archaeology; Miguel Oliva Prat, curator of the Museum of Girona; Pedro de Palol, director of the Museum of Girona; Lidia e Ester Boggia, from the Museum of Locarno; Madames Crivelli e Spinelli, from the same museum; José Tudela de la Orden, deputy director from the Museum of América; Aurélio Vinas, professor from the Sorbonne; José Maná, director of the Museum of Ibiza; Joaquín Tomás, from the Board for Prehistoric Researches of Teruel; Nino Lamboglia, director of the Institute for Ligurian Studies, from Bordighera; José Colominas, curator of the Archaeological Museum of Barcelona; Luis Amorós, director of the Museum of Artá and Island Commissioner of Excavations; Manuel Solá, director of the Institute of Ibiza; Luis Pericot, from the University of Barcelona" (Oleiro 1949a). However, it should be noted that the list includes the names of scholars who participated in the third iteration of this course but who, for unknown reasons, were not mentioned by Bairrão Oleiro (Díaz-Andreu et al. 2009: 23–24).

[5] Currently the Soares dos Reis National Museum.

[6] Bastarreche was responsible for, among other episodes, *La Desbandada*, the massacre perpetrated against thousands of civilians who were fleeing on the road from Málaga to Almería on February 8, 1937, during the Spanish Civil War (1936–1939).

encountering a female archaeologist and museologist was then still a great rarity, especially in Portugal (Martins 2016). On July 9, the participants traveled to Palma de Mallorca. Between July 10 and 13, they visited talayots (Bronze Age megaliths in Menorca and Majorca) and participated in the conferences organized by the French archaeologist Raymond Lantier (1886–1980) and his Italian peer Nino Lamboglia (1912–1977), the founder and director of Istituto Internazionale di Studi Liguri (IISL: International Institute of Ligurian Studies, founded in Bordighera in 1937). Upon arriving in Valencia on July 12, they first toured the city. On July 13 they visited the Museu de Prehistòria de València (Prehistory Museum of Valencia), before going on to the city of Sagunto, where they had the opportunity to observe the Roman theatre and museum. July 14 was dedicated to visiting the pile dwelling site of Pedregal near the city of Navarrés, where they studied a stratigraphic section and practiced excavation techniques. The course ended on July 15 with a keynote speech by the Spanish prehistorian Luis Pericot (1899–1978) (Oleiro 1949a; Gracia Alonso 2017).

On the day the course ended, Bairrão Oleiro returned to Madrid and to his normal working routine. However, the depth of experience he gained on this course, which provided knowledge of relevance to his professional development, is clearly reflected in his work from then on. In addition to all the visits and conferences, "a convivência com mestres e estudiosos, favoreceu amizades e relações de intercambio científico, duma maneira particularmente eficaz" (the interaction with scholars fostered friendships and scientific exchanges in a particularly effective way) (Oleiro 1949a). Some of those relationships became genuine exchanges; through them, Bairrão Oleiro sought to interest his colleagues in Portuguese archaeology and put them in contact with study centers in Portugal. This was undertaken in the name of developing science in his homeland with the awareness that this same goal depended on international, interinstitutional and interpersonal collaborations, as had been fostered by the courses in Ampurias (see below).

Bairrão Oleiro's participation in the course in Cartagena proved mutually advantageous both to him and to the colleagues he met on that trip. Through Pedro de Palol (1923–2005), director of the Museo Arqueológico Provincial de Gerona (Provincial Archaeological Museum of Girona), whom he met in Cartagena, he became aware of the IAC's announcement about the program of the third iteration of Cursos Internacionales de Prehistoria y Arqueología en Ampurias (International Courses on Prehistory and Archaeology in Ampurias). The course had been organized since 1947 by the MAC and the Universitat de Barcelona (University of Barcelona); for the third iteration in 1949, they also collaborated with the Université de Montpellier (University of Montpellier) and the IISL due to the personal relations of Spanish prehistorian, museologist, and university professor Martín Almagro Basch (1911–1984) and Pericot (Gracia Alonso 2012).[7] Bairrão Oleiro requested to attend the course, at least the part held in Spanish territory, between September 6 and 16, which included visits to Barcelona, Ampurias and Tarragona (Oleiro 1949g). He wished to attend the course "aproveitando as conferências, excursões e excavações dirigidas pelos professores (...) tanto mais que ao curso concorrem professores estrangeiros e teria, assim, oportunidades para estabelecer relações de intercâmbio" (to take advantage of the conferences, excursions and excavations supervised by the professors [...] all the more so since the course is attended by foreign professors and would thus have opportunities to establish exchange relationships) (Oleiro 1949g). This statement reveals his strategy of establishing scientific networks fundamental to his future research, fostering multidisciplinarity, interdisciplinarity, and internationalization in the name of developing archaeology in Portugal.

By the time of the course in Ampurias, Bairrão Oleiro was lodging at the Residencia de Estudiantes (Student Residence), dependent of the CSIC, located on Pinar Street in Barcelona, somewhat further away from the MAN. Here, he would stay until August 21 when the residence closed its doors for the summer holidays, like every Spanish teaching and research institution. This caused no disruption to his plans because he had intended to return to Portugal at this time to marry Maria Helena de Seixas Nogueira de Lemos (1924–2004) on September 1, 1949. Furthermore, on the grounds of the general dearth of academic and research activity in August, traditionally a vacation month (Oleiro 1949g), he suggested spending the entire month in his country of birth, thereby extending his grant until the end of December and not November, as was originally proposed. This amended schedule received the immediate approval of the CSIC and ensured a positive response from Lisbon (Oleiro 1949h).

Before the journey back to Lisbon, Bairrão Oleiro pushed on with his working plans and visited Toledo to study the Roman mosaics discovered in the Fábrica de Armas, as well as the Roman ruins that had been unearthed on the site of Vega Baja. He also strove to update and organize his bibliographic files "para trabalhos de monografia que penso realizar" (for the monographic works I am thinking of doing), while simultaneously studying "alguma coisa a respeito de técnica de excavações, cerâmica, museografia, etc." (something about excavation techniques, ceramics, museography, etc.) (Oleiro 1949a) which reflected his ongoing interest in

[7]While Portugal was lined up with the Allied powers, Bairrao Oleiro's period in Madrid and Barcelona coincides with the international diplomatic isolation of Spain decreed by the United Nations General Assembly (Sánchez González 2015), broken on rare occasions, such as the International Courses of Ampurias that facilitated the presence of foreigner researchers in Spain, as an example of the role played by scientific diplomacy.

the relationship between theory and practice. These activities were fundamental to the research he would subsequently carry out when back on Portuguese soil, especially as regards the conservation and restoration of archaeological materials:

> Foi sempre preocupação dominante no meu espírito o aprofundar das questões que interessam particularmente à Península Ibérica, em especial a Portugal; assim como procurei sempre completar o estudo feito nos livros, com o conhecimento dos objectos, já nos museus, já "in loco". (…) // Por exemplo: estudei, em revistas da especialidade, a maneira de levantar, trasladar, restaurar e conservar os mosaicos romanos; mas, depois, durante vários dias trabalhei, no Museo Arqueologico, com os restauradores arqueológicos da Dirección General de Bellas Artes de Espana, que me ensinaram os seus processos de trabalho valorizados por uma longa experiência.
> (It has always been the dominant concern in my spirit to deepen questions that are of particular interest to the Iberian Peninsula, especially to Portugal; just as I have always striven to complete the study made in the books, with the knowledge of objects, already in museums, already "in loco". […] // For example: I studied, in the specialist journals, the means of surveying, moving, restoring, and conserving Roman mosaics; but, subsequently, during various days I worked at the Archaeological Museum, with the archaeological restorers of Dirección General de Bellas Artes de España, who taught me their working processes of such value due to their long experiences.) (Oleiro 1949a)

Anticipating his future career trajectory, Bairrão Oleiro analyzed in detail the type of file organization in effect at the IDV while also finalizing his study of aerial photography applied to archaeology.[8] As aerial photography at that time had only rarely been carried out in Portugal,[9] he hoped that he could "aplicar, com vantagem para a arqueologia portuguesa, os conhecimentos que estou adquirindo" (apply the knowledge that he acquired in this field in Portuguese archaeology). (Oleiro 1949a).

[8]He completed this study through the observation of examples obtained by the Spanish airforce on the direct request of the IDV. It is interesting to note that the introduction of aerial photography in Spain after the Civil War marks a new confrontation between the IDV of the CSIC, led by Bellido, and the General Commissariat of Archaeological Excavations, headed by Julio Martínez Santa-Olalla (1905–1972), whom Bairrão Oleiro does not seem to have approached.

[9]Aerial photography appears to have been first applied to archaeology in Portugal by Manuel Heleno (1894–1970), professor of the Faculty of Letters at the Universidade de Lisboa (University of Lisbon) and director of the Museu Nacional de Arqueologia (National Archaeological Museum) in the same city. He did this as early as 1934 to survey the Visigoth necropolis in Silveirona (Santo Estevão, Estremoz) and later the Roman villa of Torre de Palma (Monforte), based on a report he gave to his students in the academic year 1948–1949 (Cardoso 2021: 136).

10.4 Barcelona and the Third Course on the Prehistory and Archaeology of Ampurias

On his return from Lisbon, Bairrão Oleiro traveled directly to Barcelona, where he arrived on the morning of September 5. He booked a room at the Hotel Mediterráneo, on the Rambla de Cataluña (Oleiro 1949j), and immediately contacted the secretary of the third Course on the Prehistory and Archaeology of Ampurias, Juan Maluquer de Motes (1915–1988). On this same day, he was received along with his fellow course participants at the Diputación Palace. The French delegation was led by the historian and archaeologist Albert Grenier (1878–1961) and the Italian delegation by Pietro Romanelli (1889–1981), the director of excavations at the Roman Forum and the Palatine in Rome. Bairrão Oleiro was the only Portuguese scholar attending the course, a circumstance that did not hinder his active participation and, as he noted in his account of the course,

> interesse que as instituições científicas portuguesas têm pelo progresso da ciência, a perfeita harmonia e intercambio entre Portugal e Espanha e o desejo de que a aproximação e o conhecimento de investigadores de países vários, tragam uma compreensão e uma unidade cada vez maiores, fortalecendo as relações, sempre mais íntimas e mais proveitosas.
> (The interest that the Portuguese scientific institutions have in the progress of science, the perfect harmony and exchange between Portugal and Spain and the desire that the proximity and knowledge of researchers of various countries might result in an ever-greater understanding and unity, strengthening the relationships between these countries, is always increasingly close and beneficial.) (Oleiro 1949n)

With these words, Bairrão Oleiro expanded the goal of the study grant awarded to him and highlighted his position as a producer of archaeological and heritage knowledge. In effect, such knowledge mattered especially when approaching people and geographies. It resulted in collaborations that contributed to a better understanding of the archaeological artifacts that he studied. Conversations with mentors and colleagues were highly productive; Almagro Basch and Pericot became his close colleagues and friends (Oleiro 1949n).

On the same day as the course inauguration, the participants visited the University of Barcelona in the late afternoon, where they were welcomed by the rector and other professors who attended the inaugural conference by Albert Grenier in his capacity as director of the French School in Rome. Other participants included, from the French delegation,[10] Maurice Louis (1892–1966), responsible for

[10]Members of the French delegation were Jean Jannoray (1909–1958), professor at the University of Montpellier and the excavation director of the settlement of Ensérune (Occitania); Henri Rolland (1887–1970), director of excavations at Glanum (Saint-Rémy-de-Provence) and Saint Blaise (Auvérnia-Ródano-Alpes); and Richard Arselin of the Bórely Museum, in Marseille (Oleiro 1949n).

the prehistory degree course of the Faculty of Letters at the University of Montpellier. As regards the Italian delegation, this included, among others,[11] Romanelli and Lamboglia. Finally, the Spanish contingent consisted of Almagro and Pericot as degree program directors of the International Summer Courses of Ampurias; Maluquer de Motes, de Palol and Miguel Oliva Prat (1922–1974) from the Provincial Archaeological Museum of Girona; as well as Beltrán from Zaragoza, and Alberto del Castillo Yurrita (1899–1976) from the University of Barcelona. This was a list of professionals with proven track records in their respective fields, with broad theoretical knowledge and practical experience. Bairrão Oleiro could hardly have been more satisfied.

On the following morning, September 6, they set off for Tarragona by coach. On the route, they stopped off in Sitges to visit the museums of Cau Ferrat and Maricel. They then continued onwards to Olèrdola, with its fortified square and ruins prior to arriving in Vilanova i la Geltrú where they were welcomed by the local authorities and visited the museum there and in Vilafranca del Penedès. At the end of the afternoon, following a visit to the excavation site of a prehistoric silo, they reached their destination, Ampurias.

On September 7, the course participants visited the Les Farreras Roman aqueduct, the necropolis, and the Museo Arqueológico Provincial (Provincial Archaeological Museum) where they were welcomed by its director, Samuel Ventura Solsona (1896–1972), by then already a friend of Bairrão Oleiro. Prior to departing for Reus (Baix Camp Comarca) to visit the prehistoric collection of doctor and historian Salvador Vilaseca i Anguera (1896–1975),[12] Bairrão Oleiro sought out German archaeologist Adolf Schulten (1870–1960) who was at that time working in Spain. During their brief conversation Bairrão Oleiro obtained from Schulten the Portuguese version of one of his works. On the participants' return to Barcelona, there was still an opportunity to visit the Torre dels Escipions, the Arch of Berà and a Roman quarry. The following day was entirely devoted to the Catalan capital, touring its museums, churches, and archaeological ruins, including the Roman remains over which the Museu d'Història de Barcelona (Museum of the History of Barcelona) was built and could be easily visited thanks to this ingenious architectural solution (Oleiro 1949n). The course group also visited the Centre Excursionista de Catalunya (Hiking Club of Catalonia, founded in 1876), where the surviving evidence of a Roman temple was preserved. There was also time to tour the royal palace and its chapel.

September 9 and 10 were dedicated to the city of Girona with a prior stopover in Caldes de Malavella to view the Roman spa there. The course attendees also visited the archaeological museum, the historical neighborhoods, a section of the city wall known as the 'Arabic bathhouse', and the Church of Sant Feliu with its eight Roman sarcophagi embedded in the walls. On the afternoon of September 10, the participants proceeded to the ancient city of Ampurias with a pause in Palamós and the Iberian ruins of Castell. On September 11, there was a visit to San Martin de Ampurias and a topographic study of the Palaiopolis and the Neápolis, followed by a lengthy tour of the already excavated ruins and those still awaiting excavation.

The above-mentioned visits were interspersed with lectures given by the professors participating in the course. Of particular note among these was a presentation by Lamboglia, who was responsible for supervising the practical excavation course and to this end completing a stratigraphic cross-section of the *decumani* area of Ampurias. On September 13, the participants headed on to Ullastret to practice excavating in this Iberian settlement until midafternoon under the guidance of Pericot (Oleiro 1949n) before examining, on the afternoon of September 14, the stratigraphic cross-section of the Roman wall made by Almagro in Ampurias. However, for unknown reasons, very few participants took part in these excavations: Bairrão Oleiro, an Italian and two Spanish colleagues. Bairrão Oleiro produced a diary of his excavation work, which was complemented by photographs and sketches made up of the special sheets distributed by Almagro that highlighted the objects that were collected, cleaned, described, classified, and ordered during these periods of fieldwork. In conjunction with many others, these artifacts were dispatched to the warehouses of the Museu Monogràfic d'Empúries (Monographic Museum of Ampurias),[13] where they were studied in detail. Finally, Bairrão Oleiro was entrusted with the supervision of the excavation on which the participants were left to work on their own and, as it happened, Lamboglia was satisfied with the results (Oleiro 1949n).

After the course finished, Bairrão Oleiro remained in the Catalan capital until September 29, continuing with his work courtesy of the facilities provided to him by professors, colleagues, and friends (Oleiro 1949n). He highlighted the freedom of movement that he had in every institution, accessing their libraries and files without any need for special authorization or monitoring of any kind.

[11] Other Italian participants included Giovanni Becatti (1912–1973), director of excavations in Ostia; Father Blondi; and the assistant of Lamboglia, Lidia Panizzi.

[12] His collection was part of Museo de Reus, founded in 1933 by Vilaseca i Anguera himself.

[13] Currently, Museo de Arqueología de Cataluña en Empúries, the institution that convinced Bairrão Oleiro to build a monographic museum in Conimbriga (Museu Monográfico de Conimbriga).

10.5 Madrid Once Again

Soon after the course ended, Bairrão Oleiro returned to Madrid (Oleiro 1949n). However, on the advice of Taracena, he first embarked on a tour of Zaragoza, Pamplona, Sória and Burgos while postponing visits to Mérida, Itálica, Carmona, Cordoba, and Granada (Oleiro 1949a). Unfortunately, the theft of his personal documentation significantly delayed these plans (Oleiro 1949l). Installed at his former address on Calle Serrano, Bairrão Oleiro submitted a new working plan to the IAC, at the recommendation of Taracena and Bellido. This now extended to attending, beginning in October, the classical archaeology classes given by Bellido at the Central University of Madrid. These classes were only then beginning due to the participation of Bellido in the tenure jury for full professors in archaeology, numismatics and epigraphy at the Universidad de Zaragoza (University of Zaragoza) and the Universidad de Salamanca (University of Salamanca). The recently named full professors included two of Bairrão Oleiro's friends: Antonio Beltrán, who became the chair of Archaeology at the University of Zaragoza, and Maluquer de Motes, the chair of Archaeology at the University of Salamanca (Oleiro 1949n).

Bairrão Oleiro had to coordinate his classes with research work on behalf of MAN and the IDV, with a view to writing one or two monographs. One of these works was about oil lamps, analyzing "tipos, variantes, cronologia, ensaios de sistematização, etc., reunindo dados, desenhando tipos, recolhendo tábuas de formas, como, por exemplo, as de Walters e Dressel, etc." (types, variants, the chronology, schemes for systematization, etc., gathering data, drawing types, collecting tables of forms, such as those by Walters and Dressel, etc.) (Oleiro 1949n). The other publication dealt with *terra sigillata* (Oleiro 1951a), with the groundwork involving contact, through Lamboglia, with the person deemed the leading authority on this subject, the British academic Felix Oswald (1866–1958). The justification for this research into ceramics seemed clear to Bairrão Oleiro, as establishing pottery sequences enabled him to "datar estações e monumentos, com uma grande margem de segurança, e de não se ter aprofundado muito a questão em Portugal, onde a tendência foi sempre o estudar, de preferência, os monumentos e não o seu conteúdo" (accurately date sites and monuments, given that in Portugal the tendency has always been to study, by preference, the monuments rather than their contents) (Oleiro 1949n). Committed to changing this generalized research framework for archaeology in Portugal, he sought to review the ceramic materials collected in national museums that had not hitherto been appropriately studied and which were still largely unpublished:

> pois à medida que o tempo passe sobre ele, mais trabalhosa, mais difícil e mais insegura se tornará a sua sistematização e publicação, que, certamente, traria novos e preciosos elementos de juízo tanto aos estudiosos portugueses como aos estrangeiros. // Nesse sentido prático se tem orientado parte da minha atividade.
> (Indeed, in keeping with the passage of time, their systematization and publication will lead to more work that might be more challenging and uncertain but would certainly make new and valuable data available to both Portuguese and international academics. // I have structured some of my activities in this practical sense.) (Oleiro 1949n)

In keeping with the complexity of the materials to be examined and the research that would be required, Bairrão Oleiro requested the IAC to extend his grant for another six-month period, in order to gain a better understanding of the topics he studied (Oleiro 1949a), especially because,

> a arqueologia romana tem hoje, em Portugal, poucos cultores e precisa ser estudada com conhecimentos o mais possível aprofundados, para que se continue uma tradição de glória, que importa reviver. // Sei que Vossas Excelências tomarão este pedido com a certeza de que o faço, não só no desejo de completar os meus conhecimentos, mas, e principalmente, com o de ser útil ao País e à Ciência portuguesa.
> (Roman archaeology in Portugal has few scholarly investigators today and needs to be studied with the deepest possible knowledge in order to continue a glorious tradition[14] that requires revitalizing. // I know that Your Excellencies shall accept this request in the certainty that I make it not only out of the desire to complete my knowledge but with the particular objective of being useful to the country and to Portuguese Science.) (Oleiro 1949a)

One of Bairrão Oleiro's supervisors in Madrid, Bellido, supported this request because he had been the one to suggest that Bairrão Oleiro concentrate his research in two unpublished areas: Roman mosaics[15] and Roman sculpture in Portugal (Oleiro 1949i). However, because these were unpublished topics, there were certain procedures requiring compliance. Among these was making a comprehensive photographic record of the examples available in the country for subsequent analysis in the Madrid institutions where he would be working as an intern. Portugal was thus granted the opportunity to obtain two comparative studies carried out "con toda la técnica científica y en disposición de constituir aportaciones de valor internacional. Estas dos tareas son de máxima urgencia en el estudio de las antigüedades de la Península" (using all available scientific techniques and while making a contribution of international value. These

[14] This refers to the studies already carried out by F. Martins Sarmento, J. Leite de Vasconcelos and V. Correia, among others.

[15] In 1951, he published a monograph on the 'Mosaico do Labirinto' in the journal *Archivo Español de Arqueología*. The mosaic was discovered in 1899 in the intra-wall zone of Conimbriga (see Oleiro 1951b).

two tasks are of maximum priority due to their importance to the study of antiquity on the Peninsula) (Oleiro 1949i). This conviction was only possible because, in Bairrão Oleiro's own words, Spain was "dotado de faculdades de primeira ordem" (endowed with faculties of the first order) (Oleiro 1949i) to undertake such assignments. In addition, researchers such as Bellido were dedicated to internationalizing the production of archaeological knowledge and supported a broad and inclusive vision of the concept of heritage, in this case Roman origins (see Gracia Alonso 2009, 2013),

> seria una pena grande para todos nosotros, portugueses y españoles, que dejásemos pasar esta excelente oportunidad de una colaboración íntima en una materia de carácter universal ya que el patrimonio romano antiguo pertenece en realidad al patrimonio de todo el Occidente, sin distinción de nacionalidad.
> (It would be a great pity for us all to stop these partnerships, Portuguese and Spanish who wish to take up this excellent opportunity for an intimate collaboration in a field with such a universal character as that of the classical Roman heritage that in reality belongs to all of the West, without any distinction between nationalities.) (Oleiro 1949i)

The scientific expectations placed on Bairrão Oleiro seem neither excessive nor without consequence. This is duly conveyed by the opinion issued, on the request of IAC, by João Couto (1892–1968), historian of art, museologist, and director of the Museu Nacional de Arte Antiga (National Museum of Ancient Art) founded in Lisbon in 1884. According to this landmark figure in the history of Portuguese museology, who had such profound knowledge and a forthright and critical stance, Bairrão Oleiro "deve ser aproveitado em Portugal" (should be of benefit to Portugal),

> Ainda agora estive em Conimbriga e verifiquei o estado de abandono em que esses preciosos restos se encontram. Há bem pouco tempo também ouvi comentários desagradáveis dos componentes do XVI Congresso Internacional de História de Arte que ali foram. // O Museu de Coimbra, que de certo modo olha por Conimbriga, está há não sei quantos anos sem diretor. Não deve este estudioso fazer parte do seu pessoal técnico? // Quando nos decidiremos a olhar com seriedade e saber para estes problemas que ultrapassam o interesse nacional?
> (Even just recently I was in Conimbriga and verified the state of abandonment in which these precious remains find themselves. I have also very recently heard unpleasant commentaries from the members making up the 16th International History of Art Congress[16] who visited there. // The Museum of Coimbra, which to a certain extent looks out for Conimbriga, has been without a director for I do not know how many years. Should this scholar not make up part of its technical staff? // When are we going to decide to consider seriously and with wisdom these problems that extend beyond the national interest?) (Oleiro 1949m)

Couto's thoughts were also reflected in the detailed report delivered by Bairrão Oleiro to the IAC at the end of 1949. Evaluating the general state of archaeology in Portugal, in particular focusing on the Roman period, the young grant holder and rising scholar stated there was a need to undertake new excavations and carry out comprehensive studies of certain archaeological sites using modern techniques (Oleiro 1949n). Only thus would it prove possible to resolve some of the underlying problems whose resolution "nos porão em pé de igualdade com outros países onde muito e bem se vem trabalhando desde há tempos" (would place us on an equal footing with other countries where they have long since been working well) (Oleiro 1949n). To this end, he decided to organize a file of bibliographic references on the archaeological sites that he deemed a priority. This work,

> modesto, mas lento e ingrato, tem uma importância que me parece justo encarecer. A falta de uma publicação especial onde se encontrem notícias da arqueologia portuguesa e estrangeira, a falta ou escasso número de bibliotecas especializadas e de contacto com centros de estudo e investigação estrangeiros, pode conduzir a uma estagnação de conhecimentos, precipitados em redor dos poucos elementos que se possuem, muitas vezes antiquados e ultrapassados por novas descobertas e achados. // Em arqueologia o caso é frequente e assim vemos que obras que gozaram de justa fama há vinte ou trinta anos, estão hoje postas de lado ou são citadas para efeitos de mera comparação ou como curiosidades históricas.
> (modest, but slow and thankless, holds an importance that it seems appropriate to emphasize. The lack of any special publication that details news of Portuguese and foreign archaeology, the lack or scarcity in number of specialist libraries and the contacts with international study and research centers, may lead to a stagnation in knowledge, precipitated as a facet of the lack of elements in our possession, very often outdated and overtaken by new discoveries and finds. // In archaeology, such cases are frequent, and we thus see that the works that gained due fame twenty or thirty years ago have today been cast aside or are referenced merely for purposes of comparison or as historical curiosities.) (Oleiro 1949n)

Bairrão Oleiro's database of bibliographic references had the additional advantage of enabling the structuring of university study programs. His initiative responded to the needs of those Portuguese scholars who felt that archaeological classes were still scarce in Portugal. The references would become useful to Bairrão Oleiro on his own return to the country, more specifically to Coimbra where he would lecture at the university. It was not by chance that Almagro and Pericot sought to ascertain his opinion as regards establishing a Portuguese lectureship at the University of Barcelona in an exchange system with a Spanish scholar in Portugal. In any case, Bairrão Oleiro was by then more closely focused on the work of cultural resurgence led by the IAC, to which he contributed with his own knowledge and experience gained in Spain (Oleiro 1949n). His knowledge, experiences, networks, and commitment explain the approval of his grant extension request in Madrid on January 1, 1950 (Oleiro 1949k). A new chapter thus opened up for him, deepening the path hitherto taken up until his professional duties, as museologist and professor, brought him back to Portuguese territory permanently.

[16] The congress took place in the Museu Nacional de Arte Antiga in Lisbon, on April 18–24, 1949. During the congress, exhibitions and monuments were visited, both in and outside the capital.

10.6 Conclusions

The future path of Bairrão Oleiro, focused on the archaeology of the Roman period, particularly mosaics, ceramics, and statuary, was beginning to become clear during his stay in Madrid (Martins 2022). This conclusion is supported by the first report on his activities presented to the IAC as well as the letters of recommendation written by his supervisors in Spain and the opinions of his previously mentioned Portuguese evaluators. This is not surprising considering the care with which he had prepared his path since the moment he decided to deepen his knowledge alongside those who were the leading figures in Roman archaeology in Spain. His choice of Spain for international academic and research experiences was understandable—it was the closest territory with which there was common ground on the archaeological subjects and materials he aimed to study.

For Bairrão Oleiro it was clear that there was a need to use developments in Spanish archaeology as a starting point so that, step by step, he might expand the scientific collaboration networks (see Newman 2001) that he was then building up, thus integrating a peripheral archaeology—that of Portugal—into a global one, through Spain. In summary, this reflects a very well thought out plan, in which Bairrão Oleiro established a phased approach to accessing the knowledge and methodologies that could only be gained while abroad, working with some of the leading figures in the research field that he was seeking to develop. The first experience as an IAC and CSIC grant holder was perfectly in line with his objectives from the outset, due to the supervisors that were assigned to him. There then came the work he carried out in museums and libraries. Finally, there were all the classes and courses that he attended. In truth, these combined experiences would seem to have endowed him with all the tools necessary for advancing his plan:

> Não poderei, em realidade, olvidar a hospitalidade inexcedível, a amizade real, a confiança sem limites, as facilidades do trabalho e do estudo, tudo o que vi, e tudo o que senti, durante este magnífico período da minha vida, durante o qual cimentei a minha formação professional.
> (I cannot, in fact, forget the unsurpassed hospitality, the real friendship, the boundless trust, the ease of work and study, everything I saw and felt during this magnificent period of my life, during which I cemented my professional training.) (Oleiro 1950)

Classical and especially Roman archaeology had been neglected in Portugal due to the absence of its original supporters (Oleiro 1947; Cardoso 2021; Martins 2022). Bairrão Oleiro, trained in Coimbra at a university that retained a strong tradition in this field, gained the opportunity to internationalize his work and, through these experiences, Portuguese archaeology. Such internationalization enabled the development of the field in the country and placed it on the same level as much else that was produced internationally. A feasible development, in turn, thanks to his interpersonal and inter-institutional collaboration. Only thus would he have been able to break through the borders separating scientific communities in the name of producing archaeological knowledge and heritage in Portugal. This was not an entirely novel approach within the Portuguese scientific framework or even in the history of the UC with its ongoing contacts with the leading European academic circles. However, the IAC had been created with the purpose of managing and financing science in Portugal, and the support he received from this institution allowed Bairrão Oleiro to demonstrate the relevance of his studies especially in a country whose archaeological activities were designed to affirm the agendas and the political goals of the Estado Novo dictatorship (Martins 2011; Gomes 2012).

It thus comes as no surprise that Bairrão Oleiro was not entirely satisfied by the first few months he spent in Spain. In effect, there was still a great deal left to learn and to experience. He especially wished to expand his reach beyond the borders of the Iberian Peninsula. Aware of this, specifically in terms of the entities he encountered through the mediation of his Spanish supervisors, whether as authors or as professors participating in the lecture courses and programs he attended, he ensured that he would obtain in the years to come the continuation of his first grant and internships for research at Italian and German institutes as well as at various museums and archaeological sites across Italy and France. These efforts resulted in the creation of various study units at the UC where he would establish, organize and run the Instituto de Arqueologia (Institute of Archaeology, founded in 1954) and the journal *Conimbriga* (1959) which was probably inspired by the journal *Ampurias. Revista de arqueología, prehistoria y etnografía* (1939). In addition, he took up the position of assistant conservator at the Machado de Castro Museum between 1951 and 1962, returning to the excavations in Conimbriga and founding the Museu Monográfico (Monographic Museum) (Fabião 1997) there:

> opinião que sempre tenho defendido (...) é a da criação de um Museu Monográfico a instalar em edifício especial (...) junto das ruínas, à semelhança do que sucede noutras estações arqueológicas, como Pompeia, Volubilis e Ampurias [até porque] // depois do estágio de especialização em arqueologia romana que acabara de fazer em Espanha, nenhum tema de investigação me pareceu mais sugestivo, prometedor e necessário do que o problema de Conimbriga.
> (An opinion that I have always defended [...] is the setting up of a Monographic Museum for installation in a special building [...] alongside the ruins, similar to the practice of other archaeological sites, such as Pompeii, Volubilis and Ampurias [especially because] // after the specialist internship in Roman archaeology that I have finished in Spain, no research theme seems more suggestive, promising and necessary than the problem of Conimbriga.) (Oleiro 1952: 32, 35)

The activities described above are pages from a new chapter in the life of Bairrão Oleiro with immediate consequences for the development of archaeology in general, and Roman studies in particular, in Portugal, especially at the UC. These activities emerged from Bairrão Oleiro's persistent efforts at international collaboration and his continuous contribution of novel ideas and methods to Roman archaeology. All these enabled Bairrão Oleiro to become the acknowledged expert in this field in Portugal (Maciel 1996; Martins 2005, 2021).[17] It would not be an overstated to conclude that Bairrão Oleiro himself represents a landmark in Roman archaeology in Portugal. He transformed the field above all through his international educational and research experiences and his commitment to the internationalization of science through the exchange of people and ideas who constructed an authentic scientific collaboration network in the period after the Second World War.

Acknowledgments I would like to thank Bettina Arnold, Laura Coltofean, and László Bartosiewicz, organizers of the session 'Connecting people and ideas: Networks and networking in the history of archeology' at the 26th (Virtual) Annual Meeting of the European Association of Archaeologists in 2020, for their editorial work on the text as well as for their patience and understanding. This chapter was written thanks to the Institute of Contemporary History (IHC) which is funded by national funds through FCT - Fundação para a Ciência e a Tecnologia, I.P., under the projects UID/04209 and LA/P/0132/2020 (DOI 10.54499/LA/P/0132/2020).

References

Primary Sources

Archival Material

Arthur, M. de L. C. (1953). Proc. No. 1635/28.10.1953, Archives of the Instituto para a Alta Cultura, Instituto Camões, Ministério dos Negócios Estrangeiros, Lisbon, Portugal.

Oleiro, J. M. B. (n.d.). Book No. 3, Fls. 80, Proc. No. 4515/[????].19[??]/0469/3-5, Archives of the Instituto para a Alta Cultura, Instituto Camões, Ministério dos Negócios Estrangeiros, Lisbon, Portugal.

Oleiro, J. M. B. (1947). Book No. 3, Fls. 80, Proc. No. 4515/22.12.1947/0469/1, Archives of the Instituto para a Alta Cultura, Instituto Camões, Ministério dos Negócios Estrangeiros, Lisbon, Portugal.

Oleiro, J. M. B. (1949a). Book No. 3, Fls. 80, Proc. No. 4515/[??].[??].1949/0469/62/3, Archives of the Instituto para a Alta Cultura, Instituto Camões, Ministério dos Negócios Estrangeiros, Lisbon, Portugal.

Oleiro, J. M. B. (1949b). Book No. 3, Fls. 80, Proc. No. 4515/15.03.1949/0469/45, Archives of the Instituto para a Alta Cultura, Instituto Camões, Ministério dos Negócios Estrangeiros, Lisbon, Portugal.

Oleiro, J. M. B. (1949c). Book No. 3, Fls. 80, Proc. No. 4515/14.05.1949/0469/50, Archives of the Instituto para a Alta Cultura, Instituto Camões, Ministério dos Negócios Estrangeiros, Lisbon, Portugal.

Oleiro, J. M. B. (1949d). Book No. 3, Fls. 80, Proc. No. 4515/19.05.1949/0469/51, Archives of the Instituto para a Alta Cultura, Instituto Camões, Ministério dos Negócios Estrangeiros, Lisbon, Portugal.

Oleiro, J. M. B. (1949e). Book No. 3, Fls. 80, Proc. No. 4515/19.05.1949/0469/53/2, Archives of the Instituto para a Alta Cultura, Instituto Camões, Ministério dos Negócios Estrangeiros, Lisbon, Portugal.

Oleiro, J. M. B. (1949f). Book No. 3, Fls. 80, Proc. No. 4515/22.06.1949/0469/57, Archives of the Instituto para a Alta Cultura, Instituto Camões, Ministério dos Negócios Estrangeiros, Lisbon, Portugal.

Oleiro, J. M. B. (1949g). Book No. 3, Fls. 80, Proc. No. 4515/03.07.1949/0469/60, Archives of the Instituto para a Alta Cultura, Instituto Camões, Ministério dos Negócios Estrangeiros, Lisbon, Portugal.

Oleiro, J. M. B. (1949h). Book No. 3, Fls. 80, Proc. No. 4515/08.07.1949/0469/61, Archives of the Instituto para a Alta Cultura, Instituto Camões, Ministério dos Negócios Estrangeiros, Lisbon, Portugal.

Oleiro, J. M. B. (1949i). Book No. 3, Fls. 80, Proc. No. 4515/30.07.1949/0469/65, Archives of the Instituto para a Alta Cultura, Instituto Camões, Ministério dos Negócios Estrangeiros, Lisbon, Portugal.

Oleiro, J. M. B. (1949j). Book No. 3, Fls. 80, Proc. No. 4515/16.09.1949/0469/70, Archives of the Instituto para a Alta Cultura, Instituto Camões, Ministério dos Negócios Estrangeiros, Lisbon, Portugal.

Oleiro, J. M. B. (1949k). Book No. 3, Fls. 80, Proc. No. 4515/07.10.1949/0469/77, Archives of the Instituto para a Alta Cultura, Instituto Camões, Ministério dos Negócios Estrangeiros, Lisbon, Portugal.

Oleiro, J. M. B. (1949l). Book No. 3, Fls. 80, Proc. No. 4515/11.10.1949/0469/78, Archives of the Instituto para a Alta Cultura, Instituto Camões, Ministério dos Negócios Estrangeiros, Lisbon, Portugal.

Oleiro, J. M. B. (1949m). Book No. 3, Fls. 80, Proc. No. 4515/19.10.1949/0469/80, Archives of the Instituto para a Alta Cultura, Instituto Camões, Ministério dos Negócios Estrangeiros, Lisbon, Portugal.

Oleiro, J. M. B. (1949n). Book No. 3, Fls. 80, Proc. No. 4515/[??].12.1949/0469/86/3, Archives of the Instituto para a Alta Cultura, Instituto Camões, Ministério dos Negócios Estrangeiros, Lisbon, Portugal.

Oleiro, J. M. B. (1950, June 11). Archives of the Museo Arqueológico Nacional (Correspondencia General, 1950, C. 990), Ministerio de Cultura y Deporte, Madrid, Spain.

Secondary Sources

Bendala, M., Fernández Ochoa, C., Durán Cabello, R., & Morillo, Á. (Eds.). (2005). *La Arqueología clásica peninsular ante el tercer milenio: en el centenario de Antonio García y Bellido (1903–1972)* (Anejos de Archivo Español de Arqueología, 34). Madrid: Consejo Superior de Investigaciones Científicas (CSIC), Instituto de Historia.

Cardoso, J. L. (2021). Arqueologia, investigação e formação de arqueólogos em Portugal na década de 1950: um diagnóstico de João Manuel Bairrão Oleiro. *Al-Madan, 24*, 123–131.

[17] In 1951, the young Maria de Lourdes Costa Arthur (1924–2002), also an IAC grant holder in Madrid, requested Bairrão Oleiro's support during the writing of her undergraduate dissertation on the classification of Roman oil lamps, "visto ser ele o único que se poderia pronunciar" (given that he was the only one able to provide an informed opinion [on them]) (Arthur 1953).

Charneca, M. A. P. dos S. (2005). *Uma rede de museus para Abrantes. Museu D. Lopo de Almeida—passado, presente e futuro* [Master's thesis, Faculdade de Belas Artes de Lisboa, Portugal].

Díaz-Andreu, M., Mora-Rodríguez, G., & Cortadella-Moral, J. (Eds.). (2009). *Diccionario histórico de la arqueología de España (siglos XV–XX)*. Madrid: Marcial Pons.

Fabião, C. (1997). Percursos da Arqueologia clássica em Portugal: da Sociedade Archeologica Lusitana (1849–1857) ao moderno projecto de Conimbriga (1962–1979). In G. Mora, & M. Díaz-Andreu (Eds.), *La cristalización del pasado: génesis y desarrollo del marco institucional de la Arqueología en España* (pp. 105–123). Málaga: Servicio de Publicaciones de la Universidad de Málaga. https://repositorio.ul.pt/handle/10451/10036

Ferreira, L. R. (2011). O Instituto de Coimbra e a Universidade. *Rua Larga, 32*, 17–19. http://hdl.handle.net/10316/17981

Ferreira, L. R. (2012). *Instituto de Coimbra. O percurso de uma Academia*. Coimbra: Imprensa da Universidade de Coimbra. https://doi.org/10.14195/978-989-26-0704-7

García y Bellido, A. (1951). El Instituto de Arqueología y Prehistoria 'Rodrigo Caro'. *Archivo Español de Arqueología, 24*(83–84), 161–168.

Gomes, S. (2012). *Past, identity, and webs of government: Studies on the intertwining of the production of archaeological knowledge and the construction of Salazar's Portuguese national identity* [Unpublished doctoral dissertation]. University of Oporto, Portugal.

Gómez Barreda, J. A. (2016). *Blas Taracena Aguirre (1895–1951)*. Soria: Ayuntamiento de Soria/Diputación Provincial de Soria/ Ministerio de Cultura.

Gracia Alonso, F. (2009). *La arqueología durante el primer franquismo (1939–1956)*. Barcelona: Edicions Bellaterra.

Gracia Alonso, F. (2012). *Arqueologia i política. La gestió de Martín Almagro Basch al front del Museu Arqueològic Provincial de Barcelona (1939–1960)*. Barcelona: Universitat de Barcelona.

Gracia Alonso, F. (2013). La arqueología durante el franquismo. Instrumentalización identitaria. In A. Altarriba, S. Buj, R. Campos, E. Castillejo Cambra, F. Gracia Alonso, R. Gubern, & R. P. Trigueros (Eds.), *Los intelectuales y la dictadura franquista. Cultura y poder en España de 1939 a 1975* (pp. 47–76). Madrid: Editorial Pablo Iglesias.

Gracia Alonso, F. (2017). *Lluís Pericot. Un prehistoriador entre dos épocas*. Pamplona: Urgoiti Editores.

Gracia Alonso, F. (2021). Archaeology and the Armed Forces in Spain from the nineteenth to the twenty-first century. In L. Coltofean-Arizancu, & M. Díaz-Andreu (Eds.), *Interdisciplinarity and archaeology: Scientific interactions in nineteenth and twentieth-century archaeology* (pp. 181–202). Oxford and Philadelphia: Oxbow Books.

Leonardo, A. J. F., Martins, D. R., & Fiolhais, C. (2009). *O Instituto de Coimbra: breve história de uma academia científica, literária e artística*. In A. E. Maia do Amaral (Ed.), *Tesouros da Biblioteca Geral da Universidade de Coimbra* (pp. 115–125). Coimbra: Imprensa da Universidade de Coimbra.

Maciel, M. J. (Ed.). (1996). *Miscellanea em homenagem ao Professor Bairrão Oleiro*. Lisboa: Edições Colibri.

Martins, A. C. (2005). *A Associação dos Arqueólogos Portugueses na senda da salvaguarda patrimonial. 100 anos de (trans)formação (1863–1963)* [Unpublished doctoral dissertation]. Universidade de Lisboa, Portugal.

Martins, A. C. (2011). Mendes Corrêa, Manuel Heleno and Portuguese archaeology during the Estado Novo: Two names, two cities, two perspectives, one nationalism. *Archaeologia Polona, 47*, 155–178.

Martins, A. C. (2016). Pioneiras da Arqueologia em Portugal: «Another brick» against «the wall» of indifference. Maria de Lourdes Costa Arthur (1924–2003). *Clepsydra. Revista de Estudios del Género y Teoría Feminista, 15*, 77–100.

Martins, A. C. (2021). In search of interdisciplinarity in Portuguese archaeology: Notes on the 1960s. In L. Coltofean-Arizancu, & M. Díaz-Andreu (Eds.), *Interdisciplinarity and archaeology: Scientific interactions in nineteenth and twentieth-century archaeology* (pp. 133–155). Oxford and Philadelphia: Oxbow Books.

Martins, A. C. (2022). Oleiro, João Manuel Bairrão. In R. H. da Silva, J. d'Oliva Monteiro, E. Ferreira, & E. Pereira (Eds.), *Dicionário: Quem é quem na museologia portuguesa* (2nd ed.) (pp. 409–422). Lisbon: Instituto de História da Arte and Faculdade de Ciências Sociais e Humanas, Universidade NOVA de Lisboa. https://doi.org/10.34619/oelt-t7xq

Martins, A. C. (2023). A importância de estagiar em Madrid: uma nova visão sobre os museus de arqueologia em Portugal (anos 50 do séc. XX). *MIDAS. Museus e Estudos Interdisciplinares, 16*. https://doi.org/10.4000/midas.4460

Mederos, A. (2011–2012). Martín Almagro Basch, formación y consolidación como catedrático de Prehistoria (1911–1943). *Boletín del Seminario de Estudios de Arqueología, 77–78*, 335–416. https://uvadoc.uva.es/handle/10324/11567

Mederos, A. (2017). Martín Almagro Basch, un balance de su trayectoria científica (1934–1984). *Cuadernos de Prehistoria y Arqueología de la Universidad Autónoma de Madrid, 43*, 251–289. https://doi.org/10.15366/cupauam2017.43.010

Newman, M. E. J. (2001). The structure of scientific collaboration networks. *Proceedings of the National Academy of Sciences of the United States of America, 98*(2), 404–409. https://doi.org/10.1073/pnas.98.2.404

Oleiro, J. M. B. (1951a). Elementos para o estudo da 'terra sigillata' em Portugal I. Marcas de oleiro encontradas no País. *Revista de Guimarães, LXI*(1–2) 16, 81–111.

Oleiro, J. M. B. (1951b). Materiales arqueológicos de Conimbriga. El Mosaico del Laberinto. *Archivo Español de Arqueología, XXIV*(83–84), 47–52.

Oleiro, J. M. B. (1952). Conimbriga e alguns dos seus problemas. *Humanitas, IV*, 32–42.

Olmos, R. (1994). Antonio García y Bellido y su época: una posible lectura. *Archivo Español de Arqueología, 67*(169–170), 293–308. https://doi.org/10.3989/aespa.1994.v67.449

Real, M. (2011). Carolina Michaëlis de Vasconcelos (1851–1925). A primeira professora da Faculdade de Letras de Lisboa. *Revista Letras Com Vida – Literatura, Cultura e Arte, 3*(1), 51–54. http://hdl.handle.net/10451/28868

Rollo, M. F., Queiroz, M. I., Brandão, T., & Salgueiro, Â. (2012). *Ciência, cultura e língua em Portugal no século XX. Da Junta de Educação Nacional ao Instituto Camões*. Lisbon: Instituto Camões and Imprensa Nacional-Casa da Moeda.

Sánchez González, I. (2015). *Diez años de soledad: España, la ONU y la dictadura franquista 1945–1955*. Sevilla: Editorial Universidad de Sevilla-Secretariado de Publicaciones.

Schattner, Th. G. (2005). García y Bellido y la arqueología clásica portuguesa. In M. Bendala, C. Fernández Ochoa, R. Durán Cabello, & Á. Morillo (Eds.), *La arqueología clásica peninsular ante el tercer milénio: en el centenario de A. García y Bellido (1903–1972)* (Anejos de Archivo Español de Arqueología, 34) (pp. 75–80). Madrid: Consejo Superior de Investigaciones Científicas (CSIC), Instituto de Historia.

Creating Authority in South Asian Archaeology: Mortimer Wheeler, Stuart Piggott, and Archaeological Knowledge in India

11

Heidi J. Miller

Abstract

Advances in archaeological knowledge are achieved through the sharing of facts and interpretations, especially through networks amongst scholars building on each other's research. If past research is ignored, then it seems what comes after may be doomed to fail. Archaeologists Mortimer Wheeler and Stuart Piggott were both part of the Golden Generation that worked towards a professionalization of archaeological practice in Britain in the early decades of the twentieth century. Both also dabbled in the archaeology of South Asia just prior to the end of the Colonial State. This chapter explores the basis of Piggott's and Wheeler's shared view of how to practice archaeology, and how their approach and politicized interpretations were imposed onto Indian archaeology. They formed a network of two, trusting in each other and imbued with colonial power. Their ability to silence criticism stems from their belief that only the colonial power knows best, and that their way of excavation was superior. In the end, their interpretations were disproven, yet their influence on the development of excavation techniques today in the subcontinent is still given credit, when in fact, credit should be given to John Marshall and the indigenous archaeologists he trained.

Keywords

Mortimer Wheeler · Stuart Piggott · John Marshall · South Asian archaeology · Imperialist archaeology · Authoritative knowledge · Network

H. J. Miller (✉)
Division of Liberal Arts, Middlesex Community College, Bedford, MA, USA

11.1 Introduction

In order for archaeological knowledge to advance, ideas and interpretations must be built upon, and no individual is a proverbial island: "(...) if one sees the development of a particular aspect, one should consider these giants and heroes in terms of their associates, ideally inquiring after *patterns of interconnection*, rather than catalogues of individual achievements" (Ashbee 1972: 39; emphasis added). It is these interconnections between people and ideas that form the scaffolding of our knowledge base, and without corroboration between scholars or a challenge to an interpretation, our understanding of the ancient past would not develop. This is not to say that all knowledge is correct, or equal, but for an interpretation of archaeological finds to become accepted, it must be acknowledged by a group of trusted and interconnected individuals. Such was the case of archaeologists Mortimer Wheeler (1890–1976) and Stuart Piggott (1910–1996) who both wrote about the archaeology of the Indian Subcontinent; with each interaction and publication within the colonial state of India, they enhanced and legitimized the other's interpretations, leading to the establishment of disciplinary authority for aspects of the archaeology of South Asia.

During the Second World War, Piggott served in the Aerial Photographic Interpretation Unit, stationed in New Delhi (Sharples 2001: 798), when Wheeler was appointed Director General of the Archaeological Survey of India in 1944, a position he held until Independence and Partition in 1948. Although their physical presence overlapped for less than a year in India, their respective intellectual paths and publications vis à vis the archaeology of South Asia are intertwined, each agreeing with the other in publications and actively creating authority in their view of how archaeology should be practiced and interpreted in the Subcontinent. This chapter explores how their disciplinary views and lives were interconnected, demonstrating the role of trust in the creation of knowledge (Hardwig 1991: 693).

L. Coltofean et al. (eds.), *Connecting People and Ideas*, Themes in Contemporary Archaeology,
https://doi.org/10.1007/978-3-031-81006-0_11

The chapter begins with a short overview of Piggott's and Wheeler's roles in the development of the discipline of archaeology in Britain, and the primary aspects that were important to them regarding the professionalization of archaeological practice. This is the intellectual context and network that they shared. They interacted in a variety of ways over two decades prior to the Second World War, and it was evident that they shared the same goals in archaeological research. A brief history of the establishment of a state-sponsored archaeology department for Colonial India is then described, with a focus on the extraordinary work of John Marshall (1876–1958) in professionalizing archaeological practice in the Subcontinent. Regardless of Marshall's efforts, the Colonial State did not ensure a stable work environment for archaeological research, and when Wheeler was appointed Director General in 1944, the department was understaffed and underfunded. Wheeler, with Piggott's help and support, created an illusional break with Marshall's work in order to claim a revolutionary advance in scientific archaeology in South Asia. The remainder of the chapter describes various problems with this, and then notes that the work of Piggott and Wheeler actually reflects the unequal power structure inherent in a colonial state. They did not encourage indigenous scholars and by relying only on each other for validation of their research, the legacy of their interpretations has not lasted.

11.2 Piggott, Wheeler, and the Professionalization of Archaeology in Britain

The development of the discipline of archaeology in Britain has a long history, and both Piggott and Wheeler were contemporaries in its professionalization during the first third of the twentieth century. They viewed archaeology as a scientific pursuit that required rigorous standards of fieldwork and interpretation, although each travelled a different learning path to practice the same type of archaeology.

Piggott became fascinated with archaeology as a boy, and by his early teens had developed great observational skills tracing archaeological remains in the landscape (Mercer 1998: 414). This was recognized in 1926 by archaeologist O. G. S. Crawford (1886–1957) who wrote to Piggott asking for more details about some Roman period remains Piggott had found (Crawford 1926). This led to Piggott becoming one of Crawford's ferrets[1] and a lifelong friendship developed between them, with Piggott referring to Crawford as his archaeological godfather (Piggott 1983: 29). Piggott had no academic training in archaeology or in a related discipline, and was self-taught, in contrast to Wheeler, who completed multiple degrees at the University of London, including a Doctorate of Letters (Hawkes 1982: 78–79). By 1930 each was enmeshed in the transformation of archaeology in Britain, which has been described as a shift from an antiquarian pastime to a professional enterprise and academic endeavor (Stout 2008: 17–18).

The years between the World Wars in Britain, especially from the 1920s through the 1930s, witnessed the emergence of archaeology as a profession due to the diligence of a dedicated group of individuals, including Piggott, Crawford, and Wheeler. This all-male group shared the same goals of creating a "position of disciplinary authority" predicated on acceptable ways of doing archaeology (Stout 2008: 4), and this group has been referred to as the Heroic Band (Stout 2008: 18) or the Golden Generation (Díaz-Andreu et al. 2009: 3). The members of this group established standards of acceptable archaeological practice in a number of areas (see Wheeler 1956; Piggott 1983; Stout 2008) and one primary tool for establishing a network of authority and communicating correct practice was the journal *Antiquity*, begun by Crawford in 1927. In the editorial of the very first issue, Crawford wrote:

> *Antiquity* will attempt to summarize and criticize the work of those who are recreating the past. Archaeology is a branch of science which achieves its results by means of excavation, fieldwork and comparative studies; it is founded upon the observation and record of facts. (Crawford 1927: 1)

Being a scientific discipline as defined by Crawford & Co. appears to have meant embracing the goals of correct field practice, interdisciplinary study, and robust publication according to the standards created by a community of self-appointed experts belonging to the Golden Generation.

Wheeler was an early advocate for systematic and well-recorded excavation. As early as 1926, in a lecture before the Royal Society of Arts, he spoke about the proper practice of fieldwork, emphasizing the importance of recording stratigraphic observations. At the end of the talk, he declared that if unscientific excavation occurred, then this was destruction of British Heritage done by the very people who were responsible for safeguarding it (Wheeler 1927: 833). Crawford shared this view and lamented: "When will the British public learn that it is not only what you find but where (vertically and horizontally) that matters (…)" (Crawford 1930a: 384). This focus on the context of archaeological finds was further underscored by advocating for problem-oriented excavations, which Wheeler rigorously defined in the pages of *Antiquity* (Wheeler 1929: 182–87), a position also strongly supported by Crawford and the Society of Antiquaries (Crawford 1930b: 3).

At about the same time that Wheeler was pronouncing about proper excavation techniques, Piggott was improving his excavation skills by observing archaeologist Alexander Keiller's (1889–1955) excavation at Windmill Hill; Piggott

[1] As in ferreting out the details.

considered both Wheeler and Keiller as "setting new standards of technical performance in the field" (Piggott 1983: 30). Just prior to observing Keiller, Piggott began his first significant excavation experience at The Trundle site. This excavation was directed by Eliot Cecil Curwen (1895–1967) and Piggott participated in both seasons, in 1928 as well as 1930. In between the excavation seasons Piggott visited Windmill Hill and was impressed by what he saw. Upon his return to The Trundle for further excavation in 1930, he initiated a new layout of the excavation trenches, to reflect "the new precepts of order and precision" he had observed at Keiller's excavation (Mercer 1998: 417).

Along with more precise excavation and recording, the scientific approach to archaeology in the 1920s and 1930s included new methods of study, such as the use of aerial photographs and the close examination of distribution patterns in the landscape. Piggott, Wheeler, and other members of the Golden Generation also advocated for the place of interdisciplinary analyses within archaeological interpretation.[2] Piggott was a founding member, along with Keiller in 1936, of the "Sub-Committee on the Southwestern Group of Museums and Art Galleries on the Petrological Identification of Stone Axes" (Mercer 1998: 422). As for Wheeler, in his autobiography, *Still digging*, he observed that at the end of the First World War, when he returned to civilian life, archaeological technique was primitive (his word) and thus needed to expand to include analyses from related disciplines: "(…) a multitude of exact and less exact sciences is now, very properly, expected to contribute [to archaeology], and will increasingly contribute in the future" (Wheeler 1956: 91).

In all, this multifaceted research by the members of the Golden Generation created a new focus on professional practice that has been viewed as an assault on the previous methods used in archaeology (Mercer 1998: 420). The emphasis on orderly layout and stratigraphic control in excavation, problem-oriented research, interdisciplinary analysis of materials, and clear presentation of results in relation to a research question became the basic substance of the professionalized scientific archaeology in which both Piggott and Wheeler were engaged. These factors were the focal point of acceptable practice: to identify as a member of this professional group one abided by these standards and thus there was a solidification on how proper archaeology should be practiced (Roberts 2012: 213–14). In other words, these practices were the benchmarks for the identity of the new scientific archaeology in Britain, and this view was the doctrine of the journal *Antiquity* which served as the purveyor of the new discipline (Crawford 1955: 177–82; Hauser 2007: 107f.; Stout 2008: 19f). Yet, there was no objective test or way to assess what exactly fit these requirements, except for acceptance of the research by the community who set the rules. In an editorial reflecting on the first 75 years of the journal *Antiquity*, the editors' note: "In its early years, it is probably true to say there was only one archaeology (…)" and it was not a period of reflection but one of judgement, of separating the good archaeology from the bad (Malone et al. 1999: 723).

This *good archaeology* was determined by the Golden Generation according to their standards, and many members of this group were insistent about separating professional practice from amateur status. In 1948 Piggott published an editorial emphasizing this view (Piggott 1948a). Entitled "Archaeology and the Amateur", Piggott described archaeology as a complex profession that amateurs can no longer fully understand, and that the best way for an amateur to be helpful is to provide funding in the form of dues to professional organizations, subscriptions to reputable journals, and in general to help finance proper excavations. This created an uproar and two issues later Piggott tried to clarify his original comments. This was not a retraction nor an apology, rather Piggott reinforced his opinion:

> If anyone, whatever his job, is prepared to equip himself with the necessary theoretical and practical knowledge of archaeology required by modern standards in the study, and to serve apprenticeship of several excavations, he ceases to be an amateur in my sense of the word. (Piggott 1948b: 10)

And,

> Nothing is further from my mind than a 'closed shop' of archaeological pundits. The field is open to everyone, and there is no fear that unorthodox opinion will be suppressed or disregarded provided always that it fulfills the high standards of scholarship now demanded in our study. (Piggott 1948b: 10)

In actual fact, the Golden Generation was a "closed shop of archaeological pundits" with members supporting each other, regardless of whether they actually agreed on an interpretation. This is illustrated by some of the interactions between Wheeler and Piggott. For example, in his obituary of Wheeler, Piggott alluded to the fact that some of Wheeler's writings were flamboyant and not necessarily fact-reliant; that Wheeler had a "flair for interpretation" which sometimes involved incorrect data and interpretation, but Piggott excused it away as follows:

> At this juncture [late 1920s and 1930s] some sort of interpretation, however provisional, was needed to provide a conceptual framework within which excavation could be used as a practical means of attaining intellectual ends, rather than a hazily envisaged ransacking of a misunderstood past. (Piggott 1977: 627)

This side-stepping of factual interpretation is also evident in a letter from Wheeler to Piggott:

> My dear Stuart, The manuscript of your article on the metal pins and mace-heads has reached me (…) I am delighted to have it and will send it on to press as soon as I return to Delhi. It is grand

[2] For an overview of the history of interdisciplinary studies in archaeology during the nineteenth and twentieth centuries, see Díaz-Andreu and Coltofean-Arizancu 2021.

> to have these important pieces of cross-evidence so clearly and convincingly stated. It so happens that in my summary reconstruction of the Harappa chronology (…) I have rejected the evidence of the double spiral pins on the grounds that a type which occurred on the one hand in Sialk IV and on the other hand in the Terramare is useless for dating purposes without closer evidence for local values. I am however delighted with the general tenor of your evidence, since it accords with my own conclusions and feelings in the matter on the grounds. (Wheeler 1947a, his emphasis).

This article "Notes on Certain Pins and a Mace-head from Harappa" was published in Wheeler's final issue of *Ancient India* (no. 4) (Piggott 1947a). The double spiral pins are an essential part of Piggott's argument in this article as they are the lynchpins of tying the chronology of the Harappan Civilization to the archaeological evidence beyond India. In his introduction to Piggott's article, Wheeler noted how important such material connections are across the world, in direct opposition to what he said to Piggott in the letter (cf. Wheeler's introduction to Piggott 1947a: 37). What was important to Wheeler, in this instance, was that the research was done in the correct manner, by the right person, in other words, method was more important that interpretation, at least in this example from their networked research work in South Asia.

In sum, Wheeler, Piggott, and other members of the Golden Generation embraced the standards and characteristics of scientific archaeology as *they* defined them: orderly layout and stratigraphic control in excavation, problem-oriented research, and interdisciplinary analyses of materials. It was their views that established standards for practice and interpretation, and their social and professional networks worked to reinforce such views and determine what was acceptable practice. During the 1940s Piggott and Wheeler brought these ideals to bear on their work in Colonial India, and as in the example cited above, on the spiral headed pins, a united front in terms of archaeological standards was their shared priority, not necessarily correct context nor interpretation. They formed their own network for assessing the 'correct' way of doing archaeology in the Subcontinent, and this connection was created by their multiple encounters in Britain before the Second World War.

11.3 Early Interconnections Between Piggott and Wheeler

Although Wheeler and Piggott crossed paths repeatedly in the 1920s and 1930s it seems that these paths were parallel with each working on their own projects. As Wheeler was establishing himself as an excavation authority and administrative leader in the museum world, Piggott was gaining experience and steadily publishing. In the second issue of the first volume of *Antiquity*, a scathing book review of a volume by the Royal Commission on the Ancient and Historical Monuments and Constructions in Wales led to the resignation of the secretary of the society and the appointment of a new leader (O. E. 1927). Archaeologist Roger Mercer (1998) has attributed the review to Wheeler, however there is no direct evidence to support this attribution. Regardless, Crawford was close with Wheeler as well as the Royal Commission, and Piggott was nominated to serve as typist and investigator with the new director. Thus, Piggott began to work for the Wales Commission in 1928 when he was 18 years old, while Wheeler was director of the London Museum.

Wheeler and Piggott directly interacted in 1930 when Piggott was employed at the Reading Museum working on Neolithic pottery (amongst other topics) and Wheeler was editor of the *Archaeological Journal*. In an autobiographical essay Piggott noted Wheeler "offered me publication if I produced a finished paper to a rather short deadline" (Piggott 1983: 31) which he did. Piggott's work in archaeology continued with additional single author publications in *Antiquity* (e.g., Piggott 1930) as well as participation in interdisciplinary excavation teams (cf. Smith 1997). During this time, Wheeler was writing books about London, participating in the council of the Society of Antiquaries, and excavating at Lydney Park (1928–1929), Verulamium (1930–1933) and Maiden Castle (1934–1935) alongside his first wife, Tessa Verney Wheeler (1893–1936) (Carr 2012). Piggott moved from the Reading Museum to working for Keiller in 1933, gaining more experience and exploring Avebury and other megaliths as well as following in antiquarian William Stukeley's (1687–1765) footsteps (Mercer 1998: 422–30).

Undoubtedly Piggott and Wheeler encountered each other at various professional meetings and knew of each other's work. In 1935 Piggott enrolled in Wheeler's new Institute of Archaeology in London, desiring to gain accreditation as an archaeologist. This was a one-year certificate program and Piggott recalls

> Wheeler did most of the lecturing, and I am one of the few who have heard him lecture on the Lower Paleolithic. After I had attended a couple of times he begged me not to appear again, but 'mug it up for yourself'. (Piggott 1983: 32)

While Piggott does not explain why Wheeler did not want him in class, perhaps it is because he knew Piggott personally and respected his research, and thus may have felt awkward having him there, as Wheeler lectured on topics that were not his forté. Piggott and Wheeler were part of the same movement of scientific archaeology, they published in the same journals, were members of the same societies, and appear to have been well aware of each other. This is supported by the fact that their personal correspondence blossomed towards the end of the Second World War, just before Wheeler became Director General of the Archaeological

Survey of India (ASI). A significant letter from Piggott to Wheeler, dated December 9, 1943, and labeled 'confidential', begins familiarly, as in "My Dear REMW", and ends with "yours ever" (Piggott 1943). In between are multiple pages relating Piggott's opinions of the current ASI; it seems that he felt familiar enough with Wheeler to freely express his thoughts about people, projects, and goals for Wheeler to embrace as he took over ASI. During most of the Second World War, Piggott was stationed in Delhi and used the study of South Asian Archaeology to distract him from the war, as Piggott writes to Wheeler: "(…) Indian archeology (…) a welcome intellectual refreshment after the arid mental deserts of official duties" (Piggott 1943: 4). At the same time, Wheeler was about to be decommissioned to England to prepare for the role of Director General of ASI. From this point in time their correspondence greatly increased and became very familiar in tone. That they trusted each other's judgement in terms of archaeological practice and interpretation becomes even more obvious after Wheeler arrived in India and began his command of ASI. Wheeler relied on Piggott's assessments and intellectual support for the changes he implemented as they validated each other's research in India. However, archaeology had been ongoing, although intermittently, in the Subcontinent since the eighteenth century. In order to understand the impact of Piggott and Wheeler's connection and support for each other, and its effect on their archaeological practice in the Subcontinent, a review of archaeology in Colonial India, prior to their arrival, is warranted.

11.4 A Brief History of Official Archaeological Practice in Colonial India

There is a long history of British interest in the antiquities and archaeology of India prior to the arrival of Piggott and Wheeler, and by the second half of the nineteenth and early twentieth centuries archaeological practice was primarily a colonial construct, not directly related to the development of the discipline of archaeology in England. Archaeology in South Asia began officially in 1863, driven by a desire to have a record of historical art and architecture related to western models, along with conserving some monuments which were believed to be important enough to maintain (Ghosh 1953: 29; Roy 1996: 91). The Colonial Government did not envision a long-term need for archaeological research, and instead focused on five-year contracts believing that with the completion of each term, the need for archaeology would be concluded and then shut down for financial savings (Marshall 1939: 6; Lahiri 2012: 304). The continually adjusted budget and limited scope and staffing of the department curtailed the development of an extensive cohort of practitioners, although there were dedicated professional archaeologists in India who persevered and accomplished exceptional work (e.g., Ghosh and Panigrahi 1946, discussed below). Consequently, the development of archaeology in India is one of a lack of institutional permanence, and the uneven production of knowledge within the intellectual bias of the colonial state. The first two officially appointed surveyors, Alexander Cunningham (1814–1893) and James Burgess (1832–1916), focused on their individual interests, and when John Marshall (1876–1958) was appointed in 1902 as Director General of ASI, only then did the actual work of creating a professional archaeological department begin.

The beginning of survey work started in 1861 when Cunningham was appointed as first official Archaeological Surveyor for Northern India, specifically to explore the Buddhist sites recorded by the early Chinese pilgrims Fa-Hien (c. 400 AD) and Hiuen-Tsang (c. seventh century AD) (Imam 1963: 199, 200; Chakrabarti 2001: 56, 58; Lahiri 2013: 129). In 1863 this position became the government's official Director General of the ASI, but funding and staff were severely limited. Cunningham surveyed until 1865, when he returned to England, and then traveled back to the Subcontinent six years later in 1871, to resume his work until 1885 when he officially retired. Concurrent with Cunningham's work in the northern colonial state, Burgess was surveying in the southern and western regions of India until he was asked to assume the post of Director General in 1886 when Cunningham retired. Burgess was only in this position until his retirement in 1889 and during these four years he took on the additional responsibilities for the survey work in the north of the Subcontinent, in addition to the western and southern regions, with no significant addition of staff or other resources (Chakrabarti 1982: 334; Roy 1996: 66).

During the second half of the nineteenth century, the colonial government had no comprehensive plan for the survey, conservation, excavation or research into the ancient history of India. In his review of this early phase of ASI, Amalananda Ghosh, Director General of ASI from 1953 until 1968, describes the official and vacillating guidelines regarding ancient remains in Colonial India:

> (…) at times it was thought that it would be sufficient if photographic or other copies of a few monuments were prepared. In 1862 the objective of the new Archaeological Survey was 'an accurate description—illustrated by plans, measurements, drawings or photographs and by copies of inscriptions—of such remains as deserve notice'. In 1870 the same definition was elaborated as 'a complete search over the whole country and a systematic record and the description of all architectural and other remains that are remarkable for their antiquity or their beauty or their historical interest.' (Ghosh 1953: 29)

Yet what should be considered as a candidate for beauty or historical interest was left up to the individual surveyor. Archaeological research as viewed by Cunningham, Burgess and other early surveyors and antiquity hunters was seen as a tool for creating a history of art (Roy 1953: 11, 23; 1996: 47, 66) and a linear timeline comparable to western historical chronologies (Cohn 1996: 46, 54). It is noteworthy that Cunningham and Burgess worked "at the same time as Schliemann was preparing for his epoch-making excavations in Hissarlik" (Roy 1996: 47), which archaeologist Sourindranath Roy notes was an exciting time for the advancement of archaeological practice, and yet they were not influenced by archaeological work in other world areas.

When Burgess retired in 1889, he and Cunningham urged the government to decentralize archaeology and place responsibility for survey, excavation, and conservation in the hands of local leaders (Ghosh 1953: 29; Chakrabarti 1988: 7–8; Roy 1996: 47, 66) and thus there was no national leadership or guidance from 1889 until 1902 when John Marshall became Director General. These years are marked by a lack of centralization and disorganization, and thus no substantial work of any kind was undertaken. The entire Subcontinent was left with three surveyors, who were not required to report directly to anyone, and thus each followed their own interests (Marshall 1939: 6–7; Roy 1953: 23–24; Chakrabarti 1988: 6; Roy 1996: 72). This period has been referred to as a "Bleak Interlude" when government officials believed that all archaeological work in the Subcontinent was completed, and there was no longer any need for additional fieldwork or conservation (Roy 1996: 72–73). However, by 1895, the Colonial Government began to realize that ignoring monuments and history was damaging to its world reputation, and thus they called for a re-evaluation of the need for and purpose of a national archaeological department (Marshall 1939: 8). Various leaders and local governments were asked for their opinions, as well as a number of well-established learned societies. In fact, the colonial government tried to persuade the Asiatic Society of Bengal (established in 1774 and highly regarded) to assume responsibility for any and all archaeological work in the Subcontinent, but the organization refused this offer (Roy 1953: 24). By 1898 the government concluded that they could not "disassociate itself from archaeology without inviting general condemnation (…)" (Chakrabarti 2001: 105), and in this year another plan for a national archaeological survey was submitted to and then sanctioned by the Secretary of State. This new plan did not include any provisions for excavation and left all actual direction and work in the hands of the local governments, as it had been since the retirement of James Burgess. Thus, it was a plan without a leader nor a means to enact it (Marshall 1939: 8; Roy 1953: 25).

The leaderless and directionless archaeological plan, as well as the lack of status of archaeological research in the Subcontinent began to change in 1899, when Lord George Curzon (1859–1925) was appointed as Governor-General of India. With his dedicated interest in the histories of ancient India, Curzon was the impetus behind returning the ASI to national importance (Roy 1953: 30–31; Chakrabarti 2001: 120). Curzon searched for an appropriate Director General to revitalize the Survey and he chose John Marshall, 26 years old, a scholar of ancient Greek history and language educated at King's College, Cambridge, who had trained with Arthur Evans (1851–1941) at Knossos (Lahiri 2012: 46; 2013: 317). Indeed, Curzon chose well and the status of ASI under Marshall's leadership was exemplary, as described by archaeologist Dilip Chakrabarti:

> (…) 'the John Marshall period' (…) It was an orderly and secure archaeological universe which, despite threats of retrenchment and financial stringency, went about its own way, building up the archaeological image of ancient India in its manifold colours and nuances. (Chakrabarti 2001: 169–70)

Curzon left India in 1905, and Marshall carried on with the mission to modernize the practice of archaeology in the subcontinent by introducing scientific methods in all four areas of ASI's responsibility: Conservation, Exploration, Research and Epigraphy. Marshall began by defining these practices and introducing explicit policies for all of them (Chakrabarti 1988: 11; 2001: 169; Roy 1996: 91).[3] He worked tirelessly, creating an organizational structure, establishing training programs and fellowships to encourage indigenous scholars all the while defining clear, succinct policy guidelines for all aspects of the Survey's work (Chakrabarti 1988: 11; 2001: 169; Roy 1996: 93).

Marshall's first policy directive was published in 1904 as the first *Annual report of the Director General of Archaeology, from 22 February 1902 to 31 March 1903* (Marshall 1904). This report records all of Marshall's travels, actions, publications, and expenditures for his first year as Director General. He also set about defining guidelines for the Survey staff and enforcing laws that had been enacted but had not been implemented in previous decades (Marshall 1904: 1–2). His many actions during his first year as Director General had the effect of solidifying and fortifying the Department and, especially important, in defining the significant duties of the officers of ASI. In defining these duties, he emphasized their primary role in listing and conserving all ancient monuments, as well as the importance of following the 1895 guidelines (about preserving monuments and preventing looting), which had been ignored until Marshall became Director General. He affirmed the crucial role of the archaeological officers as active and accessible field representatives of the department and encouraged them to take initiative in reporting finds and sites as well as any conservation work that was required to address imminent danger to a monument, and to enforce the laws protecting all heritage remains from looting.

[3]For an excellent overview of John Marshall and Indian archaeology, see the extensive work of Nayanjot Lahiri (1997, 1998, 2000, republished in a single volume in 2012).

Additionally, he established specific responsibilities between the archaeological officers and the provincial museums in their respective territories, to encourage integration between different parts of the Survey and to aid in conveying information to the public (Marshall 1904: 1–2).

It appears that his directives had an immediate effect because in his second annual report for the year 1903–1904 (Marshall 1905), Marshall writes: "The progress of the Archaeological Survey during 1903–04 has been steady in all its Branches (…). The responsibilities of its affairs generally have been more clearly defined (…)" (Marshall 1905: 1). Over the next two decades, Marshall focused on making the ASI a permanent government department, as well as training personnel, and delving into the ancient history of the Subcontinent.

Establishing the post of Director General as a permanent advisor overseeing the myriad aspects of archaeological research in India was an early and primary focus for him because like his predecessors, Marshall was appointed to a five-year term. In 1904 he submitted a request to the government to make the post permanent and after various government machinations, the post became so in 1906 (Ghosh 1953: 33; Roy 1996: 88; Chakrabarti 2001: 120). However, the government continually tried to save money in areas it deemed less important, especially from ASI, and in 1912 there was an attempt to abolish the post of Director General, and only after much outcry, the post was left to continue (Ghosh 1953: 35; Roy 1996: 93). There were severe financial cuts to ASI during the First World War (Roy 1996: 95), and in 1923 the government tried to cut 90% of the ASI budget. After intervention by various individuals, only a cut of 22.5% was enacted (Ghosh 1953: 37; Roy 1996: 94–95; Lahiri 2013: 300). Through all these repeated attempts by the government to curtail the work of the department, Marshall persevered in establishing policy, methodologies, and scientific field practices for all aspects of the Survey's work. Most importantly, Marshall created a sizeable and integrated community of professionals, to carry out archaeological research in the Subcontinent. What follows is a description of Marshall's dedication to best scientific practices as known by him during his years as Director General. Many of these are reminiscent of the disciplinary goals of the Golden Generation in Britain, yet they occurred a bit earlier, and independently.

In 1916, Marshall published authoritative guidelines for all aspects of the Survey's work. This publication, *Indian archaeological policy 1915*, begins with a "Resolution by the Government of India, Department of Education, nos. 353–370 (Archaeology). Dated the 22nd of October 1915" regarding the work of the Archaeology Department under Marshall, and the permanent establishment of ASI (Marshall 1916). Following this, Marshall elaborates on every aspect of the Survey (e.g., history, organization) as well as all practices (e.g., the listing as well as the conservation of monuments, epigraphy). He is very detailed here, and this has to do with the fact that he believed in and practiced a scientific approach to all aspects of archaeological research. In the 1916 volume he writes:

> (…) excavation (…) has developed into an exact science, demanding expert knowledge and control. This advance has been mainly due to the labours of archaeologists in Greece, Italy and other Western countries, whose scientific methods of excavation I was instrumental in introducing into India. What success these methods have achieved, may be seen in the harvest of monumental finds and the wealth of information regarding ancient culture that has since been obtained (…). (Marshall 1916: 23–24)

Being Director General of ASI was all consuming for Marshall, although he managed a short leave in 1923 and took this time to travel to Palestine, Syria and Egypt "acquainting myself at first hand with the archaeological work in progress there" (Marshall 1923–1924: 142–43). He notes how important this time was, visiting sites, observing current field practices, and talking with colleagues,

> Archaeology is not a science to be pursued in a watertight compartment nor within the limits of any one country (…) the methods and aims that are being pursued by our fellow archaeologists in other countries are also of no small concern to us in India. (Marshall 1923–1924: 142–43)

Marshall had a systemic and scientific approach to all matters inclusive to the ASI purview. For example, he required proper equipment as well as detailed record keeping of finds and observations in all aspects of fieldwork (Lahiri 2013: 357–58). He also relied on scientific analyses of materials and objects, for example, the many compositional and source studies of metal objects from the site of Harappa (Ullah 1940) and the study of animal remains found during the early excavations at Mohenjo-daro (Sewell and Guha 1931), in addition to the minerals found at that site (Pascoe 1931). Perhaps most importantly, after detailing policies, methods and the need for analyses, Marshall created a network of practitioners, that is archaeologists, scientists, epigraphers, and conservators that were well trained and thus could work independently of his direct supervision, and yet all reported to him. He believed that the Director General should be an advisor giving general direction, and he was explicit with the officers and superintendents of the Survey about what they should be doing, and how they should be doing it (Marshall 1916). Examples of the independent and reliable work done by the officers in the Survey include a wide range of activities, such as Nani Gopal Majumdar's (1897–1938) extensive surveys and excavations in Sindh Province (Majumdar 1934) as well as the early excavations at the sites of Mohenjo-daro by Rakhal Das Banerji (1886–1930) (Lahiri 2013: 273–74, 286) and at the site of Harappa by Madho Sarup Vats (1896–1955) (Vats 1940) and Daya Ram Sahni (1931–1935) (Lahiri 2013: 214–21, 298).

Marshall read the reports sent to him by Banerji, Vats and Sahni, and noted Vats' observation of the similarity between the objects found at both sites (Lahiri 2013: 309, 315). Therefore, Marshall organized a conclave in Simla, India in 1924 with all of the excavators as well as all their documentation and the portable finds (Lahiri 2013: 311–13). Thus, working with his network and relying on their research, he was able to define a new Bronze Age civilization (the Indus or Harappan Civilization) without ever setting foot on either site (Lahiri 2013: 300). He trained and trusted his deputies, and consequently, this left him to oversee the emergence of a complex and multifaceted institution within a period of incessant administrative and financial uncertainty. Marshall had created his own version of a Golden Generation, a community of practiced and reliable researchers conducting archaeological investigations across South Asia.

Marshall retired as Director General in 1928, but stayed in India to write up his fieldwork, and he returned to England in 1934 (Ghosh 1953: 38; Roy 1996: 110). As soon as he retired, the government stepped in to reduce expenditures and in 1931 the government eviscerated the Excavation Branch of ASI. Over the next few years, it dismissed or did not renew staff contracts, stopped funding scholarships and continued to cut funding (Ghosh 1953: 38; Roy 1996: 115–16; Ray 2008: 15). This left the survey "on the verge of a breakdown" (Roy 1996: 116). The 16 years, from 1928 when Marshall retired until 1944 when Wheeler arrived, saw the appointment of four Director Generals, all of whom were trained by and worked alongside Marshall: Harold Hargreaves (1928–1931), Daya Ram Sahni, John Francis Blakiston (1935–1937) and Kashinath Narayan Dikshit (1937–1944) (Roy 1996: 114; Ray 2008: 19). All were engaged under dire funding circumstances and served short tenures, yet these individuals strived to live up to the standards established by Marshall and maintain some semblance of the archaeological department he had created. Hargreaves oversaw "fruitful explorations in Sind" (Ghosh 1953: 39), while Sahni held the Department together during the Great Depression on a subsistence allowance from the government (Ghosh 1953: 39–40). When Blakiston took over, there was no funding for any exploration, but he made it possible for a foreign team, financed by the Museum of Fine Arts Boston and organized by the American Oriental Society, to excavate at the site of Chanhu-daro (Mackay 1943). Dikshit was the longest serving Director General of this group, and when he took over in 1937 there was some improvement in the department's finances. Although money and staffing were still meager, he managed to accomplish a great deal maintaining the scientific standards set by Marshall. He oversaw the large-scale excavation at the site of Ahichchhatra from 1940–1944 specifically to document the stratification of pottery types over 15 centuries of settlement (Ghosh 1953: 42) This excavation was conducted well enough even for Wheeler, such that the seminal pottery study was included in the first issue of Wheeler's new journal series *Ancient India* (Ghosh and Panigrahi 1946: 37–59). Dikshit also implemented the first prehistoric explorations in Gujarat led by an Indian archaeologist and revived a few of Marshall's policies that had been decimated by the colonial government, especially the responsibilities between the Survey officers and the provincial museums. He is also responsible for encouraging the University of Calcutta to excavate in the state of Bengal, the first Indian university to embrace field excavation (Ghosh 1953: 42).

Marshall had a long-term impact on the practice of archaeology in Colonial India and followed the same standards as members of the Golden Generation in Britain. When Wheeler took over in 1944, the ASI budget had been decreased once again and most of Marshall's organizational and structural work had fallen apart due to lack of funding and staff (Ghosh 1953: 40). Yet there were experienced archaeologists in the subcontinent, aware of scientific practices, which, as will be discussed in the next section, Wheeler alienated, instead focusing on recreating the department solely in his image, relying on young, untrained officers and demeaning the work of Marshall. His purpose appears to have been to create an illusion of a clean break in the archaeology of India, with the introduction of revolutionary new methods by him (Guha 2015: 141–83). Wheeler could not do this alone and used his connection with Piggott to reinforce and give authority to his views. This connection was also very useful for Piggott, who shared many of the same colonial attitudes that Wheeler had, and helped to confirm his belief in the superiority of the archaeological ways they practiced in Britain, ignoring the history of archaeology in Colonial India and belittling Marshall's many accomplishments.

11.5 Wheeler, Piggott, and Creating Authoritative Archaeological Knowledge in Colonial India

In an earlier section of this chapter, details of the professionalization of archaeology in Britain have been described, and it is this context that molded both Piggott's and Wheeler's views about how archaeology should be practiced. They were both part of the Golden Generation that distinguished between proper archaeology in contrast to the amateur, defining the rules for best practice and judging what was acceptable. The many connections between the members of this group created the authoritative voice for the discipline in Britain. However, only Wheeler and Piggott pursued archaeological research in India, and thus it was just the two of them with the same views and opinions on what is important in archaeology. They also shared a lack of background on ancient India and were strongly imperialist, seeing themselves and their work as superior to the colonized.[4] Hence their connection became their authoritative basis on what constitutes good archaeology in India. They believed, from their shared experiences in Britain, that archaeology in the Subcontinent needed to be compelled into a proper discipline, and that improper and amateur practice needed to be removed. Thus, *their* judgement became paramount, and in order to create this authority, they would rely only on each other. They therefore agreed with each other's interpretations, repeatedly cited each other in a circular fashion, and believed that a complete break with all previous archaeological work in the Subcontinent was required in order to establish a new era of archaeological practice. Creating a distance from all previous ASI endeavors was crucial to creating the appearance of authority, and it is clear from their letters and publications that this began with Wheeler's appointment as Director General.

When Piggott heard about Wheeler's appointment, he was overjoyed and felt certain that Wheeler would work to establish the archaeology of South Asia on a firm footing, as they had done in Britain. From a letter Piggott sent to Wheeler in 1943, he wrote: "The news of your appointment (…) has given me enormous pleasure and I have been going round proclaiming the advent of the Golden Age in terms that make the Ninth Eclogue a modest understatement" (Piggott 1943). In another letter, to archaeologist Cyril Fox (1882–1967)—a close friend of Piggott's, Keeper of Archaeology and later Director at the National Museum of Wales—Piggott noted that what archaeology in India needed was to completely do away with how things had been done, to create new methods of practice and interpretation, and that Wheeler was the perfect person for this role (cited in Guha 2015: 170[5]). This is an interesting parallel with how Piggott and Wheeler saw their work in Britain, to dismantle the old ways and put in place a professionalized practice that they defined. Breaking with the past was done primarily by disparaging the excavations of John Marshall. By critiquing Marshall's excavation method, they implied that the data was erroneous and thus his interpretations could not be true.

Wheeler's harsh critiques of Marshall's excavation technique include his statement that Marshall's use of a bench level was "the very parody of scientific method" and its use was "absurd" (Wheeler 1947b: 144–45). Beyond Wheeler's hyperbole, this is not true; a bench-level is merely another tool for assessing vertical findspots (see Miller, in preparation 1, n.d.-a). In another attack on Marshall, Wheeler wrote: "Marshall was of a temperament which hinders the confident delegation of responsibility and hinders therefore the adequate training of subordinates to assume responsibility" (Wheeler 1956: 185). The detailed information on Marshall's work in the Subcontinent in the previous section of this chapter proves this point is also wrong. Wheeler's method differed from Marshall's primarily in that Wheeler's included more detail of soil changes and apparent cultural levels, yet both methods focused on recording the vertical and horizontal contexts of finds in relation to architecture. Wheeler was quick to promote *his* technique of excavation as better and thus so were his interpretations. It seems his purpose in discrediting all Marshall had done was in order to place himself in the central role of most important archaeologist (cf. Guha 2003b: 45) and he relished the idea of 'fighting' Marshall. For example, Wheeler critiqued Marshall's work at the site of Taxila *before* the final publication was printed, and when the volumes were published in 1951,

> (…) Wheeler tried to entice the editor of the *Journal of the Royal Asiatic Society of Great Britain and Ireland* into seeking a short article from Marshall regarding the latter's method of dating Sirkap [an area at the site of Taxila], which he hoped could be 'published together' with his own critical views of Marshall's 'facts' within the same volume. Marshall withdrew from the combat despite having written a draft, stating that he was 'too old and too ill to engage in controversy'. (Guha 2015: 137)

Guha also cites a letter from Hargreaves (who succeeded Marshall as Director General of ASI) to Richard Bertram Whitehead (1879–1967, a numismatist, specialist in the Indo-Greek periods) about how Wheeler wanted to figuratively fight Marshall, and Hargreaves believed that this was because Wheeler was jealous of Marshall's many accomplishments (Guha 2015: 137).

Wheeler's incessant critique of Marshall's work at all the sites he excavated created a division in ASI, alienating the experienced officers who had served under and trained with Marshall (Guha 2015: 136). This is exactly the prime opportunity Wheeler and Piggott wanted, a young, inexperienced cadre in which to instill the idea that their methods were superior and thus their interpretations.

[4] This is clear from their respective writings. For example, "(…) we do not find, and should not look for, an inherent element of progress in Indian history (…) India presents itself to us today as a country medieval where it is not prehistoric" (Piggott 1945: 1–2). Also, in her biography of Wheeler, Jacquetta Hawkes, who was well acquainted with him, notes that it "(…) suited him to have authority among people who were not ashamed to accept and admire authority" (Hawkes 1982: 234). Hawkes quotes at length a letter Wheeler sent to Cyril Fox, although she does not cite its source: "(…) this is an astonishing country, rotten with the vilest intrigue, riddled with inferiority (…)" yet later on he notes how he "feels shame" when he encounters the kindness of Indians (Hawkes 1982: 234). And H.P. Ray observes: "A constant refrain in Wheeler's writings after 1947 when South Asia won its independence from Britain was the need to use archaeology as a tool where Britain could retain its intellectual superiority over Asia" (Ray 2008: 69). Wheeler believed that only the colonial authority could safeguard the heritage of colonized nations, and declared: "the very urgent need, for the safe guarding of cultural and, in particular, historical and archaeological evidence in and relating to our colonial territories" (Wheeler 1954).

[5] In a letter from Piggott to Fox, dated February 16, 1944, he writes that "only by a ruthless scraping of old ideas and instilling of new can anything be done' to improve Indian Archaeology" (cited in Guha 2015: 170; source Fox Collection, W05/5/5, Cambridge University, Cambridge, United Kingdom).

In fact, stratigraphic excavation was being practiced in India before Wheeler's arrival thanks to Marshall, and in spite of severe cutbacks in funds and staffing during the 1930s. The individuals who had worked alongside Marshall continued his practices, for example the excavation at the site of Ahichchhatra (1940–1944), mentioned earlier. Piggott[6] and Wheeler accepted the methods and results of this excavation by archaeologists trained by Marshall, and Wheeler published the results in the first issue of *Ancient India* (cf. Ghosh and Panigrahi 1946):

> What is important is that Wheeler seems to have concurred with the broad stratification of the site. Even if the analysis was done after Wheeler joined the Survey, the records were kept before his time and considered adequate enough to lead to a pottery sequence. This shows that in the early forties the young officers of the survey had a general understanding of the principle of stratification and that they could relate it to the pottery sequence of a site. (Chakrabarti 2001: 176)

Although Wheeler and his supporters claim credit for the introduction of stratigraphic methodology into South Asian archaeology (e.g., Hawkes 1982: 232; Piggott 1977: 636–37), this is an exaggeration. As the example of Ahichchhatra illustrates, there was an awareness and practice of stratigraphic excavation introduced by Marshall. Additionally, it appears that Wheeler was not directly involved in imparting knowledge about stratigraphic excavation techniques. He delegated all teaching and recording activities, and was often absent from the field (e.g., Carr 2012: 111, 115, 181, 185). Even in India, he was not directly involved in teaching, recording or direct supervision. For example, the field school held at the site of Taxila in 1944, it was Dermot Casey (1897–1977) who taught the "mysteries of stratification" (Hawkes 1982: 240). Casey was an Australian archaeologist who had worked with Wheeler at Maiden Castle and whose elder brother was Richard B. Casey (1890–1976), Governor of Bengal (1943–1946) (Hudson 1993). Also, Wheeler's excavation at the site of Mohenjo-daro in 1950, for the newly independent nation of Pakistan, followed a well-known practice of his, appointing area supervisors tasked with the day-to-day excavation and recording, and writing up of summary reports (see Miller, in preparation 2, n.d.-b). According to Piggott, the reason Wheeler never published a final report on his work at Mohenjo-daro was due to the site supervisors, to whom he delegated responsibility for the preliminary report (Piggott 1977: 637). Wheeler's work at Taxila has also never been published.

According to information scientist Robin Boast, Wheeler's field methods were in actual fact "not radically different, or even significantly different, from those he challenged" (Boast 2002: 169)—especially at the site of Arikamedu, where Wheeler was trying to define a chronological baseline for South India relying on Roman-style pottery. Wheeler's excavation provided information on horizontal context and vertical sequence, but what Wheeler needed in order to create a chronology across South India was a stylistic comparison of various ceramic wares, a procedure that has remained unchanged regardless of the development of field techniques (Boast 2002: 170). Boast observes that what Wheeler really added to archaeological practice in the Subcontinent was in "regimenting" and not "revolutionizing the approach to archaeology" (Boast 2002: 169).

Thus, for all the fanfare of his supposed superior field methodology, Wheeler's methods were not revolutionary in South Asia and furthermore Wheeler's interpretations were not necessarily based on any of his field data, and hence were disproven within a few decades of their publication (Guha 2003a: 8; 2003b: 46; Ray 2008: 152). Wheeler imposed his interpretations prior to, or without consideration of the actual field data, and there are numerous examples that illustrate this. In a review of Wheeler's book "Early India and Pakistan", Braj Basi Lal—a Wheeler trainee, Director General of ASI in 1968–1972, and Director of the Indian Institute of Advanced Studies—notes: "Sir Mortimer's tendency to rush to conclusions, or throw out suggestions of far-reaching consequence, although in reality there may not exist any basis for them" (Lal 1960: 225). Lal cites four specific instances in the book that are based on Wheeler's preconceptions and in each instance, with the use of actual data and conclusive reasoning, Lal disproves all four. And as Ray observes, Wheeler's interpretations were often based on "extraneous factors, rather than emerging out of an analysis of the data" (Ray 2008: 208).

Yet, Wheeler and Piggott's interpretations held sway for decades, and even today there is an aura of authority about their work. This has been created by imposing the illusion of new and improved techniques on a cadre of young trainees, and especially their respective publications that claimed advances in understanding the past, that relied on circular references to each other. Piggott and Wheeler created authoritative archaeology by validating each other's interpretations; each saw the other as the only true scientific archaeologist working on South Asian materials at that time. This is exemplified by a story told by Piggott about Wheeler and the finding of terra sigillata in India:

> Wheeler kept me closely in touch with his activities (…) [he] rang me up in Delhi to say he had something to show me. I went round to his office, where he sat with a tigerish grin on his face and a little pile of red sherds on the desk in front of him, which I saw and recognized as I came in the door, saying 'Good heavens, Rik, you've found it. That's first century Arretine!' 'Hooray! Of course it is. That's why I got you round—there are only two people in India who'd know what that was, and that's you and me'. (Piggott 1983: 34)

[6]In his 1943 congratulatory letter to Wheeler, Piggott describes how he had visited the excavations at Ahichchhatra that year and found that "a satisfactory pottery sequence had been established and could be used to correlate with the layers" (Piggott 1943).

Wheeler trusted Piggott to confirm the identity of the sherds and in other instances Piggott trusted Wheeler to acknowledge his findings, which Wheeler did by publishing them in *Ancient India* and regularly referring to Piggott's observations in his own publications. Together they shared a singular view of how archaeology should be practiced in India and believed that they were the only ones who knew and followed the rules, as the Golden Generation in Britain had done. This included the acknowledgement of findings judged to be reliable, or good, or dismissal of research judged to be bad. In other words, by accepting what the other was concluding, Piggott and Wheeler created what was authoritative knowledge. In most instances their citations of each other were circular and they never questioned or verified what the other was writing, as the examples below will illustrate. Rather there was an implicit trust in how each conducted archaeology, based on their shared experiences in Britain, and this trust was used as a means of creating authoritative knowledge (Hardwig 1991: 702). They seem to have believed that only the two of them in the whole of South Asia understood what it meant to practice a scientific archaeology and thus each had the other's unconditional support regardless of accuracy.

For example, in his 1943 letter to Wheeler with observations from his visit to the site of Ahichchhatra, wherein Piggott concluded that the pottery sequence was acceptable (see footnote 6 above) and he also criticizes the bench-level technique (Piggott 1943). Wheeler adopts Piggott's critique wholeheartedly and cites it as a publication aimed at demeaning Marshall's field technique (Wheeler 1947b). Following this instance of Wheeler citing Piggott, Piggott then cites Wheeler, for this 'fact': 'The inadequacy of the recording at Mohenjo-daro, Chanhu-daro, and Harappa has been commented on by Wheeler in Ancient India, no. 3, 1947: 144" (Piggott 1947a: 27). Indeed, however, Wheeler's statement is based on Piggott's own earlier observation described in a personal letter to Wheeler. Thus, this circular argument starts with Piggott's 1943 letter to Wheeler (Piggott 1943) who accepts Piggott's critique, cites a version of it in his 1947 article (Wheeler 1947b), and then in a later article Piggott cites Wheeler as the authority (Piggott 1947a) which had originally cited Piggott.

Another example of this circular referencing has to do with the fact that both Wheeler and Piggott believed that the Indus cities must have been fortified. Wheeler apparently was able to just look at a site and see a fortified layout, without excavating it (Piggott 1945: 15). In Piggott's 1945 book, *Some ancient cities of India*, published as he returned to England from India and as Wheeler was settling in as Director General, Piggott notes:

> (…) the Director-General of archaeology in India, Dr. R. E. Mortimer Wheeler, has in a recent re-examination of the sites [Mohenjo-daro and Harappan], come to a conclusion, which I share, and recognized in the great Stupa Mound of Mohenjo-daro, on a platform of brick twenty feet high, and in the AB Mound at Harappa with its traces of a defensive wall and probable bastion turrets, the citadels of a ruler, be he emperor or priest-king. (Piggott 1945: 15)

In a footnote to this Piggott adds, "I am indebted to Dr. Wheeler for permission to mention this extremely important observation made to me in conversation in 1944, and not yet published by him" (Piggott 1945: 15). Then, in an article in the January 1947 issue of *Ancient India*, which was a report on his excavations at the site of Harappa, Wheeler wrote how when he first visited the site in 1944, on his inaugural tour as Director General, he immediately saw the missing defenses around Mound AB (Wheeler 1947c: 62). Then in a footnote he added: "This identification was subsequently supported by Stuart Piggott, *Some Ancient Cities of India* 1945: 15" (Wheeler 1947c: 62). Additionally, in an article by Piggott published in 1947 he writes:

> Doubts had entered the mind of the writer, and that of Dr. Mortimer Wheeler, independently, and before visiting either Harappa or Mohenjo-daro—on the ground these doubts were converted to virtual certainty, [i.e., by Wheeler's observations of 'fortifications' when visiting the sites] (…) Dr. Wheeler's subsequent work at Harappa has triumphantly confirmed his diagnosis of the site, which we now know to have been dominated by a fortified citadel (…). (Piggott 1947b: 43)

In his later book, Piggott noted: "At Mohenjo-daro, Wheeler was able to show, by a little fieldwork rather than by actual excavation, that similar defences must have surrounded the westernmost mound on that site (…)" (Piggott 1962: 163). As we know today, Wheeler and Piggott's interpretations of citadels, fortifications and turrets were wrong, but that is not the focus of this chapter. Rather, the circular citations and trust they had only in each other as archaeologists was amplified by the colonial context of their work. As part of the colonial ruling structure, they were automatically seen as authorities to be believed, although some indigenous archaeologists questioned Wheeler's interpretations and methods (e.g., see Lal's review of Wheeler's book, Lal 1960) and many of the trainees from Wheeler's Taxila field school, distanced themselves from ASI and its dedication to Wheelarian ways (see Avikunthak 2021). This network of two individuals lacking a community of practitioners, operated within the unequal power structure of a colonial state and thus their word was absolute.

11.6 Conclusions

Piggott and Wheeler inherently trusted each other's archaeological practice because they were both part of the movement to professionalize archaeology in Britain in the 1920s and 1930s. They had the same belief in the correct way to excavate, and abhorred any practice that they thought was amateur which essentially meant any excavation not conducted by the Golden Generation. Their many interactions in Britain developed into a professional affinity, which became a crucial factor in defining acceptable archaeological research, from their viewpoint, in India between 1944–1948, the last years of the colonial state. This colonial context, with its unequal power structure, is part of the explanation of why they succeeded in creating the illusion of a new and revolutionary technique.

From the start of their time together in the Subcontinent, they wanted there to be a visible break with any previous archaeology conducted by ASI in order to create the impression of a revolutionary advance in what they believed to be the true scientific practice of archaeology. Yet, archaeology in India had a very long history, and was professionalized by Marshall, during his over 20 years leading ASI. As earlier sections have described, Marshall codified archaeological work and introduced scientific methods including stratigraphic observation, use of chemical analyses on materials, as well as the study of faunal remains. He did this independently of what was happening in archaeology in Britain, and within a repeatedly challenging fiscal environment and constantly wavering government support.

In order to create a break with this past, Piggott and Wheeler needed to separate themselves from earlier archaeological endeavors, and they did this by denigrating Marshall and his research. This created a schism within ASI and allowed Wheeler to harness a group of young and inexperienced officers, and to convince them of the superiority of his (and Piggott's) methods. Piggott and Wheeler did not build on Marshall's work, which was being continued by the experienced generation, but instead asserted their views and ways on a new generation while silencing the older voices and practices. Thus, criticism was ignored, and those voices were not given an active role in ASI (Guha 2003a: 7).

The new field method Wheeler supposedly introduced, supported by Piggott (in that he gave Wheeler the critique of the bench-level method that was in practice at that time, see Piggott 1943) was not revolutionary in the way they believed it was. This is clear from the excavation results from the site of Ahichchhatra. As for Wheeler's interpretations, "many of Wheeler's theories were declared non-valid within less than two decades after their conception" (Guha 2003a: 8).

In his book on the trope of archaeologist as hero, historian and archaeologist Robin Hoeks observes that within a colonialist narrative, such a view of the archaeologist "embeds a metaphor of progress into the narrative" (Hoeks 2017: 32). And indeed, Wheeler saw himself as such a hero, especially in his autobiography, *Still Digging*, where he describes the great difficulties he had to overcome in India in order to create a true and scientific program (Wheeler 1956: 10f; 1976, 9–12). As for Wheeler and Piggott's interpretations of the Indus Civilization, with its militarily fortified cities of Mohenjo-daro and Harappa, they saw such things in their mind's eye, because they wanted to see evidence of powerful leadership and imperialism as existed in the colonial state, mirroring their own political bias within the archaeology of India (cf. Berbeck 2012).

Historian Nayanjot Lahiri (2000) observes that when the history of archaeology in South Asia is studied, there is a tendency to assume that there was a positive evolution of techniques, with improvements building on each other. However, such a view disregards the inherent colonial context of such studies. Rather, she emphasizes, that archaeology was part of "the imperial process and its politics," and the Wheeler-Piggott interpretation of the Indus Civilization reflects this (Lahiri 2000: 24). Thus, the illusion of one true scientific practice, as believed by Wheeler and Piggott, is embedded within the unequal power relations between indigenous archaeologists and the colonial authorities. The disciplinary authority wielded by Wheeler and Piggott in South Asia is predicated on their blind support for each other and no one else, and reinforced by the power structure of the colonial state.

Acknowledgements Thank you to the editors of this volume for the feedback and patience as this chapter was written. I am also indebted to the anonymous reviewers who provided thought provoking observations. Also, a shout out to those unsung heroes that make archival research possible: the archivists at the British Academy, University College London, and the School of Archaeology Oxford University who encouraged and made possible this research!

References

Primary Sources

Archival Material

Crawford, O. G. S. (1926, 26 June). [Letter to Stuart Piggott]. Piggott Archives, Box 36a, Institute of Archaeology, Oxford University, Oxford, United Kingdom.

Piggott, S. (1943, 9 December). [Letter to R. E. M. Wheeler]. Wheeler Papers, MWA/5, British Academy, London, United Kingdom.

Wheeler, R. E. M. (1947a, 22 April). [Letter to Stuart Piggott]. Piggott Archives, Box 36a, Institute of Archaeology, Oxford University, Oxford, United Kingdom.

Wheeler, R. E. M. (1954). *Colonial Archaeology. Address by Sir Mortimer Wheeler. Delivered at the Oxford Meeting of the British Association for the Advancement of Science, No. 42.* Wheeler Archives, Box Q, University College London, London, United Kingdom.

Secondary Sources

Ashbee, P. (1972). Field archaeology: Its origins and development. In P. J. Fowler (Ed.), *Archaeology in the landscape* (pp. 38–74). London: John Baker.

Avikunthak, A. (2021). *Bureaucratic archaeology: State, science, and past in postcolonial India*. Cambridge and New York: Cambridge University Press.

Bernbeck, R. (2012). The political dimension of archeological practices. In D.T. Potts (Ed.), *A companion to the archaeology of the ancient near east* (pp. 87–105). Wiley-Blackwell.

Boast, R. (2002). Mortimer Wheeler's science of order: The tradition of accuracy at Arikamedu. *Antiquity, 76*(291), 165–170.

Carr, L. C. (2012). *Tessa Verney Wheeler. Women and archaeology before World War Two*. Oxford: Oxford University Press.

Chakrabarti, D. K. (1982). The development of archaeology in the Indian subcontinent. *World Archaeology, 13*(3), 326–344. http://www.jstor.org/stable/124387

Chakrabarti, D. K. (1988). *Theoretical issues in Indian archaeology*. Delhi: Munshiram Mancharal Publishers Pvt. Ltd.

Chakrabarti, D. K. (2001). *A history of Indian archaeology from the beginning to 1947*. Delhi: Munishiram Manoharlal Publishers Pvt. Ltd.

Cohn, B. S. (1996). *Colonialism and its forms of knowledge: The British in India*. Princeton: Princeton University Press.

Crawford, O. G. S. (1927). Editorial notes. *Antiquity, 1*(1), 1–4. https://doi.org/10.1017/S0003598X00000016

Crawford, O. G. S. (1930a). Book review, London in Roman times by R. E. M. Wheeler, with contributions by T. Davies Pryce, T. V. Wheeler and E. B. Birley. London Museum Catalogues, no. 3. Lancaster House, St. James's, S.W. I. 1930. Paper covers 2s., bound 2s. 6d. pp. 211, 51 plates (including 2 maps), and 58 figures in text. *Antiquity, 4*(15), 383–384. https://doi.org/10.1017/S0003598X00005020

Crawford, O. G. S. (1930b). Editorial notes. *Antiquity, 4*(13), 1–4. https://doi.org/10.1017/S0003598X00004142

Crawford, O. G. S. (1955). *Said and done: The autobiography of an archaeologist*. London: Weidenfeld and Nicolson.

Díaz-Andreu, M., & Coltofean-Arizancu, L. (2021). Interdisciplinarity in archaeology—A historical introduction. In L. Coltofean-Arizancu, & M. Díaz-Andreu (Eds.), *Interdisciplinarity and archaeology: Scientific interactions in nineteenth- and twentieth-century archaeology* (pp. 1–21). Oxford and Philadelphia: Oxbow Books.

Díaz-Andreu, M., Price M., & Gosden, C. (2009). Christopher Hawkes: His archive and networks in British and European Archaeology. *The Antiquaries Journal, 89*, 405–426. https://doi.org/10.1017/S0003581509000080

Ghosh, A. (1953). Fifty years of the Archaeological Survey of India. *Ancient India, 9*, 29–52.

Ghosh, A., & Panigrahi, K. C. (1946). The pottery of Ahichchhatra, District Bareilly, U. P. *Ancient India, 1*, 37–59.

Guha, S. (2003a). Imposing the habit of science: Sir Mortimer Wheeler and Indian archaeology. *Bulletin of the History of Archaeology, 13*(1), 4–10. https://doi.org/10.5334/bha.13102

Guha, S. (2003b). Mortimer Wheeler's archaeology in South Asia and its photographic presentation. *South Asian Studies, 19*(1), 43–55.

Guha, S. (2015). *Artefacts of history: Archaeology, historiography and Indian pasts*. New Delhi: Sage Publications.

Hardwig, J. (1991). The role of trust in knowledge. *The Journal of Philosophy, 88*(12), 693 https://doi.org/10.2307/2027007

Hauser, K. (2007). *Shadow sites: Photography, archaeology and the British landscape 1927–1955*. Oxford: Oxford University Press.

Hawkes, J. (1982). *Mortimer Wheeler: Adventurer in archaeology*. London: Weidenfeld and Nicolson.

Hoeks, R. (2017). *Archaeologists and autobiography: (Self-)fashioning in the public autobiographical writings of Austen Henry Layard (1817–1894), William Flinders Petrie (1853–1942), and Mortimer Wheeler (1890–1976)*. Uitgeverij Jonge Historici.

Hudson, W. J. (1993). Casey, Dermot Armstrong (1897–1977). In *Australian dictionary of biography*. Canberra: National Centre of Biography, Australian National University. https://adb.anu.edu.au/biography/casey-dermot-armstrong-10007/text17135

Imam, A. (1963). Sir Alexander Cunningham (1814–1893): The first phase of Indian archaeology. *Journal of the Royal Asiatic Society, 95*(3–4), 194–207.

Lahiri, N. (1997). John Marshall's appointment as Director-General of the Archaeological Survey of India: A survey of the papers pertaining to his selection. *South Asian Studies, 13*(1), 127–139. https://doi.org/10.1080/02666030.1997.9628531

Lahiri, N. (1998). Coming to grips with the Indian past: John Marshall's early years as Lord Curzon's Director-General of Archaeology in India—Part I. *South Asian Studies, 14*(1), 1–23. https://doi.org/10.1080/02666030.1998.9628548

Lahiri, N. (2000). Coming to grips with India's past and her 'living present': John Marshall's early years (1902–06)—Part II. *South Asian Studies, 16*(1), 89–107. https://doi.org/10.1080/02666030.2000.9628582

Lahiri, N. (2012). *Marshalling the past: Ancient India and its modern histories*. Ranikhet: Permanent Black.

Lahiri, N. (2013). *Finding forgotten cities: How the Indus Civilization was discovered*. London and New York: Seagull Books.

Lal, B. B. (1960). Early India and Pakistan. By Sir Mortimer Wheeler. (Vol. 12 in the series 'Ancient Peoples and Places.') Thames and Hudson, London, 1959. pp. 241, 57 photographs, 25 line drawings and 7 maps. 25s. *Antiquity, 34*(135), 224–227. https://doi.org/10.1017/S0003598X00029410

Mackay, E. J. H. (1943). *Chanhu-daro excavations 1935–36*. New Haven, CT: American Oriental Society.

Majumdar, N. G. (1934). *Explorations in Sind* (Memoirs of the Archaeological Survey of India, 48). Delhi: Indus Publications.

Malone, C., Stoddart, S., & James, N. (1999). Editorial. *Antiquity, 73*(282), 723–734. https://doi.org/10.1017/S0003598X00065479

Marshall, J. (1904). *Annual report of the Director General of Archaeology, Part I, from 22 February 1902 to 31 March 1903*. Calcutta: Superintendent of Government Printing.

Marshall, J. (1905). *Annual report of the Director General of Archaeology for the year 1903–4*. Calcutta: Office of the Superintendent of Government Printing, India.

Marshall, J. (1916). *Indian archaeological policy, 1915*. Calcutta: Office of the Superintendent of Government Printing, India.

Marshall J. (1923–1924). *Annual report of the Archaeological Survey of India for 1923–24*. Calcutta: Government of India, Central Publication Branch.

Marshall, J. (1939). The story of the Archaeological Department in India. In J. Cumming (Ed.), *Revealing India's past. A co-operative record of archaeological conservation and exploration in India and beyond* (pp. 13–33). London: The India Society.

Mercer, R. (1998). Stuart Piggott 1910–1996. *Proceedings of the British Academy, 97*, 413–442.

Miller, H. J. (in preparation 1, n.d.-a). Embedded misconceptions: Archaeological method and stratigraphy at Chanhu-daro.

Miller, H. J. (in preparation 2, n.d.-b). Wheeler's missing report from the 1950 excavation at Mohenjo-daro.

O. E. (1927). Review of Royal Commission on the Ancient and Historical Monuments and Constructions in Wales and Monmouthshire VII, County of Pembroke. *Antiquity, 1*(2), 245–247. https://doi.org/10.1017/S0003598X00000521

Pascoe, E. (1931). Minerals and metals. In J. Marshall (Ed.), *Mohenjo-Daro and the Indus Civilization* (pp. 674–685). London: Arthur Probsthain.

Piggott, S. (1930). Butser Hill. *Antiquity*, *4*(14), 187–200. https://doi.org/10.1017/S0003598X00004506
Piggott, S. (1945). *Some ancient cities of India*. Oxford: Geoffrey Cumberlege/Oxford University Press.
Piggott, S. (1947a). Notes on certain metal pins and a mace-head in the Harappa Culture. *Ancient India*, *4*, 26–40.
Piggott, S. (1947b). India and the Bronze Age Orient. *Institute of Archaeology Annual Report 1946, London*, 37–48.
Piggott, S. (1948a). Archaeology and the amateur. *Archaeological News Letter*, *1*(1), 1–2.
Piggott, S. (1948b). Letters, archaeology and the amateur. *Archaeological News Letter*, *1*(4), 10.
Piggott, S. (1962). *Prehistoric India to 1000 B.C.* (The Belle Sauvage Library). London: Cassell & Company Ltd.
Piggott, S. (1977). Robert Eric Mortimer Wheeler, 10 September 1890–22 July 1976. *Biographical Memoirs of Fellows of the Royal Society*, *23*, 623–642.
Piggott, S. (1983). Archaeological retrospect 5. *Antiquity*, *57*(219), 28–37. https://doi.org/10.1017/S0003598X00054958
Ray, H. P. (2008). *Colonial archaeology in South Asia: The legacy of Sir Mortimer Wheeler*. New Delhi: Oxford University Press India.
Roberts, J. (2012). Excavating an identity. British fieldwork in the first half of the 20th century. In O. Wolfhechel Jensen (Ed.), *Histories of archeological practices. Reflections on methods, strategies and social organisation in past fieldwork* (pp. 211–240). Stockholm: The National Historical Museum.
Roy, S. (1953). Indian archaeology from Jones to Marshall (1784–1902). *Ancient India*, *9*, 4–28.
Roy, S. (1996). *The story of Indian archaeology 1784–1947*. New Delhi: Archaeological Survey of India.
Sewell, R. B. S., & Guha, B. S. (1931). Zoological remains. In J. Marshall (Ed.), *Mohenjo-Daro and the Indus Civilization* (pp. 649–673). London: Arthur Probsthain.
Sharples, N. M. (2001). Stuart Piggott (1910–1996). *American Anthropologist*, *103*(3), 798–800. https://doi.org/10.1525/aa.2001.103.3.798
Smith, P. J. (1997). Grahame Clark's new archaeology: The Fenland Research Committee and Cambridge prehistory in the 1930s. *Antiquity*, *71*(271), 11–30. https://doi.org/10.1017/S0003598X00084490
Stout, A. (2008). *Creating prehistory: Druids, ley hunters and archaeologists in pre-war Britain*. Malden, MA and Oxford: Blackwell Publishing.
Ullah, K. B. M. S. (1940). The sources, composition and technique of copper and its all alloys. In M. S. Vats, *Excavations at Harappa. Being an account of archaeological excavations at Harappa carried out between the years 1920–21 and 1933–34* (pp. 378–383). Delhi: Manager of Publications.
Vats, M. S. (1940). *Excavations at Harappa. Being an account of archaeological excavations at Harappa carried out between the years 1920–21 and 1933–34*. Delhi: Manager of Publications.
Wheeler, R. E. M. (1927). History by excavation. *Journal of the Royal Society of Arts*, *75*, 814–834.
Wheeler, R. E. M. (1929). Caistor, and a comment. *Antiquity*, *3*(10), 182–187. https://doi.org/10.1017/S0003598X00003240
Wheeler, R. E. M. (1947b). Technical section: 3. The recording of archaeological strata. *Ancient India*, *3*, 143–150.
Wheeler, R. E. M. (1947c). Harappa 1946: The defences and cemetery R37. *Ancient India*, *3*, 58–130.
Wheeler, R. E. M. (1956). *Still digging: Interleaves from an antiquary's notebook*. London: Readers Union.
Wheeler, R. E. M. (1976). *My archaeological mission to India and Pakistan*. London: Thames and Hudson.

12 Archaeology and Networks of Authorities in a State of War: The Four-Level System in Sicily (1939–1943)

Antonino Crisà

Abstract

Sicily was probably one of the most bombarded Italian regions, suffering much destruction in urban areas, ports, and infrastructure during the Second World War. Since archaeological sites and museums were in danger, national and regional authorities carried out essential plans to protect the island's antiquities. In this context, archaeologists Pietro Griffo (1911–2007) and Jole Bovio Marconi (1897–1986) played essential roles in the protection of antiquities in Agrigento and Palermo. This chapter assesses the activity of Griffo and Bovio Marconi as primary actors in local, regional, and national networks, which comprised a series of units/members protecting antiquities during the war. These networks included safeguarding authorities, custodians, museum directors, workers, city mayors and military personnel. Griffo strongly opposed a series of unauthorized excavations and activities by the Italian army at Agrigento, which put archaeological buildings at serious risk. He was therefore part of a wider network that connected civilian and military authorities, which clashed with each other. In her role as the director of the Palermo Museum and superintendent, Bovio Marconi dealt with the construction of air raid shelters in archaeological urban areas. Thus, she operated within a civilian network, interacting with council, regional and state authorities.

Keywords

Jole Bovio Marconi · Pietro Griffo · Second World War · Authorities · Cultural heritage · Networks · Sicily

12.1 Introduction

The Second World War was one of the major events of the twentieth century. The conflict caused much death and destruction on a European and world-wide scale. When Adolf Hitler (1889–1945) invaded Poland on September 1, 1939, no one expected a war which would last for more than five years. The Axis and Allied armies crossed nations and occupied urban and countryside areas. Entire cities were devastated by aircraft bombing and military operations, and millions of civilians suffered deprivation and considerable casualties. Berlin, Coventry, Dresden, London, Milan, and Palermo were among the European cities subjected to substantial destruction of private and public buildings, including monuments. Much research has been carried out on the historical, political, and military aspects of the conflict (e.g., Snyder 1982; Gilbert 2010; Geyer and Tooze 2015), but the effects of war on cultural heritage, and on the people and communities who were involved in safeguarding archaeological sites and museums, have been insufficiently studied.

This chapter examines how the networks of archaeologists, local people and authorities in Sicily mobilized to protect antiquities while the region was in a state of war. The subject is certainly worth studying for many reasons. Firstly, it has not been previously explored by scholars in the framework of Italy as a whole. Secondly, the situation in Sicily perfectly fits in the wider context of the European war theater, creating an effective model for wider investigations in

This contribution is derived from my paper given at the Annual Meeting of the European Association of Archaeologists in Budapest (August 24–30, 2020), which was held as a virtual event due to the COVID-19 pandemic. It is an embryonic and theoretical preview of a wider work which will be published as a forthcoming monograph. The research for this chapter was carried out within the *Cultural Heritage in Danger: Archaeology and Communities in Sicily during the Second World War (1940–45)* project (Marie Skłodowska-Curie Individual Fellowship), which is supported by the European Research Council (ERC) under the European Union's Horizon 2020 research and innovation programme (Grant Agreement No. 835876).

A. Crisà (✉)
Department of Humanities and Social Sciences, Prince Mohammad Bin Fahd University, Al Khobar, Saudi Arabia
e-mail: acrisa@pmu.edu.sa

L. Coltofean et al. (eds.), *Connecting People and Ideas*, Themes in Contemporary Archaeology,
https://doi.org/10.1007/978-3-031-81006-0_12

the field of cultural heritage and social studies. Nevertheless, since the topic can be seen as broad and complex, this chapter assesses only selected case studies, providing a fresh theoretical approach to investigating and reconstructing social networks.

The chapter begins with a short introduction to Sicily during the Second World War, contextualizing this research within a larger historical framework and followed by a general outline of the protection of antiquities in Italy and on the island before and during the conflict. Through the lens of a new theoretical approach, the third section seeks to define a social system to better investigate the role of networks in Sicily, formed by essential characters acting to defend (or threaten) antiquities at risk on the island. The fourth and fifth sections consider the activities of archaeologists Jole Bovio Marconi and Pietro Griffo, safeguarding authorities in Sicily at the time, who were both active participants in these networks. The concluding section provides some final remarks on these subjects and presents suggestions for future research.

12.2 Historical Context: War and Antiquities Protection in Italy and Sicily

After the beginning of the Second World War, several European states enacted a series of measures to protect antiquities during the conflict by putting in force new emergency laws. On a wider scale, some common ground can be observed. For instance, the national governments of Greece, Great Britain and France arranged special shelters into which the most important collections of fine art objects and antiquities were transferred for safekeeping (Pollard 2019: 1–17). In Athens, curators even excavated pits inside the National Museum to bury a selection of Classical Greek artifacts, while the painting collections of the National Gallery in London were moved to a safe shelter far from the capital city (VV.AA. 1939; Bosman 2014: 79–100). Similar measures were taken in Italy as well.

Italy entered the Second World War on June 10, 1940, joining the Axis forces, represented by Germany and Japan. As a direct consequence, following the declaration of war signed by Prime Minister Benito Mussolini (1883–1945) against France and Great Britain, enemy aircraft bombed Italian cities. Great Britain's Royal Air Force and later the United States Army's Air Force devised a key plan to hit the most strategic urban centers equipped with ports, factories, or main railways, to weaken infrastructure and demoralize civilians. Bombing became a major concern for national authorities dealing with the protection of cultural heritage and monuments (Baldoli 2010: 34–49; Coccoli 2011, 2017). Due to its rich cultural heritage, Italy was a complex area, where substantial numbers of monuments, churches, archaeological sites, and museums were under threat from enemy aircraft bombing and military operations on the ground (e.g., landing, invasions, the advance of troops, occupation of sites). In the very early stages of the conflict, Giuseppe Bottai (1895–1959), Minister of National Education, sent an urgent dispatch to all national safeguarding Superintendencies on June 12, 1940. In this document, he ordered an emergency plan to be quickly implemented to secure historical buildings and museums (Bottai 1938–1939: 42–52; 1992; Zagarrio 1976: 267–71).

Before the war, Benito Mussolini, *Dux* of Fascism, formalized a set of vital procedures to protect cultural heritage in case of imminent conflict. The document, n. 104800 of April 30, 1936, entitled *Protezione antiaerea: Servizio di primo intervento* (Anti-aircraft protection: First intervention service), is a detailed, confidential dispatch written by Mussolini and sent to Italian state administrative offices, including prefectures and all Commissioni provinciali per la protezione antiaerea (C.P.P.A.A.: Provincial Committees for Antiaircraft Protection), later known as Unione Nazionale Protezione Antiaerea (U.N.P.A.: National Union for Antiaircraft Protection). It established teams of volunteers who had to intervene in case of enemy attack to protect local offices, public buildings, and museums. These officers were conveniently equipped and trained to deal with bombing and fire damage. More importantly, these first-intervention teams were created nationwide.

The Italian government took several additional measures once the war began; Victor Emmanuel III (1869–1947), the King of Italy, was only able to approve decrees proclaimed by Mussolini. Procedures included, for instance, Royal Decree n. 1415 of July 8, 1938 (*Legge di guerra*; War Law) and Decree n. 367 of June 17, 1940 (*Determinazione di segni distintivi per la protezione di edifici e monumenti dai bombardamenti*; Regulation of distinctive signals for the protection of buildings and monuments against bombing), which regulated the use of distinctive signals on historical buildings. Another procedure was Law n. 1041 of July 6, 1940 (*Protezione delle cose d'interesse artistico, storico, bibliografico e culturale della Nazione in caso di guerra*; Protection of goods of artistic, historical, bibliographical, and cultural interest of the Nation in case of war) for the protection of cultural heritage in time of conflict. The Italian government also provided a series of guidelines to protect historical buildings, museums, and churches. For instance, monuments when possible were to be secured with protective frameworks, scaffolding, and sandbags, for the purpose of containing damage caused by bombs and blasts. However, such structures were not completely safe: depending on their caliber and power, bombs could destroy them anyway (Boi 1986: 2–3; Coccoli 2008: 304–5; 2010: 411–13; Barone 2011: 21–22).

From the Italian state's perspective, Sicily was a culturally significant region, with plentiful monuments, archaeological sites, and historical buildings that had to be protected in case of enemy attack. The island's substantial cultural heritage derived from a long-standing and elaborate stratification of ancient and modern invasions and occupations (e.g., Greek, Roman, Arabic, Norman, Spanish), which left substantial and tangible traces in urban areas and the countryside. Furthermore, the island was of vital strategic potential, being at the center of the Mediterranean Sea which became a major theater of war between 1940 and 1943. Mussolini and the Italian military authorities were therefore concerned about a potential enemy invasion in Sicily from the very early stages of the conflict (Pack 1977: 30–32; Zaloga 2013: 6–9).

Enemy attacks began in Sicily as soon as Mussolini signed the declaration of war. In fact, bombs hit Messina and Palermo as early as June 23, 1940. Both cities were certainly of strategic significance, benefitting from vast ports, railway stations, and factories. For the same reasons, the port city of Catania would also become a major target of Allied aircraft in the years to come. In the early phases of the conflict, the intensity of bombing varied and was irregular, but the Allies markedly increased their attacks on Sicily between January and June 1943 as the conflict intensified (Romeo and Rothier 2017: 12–26; Albergoni 2018).

The Sicilian authorities had designed plans to protect antiquities by June 1940, which, however, were not fully realized. All archaeological sites were hypothetically under threat in Sicily—especially those in urban zones—but could not be fully secured. For instance, cities like Agrigento, Segesta, and Selinunte had impressive monuments and ancient ruins, but they were located in rural areas that seemed less at risk from damage by enemy bombing raids. Protecting a majestic Greek temple was a demanding operation and material was at a premium: scaffolding, sandbags, and wooden frameworks were expensive due to the rationing imposed in war time. Authorities were therefore forced to prioritize certain monuments over others and offered partial protection for the select few. As explained below, Agrigento was especially problematic among these sites, due to the presence of the Italian army high command.

The protection of Sicilian museums was another major concern for the safeguarding authorities. Many of these institutions were located in urban centers whose strategic infrastructure (e.g., ports and railway stations) became military targets during the war. This occurred in Agrigento, Catania, Messina, and Palermo, Sicilian cities that suffered the most intensive enemy bombing between 1940 and 1943. Air raid shelters were an effective and rational way to protect fine art objects and antiquities in a safe place far from the dangerous urban environment. Authorities arranged the main shelters at Syracuse, Bivona (Agrigento), and San Martino delle Scale (Palermo) to store large collections of artifacts; minor refuges and armored rooms were also constructed in the basements of museums to secure less valuable or unmovable finds (VV.AA. 1942: 335–40; Boi 1986: 28–29; Oteri 2005–2006: 293–94; Barone 2011: 21–28; Pollard 2019: 1–17).

Once the Allied forces had defeated the Italian army in northern Africa on May 13, 1943, the prospect of an enemy landing from Africa to Sicily became more likely. Intensive bombing on the island, especially of major strategic cities (in particular, Catania, Messina, and Palermo) turned into a massive aircraft action in preparation for Operation Husky, which took place on the southern coastline of Sicily on July 9, 1943. The Allied landing was carried out by Generals George Smith Patton (1885–1945) (US 7th Army) and Bernard Law Montgomery (1887–1976) (British 8th Army), while Alfredo Guzzoni (6th Italian Army) and Hans-Valentin Hube (1890–1944) (14th Armoured Corps of Germany) defended the island. The military operations were extremely rapid, and some battles were particularly intense, resulting in much destruction in small cities and towns (e.g., Gela, Randazzo, and Troina). After an initial fierce resistance by Italian and German forces, the Allies advanced towards the inner areas of Sicily, conquering Palermo on July 22, Catania on August 5, and Messina on August 17, by which time German troops had already abandoned the island, quickly crossing the Strait of Messina towards Calabria (Pack 1977: 73–159; Zaloga 2013: 33–88; Fitzgerald-Black 2018: 39–132).

At that point, in order to consolidate its headquarters and prepare for the new invasion of the Italian peninsula, the Allies established a military command center in Palermo. Thus, Sicily became Region I of the Governo Militare Alleato dei Territori Occupati (AMGOT: Allied Military Government of Occupied Territories; later known as AMG). As a military occupying force and authority, the AMGOT had different tasks in Sicily. For instance, it was responsible for easing the transitional political and social process from fascism after the fall of the regime on July 25, 1943, to the new provisional national government; this also included the punishment and the purging of ex-fascist authorities. The AMGOT also managed local justice—a key aspect considering the complex role of mafia in Sicily—involving food distribution, public health, and security, with the help of national forces (the *Carabinieri*) (Boi 1986: 11–19; Baldriga 1999: 87–93).

Some Allied officials were also in charge of supporting local authorities in their efforts to safeguard cultural heritage, providing advice to renovate damaged monuments, churches, and museums, and allocating all relevant funds for that purpose. These officials operated in synergy with other military and civilian corps. Made up of professors, art historians, and researchers mostly hired from American universities and fully trained to act in warzones, the so-called 'Monuments Officers'— known to the general public from their role in the film *The Monuments Men* (2014)—offered their expertise and experience to protect at risk antiquities and to trace plundered fine art objects. These officers included Perry Blythe Cott (1909–1998), Joseph Paul Gardner (1894–1972), and Mason Hammond (1903–2002). British officers such as Fred H. J. Maxse (1904–1970) and John Bryan Ward-Perkins (1912–1981), the future director of the British School at Rome, were also employed in these roles (Barone 2011: 21–23; Dagnini Brey 2015: 311–20; Coccoli 2017: 33–52; Pollard 2020: 107–8, 182–83).

12.3 Networks of Authorities: The Four-Level System

The previous section outlined the historical context of the Second World War in Italy and Sicily, with a specific focus on government plans to protect cultural heritage and museums against enemy attacks and military operations. This national scheme was extensive and complex: it in fact required a massive deployment of administrators, coordinators, authorities, and personnel acting in concert to defend monuments and historical buildings. How did these characters operate in Sicily in the civilian and military spheres? Can we fully assess their roles within well-defined social networks? This section outlines how authorities and employees interacted within the predetermined social levels of the Sicilian context.

According to the *Collins dictionary* (n.d.), "a network of people is a large number of them that have a connection with each other and work together as a system". This definition is certainly relevant to this volume in its reference to major and extensive social webs. Scholarship on social networks is well-established and can offer valid terms of comparison for this contribution to the history of Sicilian archaeology. In particular, this chapter benefits from essential theoretical contributions, such as *Social networks* by the social anthropologist John Arundel Barnes (1972); *Science in action: How to follow scientists and engineers through society* (1987); and *Reassembling the social* (2005) by philosopher and sociologist Bruno Latour. Two cardinal works by Carl Knappett, *An archaeology of interaction* (2011) and *Network analysis in archaeology* (2013), offer novel approaches to the study of archaeological networks, especially regarding artifact production and distribution in human communities. There are also valid sociodemographic models (e.g., on human mobility and urban interactions) which scholars have designed to investigate predetermined social spheres. Taking a broad point of view, anthropologists and sociologists often analyze large scale and extensive networks in a variety of contexts, epochs and fluid circumstances. Networks can be substantially vast: in fact, "networks may be composed of just a handful of nodes connected by short links, or millions of nodes connected by distant links" (Knappett 2011: 39). Scholars engaged in network analyses consider and assess substantial groups of similar people who tend to interact with each other. When these groups approach each other reciprocally, the interaction of social units is potentially significant (Barnes 1972; Clyde Mitchel 1974: 279–99; Latour 1987, 2005; Miller McPherson et al. 1992: 153–70; Knappett 2011, 2013). These works are complemented by recent studies on networks within nineteenth-century museum institutions, like the Pitt Rivers in the United Kingdom or the Smithsonian in the United States (Larson et al. 2017: 211–39; Snead 2020: 34–46).

In the context of Sicilian archaeology during the Second World War, a social network was not necessarily formed of a "large number" of units; in fact, smaller groups of people can also represent a network. This also in turn implies that multiple networks may interact with each other, generating interconnections among social units. In this chapter, in order to analyze social dynamics in the Sicilian theater of war context, a network is defined as a unitary group of people acting to safeguard antiquities and protect museums against enemy attacks. As part of a group, a variety of characters or social units interact with each other to achieve the same objective (or even impede it). Nonetheless, networks are influenced by social norms, or even legal and bureaucratic regulations. They also establish the parameters of effective hierarchic systems, as explained below.

The network model presented in this chapter arises from a theoretical system previously applied by the author to the history of archaeology in Sicily during the post-Unification period (1861–1918) (Crisà 2018: 30–31). The author's recent investigation in Italian archives[1] has provided substantial documentary data to reconstruct social dynamics in Sicily after the island was annexed as a region of the newly founded Kingdom of Italy in 1861. At that time, archaeology had evolved into a more professional discipline and archaeologists became state officers in the Italian government. Antonino Salinas (1841–1914), director of the Museo Nazionale di Palermo (National Museum of Palermo), played a leading role in shaping a modern institution representing all epochs of the island's history. In this context, a variety of social units (e.g., curators, custodians, antiquities safeguarding officers) interacted within a Three-level system formed of three-dimensional scales: (1) the Italian state: represented by the Ministry of Public Education; (2) the regions: represented by authorities operating in Sicily to supervise archaeological heritage on behalf of the Level One national body; and (3) the local: represented by other social units

[1] For instance, the Archivio Centrale dello Stato (Central State Archive) and The British School in Rome.

(e.g., mayors, prefects, and police officers) acting in a narrower geographic and administrative context such as a small town or village.

The above-described structure can be transferred and applied wholesale to war-time Sicily, generating a new Four-level system comprising the state, the region, the local and the supranational. The additional fourth, supranational level, which was not present in the original system, encompasses the other dimensions. The significant features of the Four-level system (and, of course, the previously conceived Three-level scheme) can be revealed by a direct social approach to assessing networks. Previous models have utilized networks to reconstruct interactions among communities starting from the archaeological data collected "in the field" (Knappett 2011: 149–71). The Four-level system, by contrast, depends on records in the archives that allow characters to be identified who were acting in social contexts that are essential for acquiring a more complete understanding of the history of archaeology in Sicily.

The first level, the state, is principally represented by the Ministry of National Education, the main government body which supervised the safeguarding of antiquities on a national scale. Giuseppe Bottai headed this state body between 1936 and 1943. Established from the traditional Ministry of Public Education on September 12, 1929, by Mussolini, and finally abolished by the Badoglio II Cabinet on May 29, 1944, the new fascist Ministry included the Direzione di Antichità e Belle Arti (Direction of Antiquities and Fine Arts) which supervised sites, museums, and galleries. It also coordinated all regional bodies, providing essential guidelines for synergistic operations of all relevant administrative entities (Dell'Era 2003: 145–46). In the period between 1940 and 1945, the Ministry of War was an essential part of the state, Level One, operations. This national body, founded in 1861, played a leading role in planning and managing all Italian conflicts until 1947, when it was disbanded. The Ministry of War occasionally clashed with the Ministry of National Education when military authorities interfered with the required protection of antiquities on a regional or local scale. In wartime, military forces, represented by soldiers, lieutenants, commanders, and generals, were primarily operational on a national scale to defend Italy against any potential enemy attack or invasion. As a part of the Italian state (but separate from the civilian authorities), military forces can therefore be properly associated with Level One in the network model developed in this chapter. Theoretically, German forces that joined the Italian army in Sicily to defend the island against the enemy invasion could be considered part of Level Four; however, it is more appropriate to consider them a parallel entity to Level One, being allied with the Italian army on the ground.

The Level Two network is represented by Sicily as a region, which had a well-organized system for supervising antiquities beginning in the post-Unification period, when a special regional Commissione di Antichità e Belle Arti (Commission for Antiquities and Fine Arts, 1861–1875) was created. Museum directors were also actively involved in overseeing artifact collections, excavating at museums and providing technical advice on fine art objects. The versatile archaeologists Antonino Salinas in Palermo, who directed the first regional Superintendency, and Paolo Orsi (1859–1935) in Syracuse (Pelagatti 2001: 599–621; Crisà 2018: 22–30), are examples of this system. Following a series of national administrative reforms carried out in the 1920s and 1930s, Sicily continued to maintain regional bodies to monitor cultural heritage and its distinct branches (e.g., sites, museums, and galleries). In particular, the Superintendencies for Antiquities included three main regional bodies: one for Eastern Sicily (1941–1973), one for Palermo and Trapani (1939–1963) and one for Agrigento and Caltanissetta (1939–1961) (Matarazzo 1938–1939: 415–18; Vitale 2011: 12–13; De Caro 2017: 240–43). Bovio Marconi and Griffo were superintendents in Palermo, Trapani and Agrigento during the war and acted as regional authorities under the direct supervision of the Ministry in Rome (Vassallo and Villa 2017: 381–82).

Level Three includes a variety of social units acting on a local scale. The fascist *Podestà*, who replaced the canonical city mayors after Mussolini promulgated Law n. 237 of February 4, 1926, was responsible for administrating each town district. Sometimes they collaborated or even interfered with the management of antiquities and sites. Local police forces and Carabinieri were also occasionally involved in antiquities safeguarding. They could investigate illegal trafficking or seize finds if not properly reported to the regional authorities. For instance, this occurred in post-Unification Sicily when police forces seized finds from a local canon in Lipari in 1883 (Crisà 2018: 194–98).

When comparing the Four-level network system to the Three-level model that was in operation in pre-war, post-Unification Sicily (Table 12.1), the fourth supranational level is clearly different from the others. In fact, the supranational dimension refers to the occupying military forces which acted in Sicily during the Second World War. They are represented by the AMGOT and the Monuments Officers who collaborated with regional and local authorities to safeguard antiquities and provided support for the renovation of museums and monuments damaged by military operations. If the interactions between Level Four and the previous ranks were essential and profitable, the AMGOT represented an unaffiliated and autonomous body which operated in Sicily only for a temporary timespan between July 1943 and February 1944 (Boi 1986: 20–24, 27–28; Barone 2011: 446–48). At that point, the Allied Control Commission (ACC) replaced the AMGOT in all Italian territories. In Sicily, Captain Hammond served as a Monument Officer together with Captain Maxse, who was his Assistant Advisor (Boi 1986: 20–26).

Table 12.1 Scheme showing the Three-level and Four-level network systems in Sicilian archaeology between the late nineteenth century and the Second World War

	Three-level system (1861–1918)			Four-level system (1939–1945)			
Main body/personnel	1	2	3	1	2	3	4
AMGOT							●
City mayors/fascist *Podestà*			●			●	
Commission of Antiquities and Fine Arts		●					
Italian Army and German troops				●			
Ministry of National Education				●			
Ministry of Public Education[a]	●			●			
Museum directors		●			●		
Police forces			●			●	
Public works office			●			●	
Prefects			●			●	
Superintendencies		●			●		

[a]It is necessary to note that the Italian provisional Badoglio II Cabinet re-established the Ministry of Public Education on May 29, 1944, while the Ministry of National Education was still operational in the Fascist Republic of Salò until April 25, 1945. For more information on Italian ministries after the fall of fascism, see Borghi 2018: 1–13

Bovio Marconi and Griffo, the two key figures engaged in safeguarding antiquities in Sicily during the Second World War, are essential figures in understanding how social networks on the island operated during the war. The next two sections provide some preliminary bibliographical information on both archaeologists to better contextualize the roles they played in the history of Sicilian archaeology. Their roles in the Four-level system are also assessed, identifying any potential connection with other authorities directly involved in the safeguarding of antiquities or even threatening their preservation.

12.4 Jole Bovio Marconi: An Active Archaeologist in Palermo

Born in 1897 in Rome and dying in Rome in 1986, Jole Bovio Marconi graduated from the University of Rome 'La Sapienza' in 1921 after studying with classical archaeologist Rodolfo Lanciani (1845–1929). She continued her training in Greece, where she met the archaeologist Pirro Marconi (1897–1938) whom she married in 1926. She became inspector of the National Museum of Palermo in 1927 and then director in 1937; two years later she was appointed superintendent of Western Sicily. As the first woman superintendent in Italy, Bovio Marconi was often actively involved in promoting women's rights in local administration and Palermo's daily public life. She also taught archaeological subjects at the University of Palermo. Nevertheless, Bovio Marconi is mostly known for having managed the National Museum of Palermo during and after the Second World War. She was solely responsible for designing and carrying out a complex plan to protect the National Museum of Palermo against bombing and arranged to have all collections from Palermo moved to the abbey of San Martino delle Scale (close to Monreale) and *vice versa*. The museum was completely reorganized after the war (Almagro 1952: 243–46; Bovio Marconi 1952; 1955: 175–81). As a professional archaeologist, active both in the field and in the museum, and primarily specializing in protohistory, Bovio Marconi excavated at Boccadifalco, Cefalù, Monte Castellaccio, Palermo (Addaura Cave), and Petralia Sottana (Vecchiuzzo Cave) (Di Stefano 1996: 11–18; Tusa 1996: 19–24; Battaglia 2017: 15–24).[2] She was certainly a central character in the Four-level system through her multiple roles in managing antiquities in Palermo and western Sicily. This section will briefly analyze an important case study involving the construction of air raid shelters in Palermo, which shows distinctly how archaeologists acted in Sicily during the war.

Once Italy entered the war and all regions and cities became potential enemy aircraft targets, the construction of civilian refuges was considered an impelling priority. Following the transmission of national guidelines circulated by the Ministry of War, local authorities, specifically the *Podestà* and the public works offices, under the higher supervision of the prefects, had to plan the creation of shelters as soon as possible in order to guarantee most Italian citizens a proper place to seek refuge in case of bombing. The Italian state only provided general advice on how to build the shelters to be as safe as possible, while local authorities organized their construction and management together with the Unione nazionale protezione antiaerea (National Union for

[2]Some of the most significant works written by Bovio Marconi include: *La stazione preistorica di Serraferlicchio* (1940), *La cultura della Conca d'Oro della Sicilia Nord-occidentale* (1944), *Sui graffiti dell'Addaura* (1953), *Il Museo Nazionale di Palermo* (1956), and *Sulla diffusione del campaniforme in Sicilia* (1963).

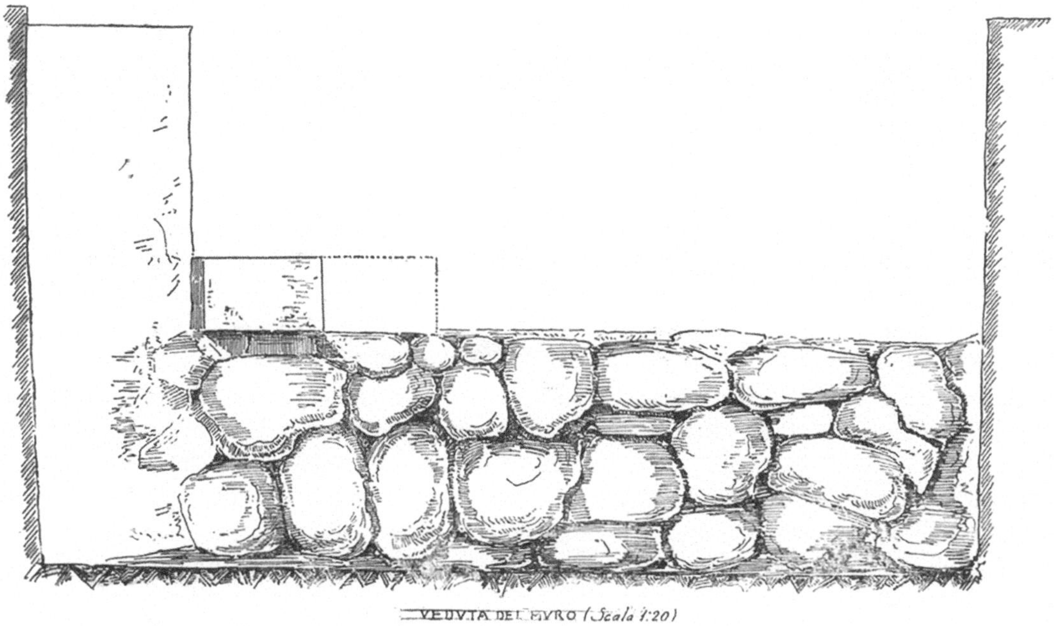

VEDUTA DEL MURO (Scala 1:20)

Fig. 4.

Fig. 12.1 Section of an ancient wall discovered at an anti-aircraft shelter construction site in Palermo. (Reproduced from Bovio Marconi 1941–1942, Fig. 4)

Anti-Aircraft Protection), which also coordinated the distribution and inflow of people inside them. Air raid shelters could be constructed *ex novo* by creating a bombproof subterranean structure reinforced with concrete, or they could be converted existing structures (e.g., a vast subterranean warehouse or the basement of a private house) or natural sites (e.g., a cave or ravine), which were reinforced to make them more secure (Chirone and Fasolini 2011: 587–89).

Palermo, especially its old town, is densely populated and includes important areas with a high archaeological potential. The city council of Palermo thus planned to build dozens of shelters, allocating substantial funds for this activity. Construction activity was on a massive scale and shelters were scattered across the city. One of the largest shelters was placed below Piazza Pretoria in the heart of Palermo's downtown. According to state guidelines, each shelter had to reserve 0.4 m^2 for each citizen sheltering in the structure. The state authorities also asked local city councils to equip public schools with shelters to protect students in case of enemy air raids. Fifty-eight refuges would have protected 27,963 school students during the war. Within a very limited timespan, the public works office planned and built numerous reinforced structures to serve as air raid shelters (Albergoni 2007: 103–10; Bertorotta 2008: 30–34; Crisà 2025, PAL.1).

In the previously described context, Bovio Marconi's specific task arose from the construction process itself. As one can readily imagine, building an air raid shelter would have required the excavation of selected urban areas. The digging process and substantial soil removal could generate casual discoveries of archaeological artifacts and structures (Fig. 12.1), especially if those shelters were located within the ancient and medieval zone of Palermo. As an expert archaeologist, Bovio Marconi knew the ancient topography and archaeology of Palermo intimately. Once she heard about the city council's plan to build the refuges, she quickly organized a monitoring system to supervise the excavation process at construction sites. The episode generated various social interactions that are documented in the city archives.[3] For example, Bovio Marconi contacted the Ufficio dei Lavori Pubblici (Public Works Office) in October 1941 asking about archaeological discoveries at the construction site of Piazza Sett'Angeli, close to the cathedral. The area contained archaeological evidence that was essential for shedding new light on the topography of ancient *Panormum*. Once the excavations had started, the archaeologists identified and documented the remains of several Roman houses decorated with mosaics. Additional important discoveries were made at the Istituto Magistrale de' Cosmi (de' Cosmi Teaching School), based on Bovio Marconi's report to the Ministry of National Education in November 1941. Substantial quantities of pottery sherds were recovered there, but local workers did not alert the authorities in time (Bovio Marconi 1941–1942: 501–12).

Bovio Marconi's monitoring of construction sites is important in understanding how the Four-level system

[3] For instance, the Archivio Comunale di Palermo (City Council Archive of Palermo).

worked in Sicily during the war. Level Two was only represented by Bovio Marconi. The regional dimension was therefore limited to a single archaeological safeguarding officer and a few expert custodians who operated in the field if the archaeologist was absent. As a key character who served as a node connecting various social units, Bovio Marconi interacted with other network dimensions. She had direct contact with Upper State Level One (Ministry of National Education), to which she reported updates on archaeological discoveries and from which she requested funds to perform further investigations or renovation of structures. However, most of Bovio Marconi's direct, personal interactions occurred within Level Three, which involved the local, circumscribed, Palermo environment. More specifically, clerks of the Public Works Office were constantly in contact with Bovio Marconi because they designed and directed all construction activities. The prefect, a local state authority in charge of public order in Palermo, was also responsible for overseeing all building operations, because shelters were related to the safety of citizens.

Sometimes the Four-level system and its interactions were limited due to *force majeure*, which did not depend on the social units involved. Although Bovio Marconi was very concerned about all potential archaeological finds, she could not oversee all sites: her personnel and resources were extremely limited due to the wartime economy. Of course, this also meant that fewer officials and workers could operate in this context. The social units were therefore restricted due to extenuating circumstances. Sometimes Bovio Marconi visited the sites personally or asked reliable and skillful custodians of the Palermo Museum to inspect them in her place. Bovio Marconi's efforts to monitor archaeological discoveries were exceptional when we consider the complicated wartime context in which defensive urban construction took place. In fact, the public works office and local building companies were under pressure and considered archaeological discoveries inconvenient accidents to be resolved quickly or even ignored.

12.5 Pietro Griffo Against Military Power in Agrigento

Pietro Griffo was born in Palermo in 1911. He obtained a degree in Classical Studies at the University of Palermo in 1935 and also studied at the Scuola Archeologica Italiana di Atene (Italian Archaeological School) in Athens. His first role as a state safeguarding officer was in Syracuse (1937–1941). The young Griffo subsequently became superintendent of Sicilian Antiquities for Agrigento and Caltanissetta in 1941, succeeding archaeologist Goffredo Ricci (1902–1984). At that point, he moved to Agrigento, where he founded a new office and library and re-arranged the collections, establishing the Museo Archeologico (Archaeological Museum) in 1967. During the war, Griffo carried out a systematic plan to protect the site of Agrigento and move archaeological collections to a safe shelter in Bivona. He was also a skillful field archaeologist, excavating at Agrigento, Eraclea Minoa and Gela. After the war, Griffo continued his job at Agrigento until 1968, when he moved to Rome to work as a superintendent of antiquities, and he died there in 2007 (Griffo 1946; Pappalardo 2012: 391–92; Gullì 2017: 133–45).[4]

As a safeguarding officer, Griffo's role often clashed with the wider and more urgent military need to defend Sicily against enemy attacks. His career represents an important case study that sheds new light on social networks in a wartime context in connection with air raid protection at Agrigento. The city and archaeological site were located in a strategic area of southern Sicily, overlooking the North African front. In case of an enemy invasion, Agrigento would have been at serious risk of bombing or being attacked by seaborne military forces coming from the southern Mediterranean. The risk was certainly tangible. For this reason, Italy installed its command for the defense of Sicily in Agrigento. The 6th Army established a military district close to the archaeological site; the 207th Coastal Division, created on November 15, 1941, and other minor corps were transferred there. Some troops even camped within the site. The command arranged an essential defensive system in the site, fortifying buildings and installing new anti-aircraft, military, and gunner positions. Military interference became more frequent and worrisome between 1942 and 1943, however, when there was the constant threat of an enemy landing.[5]

Griffo, the main archaeological safeguarding officer at Agrigento, was constantly under pressure, interacting with a variety of civilian and military authorities. Monitoring the Italian army's interference at the site was extremely problematic and difficult, especially during the last years and months of the war. Ultimately, the military authorities prohibited civilians from entering the area, and even Griffo was forced to ask for special permission to gain access. In addition to the army's general obstructionism on the site, Griffo dealt with several tangible and detrimental actions by soldiers who camped there. After January 1943, when military

[4]Griffo's major works on Sicilian archaeology include *La difesa del patrimonio archeologico agrigentino contro i pericoli della recente guerra* (1946), *Problemi di architettura greca arcaica* (1947), *Gela preistorica ed ellenica* (1949), *Breve guida del Museo Civico di Agrigento* (1954), and *Il Museo Nazionale Archeologico di Agrigento* (1967).

[5]For a preliminary outline of all events on the site of Agrigento described in this section, see Griffo 1946.

Fig. 12.2 Postcard showing the Temple of Demeter at S. Biagio in Agrigento in 1928. (Image from Antonino Crisà's private collection)

authorities aimed to strengthen the defensive system at Agrigento, interference was a major concern; soldiers built many bunkers, small anti-aircraft, and machinegun placements.

One of the most unpleasant events occurred on April 15, 1943, when Griffo contacted the Command of the 207th Coastal Division, warning the military authorities that they were to stop building on the site. Griffo reported that soldiers had blasted the rock in the northern area adjacent to the ancient sanctuary of the goddess Demeter at S. Biagio in Agrigento (Fig. 12.2) in order to install an armored shelter, without first asking him for permission. The shelter was very close to the temple and several small ancient altars. He also complained that soldiers used dynamite elsewhere on the site to quarry stone blocks as building material. More importantly, even if it was partially occupied by the army, the archaeological area was protected by state laws and any excavation had to be authorized by the Superintendency.

General Brigade Commander Ottorino Schreiber (1890–1978),[6] personally wrote to Griffo, reprimanding him vehemently. The General demanded more respect and did not tolerate the idea that a public employee was entitled to give orders to a military authority. Griffo forwarded Schreiber's strong reply to the Ministry of Public Education, which contacted the Ministry of War to resolve the matter with some urgency on April 16. A few days later, Mario Roatta (1887–1968)[7] (Fig. 12.3), the General Commander for the Defense of Sicily, wrote a harsh letter to Griffo on

Fig. 12.3 General Mario Giuseppe Leon Roatta (1887–1968). (Photo: Wikipedia; copyright free, https://en.wikipedia.org/wiki/Mario_Roatta)

[6] Born in Parma to a noble German family, Ottorino Schreiber joined the Esercito Italiano (Royal Italian Army) in 1909 and trained at the Accademia Militare (Military Academy) in Modena. He was stationed in Eritrea between the 1920s and 1930s. In 1941 he was transferred to Sicily and then became commander of the 52nd Infantry Division fighting on the Russian frontline. He returned to Sicily and defended the island during Operation Husky in July 1943 (Zingali 1962: 175–76, 318–23).

[7] Mario Giuseppe Leon Roatta was born in Modena and started his military career in 1906. He fought during the First World War, obtaining three silver medals. After having performed various military tasks in Germany, Denmark, Finland, and Lithuania, Roatta was appointed Sub-head of Command of the Italian Army in 1940 and then Head of Command between 1941 and 1942. On February 5, 1943, he was sent to Sicily as commander of the 6th Army and (again) Head of Command. He also played an essential role in the Badoglio I Cabinet, although he was tried for serious negligence in defending Rome against the German occupation in 1943 (Focardi 2016).

April 22, saying he respected not only ancient art and history, but also was responsible for protecting Italy against its enemies. The general ordered Griffo to stop issuing warnings to his subordinates; if he wanted to complain, he had to contact the Allied Air Forces directly (Griffo 1946: 23–28). The incident was unfortunate and ended in May 1943. In fact, few additional dispatches were sent by Roatta and Griffo, but similar activities and disturbances continued to threaten the Agrigento site until July 1943, when the Allies carried out Operation Husky.

Archival material demonstrates that Griffo, as the main regional authority in charge of protecting antiquities (Level Two), was fully involved in the complex reality of social interactions at Agrigento. He mostly interacted with state authorities (Level One) who were overall more powerful than he was within the wartime context. The Italian army had commandeered a vast area of the archaeological site—one of the most important Classical centers in the Mediterranean basin—usurping Griffo's tasks and position of power as a decision-maker in the field. Griffo also acted as an intermediary to connect and stimulate different units inside Level One. For example, he contacted the Ministry of National Education, which then alerted the Ministry of War. The biting letters written by the commanding military authorities in Agrigento encouraged further links between the Ministry of National Education, previously contacted by Griffo, and the Ministry of War, clearly favoring Schreiber and Roatta. They both acted to defend the Italian homeland. It seems that no direct interactions are attested between Level One, Griffo, and Level Three in this episode.

Overall, Griffo's interaction with regional authorities was strictly limited to his jurisdiction and area of competence, which only officially included the province of Agrigento. There was a regional authority managing the entirety of the archaeological heritage in Sicily in the early 1940s—namely the Royal Superintendency of Sicily—but it only coordinated the actions of the provincial bodies. Thus, Griffo had influence at Agrigento but the interference of powerful military authorities, who were also supported by the high-level Ministry of War, certainly overshadowed his ability to enforce his will.

12.6 Conclusions

This contribution has taken a theoretical approach to understanding the social dynamics of civilian and military authorities, interacting in a limited (but certainly revealing and stimulating) social context, which is not, however, typical. In fact, the state of emergency caused by the Second World War and the tangible, compelling threat of enemy attacks and military invasions, led to an effective degree of exceptional interactions which would have been unthinkable in times of peace. This section offers some final remarks and concluding thoughts on these aspects, identifying key characteristics and potential developments of the theoretical model influencing the historical analysis in this chapter.

The Four-level system (Fig. 12.4) applied in this chapter in the analysis of networks has shown that sometimes the tasks and targets of authorities that are active in separate network levels can overlap. This is certainly evident when civilian and military authorities both aim to protect antiquities and defend the Italian homeland but have different agendas and modes of operation. Both are certainly noble objectives, but the divergent opinions about the chain of command generated strong frictions, especially when military activities undermined the preservation of archaeological monuments. Agrigento is an extreme example and provides a way to understand how commanders, lieutenants, and soldiers, while posted at the archaeological site, presumed to ignore existing regulations protecting antiquities and superseded the role of civilian safeguarding authorities. The social units of Level One thus interfered with the actions of regional and

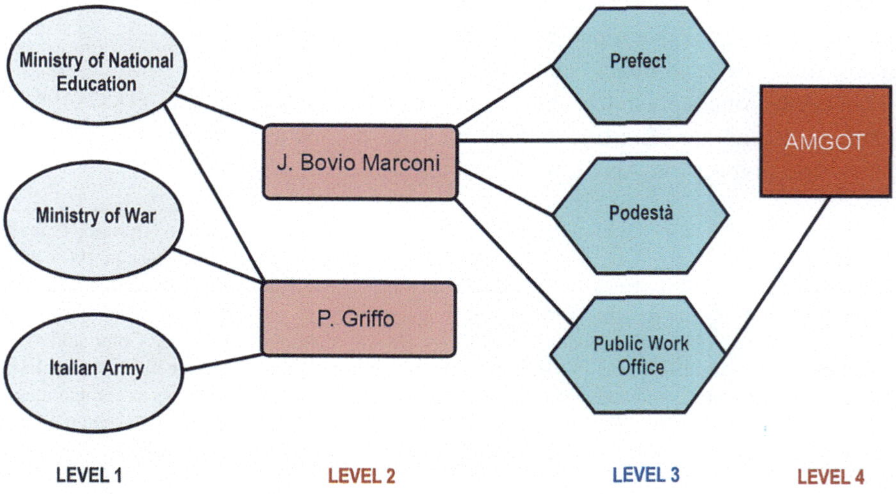

Fig. 12.4 Scheme outlining the main social interactions within the Four-level system in Sicily during the Second World War. (Diagram: Antonino Crisà)

even local authorities in Levels Two and Three, with their different dynamics.

Nevertheless, Bovio Marconi and Griffo, the two main regional authorities working in Sicily to protect archaeological heritage during the war and the central poles of the Four-level system, acted for the most part independently. Both superintendents were strictly confined to their areas of competence, the provinces of Palermo and Agrigento respectively. Opportunities to interact with each other were certainly rare. This means that horizontal interactions between Bovio Marconi and Griffo were not common. It can be argued that the limited horizontal interactions within the same level were the result of the organization of antiquities preservation in Sicily. The island had major superintendencies which were responsible for particular provincial areas of the island; they were only answerable to the Superintendency of Antiquities based in Syracuse. Therefore, both archaeologists were able to act autonomously. Furthermore, the vertical interactions linking the Ministry of National Education, the premiere national safeguarding body, with the regional superintendencies were well-established and effective.

When military units acted to protect Italy against the enemy, they not only had to deal with the canonical national authority in the form of the Ministry of National Education, which imposed strict rules to safeguard all Italian antiquities, but they had to reach direct agreements with other favorable entities as well. For instance, the Ministry of War consistently backed up the military command in Agrigento. In addition, units of Level One often clashed with Level Two at the site. Relationships between these units (military and regional authorities) were more problematic because Level One as a mediating influence was frequently limited or absent. The reasons for this can be traced to frictions within Level One itself. Ultimately, the Ministry of National Education was not able to override the conduct and demands of the Ministry of War, which played an essential role within the critical wartime context and benefitted from the strong political support of Mussolini.

The case studies discussed in this chapter exhibit limited connections with Level Four. In fact, the AMGOT was not operational in Sicily when Bovio Marconi supervised shelter construction sites in Palermo, and Griffo argued with Italian commanders at Agrigento. Nevertheless, interactions between Levels Two and Four are well documented, especially from mid-1943 on and into 1944. For instance, Bovio Marconi interacted with Captain Hammond, United States Monuments Officer, to arrange expedited renovations at the Palermo Museum, which had been partially destroyed by Allied bombs a few months before. The AMGOT was actively involved in liaising with local (Level One) and regional (Level Two) authorities to restore a democratic government after the fall of fascism.[8]

Finally, the Four-level system developed for this study could be successfully adapted and applied to other case studies, for example in Pompeii and Naples,[9] in order to understand the full spectrum of social interactions between state, regional, local and supranational authorities (both civilian and military). In those contexts, supranational authorities (Level Four), represented by the United States and British Monuments Officers acting in Naples, constantly interacted with units in Level Two (e.g., the Superintendent Amedeo Maiuri, 1886–1963), even if they opted to occupy suitable buildings in the area (e.g., the museum and the Royal Palace of Caserta) that were not as vulnerable as the site of Agrigento, for example. The author's research and archival investigations are ongoing and will certainly provide additional important evidence and documentary data to shed new light on social networks within the peculiar historical context of Sicilian archaeology—previously neglected by scholars—and of Italian archaeology in general.

References

Albergoni, A. (2007). *La Prefettura di Palermo tra potere centrale e realtà locale, 1930–1944*. Palermo: Ed. A. Albergoni.

Albergoni, A. (2018). *La guerra dell'arte. Con la cronologia dei bombardamenti e dei danni di guerra alla città di Palermo dal giugno 1940 all'agosto 1943*. Palermo: Navarra Editore.

Almagro, M. (1952). La reconstrucciòn y reinstalaciòn del Museo Nacional de Palermo. *Ampurias, 14*, 243–246.

Baldoli, C. (2010). I bombardamenti sull'Italia nella Seconda Guerra Mondiale. Strategia anglo-americana e propaganda rivolta alla popolazione civile. *DEP. Deportate, esuli, profughe, 13–14*, 34–49.

Baldriga, I. (1999). Contributo alla storia dei danni di guerra: l'opera della Allied Commission for Monuments Fine Arts and Archives (MFAA). *Ricerche di storia dell'arte, 68–69*, 87–93.

Barnes, J. A. (1972). *Social Networks*. Reading Mass: Addison Wesley.

Barone, Z. (2011). Distruzione e ricostruzione in Sicilia tra il 1943 e il 1945. *Lexicon. Storie e architettura in Sicilia e nel Mediterraneo, 12*, 21–28.

Battaglia, G. (2017). L'archeologia al femminile: la figura e l'opera di Jole Bovio Marconi. In R. Panvini, & A. Sammito (Eds.), *Archeologia in Sicilia tra le due guerre. Atti del Convegno di Studi, Modica, 5–6–7 Giugno 2014* (pp. 15–24). Modica: Ente Autonomo "Liceo Convitto".

Bertorotta, S. (2008). *Bombardate Palermo!* Palermo: Edizioni Fotograf.

[8]As with the other important case studies, the relationship between Bovio Marconi and Hammond will be fully assessed in a forthcoming monograph by the author, reporting all relevant documentation (Crisà 2025).

[9]The bombing of Pompeii and the military occupation of the National Museum of Naples are discussed in a recent essay (Pollard 2020).

Boi, M. M. (1986). *Guerra e beni culturali (1940–1945)*. Pisa: Giardini Editori e Stampatori Riuniti.

Borghi, M. (2018). L'amministrazione centrale dello Stato durante la Repubblica sociale italiana. *E-Review Dossier, 6*, 1–13.

Bosman, S. (2014). Un museo sotto una montagna: la dimora della National Gallery in tempo di guerra. In T. Calvano, & M. Forti (Eds.), *Musei e monumenti in guerra 1939–45. Londra – Parigi – Roma – Berlino. Atti del Convegno, Città del Vaticano – Musei Vaticani, Roma – Galleria Nazionale d'Arte Moderna, 15–16 novembre 2012* (pp. 79–100). Vatican City: Edizioni Musei Vaticani.

Bottai, G. (1938–1939). Direttive per la tutela dell'arte antica e moderna. *Le Arti, 1*, 42–52.

Bottai, G. (1992). *La politica delle arti: scritti 1918–1943*. Rome: Editalia.

Bovio Marconi, J. (1941–1942). Un rudere delle più antiche mura di Palermo. *Atti della Reale Accademia di Scienze, Lettere e Arti di Palermo, 3*(3), 501–512.

Bovio Marconi, J. (1952). *Il riordinamento del Museo Nazionale di Palermo dopo le distruzioni del 1940–44*. Palermo: Tip. F.lli De Magistris e C. Succ. V. Bellotti e F.

Bovio Marconi, J. (1955). Il Museo Archeologico di Palermo. *Bollettino d'Arte, 40*(2), 175–181.

Chirone, L., & Fasolini, S. (2011). La protezione antiaerea nella preparazione alla guerra. In L. De Stefani (Ed.), *Guerra monumenti ricostruzione. Architetture e centri storici italiani nel secondo conflitto mondiale* (pp. 587–589). Venice: Marsilio Editore.

Clyde Mitchel, J. (1974). Social networks. *Annual Review of Anthropology, 3*, 279–299.

Coccoli, C. (2008). Repertorio dei fondi dell'archivio centrale dello Stato relativi alla tutela dei monumenti italiani dalle offese belliche nella seconda guerra mondiale. In C. P. Treccani (Ed.), *Monumenti alla guerra. Città, danni bellici e ricostruzione nel secondo dopoguerra* (pp. 303–330). Milan: Franco Angeli.

Coccoli, C. (2010). I «fortilizi inespugnabili della civiltà italiana»: la protezione antiaerea del patrimonio monumentale italiano durante la Seconda Guerra Mondiale. In G. Biscontin, & G. Driussi (Eds.), *Pensare la prevenzione. Manufatti, usi, ambienti. Atti del XXVI Convegno di Studi "Scienza e Beni Culturali", Bressanone 13–16 luglio 2010* (pp. 409–418). Venezia: Arcadia Ricerche.

Coccoli, C. (2011). *Danni bellici e monumenti italiani durante il secondo conflitto mondiale: le fonti dell'esercito alleato*. Venice: Marsilio.

Coccoli, C. (2017). *Monumenti violati. Danni bellici e riparazioni in Italia nel 1943–1945: il ruolo degli Alleati*. Florence: Nandini Editore.

Collins dictionary. (n.d.) Network. In *Collins English Dictionary*. Retrieved January 09, 2025, from https://www.collinsdictionary.com/dictionary/english/network

Crisà, A. (2018). *When Archaeology Meets Communities: Impacting Interactions in Sicily Over Two Eras (Messina, 1861–1918)*. Oxford: Archaeopress.

Crisà, A. (2025). *Codename SICILYWAR. Archaeology, museums and social networks under threat in Sicily during World War 2*. Oxford: Archaeopress.

Dagnini Brey, I. (2015). La Roberts Commission e la formazione dei *Monuments Officers* negli Stati Uniti. In P. Dragoni, & C. Paparello (Eds.), *In difesa dell'arte. La protezione del patrimonio artistico delle Marche e dell'Umbria durante la seconda guerra mondiale* (pp. 311–324). Florence: Edifir.

De Caro, S. (2017). Le Soprintendenze alle Antichità in Italia nella prima metà del XX secolo. In C. Capaldi, O. Dally, & C. Gasparri (Eds.), *Archeologia e politica nella prima metà del XX secolo. Incontri, protagonisti e percorsi dell'archeologia italiana e tedesca nel Mediterraneo. Atti delle giornate internazionali di studio, Napoli, 24–26 Febbraio 2016* (pp. 233–324). Pozzuoli: Naus Editoria.

Dell'Era, T. (2003). Tra educazione nazionale e pubblica istruzione: le politiche ministeriali dell'istruzione pubblica dal 1943 al 1948. *Ventunesimo Secolo, 2*(4), 145–178.

Di Stefano, C. A. (1996). Jole Bovio Marconi, in memoriam. *Quaderni del Museo archeologico regionale Antonino Salinas, 2*, 11–18.

Fitzgerald-Black, A. (2018). *Eagles Over Husky: The Allied Forces in the Sicilian Campaign, 14 May to 17 August 1943*. Solihull: Helion & Company Limited.

Focardi, F. (2016). Roatta, Mario. In *Dizionario biografico degli Italiani on-line*. Retrieved January 09, 2025 from https://www.treccani.it/enciclopedia/mario-roatta_(Dizionario-Biografico)

Geyer, M., & Tooze, A. (Eds.). (2015). *The Cambridge History of the Second World War*. Cambridge: Cambridge University Press.

Gilbert, M. (2010). *Second World War*. London: Phoenix Press.

Griffo, P. (1946). *La difesa del patrimonio archeologico agrigentino contro i pericoli della recente guerra*. Agrigento: Soprintendenza alle Antichità.

Gullì, D. (2017). L'istituzione della soprintendenza di Agrigento. Pietro Griffo e le sue guerre. In R. Panvini, & A. Sammito (Eds.), *Archeologia in Sicilia tra le due guerre. Atti del Convegno di Studi, Modica, 5–6–7 Giugno 2014* (pp. 133–145). Modica: Ente Autonomo "Liceo Convitto".

Knappett, C. (2011). *An Archaeology of Interaction: Network Perspectives on Material Culture and Society*. Oxford: Oxford University Press.

Knappett, C. (2013). *Network Analysis in Archaeology: New Approaches to Regional Interaction*. Oxford: Oxford University Press.

Larson, F., Petch, A., & Zeitlyn, D. (2017). Social networks and the creation of the Pitt Rivers Museum. *Journal of Material Culture, 12*(3): 211–239.

Latour, B. (1987). *Science in Action: How to Follow Scientists and Engineers Through Society*. Cambridge: Harvard University Press.

Latour, B. (2005). *Reassembling the Social. An Introduction to Actor-Network-Theory*. Oxford: Oxford University Press.

Matarazzo, L. (1938–1939). Il nuovo ordinamento delle Soprintendenze alle opere di antichità e d'arte. *Le Arti, 1*(4): 415–418.

Miller McPherson, J., Popielarz, P. A., & Dobrnic, S. (1992). Social networks and organizational dynamics. *American Sociological Review, 57*(2), 153–170.

Oteri, A. M. (2005–2006). Riordinare e riparare. L'attività del governo alleato per la salvaguardia dei monumenti di Sicilia nell'immediato dopoguerra (1943–46). *Quaderni del Dipartimento Patrimonio architettonico e urbanistico, 29–32*, 293–302.

Pack, S. W. C. (1977). *Operation Husky: The Allied Invasion of Sicily*. Newton Abbot: David & Charles.

Pappalardo, U. (2012). Pietro Griffo. In M. Costantini (Ed.), *Dizionario biografico dei soprintendenti archeologi (1909–1974)* (pp. 391–393). Bologna: Bononia University Press.

Pelagatti, P. (2001). Dalla Commissione Antichità e Belle Arti di Sicilia (CABAS) alla Amministrazione delle Belle Arti nella Sicilia post-unitaria. Rottura e continuità amministrativa. *Mélanges de l'École Française de Rome. Italie et Méditerranée, 113*(2), 599–621.

Pollard, N. (2019). Refuges for movable cultural property in wartime: Lessons for contemporary practice from Second World War Italy. *International Journal of Heritage Studies, 2019*, 1–17.

Pollard, N. (2020). *Bombing Pompeii. World Heritage and Military Necessity*. Ann Arbor, MI: University of Michigan Press.

Romeo, S., & Rothier, S. (Eds.). (2017). *Bombardamenti su Palermo. Un racconto per immagini. Edizione illustrata*. Palermo: Istituto Poligrafico Europeo.

Snead, J. E. (2020). Circular 316: Archaeology, networks, and the Smithsonian Institution, 1876–79. In J. Roberts, K. Sheppard, U. R. Hansson, & J. R. Trigg (Eds.), *Communities and Knowledge Production in Archaeology* (pp. 34–46). Manchester: Manchester University Press.

Snyder, L. L. (1982). *Louis L. Snyder's Historical Guide to World War II*. West Port, CT: Greenwood Press.

Tusa, V. (1996). Jole Bovio Marconi Soprintendente alle Antichità della Sicilia Occidentale. *Quaderni del Museo archeologico regionale Antonino Salinas, 2*, 19–24.

Vassallo, S., & Villa, A. (2017). Ricerche archeologiche della Soprintendenza nel territorio di Palermo tra le due guerre. In R. Panvini, & A. Sammito (Eds.), *Archeologia in Sicilia tra le due guerre. Atti del Convegno di Studi, Modica, 5–6–7 Giugno 2014* (pp. 381–392). Modica: Ente Autonomo "Liceo Convitto".

Vitale, M. R. (2011). «Il tempo della solitudine». Istituzioni per la tutela dei monumenti in Sicilia prima e dopo la guerra. *Lexicon. Storie e architettura in Sicilia e nel Mediterraneo, 12*, 11–20.

VV.AA. (1939). *La protection des monuments et œuvres d'art en temps de guerre*. Paris: Institut international de coopération intellectuelle.

VV.AA. (1942). *La protezione del patrimonio artistico nazionale dalle offese della guerra aerea*. Florence: Le Monnier.

Zagarrio, V. (1976). Bottai: un fascista critico? *Studi storici, 17*(4), 267–271.

Zaloga, S. (2013). *Sicily 1943: The Debut of Allied Joint Operations*. Oxford: Osprey Publishing.

Zingali, G. (1962). *L'invasione della Sicilia (1943). Avvenimenti militari e responsabilità politiche*. Catania: G. Crisafulli.

13 Archaeological Learned Societies in the Kingdom of Yugoslavia (1918–1941): Changing States and Roles

Črtomir Lorber

Abstract

This chapter presents the work of various archaeological learned societies during the time of the Kingdom of Yugoslavia. The importance of learned societies as a networking tool utilized by Yugoslav archaeologists in the development of the discipline in the former Yugoslavia is examined, followed by an overview of the development of learned societies before 1918. Finally, the chapter assesses the evolution of archaeology in Yugoslavia and its archaeological learned societies in the period between 1918 and 1941. The few unsuccessful attempts to merge divergent archaeological traditions into a coherent whole are also discussed. These attempts were mainly motivated by the desire to create a pan-Yugoslav archaeological society and scholarly network that could include all of the existing scholarly traditions.

Keywords

Learned societies · Kingdom of Yugoslavia · National archaeologies · Chronology · Scholarly networks

13.1 Introduction

When we discuss the development of archaeology, we seldom focus on the work of learned and scholarly societies, focusing instead on the development of archaeological doctrines (e.g., Daniel 1975; Sklenář 1983; Trigger 1989), ideologies (e.g., Slapšak and Novaković 1996; Jones 2007; Slapšak 2011), major supranational and national institutional networks (e.g., Novaković 2014, 2021), and the work of individual scholars. With a few exceptions, learned societies are often overlooked and their presentation is reduced to a supplementary role in the development of archaeology. In my work (Lorber 2019, 2020, 2021, 2022) I attempt to fill in this gap in research, focusing on the importance of learned societies in the development of archaeology in southeastern Europe. This chapter deals with the countries of former Yugoslavia from the beginning of the twentieth century up until 1941, when the Axis powers invaded and carved up Yugoslavia. In wartime circumstances most, although not all, of the learned societies had their work suspended or diminished due to the occupation and many were never reformed after the war. The Second World War changed both the political and societal landscape in wider southeastern Europe as well as Yugoslavia. The post-war years were a time of change for archaeology and learned societies in Yugoslavia,[1] as a new, communist regime, took a different, more active approach in its relationship with archaeology and different institutions as well as learned societies. As these developments were accompanied by profound and complex changes, the discussion in this chapter only deals with the period until the beginning of the Second World War.

Learned societies are an aspect of wider developmental trends in archaeology. The evolution of archaeology in former Yugoslavia is complex and will not be discussed here at length, as there are other works that deal specifically with this topic (e.g., Rendić-Miočević 2009; Bandović 2014; Janković 2016; Novaković 2014). Suffice to say that we cannot offer a single development model of archaeology in the region, which is one of the main difficulties when studying the history of archaeology in former Yugoslavia. However, in

The chapter was written during the author's doctoral research at the Department of Archaeology, Faculty of Arts, University of Ljubljana, which ended in 2022 (see Lorber 2022).

Č. Lorber (✉)
Department of Archaeology, University of Ljubljana, Ljubljana, Slovenia
e-mail: crtomir@t-2.net

[1] The history of the Archaeological Society of Yugoslavia (1949–1991), which was formed after the Second World War, is the subject of the author's doctoral dissertation (Lorber 2022).

L. Coltofean et al. (eds.), *Connecting People and Ideas*, Themes in Contemporary Archaeology,
https://doi.org/10.1007/978-3-031-81006-0_13

this chapter, the emergence of archaeology in the area will be discussed in four periods defined by four major geopolitical changes in European history: before 1800, 1800–1918, 1918–1945, 1945–onwards. Each of these periods manifests itself differently in various Yugoslav regions due to differences in their political, social, and scientific trajectories.

In 1918 Yugoslavia united three regions with different histories of archaeology.[2] The first of these regions was composed of the integral parts of the former Austro-Hungarian Empire (i.e., Slovenia, Croatia, Vojvodina, parts of coastal Montenegro). Here archaeology has its roots in the antiquarian traditions of the renaissance, followed by hybridization with the continental *Landeskunde* traditions.[3] National and regional museum networks were established relatively early in this region, as were archaeological university curricula and heritage protection services. The second region, Bosnia and Herzegovina, was defined by the fact that modern archaeology there is a direct consequence of the Austrian colonial interventions aimed at the modernization of the province after 1878. This can be seen in the direct transplantation of Austrian-style institutions and models. As the most notable example of this we can point to the Zemaljski Muzej Sarajevo (Provincial Museum in Sarajevo/Landesmuseum Sarajevo), which is still one of the most important archaeological institutions in Bosnia and Herzegovina today. The third region—Serbia, inland Montenegro, North Macedonia, and Kosovo—was mostly ruled by the Ottomans until their independence or annexation by other countries (1878–1912). Here archaeology was a part of wider projects of modernization and national resurgence wherein Western models, notably those developed in Austro-Hungary, were copied in archaeology. Serbia was the most successful among the nations of the third group, forming a true nation-state with its own archaeological institutions (i.e., national museum, national archaeological society, and university department of archaeology), while the other nations in this group only began to develop national archaeologies after the Second World War.

When the Austro-Hungarian Empire collapsed after the First World War, the Kingdom of Yugoslavia united these regions into a new state. This period (1918–1941) is interpreted as a period of stagnation in archaeology (Novaković 2014: 224). Archaeology continued to exist in the already-established paradigms although this developmental phase is the least known today, mostly due to the lack of historical sources and the fact that the Yugoslav Kingdom was short lived and unstable. The modern period in archaeology begins after the Second World War, when the new socialist regime supported archaeology far more than the previous royal regime ever had. The main goal of the regime was not the modernization of different national archaeologies but rather their integration into a new societal system. Modernization happened, however, as the state became an environment in which national archaeologies developed in correlation with one another, becoming the modern national archaeologies we know today.

In each of the phases of this process different learned societies played a major role, serving as forums for discussions, as support structures for various museums, as players in the development of national movements, and as collective professional forums on a national level. This chapter discusses the contribution of learned societies to the development of archaeology in the Kingdom of Yugoslavia (1918–1945). The goal is to present their work and place within a wider archaeological infrastructural network. In the first part we shall offer a short overview of the work of different learned societies in the study area before the establishment of the kingdom aimed at showing how the diversity seen in the developmental trends in archaeology is reflected in later developments. This is followed by a short introduction to the legal structure of Yugoslavia, particularly the aspects relevant for archaeology and learned societies. The general development of archaeology in Yugoslavia during the interwar period is presented in the chapter's third part, whilst the fourth deals with the development of different archaeological/antiquarian learned and scholarly societies during this time.

13.2 Learned Societies and the Development of Archaeology in Southeastern Europe Through 1918

Throughout their history, learned societies have appeared in many different forms ranging from Renaissance academies, enlightenment scholarly circles and salons to modern professional societies. Members of these organizations reflected their respective societies; thus, the membership structure of Renaissance academies was different from the membership structure of Enlightenment scholarly circles. These organiza-

[2] For a general overview of the development of archaeology in southeastern Europe, see Novaković 2014, 2021.

[3] The German term *Landeskunde* is difficult to translate and it does not have a direct translation in English, where it is usually translated as 'historical geography' or 'regional historical geography'. Similar descriptive translations can be found in Russian and French. However, the term is well understood in some languages such as Slovene, where it is translated as 'domoznanstvo'. It should be noted that while there are no direct translations of the term in Serbian or Croatian, traditions like *Landeskunde* were important for the development of archaeology in Croatia and Serbia. The term denotes an interdisciplinary and universal approach to the regional science of a certain area, including its natural heritage, physical geography, history, archaeology, and ethnography. Some elements of this approach can be found even today, mainly in regional geography. Fore more details about the role of *Landeskunde* in the development of archaeology, see Novaković (2003).

tions were integrated within the overall social structure and perceptions of scientific inquiry more generally, most notably the question of which parts of society should actually be engaging in science. It is difficult to discuss the development of archaeological learned societies in Yugoslavia specifically, as there is no uniform developmental model that applies to their formation. Therefore, this chapter will discuss academies, antiquarian societies, historical societies, museum societies, and archaeological societies through the prism of modern nation states—previously part of Yugoslavia—for easier understanding and reference (Lorber 2019: 912).

13.2.1 Archaeological Learned Societies in Modern-Day Slovenia

I have discussed the development of archaeological learned societies in modern-day Slovenia elsewhere (Lorber 2019, 2020).[4] The earliest development of such societies is connected to two urban intellectual centers—Ljubljana (Habsburg Empire) and Koper (Venetian republic)—where two different antiquarian traditions were nourished in different academies, continental *Landeskunde* and Renaissance antiquarianism, respectively. After the Napoleonic wars (1803–1815), the territories of the Venetian republic were annexed by Austria and this was the time when scholarly societies in the modern sense of the word began to develop, most notably in Ljubljana, as Koper was slowly losing importance. Modern-day Slovenia was considered a part of Austria proper (Inner Austria/Innerösterreich) at this time and the early development of archaeology occurred in tandem with similar developments in other German-speaking areas, so much so that the genesis model of German historical societies (Heimpel 1972) was used to map the development of different historical societies in modern-day Slovenia (Janša-Zorn 1991).

The formation of a network made up of different societies affiliated with museums was at the forefront of the emergence of archaeology in Slovenia. The formation of museum networks and affiliated societies is seen especially clearly in the provinces of Krain (Kranjska), Kärnten (Koroška), and Steiermark (Štajerska). The development of archaeological learned societies began on this provincial level and later expanded to the local environment. This process effectively began with the establishment of the Kranjski deželni muzej in Ljubljana (Krainisch Ständisches Museum/Estate Museum of Carniola, 1821), which was followed by the formation of the Muzejsko društvo za Kranjsko (Museum Society of Krain, 1839) and the Historično društvo za Kranjsko (Historischer Verein für Krain/ Historical Society of Krain, 1850).[5] These two major societies united in 1881 under the name Museum Society of Krain and worked closely with the provincial museum. Similarly, historical societies in the provinces of Steiermark (Historischen Vereines für Steiermark) and Kärnten (Geschichtsverein für Kärnten) were connected to their respective provincial museums in Graz (Gradec) and Klagenfurt (Celovec). In addition to these major organizations, smaller societies were formed in towns, most notably in Trieste (Trst) on the coast and in Celje (1882), Ptuj (1893), and Maribor (1902) in the Steiermark region. These local societies were organized in cooperation with local and city museums. The development of a national sentiment played an important part in these developments, especially in Steiermark, where Slovene and German national movements competed with each other. As such, the activities of the Museal-Verein in Cilli (Celje Museum Society) were linked to the aspirations of the German community in Celje (Fugger Germadnik 2009: 356), whilst in Maribor a predominantly Slovene society, the Zgodovinsko društvo v Mariboru (Maribor Historical Society), was formed in 1903 (Žnidarič 2003: 236) *vis à vis* a predominantly German Museumverein (Museum Society) established in 1902.

13.2.2 Archaeological Learned Societies in Modern-Day Croatia

Croatia experienced a slightly different evolution of archaeological learned societies. Here we see divergent trends in coastal Croatia (Istria and Dalmatia) and continental Croatia, divided by the antiquarianism/*Landeskunde* dichotomy. In coastal Croatia the Renaissance academies of the Venetian and Dubrovnik republics in Istria and Dalmatia were a major influence in coastal towns such as Zadar, Split, and Dubrovnik. After the Austrian annexation of the former territories of the Venetian Republic (1815), the development of learned societies followed the Austrian continental models. The first major society to be created was the Kninsko starinarsko društvo (Knin Antiquarian Society, 1886), led

[4] Most of these societies were interdisciplinary, not just archaeological. They were, in fact, historical, museum, and antiquarian societies which preceded modern archaeological societies; some archaeological societies even developed directly from them. For simplicity and clarity, however, these societies will be generally referred to as archaeological learned societies throughout the chapter. This allows focusing on the role that these societies had in the development of archaeology without unnecessary excursions into the minutiae of the differences between individual societies—the latter would require additional research on topics that are beyond the scope of the chapter.

[5] This society developed after the division of the Historische Verein für Steiermark, Kärnten und Krain (Historical Society of Inner Austria, 1843) into smaller provincial societies in 1850.

by Brother (Fra) Luj Marun (1857–1939) (Jelovina 1988). This society had a nationalist component exemplified in its stated goal of forming a Muzej starohrvatskih starina (Museum of Old-Croat Antiquities, 1893) and reflected in the renaming of the Knin Society as the Hrvatsko starinarsko društvo (Croat Antiquarian Society, 1893). The second major society to be formed was the Hrvatsko društvo za istraživanje domaće povijesti u Splitu-Bihać (Croat Society for Studying Local History in Split-Bihać, 1894), formed and led by Father (Don) Frane Bulić (1846–1934). In addition to these two major societies, there were several other groups, such as the Starinarsko društvo Epidaurum (Epidaurum Antiquarian Society), which operated from Dubrovnik (Oreb 2009: 53). In continental Croatia, the development of learned societies was different and early antiquarians followed the *Landeskunde* tradition. The first modern societies developed after 1848, especially after the formation of the National Museum in Zagreb (1848). The most important of these was the Društvo za jugoslavensku povjestnicu i starine (Society for Yugoslav History and Antiquities, 1850–1878). This institution gradually transformed into the Hrvatsko arheologičko društvo (Croat Archaeological Society, 1878), the first national archaeological society in the former Yugoslavia (Rapanić 1981: 19). Its most notable leader was Father (Don) Šime Ljubić (1822–1896), one of the key figures in the development of archaeology in Croatia.

13.2.3 Archaeological Learned Societies in Modern-Day Bosnia and Herzegovina

No archaeological tradition had developed before Austria-Hungary seized Bosnia and Herzegovina from the Ottomans in 1908, thus we cannot talk about learned societies here. One notable exception was the work of Franciscan monks who curated small archaeological and antiquarian collections in their monasteries (Kaljanac and Križanović 2012); however, these were poor foundations for the development of archaeology. Modern archaeological concepts appeared only after the Austrian seizure of the province, followed by a major modernization program. The most important aspect of this program was the formation of the Zemaljski muzej u Sarajevu (Landesmuseum/Provincial Museum in Sarajevo, 1888) and the affiliated museum society (Muzejsko društvo Zemaljskog muzeja u Sarajevu, 1886) (Rizvanbegović 1976: 109). This organization also published the museum's scientific serial *Glasnik Zemaljskog muzeja*, which is also the oldest scientific serial in Bosnia and Herzegovina (established in 1889). The provincial museum and the affiliated society represented the only modern archaeological institutions in Bosnia until 1929, when a museum was formed in Banja Luka.

13.2.4 Archaeological Learned Societies in Modern-Day Serbia

In Serbia the initial emergence of archaeological and affiliated learned societies can be traced back to the work of the Orthodox monks in the monasteries of Fruška gora (Novaković 2014: 95), which played a similar role to the Franciscan monasteries in Bosnia and Herzegovina. However, the evolution of modern archaeology in Serbia and the development of modern learned societies really begins with the rise of the nationalist movement (Tasić 1983: 1). Two important regions are critical to understanding this phenomenon: Vojvodina and Northern Serbia (Belgrade). The former region was a part of the Austro-Hungarian Empire until its dissolution while the latter represented the territorial basis for the Kingdom of Serbia. In Vojvodina the early developments of archaeological and affiliated learned societies can be traced back to the cultural institution of Matica Srpska[6] (formed in Pest in 1826), which curated a collection of antiquities after moving to Novi Sad in 1864 (Tasić 1983: 2). This was later followed by the development of local learned societies in Vojvodina affiliated with smaller museums in towns such as Vršac and Zrenjanin.

In present-day northern Serbia the development of modern archaeology and affiliated organizations began after Serbia achieved its independence from the Ottoman Empire (autonomous principality in 1817; international recognition as an independent state in 1878). One of the earliest societies was the Društvo srpske slovesnosti (Society of Serbian Literature, 1841), which worked actively on one of the first national heritage protection laws. After the formation of the National Museum in Belgrade (1844) (Ham-Milovanović 2009: 90), the development trends picked up speed, culminating over time in the establishment of the Srpsko arheološko društvo (Archaeological Society of Serbia, 1883). The most notable member of this organization was the archaeologist and architect Mihajlo Valtrović (1839–1915), who also played a vital role in the formation of the society's magazine, *Starinar.* He was also a member of the Moscow Imperial Archaeological Society, the Croatian Archaeological Society, and the Imperial Archaeological Institute in Berlin (Novaković 2021: 157). In addition to the national society, several smaller regional societies developed during this time.

[6]Matica Srpska—as well as similar institutions in Slovenia (Slovenska Matica) and Croatia (Matica Hrvatska)—was one of the several cultural foundations typical for the nineteenth-century South Slavic and West Slavic peoples in Austria-Hungary. These institutions were closely connected with the nineteenth-century national movements and promoted national culture through books, events, and financing writers, among other actions.

13.2.5 Archaeological Learned Societies in Modern-Day Montenegro, North Macedonia, and Kosovo

In Montenegro, North Macedonia and Kosovo, archaeological institutions and learned societies formed much later. Except for Montenegro, during much of their early history these countries were considered provinces of other countries (Lorber 2019: 915) and archaeological activities were sporadic. The first archaeological institutions in North Macedonia were formed by the intervention of Serbian authorities. The partitioning of historical Macedonia between Serbia, Greece, and Bulgaria prevented the formation of learned societies for most of this early period, since all three states suppressed the Macedonian national movement. As such, most of the Macedonian societies that did exist were engaged in a wide range of cultural activities, mostly in the field of literature, aimed towards national resurgence. In Montenegro, most archaeological activities were carried out by Croat learned societies, especially those based in Split and Dubrovnik, with the cooperation of the Arheološki muzej Split (Split Archaeological Museum, 1820) and Croatian archaeologist and historian Frane Bulić. The first laws regarding the formation of the royal museum in Montenegro date to 1896 (Koprivica 2009: 795). The museum was formed in Cetinje, then the royal capital of Montenegro, although we have not been able to determine if an affiliated learned society existed as well. Kosovo was the least developed of these regions and no local archaeological institutions and learned societies formed there until after the Second World War.

13.2.6 Foreign Archaeological Learned Societies in Former Yugoslavia Until 1918

Before the First World War there were several foreign (e.g., Italian, Austrian, Hungarian) societies and institutions operating in the territories of former Yugoslavia. Scholars from Austrian and Hungarian institutions—such as the National History Museum in Vienna and the National Museum in Budapest, among others—were a common sight in southeastern Europe before the collapse of Austro-Hungarian Empire. These scholars represented their respective institutions and local scholarly networks (learned societies), such as the prominent Anthropologische Gesellschaft in Wien (Vienna Anthropological Society). They were members of local and regional learned societies as well, thus connecting the scholarly networks of southeastern Europe with the wider scholarly communities in the Empire. There were also several important Italian scholarly societies, such as the Istrian Society for Archaeology and National History (Società Istriana di Archeologia e Storia Patria, 1884), which published a prominent academic serial, *Memorie della Società Istriana di Archeologia e Storia Patria*. It must be noted that this society was quite nationalistic in its program. The second important society was the Società di Minerva, which focused its research on the region of Venezia Giulia, a part of which lay in modern-day Slovenia. This organization published the important academic serial *Archeografo triestino* (Pavan and Marini 2011), which was also relevant to archaeology. After the First World War, Austrian and Hungarian institutions lost much of their influence in the region, although Italian societies retained their influence as their regions of interest were absorbed into the Kingdom of Italy.

13.3 The Kingdom of Yugoslavia in Historical and Political Context

In southeastern Europe the First World War was a political cataclysm. The Austro-Hungarian Empire was dissolved, and its territories divided among different national and multinational states. One of them, the Kingdom of Serbs, Croats, and Slovenes, later renamed the Kingdom of Yugoslavia (1929), was formed in 1918. The new country was plagued by problems, as different regions of the new state were governed differently before the unification and there were also significant cultural and developmental differences among them. Yugoslavia faced the challenge of forming a uniform social, economic, and political system. The kingdom has been described as fighting with itself over the nature of the new state (Nećak and Repe 2008: 113), as there were two different political visions of its future. The first envisioned Yugoslavia as a federal state and was supported by a majority among the Slovene, Croat, Bosnian, and Montenegrin populations, whilst the other envisioned a centralist state and was championed by most Serbs and a part of the Montenegrin population. The centralist concept prevailed even though the state was organized on the basis of the concept of "compromise Unitarianism" (Nećak and Repe 2008: 124), which meant that the Serbs, Croats, and Slovenes were seen as different tribes of a single nation[7] rather than as a uniform national entity.

The tensions arising from trying to unite different traditions and views in Yugoslavia affected other social spheres as well, including economy, science, and foreign policy. The unification of these traditions proved an impossible task for the new state. Archaeology was no exception and serves as an illustration of the failures of the kingdom. In academic life Yugoslavia was attempting to form a coherent system by a top-down approach. This was done through legislative

[7]Bosnians, Macedonians, and Montenegrins were not recognized as individual entities in this view.

efforts in several fields. The state was able to integrate different civil service institutions into a uniform system, as well as to formulate a coherent financial policy to distribute funds to different state, provincial, and local institutions. However, Yugoslavia was never able to pass a comprehensive heritage protection law, and it remained one of the few countries in the study area during the interwar years (1918–1941) without national heritage protection legislation (Krstić 2006: 51). The country also chose a less than tactful approach when formulating laws regarding the functions of different non-governmental associations in Yugoslavia: it abolished all previous laws on the matter, initially instituting relatively authoritarian preliminary decrees (Zajc-Cizelj 2004: 163) that were only replaced by legislation in 1929, 11 years after the formation of the state.

This example makes clear why the Kingdom of Yugoslavia was a relatively unsuccessful country in many ways. On the one hand, it did manage to formulate policies on a state level, on the other, most of these policies were the result of half-hearted compromises that often did more harm than good. The frequent changes in the provincial structure of the country, which were most often aimed at suppressing different nationalist sentiments, backfired and had dire repercussions for related social and governmental systems such as museums. A relevant example is the division of Bosnia into several provinces, which seriously undercut the abilities of the Zemaljski muzej in Sarajevo to continue with high quality work (Novaković 2014: 143). This stems from the fact that the museum was originally meant as a provincial museum for the entire territory of Bosnia and it was financed as such. Following the division of Bosnia into provinces, the museum lost most of its territory of interest as well as a source of stable funding. In the end, the attempts to reorganize the administrative structure of Yugoslavia also failed to solve nationalist issues in Yugoslavia and the country fell rapidly when invaded by the Axis forces in 1941.

13.4 Archaeology in the Kingdom of Yugoslavia

Archaeology was one of the more vulnerable academic fields in the new state. This was the result of divergences in the development of national archaeologies. The development of archaeology in the Kingdom of Yugoslavia can be discussed in extremely limited terms. The formation of the new state was accompanied by gradual steps towards comprehensive heritage legislation, institutional organization, and the formation of a system of compatible archaeology studies at university centers (Novaković 2014: 224) but with little success. Presumably it was not easy for a new multi-national state to redirect the priorities of national archaeologies formulated in completely different political and administrative contexts. These differences led towards divergent development strategies among different national archaeologies in Yugoslavia. National organizational archaeologies that had formed in the more developed areas of the Austro-Hungarian Empire (i.e., modern-day Slovenia, modern-day Croatia, and modern-day Vojvodina) were different from those formed in Serbia, which was an actual nation state, and from approaches that had developed in Montenegro and North Macedonia. It would have been difficult even for a highly functional state to formulate a coherent whole from these practices and organizational networks, much less an unstable state such as Yugoslavia was at that time. This meant that there was no real push towards resolving the differences between competing national archaeologies, with the result that collaboration between these national systems was not very effective.

Thus, the interwar period can be interpreted as a period of institutional stagnation for archaeology in Yugoslavia (Novaković 2014: 224), even though there was institutional continuity between the pre-war period and the interwar era. Most museums, heritage protection services and learned societies continued or renewed their work after the First World War, but what is striking is the general lack of an overall national development strategy as formulated in government policies or cooperation between these institutions. This is also demonstrated by the fact that during the interwar period not a single Yugoslav specialized archaeological institution was created, even though the idea of an Archaeological Institute of Serbs, Croats, and Slovenes (Vasić 1927) had been suggested. Such ideas were projects promoted by individual scholars and were never implemented. As in the case of relevant legislation, there was no state-run heritage protection service, and no state-level research institute. A sole exception is the Muzej kneza Pavla (Museum of Prince Paul), which was formed in Belgrade by merging the Istorijsko-umetnički muzej (Art and History Museum in Belgrade), formerly the Narodni muzej (National Museum in Belgrade), and the Muzej savremene umetnosti (Museum of Contemporary Art) in 1935 (Ham-Milovanović 2009: 100). This institution was meant to serve as a central national museum for Yugoslavia, but it remained primarily a Serbian institution. The museum existed until 1944, when the new Yugoslav authorities changed its name and role. Exacerbating this lack of institutional development was the fact that no Yugoslav archaeological serial publication was established during the interwar period, a stark contrast with the publishing traditions of the more advanced national and provincial archaeologies in Serbia (e.g., *Starinar*), Croatia (e.g., *Vjesnik za arheologiju i historiju Dalmatinsku*, *Vjesnik Arheološkog muzeja u Zagrebu*), Bosnia (e.g., *Glasnik Zemaljskog muzeja u Sarajevu*) and Slovenia (e.g., *Glasnik Muzejskega društva za Slovenijo*).[8]

There were some attempts to formulate a coherent structure for practicing archaeology, mostly based upon older regional networks of scholarly cooperation. On a state level, for example, a policy of appointing university professors to posts in different areas of the country did exist. As such, archaeologists and historians Grga Novak (1888–1978) (Croatia), Ćiro Truhelka (1865–1942) (Bosnia and Herzegovina) and France Mesesnel (1894–1945) (Slovenia) were all employed at the Faculty of Arts in Skopje (North Macedonia) (Novaković 2014: 225). There were also some excavation projects, most notably in North Macedonia (Stobi) (Ninković 2009: 134–35; Novaković 2021: 392) that involved cooperation among archaeologists from different areas of the country. Even so, the level of interaction was low, as exemplified by the fact that there was almost no cooperation on projects among the three main university centers in Ljubljana, Zagreb, and Belgrade.

The lack of cooperation among archaeologists from different parts of Yugoslavia can be seen in other areas as well. Only a few major archaeological projects were initiated during the interwar years. One of them was the *Archaeological Map of Yugoslavia*, which involved the production of a cartographic representation of archaeological sites in different regions of Yugoslavia. This project was a part of a wider European initiative led by the French Union Académique Internationale. In Yugoslavia, the project was led by Slovene-born historian and archaeologist Balduin Saria (1893–1974) (Novaković 2014: 225). Two meetings were held connected to this enterprise. The first was organized in Ptuj (1923, Slovenia), where the attendees discussed joining the project (Žižek 1992: 149), and in Zagreb (1931, Croatia), where the technical side of the project was discussed (Saria 1932: 14). The second important project was the *Tabula Imperii Romani*, an international collaboration with the goal of producing the map of the Roman Empire in its entirety. This international initiative included archaeologists and historians from numerous European countries (e.g., United Kingdom, France, Spain, Germany, Italy, and Yugoslavia). The work in Yugoslavia began at a meeting in Ptuj (1937, in Slovenia), at which the participants agreed to the planned project structure and timeline. Due to the advent of the Second World War, the work on both projects was not carried out until after 1945 in Yugoslavia as well as in the other participating countries.

The international level of archaeological work in the interwar period was dominated by individual initiatives and the work of both foreign and local learned societies. There were several excursions to the Kingdom of Yugoslavia organized by foreign institutions. The most notable of these was an excursion organized by the Deutsche Gesellschaft für Vorgeschichte (German Society for Prehistory) in the Kingdom of Yugoslavia during the early 1930s (Lorber 2019: 919). However, due to the lack of reliable historical sources we are unable to say whether this was a one-time event or if there were other similar excursions. There were also several smaller one-time excursions, the first being the excursion of the London and Middlesex Archaeological Society (in Yugoslav newspapers this was referred to as the London Archaeological Society) in Dalmatia in 1929 (Anonymous 1929b). The attendees were received by the representatives of local authorities as well as by Frane Bulić (as a representative of the Bihać learned society) and Mihovil Abramić (1884–1962) (as a representative of the Archaeological Museum in Split). Alongside these there were several larger excursions organized by the Römisch-Germanische Kommission (Romano-Germanic Commission) in Frankfurt am Main as a part of the project *Studienfahrten deutscher und donauländischer Bodenforscher* (Study trips for scholars from Germany and the Danube countries). There were seven such excursions; three of them organized in Yugoslavia. The first excursion in Yugoslavia was organized in 1931 to Dalmatia and Bosnia, and Herzegovina (Anonymous 1931; Trebsche 2005: 181). The second excursion was organized in 1933 to Bosnia and Herzegovina (Trebsche 2005: 179) and the third in 1935 to Slovenia and Croatia (Saria 1935: 734–40; Trebsche 2005: 181). These visits were far more international than most as they were attended by archaeologists from Austria, France, Germany, and Hungary, among others. The final example is a French excursion to the city of Banja Luka in 1935 (Anonymous 1935), whose organizer could not be identified.

The connections of individual scholars outside the Kingdom of Yugoslavia were also important in bringing international attention to archaeological research in the country. One of the most important of these individuals was Frane Bulić, whose international work was extensive. He collaborated with the Danish architect and archaeologist Ejnar Dyggve (1887–1961) during Dyggve's excavations in Salona. He was an important corresponding member of the Deutsches Archäologisches Institut (German Archaeological Institute) and aided the institute in building a network of correspondents in Yugoslavia. The importance of this activity can also be seen in the fact that the institute invited several Yugoslav archaeologists to be present at the 100th year anniversary of the German Archaeological Institute; this visit was reported in Yugoslav newspapers as well (e.g., Anonymous 1929a). The Yugoslav delegation—represented

[8]It is important to note the lack of both a centralized Yugoslav archaeological publishing tradition and an attempt to form one in the Kingdom of Yugoslavia. Instead, there were various national publishing traditions with major differences in their international reach. Some Yugoslav archaeological serial publications had international audiences as well as international authors (or at least translations of their articles)—in addition to Yugoslav scholars—among the contributors.

by Mihovil Abramić, Viktor Hoffiler (1877–1954), Miloje Vasić (1869–1956), Nikola Vulić (1872–1945), and Vladimir Petković (1874–1956)—attended the celebratory congress and presented lectures on the development of archaeology in Dalmatia (Mihovil Abramić) and excavations in southern Serbia (modern-day North Macedonia) (Vladimir Petković).

There are other examples of such collaborations, as the joint project of Czechoslovak-born American archaeologist Vladimir J. Fewkes (1901–1941) ("American archaeologists," 1933) with Serbian archaeologists illustrates. His work with Miodrag Grbić is notable and is extensively discussed by Aleksandar D. Bandović (Chap. 14, this volume), who also notes the fact that Yugoslav archaeologists were mostly junior partners in such collaborations. Yugoslav archaeologists participated in various international meetings as well, exemplified by a special Yugoslav delegation that attended the XV. Congrès International d'Anthropologie et d'Archèologie Prèhistorique (15th International Congress of Anthropology and Prehistoric Archaeology) in Paris from September 20–27, 1931 during the *Exposition coloniale internationale* (International Colonial Exhibition). This congress opened with a speech delivered by the head of the Yugoslav delegation, the anthropologist, ethnologist and archaeologist Niko Županić (1876–1961) (Županić 1933: 303–4).

The archaeological trips, collaborations with foreign archaeologists, and participation in international scientific events create the illusion of dynamic developments in Yugoslav archaeology during this time. It is true that individual Yugoslav archaeologists were active domestically and abroad, but their activities demonstrate the importance of individual initiative rather than a systematic approach at the national level. They were outliers rather than creating a foundation for ongoing initiatives. Perhaps the most notable aspect of the above-mentioned projects is the fact that they were either planned or led from abroad. It must also be noted that Yugoslav archaeologists had no general state-wide archaeological society or a national, state-level institution that could lead such a coordinated effort (Novaković 2014: 227). Since there was no potential organizer, Yugoslav archaeologists could meet one another only during exceedingly rare national level meetings or more frequently at meetings organized by local and regional learned societies and academic institutions. This meant that these organizations were important in maintaining a certain level of academic standards and pre-existing scientific networks, but their local or regional focus limited the reach of their initiatives and collaborations during the interwar period of archaeology in Yugoslavia. Therefore, individual initiatives and achievements—most notably interactions of individual archaeologists with wider scholarly networks and international institutions—did not benefit the archaeological community at the national level.

13.5 Archaeological Learned Societies in the Kingdom of Yugoslavia

The limited institutional development of archaeology in the Kingdom of Yugoslavia in the interwar period impacted the work of learned societies as well. These organizations faced different fates during this time, sometimes at the whim of the government as some were banned as nationalist organizations (mainly in Croatia) (Rendić-Miočević 2009: 21), while others continued with their work undisturbed. In addition to museums and universities, learned societies were one of the few ways in which continuity in archaeological practice was maintained. After the First World War these societies continued to exist on a local and provincial level and ensured that pre-existing scholarly traditions and networks survived, albeit often in a diminished fashion. To better illustrate the scope of the issue, the following section offers a short overview of various learned societies in Yugoslavia following the same structure used in the overview of the work of the learned societies before 1918.

13.5.1 Archaeological Learned Societies in Modern-Day Slovenia During the Interwar Period

In modern-day Slovenia the main learned society, the Museum Society of Krain, continued to exist relatively unhindered after the First World War. In 1919 it was renamed Muzejsko društvo za Slovenijo (Museum Society of Slovenia) (Mal 1939: 22). It also formed a natural sciences section and in 1934 the society split into the Museum Society of Slovenia and the Prirodoslovno društvo za Slovenijo (Natural Sciences Society of Slovenia). The Museum Society continued to operate without legal restrictions until 1941, whilst during the Second World War it operated sporadically and under the watchful eye of the occupying Axis powers. As modern-day Slovenia had no national archaeological or historical society, the Museum Society of Slovenia grew into this role. This is demonstrated by the fact that both the modern Zveza zgodovinskih društev Slovenije (Historical Association of Slovenia, 1946) and the Slovensko muzejsko društvo (Slovene Museum Society) see themselves as direct descendants of the Museum Society of Slovenia. Alongside the central society, several other local societies either continued or renewed their work in the interwar period.

In Celje, for example, the predominantly German Museum Society was dissolved in 1922 and replaced by predominantly Slovene Muzejsko društvo v Celju (Museum Society in Celje) (Fugger Germadnik 2009: 57). The Muzejsko društvo na Ptuju (Museum Society in Ptuj) became one of the most successful local societies in modern-day Slovenia and one of the few societies of state-wide importance. Its significance can be seen in the fact that the meeting held to discuss the *Archaeological Map of Yugoslavia* project in 1923 was organized concurrent with the anniversary meeting of the Museum Society in Ptuj. Both societies in Maribor (i.e., museum society and historical society) renewed their activities after the Second World War as well. It seems that Slovene local groupings were relatively unaffected by the general instability of the country, even though their international influence was minimal. Several new local societies were formed during the later interwar period, most notably the Muzejsko društvo v Murski Soboti (Museum Society in Murska Sobota) in 1935, the Muzejsko društvo Škofja Loka (Museum Society in Škofja Loka) in 1937, the Muzejsko društvo v Krškem (Museum Society in Krško) in 1938, and the Muzejsko društvo Novo mesto (Novo mesto Museum Society) in 1941 (Mohorič Bonča 2005: 80–81). Official records show that several of these institutions—i.e., the Museum Society of Slovenia, the Museum Society in Maribor, the Museum Society in Ptuj, the Museum Society in Celje, the Museum Society in Murska Sobota, the Museum Society in Škofja Loka, and the Historical Society in Maribor—were receiving regular state (provincial) subsidies in 1940, while the Museum Society in Krško was not supported in this way. The Museum Society in Novo mesto was established just before the Second World War and therefore was unable to begin its activities. The work of most Slovene local museum societies was either suspended or severely limited during the Second World War by the occupying Axis governments.

13.5.2 Archaeological Learned Societies in Modern-Day Croatia During the Interwar Period

If modern-day Slovenia can be seen through the lens of continuity, the story of Croatia was different, as many prominent societies had their work at least temporarily suspended. This can be attributed to the prominent role that archaeological learned societies had in the Croat national movement, most notably the national Hrvatsko arheološko društvo (Croat Archaeological Society) and the Croat Antiquarian Society—the latter primarily focused on researching the early medieval heritage of Croatia. This was seen as a problem in the new kingdom, which aspired toward a single Yugoslav nationhood and saw the Croat national movement as a complicated political issue. Thus, the Croat Archaeological Society continued its activities until the beginning of the royal dictatorship (1929–1931),[9] when its work was suspended (Rendić-Miočević 2009: 21) due to its perceived connection with the Croat national movement. A similar fate befell the Croat Antiquarian Society, which suspended its activities after 1929 (Jelovina 1988: 250). The Bihać learned society was the only one of the three main Croat learned societies that was not suspended and remained active until the Italian occupation of Split in 1941.

Local societies in Dubrovnik and Kotor continued with their work and several smaller local societies were formed, such as NARONA, a society focused on studying local history in Vid (NARONA—društvo za istraživanje domaće povijesti u Vidu, 1930). In continental Croatia, the interwar years were a time of several museum initiatives, most notably in the towns of Sisak, Osijek, and Varaždin. The oldest of these was formed in Varaždin—i.e., the Muzealno društvo u Varaždinu (Museum Society in Varaždin) in 1925 (Baš 1935: 124)—while societies in Sisak and Osijek were formed later. In Osijek the Arheološki klub 'Mursa') was established in 1933 (Šimičić 2018: 267), whilst in Sisak a museum society titled Društvo sisačkih umjetnina prošlosti i starina (Society of Historical Art and Antiquities of Sisak) was organized in 1941 (Pirc 1996: 53). As the Croat territories were controlled by a Croat puppet government, the Second World War did not represent a sharp break in the activities of most continental Croat societies and the new fascist authorities allowed their work to continue undisturbed.

13.5.3 Archaeological Learned Societies in Modern-Day Serbia During the Interwar Period

Serbia was occupied by Austria-Hungary during the First World War, so the work of several societies was suspended during that time. These societies, most notably the Archaeological Society of Serbia, renewed their activities after the war ("Obnova rada," 1922: 157–60). During the

[9] The royal dictatorship or, as it is better known, the January 6 Dictatorship (1929–1931), was a royal dictatorship established by King Alexander I (r. 1921–1934). Its goal was to create a unitary Yugoslav ideology and a single Yugoslav nation. In many respects it was a response to the long-lasting political crisis and deep polarization of society in Yugoslavia (Nećak and Repe 2008: 138–39). Officially, the dictatorship continued until 1931, when a new constitution was adopted. However, the king retained extensive executive powers, thus acting more as an autocrat than a constitutional monarch. The new constitution remained in effect through the regency of Prince Paul (r. 1934–1941) until the invasion of Yugoslavia by the Axis powers in 1941.

interwar years, existing societies continued with their work, as did most national institutions—the National Museum in Belgrade and the University of Belgrade, among others.

In a wider context the period between 1918 and 1941 was a period of the integration of Vojvodina into Serbia proper and this process was also characterized by the formation of local museums in several towns such as Pančevo (1923), Novi Sad (1933), Niš (1932), and Šabac (1933) (Novaković 2021: 161). The formation of most of these institutions was due to the aspirations of local museum societies. For example, the Muzejsko društvo Niš (Museum Society in Niš) helped form and then cooperated with the local museum in Niš (Janković 2016: 856). Serbian learned societies went through the interwar years relatively undisturbed; their work even expanded into North Macedonia during that time. This development was stopped by the Second World War and the Axis occupation of Serbia.

13.5.4 Archaeological Learned Societies in Modern-Day Bosnia and Herzegovina During the Interwar Period

In Bosnia and Herzegovina, the interwar years were marked by stagnation in archaeology, which was evident in the work of the main archaeological institution in the country—i.e., the Zemaljski muzej Sarajevo (Provincial Museum in Sarajevo)—and in the activities of the Museum Society of the Provincial Museum in Sarajevo. This lull was the result of several factors. The museum and the affiliated society were *de facto* Austrian colonial institutions. A good portion of their staff—i.e., mostly highly qualified experts from more archaeologically developed parts of the empire—and the society's members came from abroad and, as the First World War ended, they returned home. After the war, the new state tried to maintain the pre-war standards of scholarly research, but the new personnel were not able to meet the high standards established before 1918 (Novaković 2014: 143) because Yugoslavia did not have enough trained experts to replace the lost personnel. This shortfall could not be resolved in a speedy manner due to other issues as well. Archaeology, as well as other academic disciplines, evolved faster in more advanced and urban regions due to the size of the local middle class that supplied most of the scholars in the nineteenth and twentieth centuries. As Bosnia was underdeveloped and not very urbanized—in comparison to Slovenia and Croatia—it did not have a local academic tradition that was strong enough to step in after the dissolution of the Austro-Hungarian Empire and the demise of its colonial archaeological tradition. This situation was further exacerbated by the new territorial division of Bosnia (Novaković 2021: 234). This further diminished the potential of Bosnian scholars to develop a national archaeological tradition in the aftermath of the First World War, as the provincial museum and society were meant to serve the territory in its entirety and not just a small part of it as in the Kingdom of Yugoslavia. This led to increasing stagnation in the evolution of archaeology in Bosnia and Herzegovina, seen in the fact that no new institutions or societies were established in the interwar years apart from the creation of a museum in Banja Luka in 1929 (Rizvanbegović 1976: 114).

13.5.5 Archaeological Learned Societies in Modern-Day Montenegro, North Macedonia, and Kosovo During the Interwar Period

In Montenegro, the development of archaeology and archaeological learned societies was limited in the interwar years. In the coastal regions several learned societies based in Croatia continued to operate, with the support of the Archaeological Museum in Split, focusing on the area of Boka Kotorska (Marin 1995: 68). The Narodni univerzitet Boke Kotorske (Public University of Boka Kotorska, 1933), whose main objective was the promotion of the cultural development of Montenegro, was also established during this period. The institution is relevant since it organized an exhibition about the cultural heritage of Boka Kotorska in Kotor in 1934 (Zloković 1980: 290). This exhibition was not just historical or archaeological; rather it represented topics from different disciplines, such as history, art history, ethnology, and archaeology. Even though the organization aimed to formulate a national strategy for the promotion of science (including archaeology), it was unsuccessful due to the lack of government support. The development of archaeology and the long road to a national archaeological society in Montenegro began only after the Second World War.

In North Macedonia, the interwar years interestingly represent a period of relative growth in archaeology. There were several major excavation projects, most notably the one at the site of the Hellenistic and Roman town of Stobi[10] in North Macedonia (1924–1936), where several notable Yugoslav archaeologists collaborated, such as Balduin Saria and Ćiro Truhelka (Novaković 2014: 170). During this period the expansion of Serbian influence in North Macedonia is notable and archaeology was, quite literally, a colonial

[10] Stobi was an ancient urban center in modern-day North Macedonia, northwest of Thessaloniki, from the third century BC until the sixth century AD. The town became especially important during the Roman Empire as a regional economic and administrative center. After the division of the Roman Empire by Emperor Theodosius I (r. 379–395), the town became the administrative center of the province of Macedonia Salutaris. The town did not survive the turmoil in the Eastern Roman Empire during the sixth century AD and was abandoned at the end of the century.

implant of the Kingdom of Serbia. This is reflected in the fact that most of the archaeological institutions and learned societies were branches of Serbian institutions (Novaković 2014: 172). In the interwar years the Istorijsko-arheološki muzej (Historical and Archaeological Museum), the Filozofski fakultet (Faculty of Arts), and the Skopsko naučno društvo (Skopje Learned Society) developed in North Macedonia; however, the members and staff employed at these institutions hailed mostly from Serbia and later from other parts of Yugoslavia. The independent development of archaeology in North Macedonia, later followed by the formation of a national archaeological society, was only possible after the Second World War, when North Macedonia was recognized as one of the constituent republics of the new federal state.

As Kosovo was not an independent administrative unit and Albanians were not considered a nation within the Yugoslav kingdom, we can state that the situation for the development of archaeology in Kosovo was worse than in other parts of the country. Kosovo had no cultural or political autonomy; there were no attempts to develop any kind of archaeological tradition or a learned society. The development of archaeology in Kosovo only began after the Second World War.

13.5.6 Non-local Learned Societies in Yugoslavia During the Interwar Years

During the nineteenth century, the territory of former Yugoslavia hosted several foreign learned societies as well. Existing Italian societies remained active in Istria and parts of Dalmatia that were annexed by the Kingdom of Italy after the First World War. Another important aspect that should be noted is immigrant societies, most notably Russian societies. As Yugoslavia did not recognize the Soviet Union until 1939, it was a popular destination for various Russian exiles. After the Russian revolution there were around 75,000 Russians living in Yugoslavia (Pulko 2009: 154), many of them scholars. These immigrants organized Russian schools, Russian learned societies, and many exiled Russian professors taught at Yugoslav universities.

One of the societies relevant for this discussion was the Russian Archaeological Society in the Kingdom of Yugoslavia (1921–1940). It is the only such foreign society identified so far in Yugoslavia after 1918. Very little is known about it, as we were unable to locate its archive; it is known only through sporadic historiographic references. It was registered as one of the many cultural and learned societies of the Russian emigration (Pulko 2009: 175). It operated at least until the mid-1930s and one of its members was Vladimir Mošin (1894–1987), who was the director of the Archive of JAZU (Yugoslavian Academy of Sciences and Arts), director of the Historiographic institute of JAZU, as well as the director of the Archeographic Department of the National Library in Belgrade (Kurelac 1988: 151).

13.6 Attempts at Developing a Yugoslav State-Wide Network

Archaeological, historical, and museum-based learned societies in Yugoslavia were numerous during the interwar years. The previous sections have shown that the network of learned societies that were established before 1918 served as the foundation for later developments. We would expect that Yugoslav archaeologists would have been able to use this network to formulate a state-wide system as they were connected by similar languages and archaeological traditions that were related to one another, even if some were more advanced than others. However, as with major institutions and individual international initiatives, this step towards a state-wide system never materialized fully.

There were attempts to rise above individual initiatives and perhaps the most important of these was the meetings organized by Yugoslav archaeologists during the interwar years. This aspect of the development of archaeology in Yugoslavia was long overlooked. Until recently we knew of only one meeting, the First Archaeological Congress of Yugoslavia in Belgrade in 1922 (Novaković 2014: 224), and even this event was rather mysterious, as it was known only through a short report written by Croatian architect and conservator Ćiril Metod Iveković (1864–1933) (Iveković 1922) and a reference in the memoirs of archaeologist Vojeslav Mole (1886–1973) (Mole 1970). Contextually we knew that the attendees discussed the heritage protection legislation proposed by the government at the time. Researching further in the archives and the contemporary press (Lorber 2019: 916), we have discovered that this meeting was a result of a prior assembly of key Yugoslav archaeologists organized in Dobrna near Celje (modern-day Slovenia) in 1922. The meeting was hosted by industrialist and lay archaeologist and numismatist Leon Ružička (1866–1931), as reported in the newspaper *Jutro* from Ljubljana (Anonymous 1922a) and in Belgrade's weekly magazine *Ilustrovani list* (Anonymous 1922b), which also published some information regarding the Belgrade congress (Anonymous 1922c).

In Dobrna it was agreed that an actual congress would be organized in Belgrade in 1922, followed by a congress in Split in 1924. The latter was supposed to be an international event, as there were plans to invite representatives of foreign national archaeological societies and institutions. The meeting in Dobrna is also important because serious discussions were held there about the formation of an association of Yugoslav archaeological, museum, historical, and antiquarian learned societies as a substitute for a national archaeological society. This is important as the discussion showed the intention of connecting national traditions loosely rather than uniting them into a new coherent entity. Even though this plan was never implemented, this was a more federal approach to the issue than any being considered in the wider political sphere at the time.

These conclusions were presented in several additional contemporary sources, including reports in contemporary scholarly journals that discuss the meetings in Dobrna and Belgrade (Bulić 1922; Cankar 1922). It should be noted that these sources are not uniformly detailed, mainly discussing the future of such meetings. For example, Ćiril Iveković and contemporary newspapers stated that the second meeting of Yugoslav archaeologists was to be held in Split (1924, Croatia), which is supported by Frane Bulić's account (Bulić 1922: 1), whilst Slovene art historian, diplomat, and translator Izidor Cankar (1886–1954) (Cankar 1922: 164) stated that the second meeting was supposed to be held in Skopje (1924; North Macedonia). We were unable to ascertain the reasons behind these different accounts and the second meeting was never actually organized (neither in Split nor in Skopje) and no reasons for their cancelation were ever given. Yugoslav archaeologists only met again in 1923 in Ptuj (present-day Slovenia). This meeting was organized on the occasion of the 30th anniversary of the Museum Society in Ptuj (Žižek 1992: 149). The event seems to have been an *ad hoc* conference of the Yugoslav archaeologists attending the celebrations, who discussed the issues surrounding the *Archaeological Map of Yugoslavia* project, suggesting that there was quite a bit of overlap among the activities of different learned societies. This was the last meeting of the Yugoslav archaeologists during the interwar years. It is notable that any discussion of forming a state-wide Yugoslav association of different archaeological learned societies had died down by 1924 and was not publicly considered again until after the Second World War in completely different political circumstances.

The question arises: why was the attempt to create a Yugoslav association of learned societies so ambitious and why did it die out so quickly? It seems that, in spite of the numerous scholarly and amateur communities represented in these societies, their horizons shrank during the interwar years. We can offer two main reasons for this turn of events. The first reason is illustrated by the Muzealno društvo (Museum Society) in Varaždin, Croatia. The society marked its tenth anniversary in 1935 by organizing a major celebration, to which representatives of many other contemporary societies were invited. Franjo Baš (1899–1967), one of the most notable members of the Historical Society in Maribor, attended as a representative of the Historical Society in Maribor and the Museum Society in Murska Sobota. He observed that the society in Varaždin had approximately 500 members and supporters at this time (Baš 1935: 124), so it was by no means a small society. Visitors noted that the society was wealthy by comparison with many other societies and it was put forward as a rare example of positive developments in the time described as a period of state sanctioned "atentatov na muzeje" (assassinations of museums) (Baš 1935: 125). This description is relevant as it reveals how the lack of strategic development on the national level forced learned societies to rely on the goodwill of local communities (Baš 1935: 125), especially from a financial point of view. As such, many societies were forced to limit their work to the basics (supporting local museums, supporting local collectors, and struggling to maintain their publishing activities), rather than expanding their work. This limited their cooperation with other societies in Yugoslavia and even more so abroad. Such a system was unsustainable in the long-term even in modern-day Slovenia and Croatia, which had comparatively sizable middle classes, and even more so in other Yugoslav regions. This issue was highlighted by the work of the Museum Society of Slovenia (itself a central learned society in modern-day Slovenia at that time), which faced major financial challenges in the period from 1919 to 1926 that seriously hindered its publishing activities (Mal 1939: 21). It was only after additional financial resources were found that the society could continue with its work.

These financial issues—which can be interpreted as the result of wider political instability and the absence of state-wide support systems for these organizations in the early years of the Yugoslav Kingdom—were compounded by the fact that archaeology was not given prime importance in the construction of the new state. This is not to say that it was not considered, but it was generally seen as playing a secondary role unlike, for example, history. Although museum, archaeological, and antiquarian societies failed to form a state-wide organization, historians succeeded in doing so. The Jugoslovensko istorisko društvo (Yugoslav Historical Society, 1935–1941) was quite active and it even published a state-wide historical academic serial, *Jugoslovenski istoriski časopis* (1935–1939). It counted amongst its members many highly regarded Yugoslav archaeologists such as Balduin Saria.

The Yugoslav Historical Society's journal is one of the most important sources of information on the development of disciplines like archaeology, history, and museum studies in Yugoslavia during the interwar years. It is even more important that the society was rather generously supported by the Yugoslav government, hinting at the willingness of the state to be more generous in funding such initiatives towards the end of the interwar period. This observation seems to be supported by the successful formation of the subsidy system for local societies in the province of Dravska banovina (Drava banovina), encompassing most of modern-day Slovenia, but further developments were interrupted by the Second World War.

13.7 Conclusions

This chapter discussed several key issues associated with researching archaeological societies in former Yugoslavia, especially between 1918 and 1941. The general overview of archaeology's different developmental trajectories in the countries of former Yugoslavia shows that the diversity of national academic traditions—diverse in their conceptual histories as well as their institutional development—makes simple and overarching conclusions impossible. The major importance of the work of archaeological learned societies in the development of modern archaeology in the broader region could also be demonstrated. We have chosen to discuss the history of learned societies in terms of present-day countries (i.e., Slovenia, Croatia, Bosnia and Herzegovina, Serbia, North Macedonia, and Kosovo) in order to offer a reference framework for the widest possible audience and to show the importance of these societies in the context of the development of national archaeologies. Their nineteenth-century importance in connecting different national movements with the new academic discipline of archaeology could only be hinted at, however, and we should also note that their importance is not confined to single national archaeologies.

This chapter has also shown how the First World War ended one multi-national referential environment (Austria-Hungary) and fostered another (Yugoslavia) instead. It highlights how this referential framework would be expected to be a boon to different national archaeologies due to certain similarities in language and tradition, while acknowledging the magnitude of the differences in the development of archaeology in the individual provinces of the new state. The general discussion regarding the evolution of archaeology in Yugoslavia presented here has revealed periods of stagnation in the development of the discipline and demonstrates how individuals attempted to overcome this situation, albeit unsuccessfully due to the lack of state support and the general instability of the country.

The chapter has also offered an overview of the development of archaeological, museum, antiquarian, and historical societies in Yugoslavia during the discussed period using present-day nation states as a reference in comparison with the period prior to 1918. It has shown that the development was different in each province of the new country and that societies also had different experiences within the new state. Most carried on with their activities, some were suspended (e.g., most notably Croat societies), and a small number of learned societies were newly founded. This differentiated development was not supported by the government; rather it was the result of existing scholarly networks and work by the wider communities connected with these societies, pillars of archaeology's continuity in the period before the First World War and the interwar years.

Finally, we have discussed the few attempts to extend the pre-First World War traditions to the level of the entire country. This would have connected different national and local traditions in the new geo-political environment. The goal of the three meetings of the Yugoslav archaeologists in Dobrna near Celje (1922), Belgrade (1922), and Ptuj (1923) is discussed in detail. We have highlighted the aspirations toward founding a state-wide association of archaeological, historical, antiquarian, and museum societies, as well as the ambition to use this association to connect with wider European archaeology during the planned meeting in Split in 1924. Even though these attempts were unsuccessful they hint at the mindset of individuals involved and can be understood as acts of self-awareness required to reconceptualize one multi-national system into another. This self-awareness was not enough, however, and the key obstacles faced by the learned societies that hindered their ability to cooperate and engage in long-term planning have been outlined in this chapter. The lack of financial support and the general instability of the country were among these issues. Local societies, including larger ones, were forced to focus on their basic work and the state did not offer any support. This seems to have changed, albeit slightly, between 1935 and 1941, as shown by the existence of a state-wide historical society and the gradual implementation of subsidies for local learned societies in several provinces of Yugoslavia.

In the end, these developments were curtailed by the 1941 Axis invasion of Yugoslavia. The Second World War was a major break in the development of archaeology in former Yugoslavia, as discussed elsewhere (e.g., Novaković 2014; Lorber 2019; Lorber and Novaković 2020; Lorber 2022). It is interesting that after the Second World War a situation similar to the aftermath of the First World War was faced by the next generation of archaeologists in Yugoslavia. Comparable to the 1922 post-war meeting in Dobrna near Celje, Yugoslav archaeologists organized a major conference in Niška Banja (present-day Serbia) in 1950. It is also notable that both meetings aimed to form a state-wide archaeological organization and after the Second World War this was achieved with the formation of the Arheološko društvo Jugoslavije (Archaeological Society of Yugoslavia) in 1953, one of the pillars of post-Second World War archaeological development (Lorber 2022).

What made the establishment of the Archaeological Society of Yugoslavia possible? The socialist authorities viewed archaeology differently from the royal authorities in the aftermath of the Second World War mainly due to the excesses of the war, especially the use of archaeology and history as a way to justify political goals. In an effort to integrate the discipline in the new state order, the government took an active role in its promotion. There were several reasons for such a development. First, there was a rather sizable generational change in archaeology after the Second World War. In post-war Yugoslavia many older archaeologists either retired or emigrated, and many learned societies did not recover from losses incurred during the conflict. The resulting generational split in archaeology allowed for a smoother transition towards a new system. The younger generations of Yugoslav archaeologists were forced to rebuild the national archaeological infrastructure and formulate a more coherent all-Yugoslav networking system that was based on the previously conceived statewide archaeological society. Second, the society actively collaborated with the government and national archaeological institutions in the development of archaeology in the former Yugoslavia, which included the expansion of international networks on the past Yugoslav archaeologists. During the interwar years, the international work of Yugoslav archaeologists had mainly been an individual exercise but after the Second World War international collaborations became part of a relatively well-planned institutionalized process (Lorber and Novaković 2020). This institutional networking on an international scale led to several high profile cooperations, including those with several institutions in the United States referred to by Judith A. Rasson (Chap. 15, this volume), in which it should be noted that the cooperation was far more equal than in the interwar years. Such developments would have been impossible after the First World War, when most archaeologists maintained their already existing networks which perpetuated institutional traditions without actively adapting them to the new societal and political environment. Such an approach was logical at the time, especially considering the lack of support by the authorities and uneven generational continuity in the small Yugoslav archaeological community. These two factors mitigated against the reforms that were forced upon archaeology in Yugoslavia by outside factors after the Second World War.

The historical overview of the development of the archaeological learned societies in the Kingdom of Yugoslavia presented here suggests a somewhat ironic conclusion: that the perceived strengths of archaeology in the interwar years, namely personal and conceptual continuity, in fact limited the scope of the proposed changes and reforms, with practitioners focused instead on maintaining the *status quo*. In contrast to the situation after the Second World War, the disinterest of the state in archaeological research was one of the reasons for the stagnation of the discipline during the interwar years in Yugoslavia.

At this stage no further conclusions can be drawn since the discussion is ongoing concerning the work of the Archaeological Society of Yugoslavia and its successors after the Second World War. This chapter has provided an overview and some relevant general observations regarding an interesting period in the history of archaeological discipline in the Kingdom of Yugoslavia that is neither well known nor sufficiently researched. It represents a stepping-stone toward a more comprehensive research project that will hopefully be undertaken in the future.

References

Primary Sources

Newspaper and Magazine Articles

Anonymous. (1922a, August 29). Sestanek arheologov v Dobrni. *Jutro*, 1.

Anonymous. (1922b, November 5–12). Iz života našeg društva. *Ilustrovani list*, 13.

Anonymous. (1922c, November 23–30). Iz života našeg društva. *Ilustrovani list*, 2.

Anonymous. (1929a, April 19). Jugoslovenski arheolozi na proslavi Njemačkog arheološkog inštituta. *Novo doba*, 6.

Anonymous. (1929b, August 15). Engleski arheolozi u Dalmaciji. *Novo doba*, 5.

Anonymous. (1931, September 29). Dloazak stranih arheologa u našu zemlju. *Novo doba*, *5*.

Anonymous. (1935, May 5). Dolazak francuskih arheologa. *Jadranski dnevnik*, 3.

Obnova rada Srpskog arheološkog društva. (1922, May/June). *Prosvetni glasnik; Službeni list Ministarstva prosevete*, 157–160.

Secondary Sources

American archaeologists in Yugoslavia. (1933). *Nature*, *131*, 393. https://doi.org/10.1038/131393b0

Bandović, A. (2014). Muzejski kurs i arheologija tokom II svjetskog rata u Beogradu. *Etnoantropološki problemi*, *9*(3), 625–646.

Baš, F. (1935). Desetletnica Muzealnega društva v Varaždinu. *Časopis za zgodovino in narodopisje*, *30*(1–2), 124–125.

Bulić, F. (1922). Izvješće o djelatnosti Pokrajinskog Konservatorskog ureda za Dalmaciju za god. 1922. Općenite akcije. Sastanci arheologa i konservatora. *Vjesnik za arheologiju i historiju dalmatinsku, XLV* (*II. Prilog Vjesniku za arheologiju i historiju dalmatinsku god. 1922*), 1–3.

Cankar, I. (1922). Arheološki kongres v Beogradu. *Zbornik za umetnostno zgodovino*, *2*(3–4), 164.

Daniel, G. E. (1975). *A hundred and fifty years of archaeology* (2nd ed.). London: Duckworth.

Fugger Germadnik, R. (2009). Muzejsko društvo v Celju (Museal-Verein in Cilli), 1883–1922. *Kronika*, *57*, 355–362.

Ham-Milovanović, L. (2009). Naziv i ustrojstvo Muzeja kneza Pavla. In T. Cvjetićanin (Ed.), *Muzej kneza Pavla* (pp. 90–123). Beograd: Narodni muzej Beograd.

Heimpel, H. (1972). Geschichtsvereine einst und jetzt. In H. Boockmann (Ed.), *Geschichtswissenschaft und Vereinswesen im 19. Jahrhundert: Beiträge zur Geschichte historischer Forschung in Deutschland* (Veröffentlichungen des Max-Planck-Instituts für Geschichte, 1) (pp. 45–73). Göttingen: Vandenhoeck & Ruprecht.

Iveković, Ć. (1922). 1. Jugoslovenski arheološki kongres. *Narodna starina*, *2*, 197–199.

Janković, M. (2016). U potrazi za Adamom Oršićem kroz arheologiju XX. veka. *Etnoantropološki Problemi/Issues in Ethnology and Anthropology*, *11*(3), 853–867. https://doi.org/10.21301/eap.v11i3.10

Janša-Zorn, O. (1991). Historično društvo za Kranjsko. *Zgodovinski Časopis, 45*(2), 217–238.

Jelovina, D. (1988). Osnivanje i rad Kninskog starinarskog društva. *Starohrvatska prosvjeta, 18*, 241–251.

Jones, S. (2007). Discourses of identity in the interpretation of the past. In T. Insoll (Ed.), *The archaeology of identities: A reader* (pp. 44–58). Abingdon and New York: Routledge.

Kaljanac, A., & Križanović, T. (2012). Bosanskohercegovački antikvarizam osmanskog doba. Antikvari na razmeđu istoka i zapada. *Godišnjak/Jahrbuch, 41*, 229–253. https://doi.org/10.5644/Godisnjak.CBI.ANUBiH-40.14

Koprivica, T. (2009). Don Frano Bulić kao istraživač kulturne baštine Crnogorskog primorja. In L. Čoralić (Ed.), *Hrvatsko-crnogorski dodiri/crnogorsko-hrvatski dodiri: identitet povijesne i kulturne baštine Crnogorskog primorja* (pp. 795–800). Zagreb: Hrvatski institut za povijest and Matica Hrvatska.

Krstić, B. (2006). *Zakonodavstvo arhitektonske baštine*. Beograd: Republički zavod za zaštitu spomenika kulture Beograd.

Kurelac, M. (1988). Vladimir Mošin (1894–1987). *Zbornik Odsjeka za povijesne znanosti Zavoda za povijesne i društvene znanosti Hrvatske akademije znanosti i umjetnosti, 15*, 149–155. https://hrcak.srce.hr/21428

Lorber, Č. (2019). Uvod u proučavanje Arheološkog društva Jugoslavije (1949–1991). *Etnoantropološki Problemi/Issues in Ethnology and Anthropology, 14*(3), 909–936. https://doi.org/10.21301/eap.v14i3.8

Lorber, Č. (2020). Predhodniki Slovenskega arheološkega društva/The forebearers of the Slovene Archaeological Society. *Arheo, Glasilo Slovenskega arheološkega društva, 37*, 7–24.

Lorber, Č. (2021). Slovensko arheološko društvo od ustanovitve leta 1971 do leta 1991/The Slovene Archaeological Society between its formation in 1971 and 1991. *Arheo, Glasilo Slovenskega arheološkega društva, 38*, 7–24.

Lorber, Č. (2022). *Arheološko društvo Jugoslavije in njegov pomen pri razvoju arheološke vede v obdobju od 1949 do 1991: doktorska disertacija* [Doctoral dissertation, University of Ljubljana, Slovenia]. Repozitorij Univerze v Ljubljani. https://repozitorij.uni-lj.si/IzpisGradiva.php?id=141521&lang=slv

Lorber, Č., & Novaković, P. (2020). Internationalisation as a long-term strategic project of the post-war renewal of the Yugoslav archaeology (1950–1971). *Etnoantropološki Problemi/Issues in Ethnology and Anthropology, 15*(3), 689–715. https://doi.org/10.21301/eap.v15i3.3

Mal, J. (1939). Naše muzejsko društvo v toku 100 let. *Glasnik Muzejskega društva za Slovenijo (Zbornik ob stoletnici društva 1839–1939), 20*(1), 1–24.

Marin, E. (1995). Arheološki muzej u Splitu i Boka Kotorska. *Prilozi povijesti umjetnosti u Dalmaciji, 35*(1), 67–80.

Mohorič Bonča, K. (2005). Muzejska društva na Slovenskem pred 2. svetovno vojno. *Kronika, 53*(1), 69–90.

Mole, V. (1970). *Iz knjige spominov*. Ljubljana: Slovenska matica.

Nećak, D., & Repe, B (2008). *Kriza. Svet in Slovenci od prve svetovne vojne do sredine tridesetih let*. Ljubljana: Znanstvenoraziskovalni inštitut Filozofske fakultete Univerze v Ljubljani.

Ninković, V. (2009). Arheološko odeljenje Muzeja kneza Pavla. In T. Cvjetićanin (Ed.), *Muzej kneza Pavla* (pp. 124–141). Beograd: Narodni muzej Beograd.

Novaković, P. (2003). *Osvajanje prostora: razvoj prostorske in krajinske arheologije*. Ljubljana: Filozofska fakulteta Univerze v Ljubljani.

Novaković, P. (2014). *Historija arheologije u novim zemljama Jugoistočne Europe*. Sarajevo: Univezitet u Sarajevu.

Novaković, P. (2021). *The history of archaeology in the Western Balkans*. Ljubljana: Filozofska fakulteta Univerze v Ljubljani.

Oreb, F. (2009). Arheološka i starinarska društva na tlu južne Hrvatske. In J. Hekman, J. Balen, & B. Čečuk (Eds.), *Hrvatska arheologija u XX. stoljeću* (pp. 31–64). Zagreb: Matica Hrvatska.

Pavan, G., & Marini, G. (Eds.). (2011). *Il Bicentenario della Società di Minerva 1810–2010* (Archeografo Triestino, Extra Serie, 8). Trieste: Società di Minerva.

Pirc, P. (1996). Od Arheološkog društva "Siscia" do Muzeja Sisak. *Iz muzejske teorije i prakse, 1–2*, 53–57.

Pulko, R. (2009). Organizacije ruskih emigrantov na Slovenskem 1921–1941. *Časopis za zgodovino in narodopisje, 80/n.v. 45*(2–3), 144–178.

Rapanić, Ž. (1981). Postanak arheoloških društava, njihova uloga u razvitku arheologije u Hrvatskoj i doprinos nacionalnoj kulturi. In N. Majnarić-Pandžić (Ed.), *Izdanja Hrvatskog arheološkog društva* (pp. 15–28). Vukovar: Hrvatsko arheološko društvo.

Rendić-Miočević, A. (2009). 130 obletnica osnivanja Hrvatskog arheološkog društva. In J. Hekman, J. Balen, & B. Čečuk (Eds.), *Hrvatska arheologija u XX. stoljeću* (pp. 9–30). Zagreb: Matica Hrvatska.

Rizvanbegović, I. (1976). Razvojni put Muzejskog društva Bosne i Hercegovine. *Muzeologija, 20*, 109–135.

Saria, B. (1932). Arheološki zemljevidi. *Glasnik Muzejskega društva za Slovenijo, 13*, 5–16.

Saria, B. (1935). Šesto naučno putovanje nemačkih i podunavskih arheologa. *Jugoslovanski istoriski Časopis, I*(3–4), 734–740.

Sklenář, K. (1983). *Archaeology in Central Europe: The first 500 years*. New York: Leicester University Press and St. Martin's Press.

Slapšak, B. (2011). Prepletene zgodovine v jugovzhodni Evropi: spomin in arheologija/Entangled histories in South-East Europe: Memory and archaeology. *Arheo, 28*, 131–146.

Slapšak, B., & Novaković, P. (1996). Is there national archaeology without nationalism? Archaeological tradition in Slovenia. In M. Díaz-Andreu, & T. Champion (Eds.), *Nationalism and archaeology in Europe* (pp. 256–293). London: University College London Press.

Šimičić, A. (2018). Doprinos Društva prijatelja starina „Mursa" oblikovanju Odjela umjetničkog obrta Muzeja Slavonije. In D. Damjanović, & L. Magaš Bilandžić (Eds.), *Imago, imaginatio, imaginabile. Zbornik u čast Zvonka Makovića* (pp. 267–280). Zagreb: Filozofski fakultet Sveučilišta u Zagrebu.

Tasić, N. (1983). Arheologija u Srbiji do osnivanja Srpskog arheološkog društva 1883. In N. Tasić (Ed.), *Spomenica Srpskog arheološkog društva 1883–1983* (pp. 1–8). Beograd: Srpsko arheološko društvo.

Trebsche, P. (2005). Zu den internationalen Beziehungen der Urgeschichtsforschung in Oberösterreich während der Zwischenkriegs- und Nazizeit. In M. Chytráček, J. Michálek, M. M. Rind, & K. Schmotz (Eds.), *Archäologische Arbeitsgemeinschaft Ostbayern/West- und Südböhmen. 14. Treffen 23. bis 26. Juni 2004 in Heřmaň bei Písek* (pp. 178–188). Rahden/Westf.: Verlag Marie Leidorf.

Trigger, B. (1989). *A history of archaeological thought*. Cambridge: Cambridge University Press.

Vasić, M. (1927). Arheološki institut Srba, Hrvata i Slovenaca. *Srpski književni glasnik, 22*, 33–43.

Zajc-Cizelj, I. (2004). Pravna osnova za delovanje društev 1849–1941. *Arhivi: glasilo Arhivskega društva in arhivov Slovenije, 27*(1), 145–166.

Zloković, I. (1980). Kulturno-istorijsko naslijeđe Boke Kotorske u Glasniku Narodnog unvierziteta—Kotor. *Prilozi povijesti umjetnosti u Dalmaciji, 22*(1), 290–296.

Žižek, I. (1992). Muzejsko društvo in arheologija od 1893 do 1945. *Kronika: Časopis za slovensko krajevno zgodovino, 40*(3), 148–151.

Žnidarič, M. (2003). Ob stoletnem jubileju Zgodovinskega društva v Mariboru. *Časopis za zgodovino in narodopisje/Review for History and Ethnography, 74*(3–4), 227–246

Županić, N. (1933) Pozdravni govor delegata Kraljevine Jugoslavije, N. Županiča, na XV. Mednarodnem Kongresu antropologije in predzgodovinske arheologije v Parizu, dne 20. sept. 1931. *Etnolog, 5–6*, 303–304.

Between Two Stools: Miodrag Grbić and the Archaeological Networks Formed Around the American and German Expeditions to Yugoslavia in the 1930s

14

Aleksandar D. Bandović

Abstract

By reconsidering consequential aspects of the network created between the Yugoslav archaeologist Miodrag Grbić and members of two foreign expeditions in Yugoslavia during the 1930s it will be possible to review the meaning and significance of scholarly networking in the region as well as more generally. The German archaeological expedition led by Wilhelm Unverzagt initiated excavations of the Hellenistic settlement of Gradište near Lake Ohrid in present-day North Macedonia in 1931/1932. In the same year, the American School of Prehistoric Research, led by Vladimir Fewkes, a Czech-born American archaeologist, began excavating at Starčevo, an important early Neolithic settlement near Belgrade. As a prehistorian and curator of the National Museum in Belgrade, Grbić took part in both expeditions, serving as a guide and a mediator. Both expeditions shared a familiar set of ideas, among others the preconception that the Balkan region is a crossroads and the bridge between Asia and Europe, between East and West. Both expeditions were driven by a similar belief that the archaeological exploration of Yugoslavia could reveal crucial answers about European origins and identity. The contacts that Grbić established with the members of both expeditions illustrate the tight-knit interconnections between his personal motivations and his scientific and political ideas.

Keywords

Archaeological networks · History of archaeology · American expedition · German expedition · Yugoslavia

14.1 Introduction

As the First World War in Europe was raging, the Czech-born American archaeologist Vladimir J. Fewkes (1901–1941) wrote in 1939 that battlefronts in Europe were "predestined long ago by nature" (Fewkes 1939: 375) based on prehistoric evidence. He based this theory on the fact that it was in the Balkans that he had had the opportunity as director of archaeological expeditions to analyze the remains of many fortifications from different periods. Fewkes clearly saw an unbroken chain in European battles from prehistory up until 1939 (Fewkes 1939). There is no reason to worry, Fewkes implies, wars have been ongoing since time immemorial, including the current one. When we examine Fewkes' perception of prehistory today, we can easily identify anachronisms and geographical determinism. However, his article is an excellent example of the achievements of archaeology at the time, especially the 'neutrality' that it tries to maintain in the face of great events, and his comment speaks to us about the 'usability' for the present-day that archaeologists of the time associated with the past they studied. In the interwar period, archaeology came full circle, from interpretations of prehistory that were inspired by personal experiences during the First World War (Meinander 1981: 104; Fotiadis 2001: 130) to reconciliation with the "War to End All Wars" as only one of many that would follow.

Only a few years before Fewkes' article was published, he and his Yugoslav colleague Miodrag Grbić (1901–1969) announced the establishment of the American School of Prehistoric Research in Yugoslavia in a *Discovery* magazine article published by the University of Pennsylvania Museum. In this announcement, there was no trace of the gloomy tones of Fewkes' later article, and the authors hoped that excavations in Yugoslavia would answer many important questions about European prehistory (Fewkes and Grbič 1931: 247). This hope was quite widespread among European prehistorians of that time. "Archaeologists' new Mecca" or the "Key to European prehistory" were some of the titles associated with

A. D. Bandović (✉)
National Museum of Serbia, Belgrade, Serbia

L. Coltofean et al. (eds.), *Connecting People and Ideas*, Themes in Contemporary Archaeology,
https://doi.org/10.1007/978-3-031-81006-0_14

Yugoslavia in contemporary newspapers and magazines (Anonymous 1933a; Anonymous 1933b). Several European prehistorians (e.g., Carl Schuchhardt [1859–1943], Stanley Casson [1889–1944], and Gordon Childe [1892–1957]) considered Yugoslavia and the Balkans a transit area for migrating individuals and populations during European prehistory. Research conducted in the Balkans was to play a key role in synchronizing the prehistoric cultures of Central Europe, the Mediterranean and Asia Minor (e.g., Schuchhardt 1919: 132–33; Casson 1921: 216; Childe 1928: 45; "General programme," 1934; see also Fotiadis 2001 for Macedonia).

Situated between the two identities and theogonic myths of Europe, Ex Oriente Lux and Ex Septentrione Lux (Wiwjorra 2002; Chapoutot 2016: 33–35), Yugoslavia was perceived as a bridge—a liminal area between Europe and Asia, with primordial routes along the banks of the Danube or Morava-Vardar valley. Although Bulgarian historian Marija Todorova expressed reservations about generalizing in discussions of Balkanism and its relationship to scholarly knowledge, we see that the idea of a bridge/crossroads as one of the significant topoi of Balkanism is indeed a persistent theme in archaeological discourse (Todorova 2009: 19–20, 58–59). Some authors (Babić 2002: 310–11; Gori and Ivanova 2017) have already pointed out the relationship between Balkanism and archaeology through various examples of liminality and prejudice towards the Balkans or internal self-marginalization and reproduction of "nesting orientalism" (Bakić-Hayden 1995) of the inhabitants of the Balkans themselves. In this vein, I will consider the work of two foreign expeditions between the two World Wars in the Kingdom of Yugoslavia (Fig. 14.1), the state in Southeast and Central Europe that was formed in 1918 as the Kingdom of Serbs, Croats and Slovenes and renamed the Kingdom of Yugoslavia in 1929. The first project to be discussed is the aforementioned American expedition led by Fewkes, which carried out excavations at Starčevo—the eponymous Neolithic site east of Belgrade. The second is the German expedition led by archaeologist Wilhelm Unverzagt (1892–1971), which carried out excavations at the Gradište site near Lake Ohrid. Miodrag Grbić, a young Yugoslav prehistorian and curator of the National Museum in Belgrade at the time, took part in both expeditions.

In this chapter, archaeological networks are understood as the intertwining of different scientific, ideological, and personal motivations that condition the emergence and dispersion of archaeological ideas and practices, and which are at the same time dependent on institutional, economic or political structures. This definition of archaeological networks moves beyond personalized histories of archaeology and great narratives about the contributions to science of the individual genius that are inherent in "ancestor worship" (Hamilakis 2010: 893; cf. Palavestra 2020: 20), a form of academic hagiography. Through archaeological networks, we can examine the various asymmetries of power that existed between Yugoslav, German and American archaeology, as well as the rivalry that developed between these expeditions. Grbić's archaeological networks should not be understood as Grbić's *sensu stricto*. Appearing as one of the actors, certainly not the most important, the scope of his influence was very limited. This reminds us once again that, when we muse about biographies of individuals, no matter how significant, we need "to relativize the role of the individual," to understand the biography of archaeologists as a prism (*sensu* Kershaw 2008: 37–38) or "key" (Kaeser 2008: 12), in an effort to limn the circumstances, context and all the structures that surround and influence an archaeological career.

14.2 A World of Ideas: From Prague to Belgrade

Like many students from the Kingdom of Yugoslavia, Miodrag Grbić (Fig. 14.2) studied in the Czech Republic, where he completed his studies in archaeology at Charles University in Prague in 1925. At the time, Yugoslavia, as a member of the Little Entente, was oriented towards the Czech Republic in military and political terms, with Prague serving as an idealized center of "Slavic modernity" (Čalić 2010: 136). Grbić completed his studies under the supervision of archaeologist and ethnographer Lubor Niederle (1865–1944), while his education was also influenced by the Czech prehistorian Albin Stocký (1876–1934) with whom Grbić had the opportunity to collaborate at the National Museum in Prague. He received his PhD on the topic *Předřímské bronzové nádoby na území Československé republiky* (Pre-Roman Bronze Vessels in the Czech Republic) and became one of the first formally educated prehistorians in the Kingdom of Yugoslavia (Gačić 2005: 7; Bandović 2019a: 30). During Grbić's studies, defining prehistoric cultures, their chronology, origin, and distribution were important topics in Czech prehistoric archaeology. Since the beginning of the twentieth century, Czech archaeologists have used the concept of archaeological cultures as part of the so-called *New School* that was based on the publication of the *Rukovět české archeologie* (Handbook of Bohemian Archaeology) written by archaeologist and art historian Karel Buchtela (1864–1946) and Niederle (Buchtela and Niederle 1910; Fewkes 1932a: 365, f. n. 5; Sklenář 1983: 142; Salač 2006: 226–27). Frequent reviews of German archaeology (e.g., Stocký 1919, 1922) published in the interwar period indicate an undisguised anti-German sentiment. Underlining and criticizing the nationalist narrative of Gustaf Kossinna (1858–1931) was an important part of building the identity of Czech archaeology (e.g., Niederle 1912; Stocký 1912).

Fig. 14.1 Map of the directions of the American reconnaissance expedition (1932) and of the locations of the excavations carried out by the American and German expeditions (1931–1932). (Map: Aleksandar D. Bandović, Jelena Premović and László Bartosiewicz)

Upon his return to Yugoslavia in 1926, Grbić began a new life chapter at the National Museum in Belgrade. The small archaeological community in Belgrade consisted of Miloje Vasić (1869–1956), Nikola Vulić (1872–1945) and Vladimir Petković (1874–1956). They were part of the so-called Planned Elite, which was educated in Germany (e.g., in Munich, Halle, and Berlin) at the end of the nineteenth and the beginning of the twentieth centuries (Trgovčević 2003: 33–46). In a way, each of these actors was shaped by his own professional and personal interests: Vasić as *Altertumswissenschaftler* (Classical scholar), excavator and interpreter of Vinča, one of the most important Neolithic settlements near Belgrade, Vulić as a historian and archaeologist of the Roman provinces, and Petković as an art historian and archaeologist who dealt primarily with medieval heritage. Personal animosities and power struggles among these three had long-lasting consequences for archaeology in Belgrade (Bandović 2019a: 36–37; Palavestra 2020: 25, 142–43).

The most important and most powerful archaeological institution in Yugoslavia in the interwar period—both in terms of financial resources and the territory it covered—was the National Museum in Belgrade. Under the baton of director Vladimir Petković, the museum was actively recovering from the bombing and destruction of museum collections that accompanied the First World War. In 1924, the museum began excavations at Stobi, an ancient site on the banks of the Crna Reka River in present-day North Macedonia led by Slovene archaeologist Balduin Saria (1893–1974) (Bandović 2019b: 869–71). At the time, the popular phrase “Čija je

Fig. 14.2 Miodrag Grbić (1901–1969), curator of the National Museum in Belgrade. (Reproduced with the permission of the Archives of the National Museum in Belgrade, Serbia)

škola, toga je Maćedonija" (Macedonia belongs to whoever owns the school) (Jovanović 2014: 177–89) could be freely modified into "Macedonia belongs to those who dig there" since it was written in the press that Macedonia was "zemlja arheologa, pravi Eldorado" (the country of archaeologists, the real Eldorado), with Stobi represented as the "Jugoslovenski Pompeji" (Yugoslav Pompeii) (Anonymous 1924, 1927). The excavations at Stobi were a long-term undertaking that had propaganda value as a *mimesis/imitatio* project of the great powers. The director of the National Museum, Vladimir Petković, described the excavations at Stobi as of equal national and international importance (Bandović 2019b: 869–71).

Prehistoric archaeology was not the focus of Yugoslav cultural policy. Despite that, Grbić, as the curator of the National Museum's prehistoric and Roman department, which had modest funds, and with the support of Vladimir Petković, managed to begin explorations at several important prehistoric sites, such as Pločnik and Starčevo, among others. He published the results of the excavations at Pločnik and Starčevo in German (Grbić 1929, 1930). Although these were modest scientific achievements, even according to the standards of archaeology at the time, these works drew the attention of European archaeologists to the young prehistorian (e.g., Reinecke 1931; see also Bandović 2019a: 70, f. n. 113). His work on Starčevo, which resulted in a book chapter about painted pottery (Grbić 1930), whose origin was disputed among prehistorians at the time (e.g., Childe 1929: 98–111; Tompa 1929: 48–60)—attracted special attention.

Grbić had ambitious plans for the future. Among other things, he had the ultimate goal of becoming a university professor. However, apart from the support he received from Petković, the small community of archaeologists in Belgrade—which consisted of the already established professors and authorities Vasić and Vulić—showed indifference towards him (Bandović 2019a: 46–47). Therefore, Grbić had to look for allies elsewhere. Broadly speaking, Grbić's political attitude at that time inclined toward conservatism tempered with pragmatic Serbian nationalism, which in his case was more of a reluctant rhetorical tool than a developed ideological attitude (Bandović 2019a: 88–89, 116).

14.3 Prehistoric Archaeology as an Export Product: The American Expedition in Yugoslavia

One of the widespread practices of archaeologists in the interwar period consisted of study trips, tours of museum collections and examination of archaeological material, so-called comparative studies. Visits by foreign researchers and their examination of museum collections preparatory to more extensive research was intensively followed in the press at the time (Bandović 2019a: 97, 243). Among the prehistorians who visited Yugoslav museums in the 1920s were Albin Stocký, V. Gordon Childe, Oswald Menghin (1888–1973) and Hans Reinerth (1900–1990) (Bandović 2019a: 66–67). At the same time, the British prehistorian John L. Myres (1869–1954) was in various ways connected to the research of prehistory in the Balkans, especially through the excavations at Vinča conducted by Miloje Vasić, which benefitted from the extensive financial support of the British newspaper magnate Charles Hyde (1876–1942). Vasić was in correspondence with Myres, who helped him present the results of his research at the First International Congress of Prehistoric and Protohistoric Sciences held in London in 1932 (Palavestra 2020: 66–67). The Vinča excavations sparked the attention of numerous European archaeologists, such as Walter Abel Heurtley (1882–1955) and Kimon Grundmann (1891–1968), both of whom visited the site (Palavestra 2020: 25, 66–67, 88–89). Given this more general scientific interest in the prehistory of the Balkans, Grbić's position as curator of the prehistoric and Roman collections at one of the few Yugoslav museums was probably viewed as important, promising, and attractive to colleagues in other countries.

In the context of this lively interest in the prehistory of Yugoslavia, the engagement of the American School of

Prehistoric Research (ASPR) should also be considered. The founders of ASPR in 1921 were anthropologist George Grant MacCurdy (1863–1947) and Charles Peabody (1867–1939). The school was particularly active in France, but its headquarters then moved to Prague, where ASPR fellows Vladimir Fewkes (Fig. 14.3) and Robert W. Ehrich (1908–1992) excavated the Eneolithic site of Homolka (Fewkes 1932a; Browman and Williams 2013: 430–31). As Robert Ehrich later wrote "The school first focused its interest on France, later on all of Europe, and eventually on the whole of the Old World" (Ehrich 1948: 49). MacCurdy, a pioneering paleoanthropologist, was a professor at Yale who is credited with presenting European archaeological concepts to the American archaeological community (Bricker 2002: 284). He believed that archaeological research in Europe would provide comprehensive answers to the question of where the cradle of anatomically modern humans was to be found, which he assumed was in the northern hemisphere, in some part of the Old World. A step towards this discovery was represented by the establishment of a chronology for the Neolithic in Europe (MacCurdy 1929: 97, 103–4). MacCurdy later explained these ideas in an interview with the Belgrade daily *Vreme*:

> Jugoslavija me privlači svojom bogatom preistorijskom prošlošću. Za nas je od velike važnosti da vršimo istraživanja na licu mesta. Kako nov kontinent ne obiluje interesantnosti u arheološkom pogledu, mi svoja znanja moramo da obogaćujemo svakogodišnjim istraživanjima u Evropi.
> (Yugoslavia attracts me with its rich prehistoric past. It is of great importance for us to conduct research on the spot. As the new continent is not rich in archaeological points of interest, we must enrich our knowledge with annual research in Europe.) (Anonymous 1931a)

Ehrich, Fewkes and Hallam Movius (1907–1987), another student of the ASPR, stayed in Belgrade as early as 1930, studying museum collections and inspecting prehistoric sites. They established contact with Grbić through the Archaeological Institute in Prague, where Jaroslav Böhm (1901–1962), one of Grbić's friends from college, also worked.

In 1931, a cooperation agreement was signed between the National Museum in Belgrade on one side and the American School of Prehistoric Research, Peabody Museum and Penn Museum on the other. The contract, signed by Fewkes and Vladimir Petković, included cooperation in the excavations at Starčevo as well as sounding and reconnaissance of other Neolithic sites (i.e., Jabuka, Ljuljaci, and Resnik). Regulating the rights and obligations of the contracting parties, the American expedition sought to engage Yugoslav workers and specified that all archaeological objects were the property of the National Museum, on the condition of providing one "collection of duplicates" (Anonymous 1931) to the American partners at its discretion. The right to publish archaeological findings was mutual (Bandović 2019a: 97).

Together with MacCurdy, the patrons of the expedition, spouses Julius (1869–1931) and Lilian Haass (1869–1931), stayed in Belgrade in 1931. The Detroit banker was also a trustee of the Detroit Museum of Art Founders Society, and after his death his wife continued to engage in philanthropic work (Bandović 2019a: 98). The following year, the expedition organizers made an arrangement with the Fogg Art Museum and American archaeologist Hetty Goldman (1881–1972) joined the expedition in 1932. Paul Sachs (1878–1965), a cousin of Goldman's and director of the Fogg Art Museum, played a role in establishing this connec-

Fig. 14.3 Vladimir J. Fewkes (1901–1941), director of the American expedition to Yugoslavia (1932). (Reproduced with the permission of the Archives of the National Museum in Belgrade, Serbia)

tion (Ehrich 1974: 38). Goldman worked for a long time in Greece, as a Harvard Fellow at the American School of Classical Studies, where she excavated the sites of Halae, Colophon and Eutresis (Mellink and Quinn 2006: 301–16). In a letter to Grbić, Fewkes presented Goldman as a top expert: "Ona zna celu Grčku i egejsku oblast bolje nego Stocki svoje cipele" (she knows the whole of Greece and the Aegean region, better than Stocký his shoes) while he described her publication on Eutresis as "kolosalno" (colossal) (Fewkes 1932b; Bandović 2019a: 98).

In 1931, the first year of residence in Yugoslavia, the American expedition engaged in exploratory excavations at Starčevo and Lipovac (Dizaljka), another Neolithic site in the relative vicinity of Belgrade (Fewkes et al. 1933a). During the excavations at Starčevo promising structures and finds were discovered. In April of the following year, the expedition set sail for Europe with a much more ambitious project and preconceptions related to Starčevo itself. Based on an internal document in the archives of the Penn Museum (Anonymous 1932), it is clear that the painted pottery from Starčevo was interpreted as the oldest trace of the spread and origin of European civilization on the continent (Bandović 2019a: 98). After arriving in Yugoslavia in 1932, the expedition undertook reconnaissance across a wide area from southern Serbia to the south of what is North Macedonia today, carefully recording data on each locality and other impressions in field notebooks. On their way south, they drove a Ford Model A that became a hallmark of the expedition and an iconic aspect of the image of wealth built around the excavation projects associated with the projects conducted under its aegis (Ehrich 1974: 39; Mellink and Quinn 2006: 318; Roth 2020: 107).

In addition to his archaeological interests, Robert Ehrich wrote his doctoral dissertation on the *Racial analysis of Montenegro*, an area that he considered "racially speaking, practically unknown" (Ehrich 1933: 54). In 22 days, Ehrich managed, as described in the preliminary report of his research, to measure "852 subjects", carrying out anthropometry on people

> in our hotel bedroom, local hans, open markets, restaurants, cafes, town halls, spare rooms, corridors, rented rooms, private houses, and even in the bedrooms of three local administrators, to say nothing of schools, a military hospital, and a road gang found sleeping under an apple tree. (Ehrich 1933: 55)

Ehrich, as a 'racial thinker', can be compared to his contemporary Harvard professor Carleton S. Coon (1904–1981), who used the results of Ehrich's anthropometry in Montenegro in *The races of Europe* (Coon 1939: 591, f. n. 126) and who continued to "merge biology and culture" (see Goodman and Hammonds 2000).

The Belgrade press followed the expedition of the American archaeologists at every step. In one of these newspaper articles, Fewkes suggested that the sum of 120,000 dollars would be spent on the activities of the American expedition. In reality, archival documents indicate that the actual amount was much smaller and Fewkes' statements regarding finances were part of the American expedition's propaganda. As can be deduced from the private letters that Fewkes exchanged with Grbić, the expedition had funding problems. Apparently, the Great Depression had left a lasting mark on archaeology (Bandović 2019a: 99, 102–3).

The expedition donated a peculiar "collection of curiosities" to the National Museum which, in addition to Etruscan pottery and Native American tools, included objects from the Neolithic collected by Gabriel de Mortillet (1821–1898) in France (Bandović 2019a: 100, f. n. 204). Part of the collection of objects given in exchange included artifacts from Denmark (e.g., Neolithic flints, axes) donated by civil engineer W. J. Rose, while the other part of Rose's collection remained on display in the Peabody Museum at Harvard University (Putnam 1898: 8, 14). The practice of exchanging archaeological objects between the American Expedition and the National Museum reminds us of the nineteenth-century tradition of exchanging comparative collections and the fact that the attitude towards preserving cultural heritage had not yet been legally regulated in Yugoslavia. No law had yet been passed in the Kingdom of Yugoslavia regulating the exportation of cultural heritage. In accordance with the agreement signed between the American and Yugoslav partners, the National Museum in turn set aside a part of the finds from the Starčevo and Lipovac sites that were sent to the Peabody Museum of Archaeology and Ethnology at Harvard University (Fewkes et al. 1933a: 36; Fewkes 1936: 36, f. n. 312).

Despite high expectations and extensive preparations, the results of the American expedition's research at Starčevo did not yield the expected results, but it did generate sensationalist reports in the press (Obrovski 1932). By the standards of the time, the excavations were methodologically quite advanced. In the only published report, painted pottery as a "type fossil" was the only element Fewkes, Goldman and Ehrich used as *comparanda* with archaeological sites of Central, Eastern and Southeastern Europe. They remained cautious in determining the directions of cultural transmission, concluding that only further research would show in which direction cultural influences had been transmitted (Fewkes et al. 1933b: 51). Therefore, at that moment, the question of the origins of painted pottery (read European civilization) was not resolved by the Starčevo research project. Equally unsuccessful was the archaeological survey conducted along more than 4000 km of the banks of the Morava and Vardar Rivers, whose valleys they had described as a "highway" or "avenue of early cultural movement" (Fewkes et al. 1933a: 18).

The question of the relationship between Aegean civilization and Central Europe was included in the general program at the First International Congress of Prehistoric and Protohistoric Sciences in London in 1932 (see "General programme," 1934). One of the resolutions that was passed at the Congress was to form a committee to be organized by the Hungarian archaeologist Ferenc Tompa (1893–1945) devoted to systematically and scientifically researching the Balkan region at an international level ("General programme," 1934: 10–11). Thus, the establishment of a chronological system in the Balkans was a matter of scientific prestige, a burning topic. The engagement of the American expedition in Yugoslavia should be seen in this light. Still, on the basis of the preliminary report on Starčevo, the debate over a Nordic, Central European or Middle Eastern origin of the European Neolithic was not resolved but was postponed to the future, which in 1933 still seemed bright and shiny.

So far, this chapter has discussed the ideas that guided the activities of the American expedition and the key actors involved in it, with whom Grbić cooperated in the framework of this international project. A distinctly Eurocentric metanarrative and preconceived notions of the origins and direction of Neolithic and later cultural impulses with which the members of the expedition came to Yugoslavia characterized all aspects of the expedition members' activities. Their relationship with Grbić will be analyzed later in this chapter, after a discussion of the German expedition's archaeological work at Ohrid.

14.4 The German Expedition to Ohrid (1931 and 1932)

The ideas and discoveries that motivated the German expedition to Yugoslavia were somewhat different from those of the American project carried out at the same time. What attracted German archaeologists to Yugoslavia was one of the tropes of the First World War—the Iron Age necropolis of Trebeništa near Lake Ohrid, in today's North Macedonia. The excavations of the Trebeništa necropolis, which were undertaken by Bulgarian troops in 1918, later under the supervision of Bulgarian archaeologist Karel Škorpil (1859–1944), and the publication of seven graves from the archaic period with exceptionally rich grave goods, generated excitement among European archaeologists. A book about archaeological finds from Trebeništa by Bulgarian archaeologists Bogdan Filov (1883–1945) and Karel Škorpil was published in a luxury edition by the German publishing house Walter de Gruyter (Filow and Škorpil 1927).

The publication of Trebeništa raised several important questions among European archaeologists: the origin and ethnicity of the people buried at the necropolis, the custom of placing golden masks on the faces of the deceased (which even the first researchers compared to the Shaft Graves in Mycenae) and location of the settlement where people buried in the necropolis lived. Some authors (Casson 1928; Schuchhardt 1932) were quick to conclude that Mycenae and Trebenište were in fact part of the same culture, part of a "conservative nature of Macedonians and Illyrians and of the isolation in which they lived" (Casson 1928: 270). It is important to mention the opinion of archaeologist Carl Schuchhardt (1859–1943), Schliemann's biographer, who thought it was possible to trace the connections between Central Europe, Trebenište and Mycenae, that is, all the areas he believed had been occupied by the Illyrians. He considered them to be the bearers of the LBK pottery (Bandkeramik) tradition who had set out for Greece during the first invasion of the Indo-Germans, carrying the spiral motif on pottery vessels with them (Schuchhardt 1932: 343). In his opinion, the Illyrians were "phantasievolle, witzige, lebensfrohe Leute" (imaginative, humorous, life-affirming people), so even Odysseus, judging by his character, was supposedly an Illyrian (Schuchhardt 1934: 7–8). Although in Germany Schuchhardt is considered Kossinna's main rival, as well as an antipode, it should not be forgotten that his books enjoyed general popularity among Nordic-minded experts (Lutzhöft 1971: 124; Chapoutot 2016: 26, 60) who believed in *Ex septentrione lux*/*Der Nordische Gedanke*.

In Yugoslavia, in Belgrade, the question of those buried in the Trebenište necropolis was much more related to modern ethnicities than prehistoric ones and the appropriation of archaeological finds took a different path. When the book about Trebenište was published, Grbić and Petković demanded that Bulgaria respect the regulations of the Treaty of Neuilly-sur-Seine and return the finds to Yugoslavia. Melodramatically and with pathos, Grbić also wrote an article for the Belgrade daily newspaper *Politika*, publicly demanding the return of the archaeological finds. In a letter addressed to the Ministry of Education, which oversaw the Museum's activities, Grbić tried to attract the attention of the Ministry by insisting on the "veliki kulturni, moralni i materijalni značaj ove stvari" (great cultural, moral and material significance of these subjects) (Grbić 1928; Bandović 2019a: 88–89).

The publication of the book about Trebeništa also started a literal archaeological race in search of the associated settlement, with Nikola Vulić, a university professor and representative of the Serbian Royal Academy, and Grbić as a representative of the National Museum, as the main competitors. The concurrent engagement of Grbić on the one hand and Vulić on the other to find the settlement "gde su mrtvi živeli" (where the dead lived) (Krakov 1930) is an indicator of the general conflicts regarding jurisdiction that characterized the relationship between the National Museum and the Serbian Royal Academy (Bandović 2019a: 90–91).

One of the protagonists who was already in Yugoslavia in 1928, romantically indulging in the ruins of Lake Ohrid's past, was a young German baron named Johann Albrecht von

Fig. 14.4 Johann Albrecht von Reiswitz (1899–1962), initiator of the German expedition to Yugoslavia (c. 1927). (Reproduced with the permission of the Archives of the National Museum in Belgrade, Serbia)

Reiswitz (1899–1962). Reiswitz (Fig. 14.4) was an autodidact prehistorian, since in Berlin he had received his doctorate with a dissertation about German philosopher Arthur Schopenhauer (Roth 2020: 31–32). However, from 1924 onwards he was interested in various aspects of German-Serbian relations throughout history, the issue of the Illyrians and Bogomils, and so he became a regular visitor to Yugoslavia. Ohrid soon became his passion (Roth 2020: 69). It was Reiswitz who suggested Gradište as a likely site for future archaeological excavations of the German expedition, based on the "Zyklopenmauern" (cyclopean ramparts) he had seen there (Roth 2020: 74). He hoped, as he later wrote to Grbić, that the excavations of Gradište would contribute "zum rechten Verständnis für die Funde in 'Trebeniste'" (to a true understanding of Trebenište findings) (Reiswitz 1930; Bandović 2019a: 90). Being interested in prehistory, he soon became a great admirer of Schuchhardt's (Roth 2020: 67–68) and in private correspondence he called him "Papa Schuchhardt" (Reiswitz 1933a; Bandović 2019a: 90), implying his importance and authority. In order to jump-start research at Gradište, Reiswitz initiated a collaboration between Petković, director of the National Museum, and the Deutsches Archäologisches Institut (DAI: German Archaeological Institute). Petković, who was a corresponding member of the DAI himself, believed that the DAI would be a powerful international partner, not only as a financial sponsor of archaeological research that would contribute to cultural propaganda abroad, but also as an ally in the case of regional disputes with Bulgaria (Bandović 2019a: 91). The DAI's perspective on the matter was probably similar. According to historian Suzanne Marchand (1996: 279), in the 1920s the DAI, led by Gerhardt Rodenwaldt (1866–1945), had expanded its area of interest and influence. In this regard archaeologists played a significant role as cultural ambassadors in the service of cultural policy. The contract between the DAI and the National Museum was signed at the end of 1929 and was almost identical to the contract signed with their American partners. One of the points that differed, however, was the restriction of the National Museum regarding the publication of archaeological finds because the primacy in publication was claimed by the German side. Petković thought that the contract signed with the American expedition was more favorable (Bandović 2019a: 92, 97).

The archaeological work at Gradište was supervised by German archaeologist Wilhelm Unverzagt with the assistance of Alexander Langsdorff (1898–1946), who withdrew from the project due to illness (Roth 2020: 100–2). Unverzagt had been a member of the DAI since 1927 and was Schuchhardt's successor as Director of the Prehistoric Department in the Museum für Vor- und Frühgeschichte (Museum for Pre- and Early History) in Berlin. Like Schuchhardt, Unverzagt was considered an experienced field researcher who had excavated at Losow and Reitwein before Gradište, and together with historian Albert Brackmann (1859–1952) was an important proponent of German *Ostforschung* (East research) (Fehr 2004; Saalman 2017; Grunwald 2019). The Ostforschung refers to the research of various German scientists in the years after the First World War with the goal of finding supporting evidence for political and territorial claims at the expense of Germany's eastern neighbors. When it comes to the excavations in Ohrid, it is interesting that, in a letter that Unverzagt sent to the Notgemeinschaft der Deutschen Wissenschaft (Mutual Assistance Association of German Science) he stated that the excavation of Gradište in Ohrid was important in order to investigate the connection between the "griechisch-ägäischen Welt und dem thrakisch-illyrischen Hinterlande" (Greek-Aegean world and the Thracian-Illyrian hinterland) (see Kott 2017: 250 for the letter). In addition to the scientific goals of the project, Unverzagt emphasized the importance of cooperation between German and Yugoslav science because, in his opinion, there was a risk that American aid would arrive before German cooperation (Kott 2017: 250–51). Apparently, Unverzagt was already aware of the intentions of the American expedition to begin research and the competitiveness that he underlined played a significant role in obtaining funding for his research. Unverzagt did not hesitate to invoke nationalist and *völkische* narratives in funding applications for his excavation projects (Saalman 2017: 852; Grunwald 2019: 87).

While foreign expeditions were preparing for great archaeological discoveries at the sites for which they had obtained excavation permits, in a dramatic turn, Nikola Vulić found Tomb VIII at Trebeništa, resulting in a sensational tour of Europe. The impressive grave goods led Vulić to initially interpret the burial as the grave of a fallen warrior (Vulić 1930). Vulić's excavations and their aftermath were the reason for Gerhart Rodenwaldt's public address stating that the DAI was pleased with the discoveries by Vulić and that he did not expect the German expedition to have equally sensational results at Gradište (Milojević 1930). Rodenwaldt's reaction to Vulić's discovery can also be interpreted as DAI's concern about the fate of the excavation concession. In his diary, Reiswitz commented that Vulić had taken over the primacy at Ohrid (Reiswitz in Roth 2020: 102).

Despite all the vicissitudes, in April 1931, Unverzagt, Grbić and Reiswitz posed for journalists at the train station in Belgrade, pompously announcing the start of excavations at Gradište. Archaeologist Josef F. Keller (1902–1981) participated in the excavations with them as Unverzagt's assistant (Bandović 2019a: 92; Roth 2020: 114, 435, f. n. 466). After engaging in preliminary investigations at several sites in the vicinity of Ohrid, they finally decided that Gradište was the most promising. The Belgrade press used the presence of the German expedition for propaganda purposes, emphasizing the fact that the Yugoslav government was mainly motivated by the pacification and modernization of Macedonia and that the joint work of Yugoslav and German archaeologists would contribute to breaking down prejudices against Yugoslavs "naturene Nemcima kroz tuđe naočare" (dictated to Germans through someone else's glasses). Unverzagt, like a skilled politician, looked forward to working together with the "vedri i radni Jugosloveni" (cheerful and hard-working Yugoslavs) and stated, "da posle rimske epohe koja je vekovima održavala red i mir u Makedoniji, dolazi jugoslovenska epoha, koja će učiniti kraj nesretnim trzavicama kojim je Južna Srbija bila izložena" (that after the Roman epoch that maintained order and peace in Macedonia for centuries, the Yugoslav epoch is coming, which will put an end to the unfortunate setbacks to which Southern Serbia had been exposed for centuries) (Anonymous 1931b). By "cheerful and hard-working Yugoslavs" he probably meant cheap manual workers who were severely underpaid (cf. Roth 2020: 106).

In the end, the German archaeological team postponed the start of systematic excavations, which were to be carried out in September and in the following year. The great discoveries promised by the excavations at Gradište were not fulfilled as the excavations revealed the remains of a settlement from the early Hellenistic period that could not be linked to the Trebeništa necropolis. Despite the modest archaeological results, Reiswitz, in full accordance with Schuchhardt's and Unverzagt's ideas, in *Auf den Spuren der Illyrer* published in the newspaper *Deutsche Allgemeine Zeitung*, presented Gradište as the southernmost Illyrian fortification that could be linked with sites in east and southeastern Germany, Pomerania and Lausitz (Reiswitz 1932).

Despite these plans for excavating additional archaeological sites, the German archaeologists did not return to Yugoslavia in 1933. The lack of financial resources and the "grosser politischer Umschwung" (great political change)—as Reiswitz put it (Reiswitz 1933b), referring to the rise of Chancellor Hitler in a letter to Grbić—were the reasons why archaeological research was postponed from year to year. Grbić responded that "domaća štampa je sa simpatijama pisala o velikim političkim događanjima, verovatno zbog toga što ništa protiv jugoslovenske strane nije pisano ili urađeno" (the local press wrote with sympathy about the great political events in Germany, probably because nothing was written or done against the Yugoslav side) (Grbić 1933a; Bandović 2019a: 118). Grbić, of course, was speaking on his own behalf. The following section discusses Grbić's position in the American and German expeditions in Yugoslavia, as well as their mutual relations.

14.5 Miodrag Grbić Caught Among Benefits, Difficulties, and Intrigues

Grbić played several roles in the American and German expeditions to Yugoslavia and expressed different attitudes towards the individuals involved in them. For example, although in his letters he maintained an official and restrained attitude towards Unverzagt, in correspondence with Reiswitz and Fewkes, his tone was friendly and informal. Grbić spoke German much better than English, so he communicated with the members of the German expedition in German, while with Fewkes, who advised him that he would have to study English because of future career plans, he exchanged letters in Czech and Serbian (Bandović 2019a: 101–3).

Although he was the main and most often the only representative of the Yugoslav side in both expeditions, Grbić was not on equal footing with other team members but was instead assigned the role of mediator, interpreter and translator for the members of the foreign archaeological teams and local or regional authorities. When responding to numerous requests from the international archaeological teams, he sometimes found temporary and grotesque solutions. When Ehrich requested that Grbić provide him with police or an escort of officials during his research on the "Dinaric race" in Montenegro (Ehrich 1932) Grbić sent his own brother-in-law as an escort (Ehrich 1933: 55; Bandović 2019a: 115). Similarly, Grbić readily responded to Unverzagt's insistence that all archaeological sites in the vicinity of Ohrid, to which the Germans "have a string" (Grbić 1932), be placed under the protection of the gendarmerie in their absence (Bandović

2019a: 93). The gendarmerie played an important role in securing the activities of both expeditions (cf. Fewkes et al. 1933a: 25).

One particular event eloquently illustrates the relations between Fewkes and Grbić. At the beginning of their cooperation in 1931, Grbić wanted to publish a paper in the magazine *Discovery* with the aim of popularizing Yugoslav archaeology while announcing the arrival of the American expedition in Serbia (Fewkes and Grbič 1931). However, Fewkes submitted Grbić's paper as a co-author. Fewkes justified this by claiming that it had happened without his knowledge: "Urednik zna mene, ne tebe" (The editor knows me, not you) (Fewkes 1932b). Hence, in private letters, Grbić considered Fewkes "a first-class field expert, an appealing man and his friend" (Grbić 1933b), but Fewkes's machinations did not recommend themselves when it came to future collaborative publishing ventures (Bandović 2019a: 104). Grbić did not fare any better with Unverzagt when it came to joint publishing ventures. Grbić's attempt to publish a critique of Professor Miloje Vasić's book *Prehistoric Vinča* in the journal *Praehistorische Zeitschrift* received no support from Unverzagt, who, as the editor of the magazine, had considerable objections to Grbić's German language abilities even after proofreading. Apart from the language issues, the impression is that Unverzagt was not particularly interested in the internal conflicts and arguments in Yugoslav archaeology (Bandović 2019a: 107).

Grbić's reasons for accepting the assigned roles in the two expeditions—roles which differed from those that he wished and expected—were numerous. For instance, although he was not satisfied with the remuneration he received as an expedition team member, Grbić accepted this as he did not want to lose his privileges within the American expedition. In 1933, when Grbić could not participate in the reconnaissance survey, his museum colleague and friend Đorđe Mano-Zisi (1901–1995) replaced him. Grbić prepared a "poverljivi pravilnik rada sa Amerikancima" (confidential rulebook for working with Americans) for Mano-Zisi, in which he insisted that, among other things, one should be "rezervisan i nepoverljiv, jer su nekoliko puta hteli naklonost, da zloupotrebe. Držati se na visini, a ne pokazivati se im da nam je stalo do toga, da su ovde, jer hoće da nas degradiraju zbog toga i da se prave veliki gospodari" (reserved and distrustful because several times they [the Americans] wanted to abuse affection. To stay on top by not showing that we care about them, that they are here, because they want to degrade us pretending to be great masters) (Grbić 1933c; Bandović 2019a: 103).

After the excavations at Starčevo were completed in 1932, there was a plan to send archaeological material for analysis to the laboratory of the Peabody Museum and then to prepare a publication and a major archaeological exhibition in which Grbić would also participate. Grbić included Unverzagt as an advisor in the negotiations with the Americans. After initial negotiations, the Yugoslav Ministry of Education demanded insurance in the amount of 13,000 dollars from the American side, which Grbić considered an unrealistically high amount, and he expressed his fear that abandoning this arrangement would prevent further American investments in Yugoslav archaeology. However, as in the case of the German expedition, financing such an endeavor was not possible at the time (Bandović 2019a: 102). Speaking of financing, it should be said that although some estimates suggest the funds of the American expedition were as much as six times the size of the German ones (Roth 2020: 110), one should be careful in accepting these claims, since the media propaganda associated with the American expedition cannot be considered a reliable source of information (cf. Obrovski 1932).

Thanks to his relations with Reiswitz and Unverzagt, Grbić became a corresponding member of the DAI at their invitation. In addition, his acquaintance with German archaeologists enabled him to receive a scholarship and a study visit to the DAI in Athens at the beginning of 1934. Grbić used his visit to study comparative Neolithic material but also expanded the circle of his acquaintances and established new connections (Bandović 2019a: 119–20).

The visit of the American expedition to the German excavations of Gradište in May 1932 speaks to Grbić's position within both expeditions. From the short notes that Fewkes left (see Roth 2020: 107, f. n. 468), it is clear that the Americans also had plans in connection to Macedonia. Fewkes was delighted with the excavation technique of German archaeologists, while he described Vulić's excavations in Trebenište as catastrophic. During dinner with members of the German expedition, in spite of the delicious Ohrid fish on the table, Fewkes felt he was in an unpleasant environment. Because of the fact that the German team used the American expedition's Ford Model A for their own needs and since Grbić cut short his work with the American team, Fewkes considered Grbić's behavior to be irresponsible and reckless (Roth 2020:107–9). At the same time, due to the American archaeologists' visit to Gradište, which Grbić facilitated, he was later accused by Unverzagt of endangering the concession:

> (...) dass die Schuld an allen diesen Vorgängen letzen Endes Sie selbst trifft, da Sie die Amerikaner nach Ochrid eingeladen und unsere Grabung dem Augenblick an vernachlässigt haben, in dem die Amerikaner in Ochrid erschienen sind. (...) An der Aufrichtigkeit Ihrer zugleich im Namen von Herrn Direktor Petković abgegeben Erklärung, dass die Umgebung von Ochrid für uns reserviert bleiben, habe ich mir niemals irgendwelche Zweifel erlaubt.
> ([…] you are to blame for all of these occurrences, as you called the Americans to Ohrid and neglected our excavation from the moment the Americans appeared in Ohrid. [...] I have never allowed myself any doubts about the sincerity concerning the explanation made on behalf of Director Petković, that the surroundings of Ohrid remain reserved for us.) (Unverzagt 1932a)

Although Unverzagt later wrote from Berlin that he had become good friends with Fewkes (Unverzagt 1932b), there is no doubt that there was competition between the two expeditions (Bandović 2019a: 105). Unverzagt's furious reaction when Vulić found several more graves related to the Trebeništa necropolis in 1932 also shows his attitude towards Grbić. He could not hide his surprise and wrote in a letter to Grbić that he learned about the discovery from the newspaper and not from Grbić or Vulić, and that "dass in dem Belgrader Zeitungsbericht unsere Ausgrabung mit keinem wort erwähnt waren" (the Belgrade newspapers did not say anything about the German excavations). He also added: "Ein solches Vorgehen ist keineswegs geeignet, mir die Beschaffung der Mittel für die nächsten Ausgrabungen in Ochrid zu erleichtern" (Such an approach is by no means suitable to make it easier for me to raise funds for the next excavations in Ohrid) (Unverzagt 1932c).

Considering that "die Serben darin nicht bewandert sind" (the Serbs were not experienced) in the study of pottery (see Roth 2020: 106), Unverzagt thought that the best solution was to bring the sherds to Berlin, where Robert Zahn (1870–1945), "dem besten Kenner auf diesem Gebiete" (the best expert in that field), would analyze them (Unverzagt 1932b). While such a move was not addressed in the contract, Grbić and the museum commission, after determining "da keramički fragmenti ne predstavljaju nikakvu materijalnu ni naučnu vrednost" (that the pottery had no scientific and material value) (Grbić 1932) sent six boxes to the Museumsinsel (Museum Island, built under the Prussian rulers, encompassing five large museums) in Berlin. Although the pottery was supposed to be returned to Belgrade after processing, the transaction was forgotten, and Grbić himself contributed to that. When Milan Kašanin, the new director of the National Museum in Belgrade, made an inquiry about the finds from Gradište, Grbić explicitly denied the existence of any significant material (Grbić 1939; Bandović 2019a: 94–95). When Unverzagt published a report on the excavation of Gradište after the Second World War, he wrote that there were no movable finds which would enable dating the site (Unverzagt 1954: 21).

Even though the excavations at Gradište did not continue in 1933, it is interesting that Reiswitz still thought about the site. In the archives of the National Museum in Belgrade, the bequest of Johann Albrecht von Reiswitz includes excerpts from Belgrade newspapers about Vulić's and the German excavations that indicate how closely Reiswitz followed the news from Belgrade. Reading *Politika*, in August 1933, Reiswitz wrote to Unverzagt that he sees "manches politisch sehr lustige" (a lot of funny things politically) but that he finds "noch nichts von etwaigen neuen Vulic-Aktionen" (nothing new about Vulić's actions). However, the whole situation was getting on his nerves, and he wrote that *nervus vagus* tormented him „mit der Vorstellung, das Vulić eben jetzt unterhalb unseres Gradištes wühlt, oder dass überhaupt da unten etwas vor sich geht. Vieleicht kraxelt Grbic mit Miss Goldman auf Gradište?" (with the idea that Vulić is digging under our Gradište right now, or that something is happening down there. Perhaps Grbić climbs up to Gradište with Miss Goldman?) (Reiswitz 1933c). German intentions and hopes of continuing the excavations near Ohrid extended into the years preceding the Second World War (Kott 2017: 251–52) as well as after the division of Yugoslavia between the Axis powers. The correspondence between Reiswitz and Unverzagt clearly shows that the new political reality prevented the realization of further plans since in 1941 Lake Ohrid became the demarcation line between Albania and Bulgaria (Reiswitz 1941).

14.6 Conclusions

Grbić's position within the expeditions could be viewed in the context of a German proverb quoted by Hetty Goldman in a letter to Fewkes: "Wer A sagt, muss auch B sagen" (whoever says A must also say B) (Goldman 1932; Bandović 2019a: 96), which in English equates to: In for a penny, in for a pound. Contrary to the spirit of the saying, neither Grbić nor Yugoslav archaeology were mature enough to compete with the foreign expeditions. The Yugoslav archaeological community was strikingly small and in it Grbić did not have a support system. Bearing in mind that Grbić was the only representative of the host side, he certainly was not an equal partner in the archaeological networks formed around the expeditions. On one side stood the foreign archaeologists who brought to Yugoslavia the funds necessary to carry out large-scale archaeological excavations and reconnaissance expeditions as well as up-to-date methods of field research, while on the other side we have delegate Grbić, whose only role was to provide logistical support as an interpreter and intermediary (cf. Kott 2017: 251). The only competition in the rush for Yugoslav sources of archaeological material was between the American and German expeditions.

Within Grbić's social and archaeological networks, different 'stratigraphic layers' can be observed. Conceptually, in the first 'stratigraphic layer', we see that the German and American expeditions did not differ that much, as both were in search of origins. The German expedition was in search of a reflection of the Mycenaean golden masks, chasing "Schliemann's ghost" (Bandović 2019a: 87–95) at Ohrid and attempting to reveal the outlines of the Nordic Mediterranean (*sensu* Chapoutot 2016: 51–97). The German initiative can be seen more broadly in the context of the diplomatic rapprochement of Yugoslavia and Germany after 1931 and the establishment of various Yugoslav-German societies based on economic and ideological unity (cf. Gašić 2005: 23–24, 30–31). On the other hand, the American team was in search of Eurocentric myths about the origins of Neolithic civilizations. Robert Ehrich's 'racial' adventures in Montenegro were not the digression they may seem in this connection. Yet, the political and economic background of the American endeavors remains unexplored. In the next 'stratigraphic layer' of Grbić's network, we can see fragments of inconsistency, intrigue, "banal nationalism" (*sensu* Billig 1995), colonial arrogance and opportunism. In the same 'stratigraphic level', based on daily newspaper articles and echoes from the general public, we can observe that Yugoslav authorities used the presence of the expeditions either to promote the victory of their own politics in Macedonia or as a propaganda framework for sensational discoveries about the origins of European civilization. The Yugoslav authorities tacitly allowed and supported the archaeological competition that developed between these different players. To conclude, all sides had their own interests in the sense of a hybrid (i.e., nationalist and colonialist) form of archaeology (Hamilakis 2013). The power asymmetries can be seen on the level of the personal experience of the individuals involved as well as through the extent of the funding each expedition was willing to invest.

The archaeological networks built around and by Grbić in the 1930s had a strong influence on the future of Yugoslav archaeology as well (Bandović 2019a). For Fewkes, who in 1939 saw the beginning of the war as Europe's historical destiny, as quoted in the beginning of the chapter, the article that he published the same year (Fewkes 1939) was the last that he dedicated to Yugoslavia. He passed away during the war in 1941 (Alden Mason 1942: 114). The American interest in the prehistory of Yugoslavia was on hold until the 1960s and 1970s, when Robert Ehrich returned to Starčevo (see Rasson, Chap. 15, this volume). For Unverzagt, Grbić and Reiswitz, the Second World War provided a new opportunity to collaborate, and they were to meet once again under the walls of the Belgrade fortress in 1942 when they began excavations under the patronage of the Ahnenerbe (Kott 2017: 253–67; Bandović 2019a: 144–65; Roth 2020: 211–323).

References

Primary Sources

Archival Material

Anonymous. (1931, June 20). *Contract with American expedition*. Archives of the National Museum in Belgrade, Serbia.

Anonymous. (1932). *Harvard archaeological expedition to Jugoslavia*. Penn Museum Archives, Philadelphia, PA, United States.

Ehrich, W. R. (1932, March 2). [Letter to Miodrag Grbić]. Fond Miodrag Grbić, Archive of the Serbian Academy of Sciences and Arts, Department in Sremski Karlovci, Serbia.

Fewkes, J. V. (1932b, March 10). [Letter to Miodrag Grbić]. Fond Miodrag Grbić, Archive of the Serbian Academy of Sciences and Arts, Department in Sremski Karlovci, Serbia.

Goldman, H. (1932, March 15). [Letter to Vladimir J. Fewkes]. Penn Museum Archives, Philadelphia, PA, United States.

Grbić, M. (1928, February 16). [Letter to the Ministry of Education]. Archives of the National Museum in Belgrade, Serbia.

Grbić, M. (1932, July 6). Commission report. Archives of the National Museum in Belgrade, Serbia.

Grbić, M. (1933a, September 2). [Letter to Johann Albrecht von Reiswitz, draft]. Fond Miodrag Grbić, Archives of the Serbian Academy of Sciences and Arts, Department in Sremski Karlovci, Serbia.

Grbić, M. (1933b, October 14). [Letter to Radoslav Grujić, draft]. Fond Miodrag Grbić, Archives of the Serbian Academy of Sciences and Arts, Department in Sremski Karlovci, Serbia.

Grbić, M. (1933c, September 24). [Letter to Đorđe Mano-Zisi, draft]. Fond Miodrag Grbić, Archives of the Serbian Academy of Sciences and Arts, Department in Sremski Karlovci, Serbia.

Grbić, M. (1939, April 27). [Letter to the Ministry of Education]. Archives of the National Museum in Belgrade, Serbia.

Reiswitz, A. J. (1930, July 1). [Letter to Miodrag Grbić]. Fond Miodrag Grbić, Archives of the Serbian Academy of Sciences and Arts, Department in Sremski Karlovci, Serbia.

Reiswitz, A. J. (1933a, December 10). [Letter to Erna Bocks]. Bequest of Johann Albrecht von Reiswitz, Archives of the National Museum in Belgrade, Serbia.

Reiswitz, A. J. (1933b, August 8). [Letter to Miodrag Grbić]. Fond Miodrag Grbić, Archives of the Serbian Academy of Sciences and Arts, Department in Sremski Karlovci, Serbia.

Reiswitz, A. J. (1933c, August 6). [Letter to Wilhelm Unverzagt, draft]. Bequest of Johann Albrecht von Reiswitz, Archives of the National Museum in Belgrade, Serbia.

Reiswitz, A. J. (1941). [Letter to Unverzagt, draft]. Bequest of Johann Albrecht von Reiswitz, Archives of the National Museum in Belgrade, Serbia.

Unverzagt, W. (1932a, June 15). [Letter to Miodrag Grbić]. Fond Miodrag Grbić, Archive of the Serbian Academy of Sciences and Art in Sremski Karlovci, Serbia.

Unverzagt, W. (1932b, July 9). [Letter to Miodrag Grbić]. Fond Miodrag Grbić, Archive of the Serbian Academy of Sciences and Art in Sremski Karlovci, Serbia.

Unverzagt, W. (1932c, August 13). [Letter to Miodrag Grbić]. Fond Miodrag Grbić, Archive of the Serbian Academy of Sciences and Art in Sremski Karlovci, Serbia.

Newspaper and Magazine Articles

Anonymous. (1924, December 1). U južnoj Srbiji leži neiskorišćeno ogromno umetničko blago. *Vreme*, 3.

Anonymous. (1927, June 6). U našoj još ne otkrivenoj Pompeji. *Vreme*, 5.

Anonymous. (1931a, August 26). Jedna grupa uglednih američkih arheloga u Beogradu. *Vreme*, 7.

Anonymous. (1931b, May 6). U okolini Ohrida otkrivena su tri preistoriska grada. *Politika*, 5.

Anonymous. (1933a, September 9). Archaeologists' new Mecca. *The Nothingham Evening Post*, 7.

Anonymous. (1933b, September 29). Strani arheolozi proveli su jučerašnji dan i na preistoriskom naselju kod Vinče. *Politika*, 4.

Fewkes, J. V., & Grbič, M. (1931). The prehistory of Yugoslavia. *Discovery*, *1*(8), 247–251.
Krakov, S. (1930, June 6). Tajna sedam kneževskih grobova, punih zlatnih predmeta, kraj Ohridskog Jezera. *Vreme*, 1, 3.
Milojević, P. (1930, July 29). Veliko interesovanje nemačkih arheologa za g. Vulićeva otkrića. *Politika*, 5.
Obrovski, N. N. (1932, September 13). Senzacionalno arheološko otkriće kod Starčeva. *Vreme*, 3.
Reiswitz, A. J. (1932, August 18). Auf den Spuren der Illyrer. *Deutsche Allgemeine Zeitung*

Secondary Sources

Alden Mason, J. (1942). Vladimir J. Fewkes. *American Antiquity*, *8*(1), 114–117.
Babić, S. (2002). Still innocent after all these years?—Sketches for a social history of archaeology in Serbia. In P. F. Biehl, A. Gramsch, & A. Marciniak (Eds.), *Archäologien Europas: Geschichte, Methoden und Theorien/Archaeologies of Europe: History, methods and theories* (Tübinger Archäologische Taschenbücher, 3) (pp. 309–321). Münster, New York, München, Berlin: Waxmann.
Bakić-Hayden, M. (1995). Nesting Orientalism: The case of Former Yugoslavia. *Slavic Review*, *54*(4), 917–931.
Bandović, A. (2019a). *Miodrag Grbić i nastanak kulturno-istorijske arheologije u Srbiji* [Doctoral dissertation, University of Belgrade, Serbia].
Bandović, A. (2019b). Upotreba ljudskog rada kroz istoriju srpske arheologije: tri primera. *Issues in Ethnology Anthropology*, *14*(3), 861–887.
Billig, M. (1995). *Banal nationalism*. Los Angeles, London, New Delhi, Singapore, Washington, DC: SAGE Publications.
Bricker, H. (2002). George Grant MacCurdy: An American pioneer of palaeoanthropology. In D. L. Browman, & S. Williams (Eds.), *New perspectives on the origins of Americanist archaeology* (pp. 265–285). Tuscaloosa: University of Alabama Press.
Browman, D. L., & Williams, S. (2013). *Anthropology at Harvard: A biographical history*. Cambridge: Peabody Museum Press.
Buchtela, K., & Niederle, L. (1910). *Rukověť české archaeologie*. Praha: Jan Laichter.
Casson, S. (1921). The Dorian invasion reviewed in the light of some new evidence. *The Antiquaries Journal*, *1*(3), 199–221. https://doi.org/10.1017/S0003581500002249
Casson, S. (1928). Die archaische Nekropole von Trebenischte am Ochridasee. By Dr Bogdan D. Filow und K. Schkorpil. Pp. 110, figs. 119, Pl. 15. Berlin and Leipzig: de Gruyter, 1927. *The Journal of Hellenic Studies*, *48*(2), 267–270. https://doi.org/10.2307/624984
Čalić, Ž. M. (2010). *Istorija Jugoslavije u XX veku*. Beograd: Clio.
Chapoutot, J. (2016). *Greeks, Romans, Germans: How the Nazis usurped Europe's classical past*. Oakland, CA: University of California Press.
Childe, V. G. (1928). O značaju Srbije za preistoriju Evrope. *Starinar*, *IV*, 45–54.
Childe, V. G. (1929). *The Danube in prehistory*. Oxford: The Clarendon Press.
Coon, S. C. (1939). *The races of Europe*. New York: The Macmillan Company.
Ehrich, W. R. (1933). Preliminary report on an anthropological reconnaissance in Montenegro, season of 1932. *Bulletin of the American School of Prehistoric Research*, *9*, 55–59.
Ehrich, W. R. (1948). George Grant MacCurdy, 1863–1947. *American Antiquity*, *14*(1), 49–50.
Ehrich, W. R. (1974). The Yugoslavian Interlude. In *Symposium in memory of Hetty Goldman* (pp. 38–42). Princeton, NJ: Princeton University Press.
Fehr, H. (2004). Prehistoric archaeology and German *Ostforschung*: The case of the excavations at Zantoch. *Archaeologia Polona*, *42*, 197–228.
Fewkes, J. V. (1932a). Excavations in the Late Neolithic fortress of Homolka in Bohemia. A Preliminary Report. *Proceedings of the American Philosophical Society*, *71*(6), 357–392. http://www.jstor.org/stable/984647
Fewkes, J. V. (1936). Neolithic sites in the Moravo-Danubian area (eastern Yugoslavia). *Bulletin of the American School of Prehistoric Research*, *12*, 5–81.
Fewkes, J. V. (1939). Battle fronts of Europe's war predestined long ago. *The Science News-Letter*, *36*(24), 375–380. https://doi.org/10.2307/3915595
Fewkes, J. V., Goldman, H., & Ehrich, W. R. (1933a). Preliminary report on an archaeological reconnaissance in Yugoslavia, American expedition, season of 1932. *Bulletin of the American School of Prehistoric Research*, *9*, 17–32.
Fewkes, J. V., Goldman, H., & Ehrich, W. R. (1933b). Excavations at Starčevo, Yugoslavia, seasons 1931 and 1932. *Bulletin of American School of Prehistoric Research*, *9*, 33–54.
Filow, B., & Škorpil, K. (1927). *Die archaische Nekropole von Trebenischte am Ochridasee*. Berlin: Walter de Gruyter.
Fotiadis, M. (2001). Imagining Macedonia in prehistory, ca. 1900–1930. *Journal of Mediterranean Archaeology*, *14*(2), 115–135. https://doi.org/10.1558/jmea.v14i2.115
Gačić, D. (2005). *Miodrag Grbić. Život i delo*. Novi Sad: Muzej grada Novog Sada.
Gašić, R. (2005). *Beograd u hodu ka Evropi. Kulturni uticaji Britanije i Nemačke na beogradsku elitu 1918–1941*. Beograd: Institut za savremenu istoriju.
General programme. (1934). In *Proceedings of the First International Congress of Prehistoric and Protohistoric Sciences*, London, August 1–6, 1932 (pp. 9–13). London: Oxford University Press.
Goodman, A., & Hammonds, E. (2000). Reconciling race and human adaptability: Carleton Coon and the persistence of race in scientific discourse. *Kroeber Anthropological Society Papers*, *84*, 28–44.
Gori, M., & Ivanova, M. (2017). *Balkan dialogues: Negotiating identity between prehistory and the present*. In M. Gori, & M. Ivanova (Eds.), *Balkan dialogues: Negotiating identity between prehistory and the present* (Routledge Studies in Archaeology) (pp. 1–14). London and New York: Routledge.
Grbić, M. (1929). *Pločnik. Aeneolithische Ansiedlung, Prähistorische Denkmaler*. Belgrade: Nationalmuseum in Belgrad.
Grbić, M. (1930). Bemalte Keramik aus Starčevo im banater Donaugelände – Jugoslavien. In J. Kostrzewskiego (Ed.), *Księga pamiątkowa ku uczczeniu siedemdziesiątej rocznicy urodzin prof. dr. Włodzimierza Demetrykiewicza* (Bibljoteka Prehistoryczna, 1) (pp. 111–112). Poznan: Nakładem Polskiego Towarzystwa Prehistorycznego.
Grunwald, S. (2019). Scientific capital after 1945 in German archaeology – Wilhelm Unverzagt and the archaeology of hillforts. *Archaeologia Polona*, *50*, 85–109.
Hamilakis, Y. (2010). Nathan Schlanger & Jarl Nordbladh (ed.). Archives, ancestors, practices: archaeology in the light of its history. xx 356 pages, 89 illustrations. 2008. New York & Oxford: Berghahn; 978-1-84545-066-3 paperback £45. *Antiquity*, *84*(325), 893–894. https://doi.org/10.1017/S0003598X00100328
Hamilakis, Y. (2013). Double colonization: The story of the excavations of the Athenian Agora (1924–1931). *Hesperia: The Journal of the American School of Classical Studies at Athens*, *82*(1), 153–177.
Jovanović, V. (2014). *Slika jedne neuspele integracije: Kosovo, Makedonije, Srbija, Jugoslavija*. Beograd: Fabrika knjiga.
Kaeser, A. M. (2008). Biography as microhistory. The relevance of private archives for writing the history of archaeology. In N. Schlanger, & J. Nordbladh (Eds.), *Archives, ancestors, practices: Archaeology*

in the light of its history (pp. 9–20). New York and Oxford: Berghahn Books.

Kershaw, I. (2008). Biography and the historian: Opportunities and constraints. In V. R. Berghahn, & S. Lässig (Eds.), *Biography between structure and agency: Central European lives in international historiography* (Studies in German History, 9) (pp. 27–39). New York: Berghahn Books.

Kott, C. (2017). Kunstschutz im Zeichen des totalen Krieges. Johann Albrecht von Reiswitz und Wilhelm Unverzagt in Serbien, 1941–1944. *Acta Praehistorica et Archaeologica*, *49*, 245–270.

Lutzhöft, H. J. (1971). *Der nordische Gedanke in Deutschland 1920–1940*. Stuttgart: E. Klett.

MacCurdy, G. G. (1929). Old-World prehistory in retrospect and prospect. *Proceedings of the American Philosophical Society*, *68*(2), 95–106.

Marchand, S. (1996). *Down from Olympus: Archaeology and philhellenism in Germany, 1750–1970*. Princeton, NJ: Princeton University Press.

Meinander, C. F. (1981). The concept of culture in European archaeological literature. In G. Daniel (Ed.), *Towards a history of archaeology* (pp. 100–111). London: Thames and Hudson.

Mellink J. M, & Quinn, M. K. (2006). Hetty Goldman, 1881–1972. In G. M. Cohn, & M. S. Joukowsky (Eds.), *Breaking ground: Pioneering women archaeologists* (pp. 298–350). Ann Arbor: University of Michigan Press.

Niederle, L. (1912). G. Kossinna. Die deutsche Vorgeschichte: Eine hervorragend nationale Wissenschaft, Mit 157. Abb im Text, (Würzburg 1912. C. Kabitzsch) Mannus Bibliothek. *Pravek*, *VIII*, 120–121.

Palavestra, A. (2020). *Usamljeni arheolog. Terenski metod Miloja M. Vasića*. Beograd: Univerzitet u Beogradu, Filozofski fakultet.

Putnam, F. W. (1898). *Guide to the Peabody Museum of Harvard University, with a statement relating to instruction in anthropology*. Salem, MA: Peabody Museum of American Archaeology and Ethnology, The Salem Press Company.

Reinecke, P. (1931). Pločnik. Eine Prähistorische Ansiedlung aus der Kupferzeit (Prähistorische Denkmäler, National Museum in Belgrad I). *Wiener Prähistorische Zeitschrift*, *XVIII*, 70–72.

Roth, A. (2020). *Johann Albrecht von Reiswitz (1899–1962): Von unbequemen Südosteuropa Experten zum Kunstschützer*. Graz: Ares Verlag.

Saalman, T. (2017). Wilhelm Unverzagt. In M. Fahlbusch, I. Haar, & A. Pinwinkler (Eds.), *Handbuch der völkischen Wissenschaften. Akteure, Netzwerke, Forschungsprogramme* (pp. 850–853). Berlin and Boston: De Gruyter Oldenbourg. https://doi.org/10.1515/9783110429893

Salač, V. (2006). Abriss der Geschichte der Universitätsarchäologie in Tschechien bis zum Zweiten Weltkrieg. In J. Callmer, M. Meyer, R. Struwe, & C. Theune (Eds.), *Die Anfänge der ur- und frühgeschichtlichen Archäologie als akademisches Fach (1890–1930) im europäischen Vergleich: Internationale Tagung an der Humboldt-Universität zu Berlin vom 13.–16.03.2003/The beginnings of academic pre- and protohistoric archaeology (1980–1930) in a European perspective: International Conference at the Humboldt University of Berlin, March 2003, 13–16* (Berliner Archäologische Forschungen, 2) (pp. 221–236). Rahden/Westf.: Verlag Marie Leidorf.

Schuchhardt, C. (1919). *Alteuropa in seiner Kultur- und Stilentwicklung*. Straßburg: Trübner.

Schuchhardt, C. (1932). Georg Karo. Die Schachtgräber von Mykenai. Text II. Teil, 4°, S. 173–372. München, Bruckmann, 1933. *Praehistorische Zeitschrift*, *23*, 342–345.

Schuchhardt, C. (1934). Die frühesten Herren von Ostdeutschland. *Sitzungsberichte der Preussischen Akademie der Wissenschaften, Philosophisch-Historische Klasse*, *1934*, XXXVII–XLVIII.

Sklenář, K. (1983). *Archaeology in Central Europe: The first 500 years*. New York: Leicester University Press and St. Martin's Press.

Stocký, A. (1912). Poznámky k některým názorům na český neolith v novější německé literatuře. *Pravek*, *VIII*, 26–36.

Stocký, A. (1919). Studie o českém neolithu, I. *Památky archaeologické*, *XXXI*, 67–87, 121–134.

Stocký, A. (1922). Praehistorie a národopis. *Národopisný Věstník Českoslovanský*, *XV*, 12–22.

Todorova, M. (2009). *Imagining the Balkans* (Updated ed.). Oxford: Oxford University Press.

Tompa, F. (1929). *A szalagdíszes agyagművesség kultúrája Magyarországon. A bükki és a tiszai kultúra/Die Bandkeramik in Ungarn. Die Bükker- und die Theiss-Kultur* (Archaeologia Hungarica, 5–6). Budapest: Franklin-Társulat Nyomdája.

Trgovčević, Lj. (2003). *Planirana elita. O studentima iz Srbije na evropskim univerzitetima u 19. veku*. Beograd: Istorijski institut.

Vulić, N. (1930). Das neue Grab von Trebenischte. *Archäologischer Anzeiger*, *III/IV*, 296–299.

Unverzagt, W. (1954). Die Burganlage über dem Kloster St. Erasmo am Ochridasee. *Germania*, *32*, 19–21.

Wiwjorra, I. (2002). „Ex oriente lux"— „Ex septentrione lux": Über den Widerstreit zweier Identitätsmythen. In A. Leube, & M. Hegewisch (Eds.), *Prähistorie und Nationalsozialismus. Die mittel- und osteuropäische Ur- und Frühgeschichtsforschung in den Jahren 1933–1945* (Studie zur Wissenschafts- und Universitätsgeschichte, 2) (pp. 73–106). Heidelberg: Synchron Verlag.

When Paradigms Meet: American and Yugoslav Archaeologists and the Neolithic in the 1960s and 1970s

15

Judith A. Rasson

Abstract

In the late 1960s, American and Yugoslav archaeologists began a series of Neolithic-period excavations that opened research in the Balkans to Americans. The first projects were funded by the Foreign Currency (Blocked Funds) Program of the Smithsonian Institution. Few scientific strings were attached apart from the fact that excavations had to be joint ventures; no team could work independently. The Americans adopted the already-established Neolithic culture history approach dominant in Yugoslav archaeology, although many of the Americans (mainly graduate students) were trained in the methods of processual archaeology. Differing theoretical orientations between Americans and Yugoslavs were potential sources of conflict but had relatively little impact during the excavations. Differences of opinion arose over excavation techniques and data management, which reflected different theoretical objectives. The long-term consequences of these joint excavations were to solidify the understanding of domesticated plants and animals in the Balkans—a bridge linking Europe with earlier Neolithic cultures in Anatolia—and to increase communication between American and European archaeologists.

Keywords

Bosnia · Serbia · Macedonia · Yugoslavia · Blocked funds · Neolithic · Networks · Research paradigms

J. A. Rasson (✉)
Department of Medieval Studies, retired, Central European University, Budapest, Hungary

15.1 Introduction

In the late 1960s, the United States government offered American scholars the opportunity to conduct research in a number of countries worldwide. This was the Cold War period, when the American government invested in programs such as the Peace Corps and the Fulbright program to build friendly relations around the world. Funding for Eastern Europe came from the Blocked Funds Program administered by the Smithsonian Institution (see McPherron and Srejović 1988: xiii). This program, which required participation by researchers from both the host nation and the United States, had lasting effects on American and Yugoslav archaeologists. Impacts extend to the present and include long-lasting collegial networks across national borders. The term Yugoslavia (i.e., Socialist Federal Republic of Yugoslavia) is used here because in the 1960s it was a country comprised of six republics (Bosnia and Herzegovina, Croatia, Macedonia, Montenegro, Serbia, and Slovenia). Its political stance (non-aligned socialism) was important in the availability of research funding. This chapter notes the socio-cultural and academic background, discusses the earliest joint excavations, compares the research paradigms and data management practices, and describes the networks that resulted from these projects.

This opportunity for joint projects between the United States and Yugoslavia was rather unexpected. The archaeologists from each country contributed research questions and methods that they regarded as normal, but which ultimately required negotiation, often on several levels, from the research objectives, excavation methods, and curation of finds to managing the daily life of the excavation team. Encounters between groups generally require managing power through means such as negotiation (discussed in Faier and Rofel 2014; see further references in Coltofean et al., Chap. 1, this volume). It was a critical point that on all the joint excavations each partner had control (power) over important aspects of the excavation. The Americans brought

L. Coltofean et al. (eds.), *Connecting People and Ideas*, Themes in Contemporary Archaeology,
https://doi.org/10.1007/978-3-031-81006-0_15

the funding and access to resources like Carbon-14 dating and the Yugoslavs controlled local infrastructure like museum services, a labor pool of workmen, and arrangements for daily life. The money from the Blocked Funds program was distributed among members of the community: workmen on the excavation, payments to the families that owned the land where the sites were located, students and archaeologists participating in the excavation, museum workers, and museums as institutions.

It may be hard for scholars in the 2020s to understand that in the absence of digital media and the internet everything had to be done by hand. Communication was through regular mail services; telephone calls were prone to monitoring by the authorities in the socialist countries, and there were no cell phones. English was not widely spoken, especially among the older generation of Yugoslav archaeologists, who mainly spoke German, and few American archaeologists were able to read the languages of Eastern Europe. Yugoslavia and other Eastern European countries had a far-reaching network of institutions that exchanged publications in their own languages, thus avoiding problems of currency exchange, but few institutions in the United States participated and the language barrier impeded access to current research in Eastern Europe for American researchers. Eventually, academic institutions in the United States developed basic collections of Yugoslav publications (Rasson forthcoming).

Black and white film was available in Yugoslavia, but photography in the field required developing the film, usually at the cooperating local museum. Computer-assisted data processing required punch cards, in use into the 1990s (Fig. 15.1); each card allowed for 80 columns of compressed information. If one card were devoted to each artifact from a site thousands of cards might be required (e.g., the American excavation units at the Obre sites yielded 102,542 ceramic artifacts [Sterud and Sterud 1974: 161]). Preparing text for publication required a typewriter, laborious correction of manuscript changes, and inserting diacritical marks by hand. Perhaps to avoid self-published (in Serbo-Croatian/Russian: *samizdat*) publications or because they had to be purchased with hard currency, no photocopiers were available in Yugoslavia at the time, which meant that fieldnotes had to be copied in longhand. There was no GPS (global positioning system), no access to modern topographic maps or aerial photographs, which were controlled by the military. In my own work, I used nineteenth-century maps (1:25000 scale) of the Austro-Hungarian Military Survey (1869–1887) (since then available on-line: Arcanum Maps n.d.).

Archaeological research in Yugoslavia was conducted through museums and universities (Lorber and Novaković 2020; Milosavljević 2020; see also Lorber, Chap. 13, this volume). Funding was relatively meager in the 1960s; Yugoslavia at the time was in the process of moving from an agricultural to an industrial economy (see Rasson et al. 1999). There was a national drive to develop infrastructure and to provide consumer goods and a definite desire among the population to adopt a middle-class lifestyle with televisions, refrigerators, and automobiles (see Aranđelović 1967; Hammel 1969). Yugoslavia's non-aligned political stance allowed it to be among the first Communist-bloc countries to import Western goods, notably Pepsi Cola in the late 1960s and Levi's jeans in the 1970s.

Fig. 15.1 Timothy Kaiser keypunching data cards for later computer analysis for Selevac material in Smederevska Palanka during the excavation, fortified by coffee, beer, and a cigarette; Selavac excavation, c. 1977. (Photo: .Judith Rasson)

The issues that arose between the American and Yugoslav participants during and after the various excavation projects included research methods, data management, and protocols for sharing information. The short-term results of these encounters were cross-cultural archaeological awareness, a network of contacts in the United States and Yugoslavia and beyond into wider European archaeological scholarly (and multilingual) networks. The Blocked Funds excavations were an excellent opportunity for American students to collect data for theses and dissertations (Fig. 15.2). The long-term practical outcomes included the education of a number of American archaeologists specializing in the prehistory of Eastern Europe, the development of an English-language literature on the archaeology of the region, the application of

Anthony, David W. 1985. *The Social and Economic Implications of the Domestication of the Horse*. University of Pennsylvania.
Elster, Ernestine S. 1977. *Neolithic Technology: A Case Study in Lithic Analysis for Old Europe, 6500–4000 BC*. University of California Los Angeles.
Evans, Robert K. 1973. *Craft Specialization in the Chalcolithic Period of the Eastern Portion of the Balkan Peninsula*. University of California Los Angeles.
Gardner, Elizabeth J. 1972. *The Technological Development of Ceramics of the Neolithic Period of Macedonia*. University of California Los Angeles.
Glumac, Petar D. 1992. *The Advent of Metallurgy in Prehistoric Southeast Europe*. University of California Berkeley.
Greenfield, Haskel J. 1984. *The Paleoeconomy of the Central Balkans (Serbia): a Zooarchaeological Perspective on the Late Neolithic and Bronze Age (ca. 4500-1000 BC)*. City University of New York.
Kaiser, Timothy. 1984. *Vinča Ceramics: Economic and Technoogical Aspects of Late Neolithic Pottery Production in Southeast Europe*. University of California Berkeley.
Mallory, James P. 1975. *The Indo-European Homeland Problem: The Logic of the Inquiry*. University of California Los Angeles.
Mock, Renita. 1972. *The Neolithic of Upland Western Macedonia*. MA thesis, University of California Los Angeles.
Prinz, Beth. 1986. *Mesolithic Adaptations on the Lower Danube. Vlasac and the Iron Gates Gorge*, University of Pittsburgh.
Rasson, Judith A. 1984. *Interaction Spheres as Adaptive Mechanisms; Bosnian-Dalmatian Relations in the Neolithic*. State University of New York at Binghamton.
Russell, Nerissa. 1993. *Hunting, Herding and Feasting: Human Use of Animals in Neolithic Southeast Europe*. University of California Berkeley.
Schwartz, Charles A. 1989. *Neolithic Cattle Exploitation in the Balkans*. Hungarian Academy of Sciences, Budapest.
Stefanovich, Mark. 1989. *The Archaeology of Reification, Continuity and Interpretation*. University of California Los Angeles.
Sterud, Eugene L. 1976. *The Kakanj/Butmir Transition at Obre: A Case Study in the Use of Quantitative Analysis*. University of California Los Angeles.
Stevanović, Mirijana. 1996. *The Age of Clay: the Social Dynamics of House Destruction*. University of California Berkeley.
Voytek, Barbara. 1985. *The Exploitation of Lithic Resources in Neolithic Southeast Europe*. University of California Berkeley.

Fig. 15.2 Degrees in the United States using data from former Yugoslavia (PhD unless otherwise indicated; original titles)

anthropological theories to Eastern European prehistoric data, a better appreciation for the importance of Southeast Europe in the dispersal of agriculture, and contributions to the development of an international network of archaeologists (i.e., the European Association of Archaeologists [EAA], founded in 1994).

There was a time gap between the earliest American archaeological research in Yugoslavia and the Blocked Funds excavations. In the 1930s, Vladimir J. Fewkes (1901–1941), originally from Czechoslovakia and director of the American Archaeological Expedition to Europe (1934–1939), conducted extensive archaeological research in Yugoslavia, aided by his student, Robert W. Ehrich (1908–1992) (see more about the details and context in Bandović, Chap. 14, this volume). Among others, they excavated at the two principal Neolithic sites in Serbia, Starčevo (Early Neolithic) (Ehrich 1933; Fewkes et al. 1933; Novaković 2021: 161) and Vinča-Belo Brdo (Late Neolithic) (Fewkes 1935), but their research was not integrated into the main body of archaeological knowledge in the United States even though it was published there. Most American archaeologists were deeply involved in establishing a culture-historical framework for the New World, for instance, Willey and Phillips (1958).

After the Second World War, the political situation made international archaeology difficult. Ehrich, teaching at City University of New York, edited multiple versions of

Chronologies in Old World Archaeology (called *Relative Chronologies* before Carbon-14 dating; Ehrich 1954), showing the perennial need to construct time for the past. In 1969 and 1970, Ehrich returned to excavate at Starčevo (Ehrich 1977; Clason 1980). He was an adjunct to the developing network of archaeologists, offering advice and attending networking events in the United States like the Hleb i Vino (Bread and Wine) conferences (discussed below).

The Blocked Funds program also supported significant excavations at the Classical sites of Sirmium in Serbia (Popović and Ochenschlager 1971–1980) and Stobi in North Macedonia (Wiseman and Mano-Zissi 1973–1981). Classical archaeology was greatly informed by its own long research history, the availability of written records from the past, and the nature of the remains—large settlements, stone architecture, and infrastructure. Prehistoric archaeology addressed the past in different ways. Although the teams from the different projects knew about the others (the excavation team from Anza visited the Stobi excavations, for instance), there was almost no interaction.

15.2 Neolithic Excavations Under the Blocked Funds Program

In the United States in the 1960s, interest in the origins and spread of agriculture in the Near East had been raised by the University of Chicago archaeologist Robert Braidwood (1907–2003) (Braidwood 1960), who proposed that agriculture developed in the foothills of the Taurus and Zagros mountains (the Hilly Flanks theory). Archaeologists from outside Eastern Europe saw the potential of Neolithic archaeology for understanding the spread of early agriculture further west, especially with the addition of corrected Carbon-14 dating (C. Renfrew 1973). The Early Neolithic site of Nea Nikomedeia in Greece, for example, excavated from 1961 to 1964 (Rodden et al. 1996), sparked interest in questions about early agriculture in Europe. All of the Blocked Funds excavations in Yugoslavia and the roughly contemporaneous projects at Sitagroi and Achilleion in Greece were at Neolithic sites.

There were few specialists in European prehistoric archaeology in the United States at that time. Two of them were at the University of California Los Angeles (UCLA): James Sackett (1933–2019) in the Anthropology Department (who dealt with the Paleolithic in France), and Marija Gimbutas (1921–1994) in the Department of Indo-European Studies. Although they shared students, they approached prehistory quite differently...

Originally from Lithuania, Gimbutas earned a master's degree in Lithuania and a doctoral degree in Germany at the University of Tübingen. After she moved to the United States in 1949, she began to work with Russian-language sources at Harvard, which opened a view of Soviet archaeology in English (Gimbutas 1956, 1965). She developed the "kurgan (burial mound) hypothesis" that Eurasian steppe inhabitants had spread Indo-European languages through migrations in prehistoric times (reviewed in Anthony et al. 1986). This idea was later pursued by her student, James P. Mallory (1989), and others, like David W. Anthony (2010), both of whom participated in Neolithic excavations in Yugoslavia. Renfrew (1987) also addressed this hypothesis and more recent DNA research seems to support some aspects of this model (Jacobson 2018).

The first Blocked Funds' excavation took place in 1967 and 1968 in Bosnia at Obre (two sites usually referred to as a unit), directed by Gimbutas and Alojz Benac (1914–1992), the most important archaeologist in Bosnia at the time. These excavations were followed in 1968 and 1969 by those in Serbia at the site of Divostin, directed by Alan McPherron (1929–2020) from the University of Pittsburgh and Dragoslav Srejović (1931–1996) from the University of Belgrade. Srejović (see Novaković 2021) had begun a rescue excavation on the bank of the Danube at the site of Lepenski Vir (*vir* means whirlpool), which was slated to be flooded by a new hydroelectric dam. The surface remains indicated nothing out of the ordinary, but below lay a spectacular pre-Neolithic fishing settlement (Srejović 1969, 1972; Prinz 1987; Radovanović 1996; Radovanović and Voytek 1997; Borić 2014). In 1969 and 1970, Gimbutas co-directed excavations in Macedonia at the Neolithic site of Anza with archaeologists Milutin Garašanin (1920–2002) from Serbia and Vojislav Sanev (1938–2007) from Macedonia. In parallel, the archaeologist Sarunas Milisauskas (originally from Lithuania), whose career was spent at the State University of New York at Buffalo, began a Blocked Funds Neolithic excavation in Poland (Milisauskas 1986) and went on to work there for a number of years (Milisauskas et al. 2010).

Networking began in 1967, when British archaeologist Colin Renfrew, then teaching at the University of Sheffield, was in residence at UCLA, invited by Gimbutas. He lectured there about using obsidian analysis to reconstruct prehistoric cultural contacts (Renfrew et al. 1966). Archaeologists Eugene Sterud (who had excavated an obsidian quarry in California) and Robert K. Evans attended his class and I also visited (Evans and Rasson 1984). After the excavations at Nea Nikomedeia, Robert J. Rodden began teaching at the University of California Berkeley in the late 1960s and was in contact with Gimbutas. One of his students, Renita Mock (1946–2015; see Fig. 15.2), transferred to UCLA in 1972 to study with Gimbutas and participated in the Anza excavations.

In 1968, Renfrew, in cooperation with Gimbutas but funded by British sources, began excavations at the Neolithic site of Sitagroi in Greece (Renfrew et al. 1986; Renfrew and Elster 2003). Some of the participants in these projects, such

Fig. 15.3 A discussion of field methods. The author and Borislav Jovanović at Selevac, 1976. (Photo: Carol Spears; reproduced with the permission of Ruth Tringham)

as then-students Robert K. Evans (1984; see also Fig. 15.2), Ernestine Elster (1976; see also Fig. 15.2), and Elizabeth Gardner (1976; see also Fig. 15.2), became part of the network of those working in Yugoslavia or attached at a distance, but most of them remained "outliers" owing to their own research interests elsewhere or personal situations (spouses and children). Renfrew went on to other projects after the 1960s excavations, while Gimbutas (1986a, b) continued to work at the Neolithic site of Achilleion in Greece. At the same time, networks were also developing in the United Kingdom. Eric Higgs (1908–1976) began the Cambridge Palaeoeconomy School; Anthony Legge (1939–2013), a student of Higgs, later analyzed the faunal material from Selevac in Serbia (Legge 1990). John Chapman (1981) researched the Vinča archaeological culture in Serbia and directed the "Neothermal Dalmatia Project" in Croatia (Chapman et al. 1996). Chapman's memoir (2021) is informative about British work in the Balkans in this period.

The network of archaeologists from the Blocked Funds projects was augmented by others from joint projects at other sites in Serbia that continued after the end of the Blocked Funds Program, notably Ruth Tringham and her students, first at Selevac (Tringham and Krstić 1990) and later at Opovo (Tringham et al. 1992). Over time, other projects were conducted elsewhere, in Serbia (e.g., Bartel et al. 1979; Bankoff and Winter 1990; Arnold and Greenfield 2006), in Croatia (e.g., Forenbaher et al. 1994, 2013), and in Bosnia (Whallon 1989).

Proximity played a role in transferring information and developing networks. Most Americans had limited contact with the academic establishments in Belgrade and Skopje (there was no department of archaeology in Sarajevo in the republic of Bosnia and Herzegovina). People met each other while working at the same site; they were there during the summer when there were no classes; most participants could not speak the language of the others, depended on bilingual team members for translation, and developed their own rudimentary vocabularies for fieldwork. The exchange of knowledge was informal, incidental, and incremental. Later, the EAA enhanced communication among archaeologists from the United States, the United Kingdom, and former Yugoslavia.

15.3 Joint Neolithic Excavations in Bosnia, Macedonia, and Serbia

The earliest Blocked Fund excavations in Eastern Europe took place in Bosnia at two deeply stratified adjacent sites, Obre I (Raskršće) and Obre II (Gornje Polje). They had been tested earlier in the 1960s and were pristine in the sense that there was no later occupation over the Neolithic layers except for modern farming (with horse-drawn plows). Between them the two sites spanned the complete chronological sequence of the Neolithic in Bosnia (Benac 1973a, b).

Neither Gimbutas nor Sterud, who became involved after Renfrew's lectures, was in the United States from 1967 through 1968. Working on his PhD dissertation on the Obre material in Sarajevo (with a Ford Foundation grant), Sterud recruited most of the students at Obre by snail mail from the UCLA Archaeological Survey, the archaeological training unit of the Department of Anthropology, including the author, Joan Carpenter (Grasty), Robert K. Evans, and Charles A. Schwartz. Gimbutas also invited her student, James Mallory. Other members of the team were the Americans Deborah Gates and Mary Dawes (both from UCLA), four of Renfrew's students from Sheffield, and Anne Bohlin (Mrs. Eugene) Sterud, a Swedish-trained archeologist. Ironically, the American team consisted of Gimbutas and a group of unknown graduate students, but the Yugoslav teams for both seasons were a Who's Who of Yugoslav archaeology assembled by Benac: Djuro Basler (1917–1990) from the Zemaljski Muzej in Sarajevo, Bosnia; Bogdan Brukner (1931–2006) from the University of Novi Sad, Serbia; Branko Belić from the Zavičajni Muzej in Doboj, Bosnia; Ivan Marović (1920–2014) from the Arheološki

Muzej in Split, Croatia; Katarina Petrović, director of the museum in Slavonski Brod, Croatia; Šime Batović (1927–2016) from the Archaeological Museum in Zadar, Croatia; and Tatjana Bregant (1932–2002) from the University of Ljubljana, Slovenia. Benac also invited two students: Irena Rajterič, a student of Bregant's from the University of Ljubljana and Slobodan Čače (1946–2020) from Croatia, recommended by Batović. Čače went on to conduct research on islands in the Adriatic Sea in the Classical period; we corresponded intermittently over the years, and I was happy to meet him again in the 2000s at the University of Zadar, where he became a well-regarded professor. Two Czech archaeologists,[1] Juraj Pavúk (Pavúk and Bakamska 2018) and Viera Němejcová-Pavúk, were also invited to participate.

The Neolithic site of Divostin in Serbia also had modern occupation on the surface, but parts of it remained relatively undisturbed. It was surveyed with geomagnetic methods (the first time they were used in Yugoslavia [Novaković 2021]) that revealed the location of burnt Neolithic house remains (Mužijević and Ralph 1988). The Americans at Divostin, recruited by McPherron, were Nicholas Gessler,[2] Gloria Y'Ednyak (coming from a Stanford University Blocked Funds project in Bosnia [Belošević n.d.]), Gregory Johnson[3] (coming from Milisauskas' project in Poland), Birutė Galdikas,[4] John Kenagy, and Deborah Gates (all three graduate students in anthropology at UCLA), and the author, the last two coming from the excavations at Obre. The local museum in Kragujevac provided staff and Srejović invited his own students, among others Vladimir Stančić and Mila Popović, my personal friend, who made her career at the Narodni Muzej in Belgrade.

Fig. 15.4 Marija Gimbutas and Milutin Garašanin at Anza, 1970. (Photo: Judith Rasson)

Macedonia has a rich archaeological heritage (M. Garašanin et al. 1971; Bugaj et al. 2014). The large site of Anza, on fields belonging to the village of Amzabegovo, had both Neolithic and Classical remains. Only the Neolithic was investigated with Blocked Funds.The project was directed by Gimbutas, M. Garašanin (Fig. 15.4), and Vojislav Sanev (1938–2007), director of the museum in the nearby town of Štip. The other participants included D. Garašanin (1921–1997) and four Macedonian archaeologists: Saržo Saržoski from the Archaeological Museum in Skopje; Dragica Simoska (1931–2019) from the Naroden Muzej in Bitola, and students from Belgrade, Mark Stefanovich (see Fig. 15.2), and Milan Ivanovski (d. 2009), who went on to work for the state cultural heritage organization in Macedonia (Fig. 15.5).The Americans at Anza were all from California: Renita Mock, Joan Carpenter (Grasty), Charles A. Schwartz, Eugene L. Sterud, David Weide (geologist), and the author, all students, except Peter Banks, a cultural resource archaeologist, and Margaret (Weide) Lyneis (1938–2018),[5] then professor of anthropology at California State College Long Beach. Two of Gimbutas' students from Indo-European Studies at UCLA participated, Nancy Brandt and Linda Mount-Williams. They were joined by several British students: Gerald Ian Thomas, Geoffrey Sayre (from the Obre excavation), Robert W. Chapman (now retired), Stephen Shennan (now retired), and Susan Shennan. Two French archaeologists also participated: Jean-Paul De Moule and

[1]Czechoslovakia at the time was progressive in archaeological research (see Neustupný 2016).

[2]Currently at Duke University, United States: https://people.duke.edu/~ng46/

[3]Later a professor at Hunter College of the City University of New York; he met Jasmina Drašković at the Divostin excavation and married her; see also Chapman 2021: 37.

[4]Currently primatologist, conservationist, and educator at Orangutan Foundation International: https://orangutan.org/about/dr-birute-mary-galdikas

[5]That the Balkans continue to confuse Americans is evident from a remark in Lyneis' obituary that "She even had one publication (Lyneis 1988) on antler and bone artifacts from a Neolithic site in Siberia," which sounds like Serbia to the uninitiated (Harry et al. 2019: 283).

Fig. 15.5 The Serbian-Macedonian team taking a break at the Anza excavations, 1970. From left to right: Milutin Garašanin, Dragica Simoska, Vojislav Sanev, Milan Ivanovski, Draga Garašanin. (Photo: Judith Rasson)

Serge Cleuziou (1945–2009), who had both worked at Dikili Tash, a French-Greek project.[6]

The Americans working at Selevac (some of them intermittently) were mainly graduate students: the author, Barbara Voytek, Timothy Kaiser, Nerissa Russell, Petar Glumac (see Fig. 15.2), David W. Anthony, Cheryl Claassen, and Michael Dietler, except Eugene Sterud, then teaching at the State University of New York at Binghamton, and Carol Spears, an American cultural resource archaeologist who was the project photographer, along with a few others including British archaeologist John Chapman (1981, 2020, 2021). Dušan Krstić, co-director of the excavations at Selevac, invited participants from Serbia, including Borivoj Jovanović from the Institute of Archaeology in Belgrade and graduate students Mirijana Stevanović, Mirjana Vukmanović, and Nina Ilić, among others. Ivana Radovanović, also a graduate student in Belgrade, heard about the excavation and was accepted by Tringham. These lists of participants are reconstructed from memory. Documentary evidence consists mainly of scant notes in the final site reports.[7] After the excavations, many students went on to other professions or conducted archaeological research in other periods and places. Those who invested in learning the other's language (the author, McPherron, Tringham, Voytek, Kaiser, Radovanović, and Stevanović—Glumac and Stefanovich are native speakers of Serbian) were more likely to become part of the network of archaeologists conducting research in Yugoslavia. Those who did not are more ephemeral in the literature.

During the 1968 Obre excavation, Gimbutas organized what she called "the university of Obre," inviting the professional archaeologists at the project and visiting specialists like the Hungarian archaeozoologist Sándor Bökönyi (1926–1994), from the Archaeological Institute of the Hungarian Academy of Sciences in Budapest, to give evening lectures. These brief presentations revealed in a nutshell the different frameworks of interpretation among the professionals. Benac spoke about culture change as arising from migration and autochthonous development, while Brukner attributed culture change to diffusion. Bökönyi spoke on the origins of cattle domestication. These presentations were a quick low-key introduction to current archaeological thinking.

Depending on the availability of suitable local housing, on some excavations the teams shared accommodations (e.g., Obre and Selevac). On others (e.g., Divostin and Anza) the teams lived separately, often renting rooms in village homes, usually gathering for a shared mid-day dinner (in Serbian: *ručak*). At the Obre excavations everyone ate *ručak* together on the veranda of our residence (a worker's hostel), which encouraged collegiality.

Joint excavations had other events like parties, barbecues, and visits to other sites. Srejović, for example, invited the Divostin team to visit Lepenski Vir before it was flooded, which was a rare opportunity. There were no dedicated intellectual exchanges, however, at least partly because the teams had no central gathering place and not all participants were bilingual; the novelty of encounter had waned. Of course, informal discussions (even arguments) took place. I remember trying to translate the essence of Lewis Binford's "doc-

[6]The Dikili Tash French-Greek project: http://www.dikili-tash.fr/content_en

[7]For example, I am not listed in either Benac's (1973a: 5) publication on Obre or McPherron and Srejović's (1988: xi) work on Divostin, even though I participated in both projects.

trine" to my teammates at Divostin in broken Serbo-Croatian (as the official language of Yugoslavia was called then).

There were occasional conflicts over field methods, such as how to define an intrusive pit (Fig. 15.3) and whether to discard plain ceramic wares before tabulation (at Divostin in the first season, although this was quickly resolved). At Obre and Anza, many potential points of conflict were solved by avoidance; each team had separate excavation units with attendant artifact processing and, ultimately, separate final reports, which could be regarded as two different treatments of the same site (Benac 1973a, b; Gimbutas 1974, 1976). At Divostin, and later at Selevac, the fieldwork was conducted by mixed teams of Americans and Yugoslavs and both the artifact processing and final site reports were combined, despite different theoretical and methodological approaches:

> Dragan Srejović and I had decided that the project should be a truly joint one throughout. There was to be a single unified workforce. (...) The project was designed as a genuine effort at international cooperation. Neither Dragan nor I would claim that the task was made easier by such an aim. There are fundamental and firmly ingrained differences between the two parts of the world in the philosophy of archaeology. (McPherron and Srejović 1988: xiii)

On the early excavations, none of the American excavators except McPherron spoke or read Serbo-Croatian. There was extensive literature on the Neolithic in Serbian: Miloje Vasić (1932–1936), Milutin Garašanin (1951), and Draga Garašanin (1954); in German: Vladimir Milojčić (1949 and later); and even in English: Stuart Piggott (1965); plus Fewkes' work from the 1930s. Chronology is always an issue among archaeologists; in Europe, besides stratigraphy and typology, the utility of Carbon-14 dating was debated at the time (exemplified by Mellaart 1960 and Garašanin 1961), with some, like János Makkay (1985) rejecting all radiocarbon dates.

The pivotal importance of the Balkans as a bridge between Anatolia and Western Europe had not yet been acknowledged, which changed after the joint American-Yugoslav excavations, thanks in part to the use of Carbon-14 dating, more systematic archaeobotanical and archaeozoological studies, and the application of different models of prehistory.

15.4 Research Paradigms

All the participants in the joint excavations were interested in, even passionate about, explaining the prehistoric past, but their perspectives varied. In the United States, developmental sequences of archaeologically defined "cultures," horizons, or zones provided the main structure for research (Swartz 1996), relying on the attributes and distribution of artifacts as well as stratigraphy and ethnographic analogies to Native American cultures. Even though Marxism was the official theory in the wider socio-political establishment in Yugoslavia (and in Hungary and elsewhere in the Communist Bloc; see Neustupný 2016), archaeologists did not see that it had potential for explaining prehistory (Novaković 2011: 398); Marxist interpretations were largely pro forma (Novaković 2023). The Yugoslavs preferred to work with a traditional culture-area concept framed as core-periphery relations (Novaković 2011: 342) that was anchored in stratigraphic sequences to define prehistoric cultures by ceramic styles and other cultural features. There were relatively few archaeologists in Yugoslavia; those in academia and museums knew each other and had gone to the university or done fieldwork together (e.g., Novaković 2021: 329–31). Their shared common knowledge and communication through publications were more influential than abstract "texts focused on presenting the advances in archaeological knowledge" (Novaković 2011: 349), which would probably have been seen as pretentious.

Gimbutas (1974, 1991, see Chapman 1998, and Elster 2007), an outsider to American archaeological developments, took an individual and somewhat idiosyncratic approach to culture history. Recognizing that the archaeological record reflects supernatural beliefs as well as social and economic behavior, she developed a vision of prehistoric religion in "Old Europe" that made it logical to look for evidence through archaeology. She was an avid visitor of regional museums in Southeast Europe and, once seen, never forgot a figurine. She was enthusiastic about the promise of Carbon-14 dating and had samples using tree-ring calibrated dates analyzed from each site where she worked (Ferguson et al. 1976). She translated her passion for understanding the past into successful grant applications. The opportunities of the Blocked Funds Program, fortuitously, coincided with rising interest in the spread of domesticated plants and animals from the Near East to Europe.

Theoretical orientation became important to American archaeology in the 1960s, thanks to the vociferous proselytizing of Lewis Binford (1931–2011) (Binford 1962; Binford and Binford 1968, and others), James Hill (1934–1997), and Fred Plog (1944–1992), all teaching at UCLA. This shifted attention away from defining and refining prehistoric culture areas, so-called "space-time systematics," toward self-consciously scientific and evolutionary investigations of past cultures, self-identifying a paradigm shift following the work of "neoevolutionary" anthropologist Leslie White (1900–1975) (White 1949) and the philosopher of science Thomas Kuhn (1922–1996) (Kuhn 1962). New Archaeology (later called Processual Archaeology) posed specific anthropological questions before excavation to understand behavioral patterns in the past. Research questions were framed as hypotheses with attendant test implications. Trying to apply the principles of New Archaeology contributed to some of the tension between Americans and Yugoslavs. Many years later, in 1985, Binford did lecture at the University of Ljubljana (Slovenia being the most Western-oriented Yugoslav republic); even then "one would hardly say that Binford's lecturing had an immediate impact (...) [it was] discourse on a completely different wavelength" (Novaković 2015: 124–25). Several of the American students who went to Yugoslavia in 1968, including the author, attended Binford's classes at UCLA. I had never been attracted by culture history because it seemed to be just lists of artifacts. New Archaeology's approach was focused on people and what they did, the processes behind culture history. My experience in Neolithic archaeology in Yugoslavia would have been different (but no less enthusiastic) had I not attended his class.

Alan McPherron earned his PhD at the University of Michigan and was part of the first wave of "New" archaeologists that included Binford, Albert Spaulding (1914–1990), Robert Whallon (b. 1940), and H. Martin Wobst (b. 1943). Archaeology in the United Kingdom followed similar developmental patterns that grew out of traditional archaeology. The work of David Clarke (1937–1976) (Clarke 1968) was particularly influential and also of interest to Americans (Sterud 1978). Ruth Tringham (b. 1940) with a British background in culture history, moved from the United Kingdom to the United States (Tringham 1971). Initially she approached the past through experimental archaeology, the study of microwear on chipped stone artifacts (Tringham et al. 1974) and went on to address more theoretical topics (Tringham 1991; Tringham and Stevanović 2012), including digital archaeology and gender (e.g., Conkey and Tringham 1996).

In the late 1960s, ideas about interpreting prehistory were also changing in Yugoslavia, although M. Garašanin "and Srejović both remained quite faithful to culture history and their authority and reputation in academic archaeology had an impact on the rather late appearance of processual and postprocessual ideas in Serbian archaeology" (Novaković 2011: 398; 2021: 177; see also Lorber and Novaković 2020). Alternative views by outsiders, such as John Nandriș (who visited the Anza excavations in 1969) bio-social interpretation of Lepenski Vir (Nandris 1968) may not have had great impact. In Bosnia, where Benac was prominent,

> the study of the Neolithic Period was treated with the standard interpretative apparatus (e.g., migrationist theories to explain neolithization, pottery forms and decorations used as basis for distinguishing regional groupings (i.e., cultures), detailed chronological analyses, 'historical' interpretation of social processes, cross-cultural comparisons. (Novaković 2011: 409; 2021: 239)

The Neolithic chronology was anchored in the large, stratified Neolithic (and later) site of Vinča on the banks of the Danube in Serbia, excavated in 1908 and from 1929 to 1931 by Miloje Vasić (1869–1956). Accumulating evidence in later years suggested that Vasić's interpretation needed revision (Novaković 2011: 391). Milutin Garašanin (1951) re-evaluated the Late Neolithic (Vinča) material and Draga Garašanin (1954) the Early Neolithic (Starčevo) material. They later became part of the network that developed among the participants in the Blocked Funds excavations in former Yugoslavia.

On the Blocked Fund excavations in Yugoslavia, Americans adopted the already-defined cultural chronology and most of the local terminology as the most efficient approach. Discovery played a role in learning about the Neolithic. Excavating at Obre I, for example, I became familiar with features of the Neolithic in Bosnia, including black burnished ware, barbotine ("roughened") ceramics, and so-called rhytons (distinctive four-footed ceramic vessels with loop handles) similar to those found along the Adriatic littoral. This confirmed the already-developed Yugoslav cultural chronology, but I decided to explore the cultural mechanisms behind the chronology using the methods of processual archaeology and a model of cultural interaction (interaction spheres) from the American Midwest that also included the role of transhumant herding and exchange during the Neolithic (see Fig. 15.2).

In artifact analysis the material had to be wrenched from culture history and reassembled in a new paradigm. I adapted my research questions to the available materials, mainly ceramics from museum holdings, which did not lend themselves to accurate statistical analysis. I do not recall much discussion with my Yugoslav peers; for them the basic notion of establishing patterns of interaction may have seemed merely a refinement of the existing sequence of archaeological cultures. We argued more vigorously about whether calibrated Carbon-14 dates could be relied on.

15.5 Data Management

The basic excavation techniques were similar in the United States and Yugoslavia. Artifact recovery was controlled by standard-sized excavation units (in Serbo-Croatian: *sonda*) and by digging in artificial levels. The rationale for the American methods at the Obre excavations was the necessity to collect information to be used for "statistics." This was shorthand and everyone (Americans and Yugoslavs) had an idea of what it meant even if it was not exactly clear what operations would be performed or why. Such analyses depended on computerized data handling. Using statistics thus called for a different kind of management in collecting and analyzing information, which required a more disciplined approach to collecting and processing artifacts. I do not believe that this has yet been achieved in much of former Yugoslavia.

Ultimately, Sterud (1978) used statistics for his analysis of the Obre material and I used the non-parametric measure of Chi-square for ceramics (Rasson 1984b). This was risky given the excavation methods (see below) but was expected in a New-Archaeology-inspired dissertation. Analysis of the Divostin material used multivariate statistics such as factor analysis. Joel Gunn's factor analysis (McPherron and Gunn 1988) explained the nature of chipped stone artifacts clearly and was written so that the same system could be applied to other collections for comparison: "An operations sequence was based on a series of binary decisions that would guide the artisan through the correct behavioral steps to achieve the desired goal" (McPherron and Gunn 1988: 245). It was, however, one step removed from seeing the people behind the chipped stone cores.

15.5.1 Excavation

Practical matters of excavation differed in the United States and Yugoslavia. Control of excavated remains was well-developed in California by the 1960s. All soil from excavation units was screened and all the materials were at least washed and examined, although some were later discarded (mainly marine shell). Thus, although there was not much discussion of it, sampling of deposits was attended to as normal and systematic. Another strong feature of American archaeological practice was attention to food remains; in California, archaeozoology was well established. At the time, not much attention was devoted to botanical remains, either carbonized or pollen, but new information showed the potential of flotation methods (Struever and Vickery 1973).

On joint Blocked Funds projects, excavation was carried out in artificial levels, usually 10 to 15 cm in depth. Stratigraphy was important, but the Harris matrix method, developed only in the 1970s (Harris 1979) was unknown to everyone. In Yugoslavia there was no screening of archaeological material (for a variety of reasons, both philosophical and practical); only samples of environmental remains like bone were saved. As they dug, workmen picked up artifacts and bones and gave them to the supervisor of the *sonda* (trench). The material was then washed and sorted, discarding items they had many examples of, including animal bones and non-diagnostic body sherds. The result of this practice was that the remains that ended up in museum collections were carefully selected to be either unusual (like decorated pottery and bone tools) or "typical" pottery (either plain or decorated). The frequencies of different categories and types were not only disregarded but lost.

In contrast, American excavators used screens (e.g., at Anza and Selevac) or systematic hand collection where clayey soil was too heavy to screen (e.g., at Kakanj and Obre [Sterud and Sterud 1974: 156–57], and Divostin [McPherron and Srejović 1988: 6]). They kept all the finds, washed them, and then sorted them. The effect on sampling was that the American excavations recovered more small items (especially bone and stone). I once heard a distinguished senior East European archaeologist remark, without irony, "It's funny how all the small animals lived on your side of the site." Occasional discoveries earlier of carbonized grain by Yugoslav archaeologists had already established Neolithic reliance on domesticated plants, but macrobotanical remains had not been collected systematically. Flotation, pioneered in the American Midwest (Struever 1968), was one of the technical innovations introduced by the Americans in Yugoslavia, first at the Kakanj site, then at the Obre and Anza excavations. Although it was not implemented at Divostin (where water could have been a problem because most of the modern village relied on one spring), it later became standard practice, which added immeasurably to our understanding of the adoption of domesticated plants outside the Near East (see Hillman 1981; J. M. Renfrew 1973).

Systematic identification of animal bone was another innovation of the Blocked Funds' excavations. Bökönyi's research interest was the evolution of domesticated cattle (Bökönyi 1974, 1976, 1986, 1988, 1989). Originally invited by Gimbutas, he analyzed the faunal remains from Divostin, Achilleion, Obre, and Anza, assisted at the latter two sites by Charles A. Schwartz (1939–2016) (see Fig. 15.2), then a UCLA student experienced in zooarchaeology, who processed all the faunal remains from the American excavations at Obre and Anza (Schwartz 1976). Bökönyi played a role in developing the network of American-educated archaeologists working in the Balkans. Later, Bökönyi advised Alice M. Choyke, a student of Sterud, during the research for her dissertation on Bronze Age bone material in Hungary (Fig. 15.2). She married fellow archaeozoologist László Bartosiewicz and remained in Hungary, eventually joining other, mostly European, archaeozoologists working in Europe and the Near East, who had already formed the International Council for Archaeozoology in 1971 (see Bartosiewicz, Chap. 16, this volume), which integrated European-educated specialists into the wider field of archaeozoology.

15.5.2 Publication

In the 1960s, few journals in English offered articles on prehistoric material from Southeast Europe; some appeared in German. This may have been a lack of suitable submissions because some articles were published, such as Mellaart (1960) and Garašanin (1961). Later, the *Journal of Field Archaeology*, the *European Journal of Archaeology* and the like offered more opportunities for disseminating information and ideas. The Cotsen Insitute of Archaeology at UCLA, with which Gimbutas was associated, became important in publishing reports of excavations in former Yugoslavia. Although Gimbutas could be high-handed in editing, occasionally changing the texts of her students to fit her views, she responsibly published professional academic volumes on her excavations at Obre (Gimbutas 1974), Anza (Gimbutas 1976), Sitagroi (Gimbutas 1986a; Renfrew et al. 1986), and Achilleion (Gimbutas et al. 1989).

In Los Angeles, a circle of avocational archaeologists called the Friends of Archaeology developed around Gimbutas, sponsoring lecture series and funding scholarships. One of the Friends, Lloyd Cotsen, later endowed the archaeological institute, formerly the UCLA Archaeological Survey. Gimbutas was a magnetic speaker, but not a native speaker of English, so she relied on others when it came to editing for publication. In 1971, Giorgio Buccellati, then director of the institute, hired UCLA-graduate Ernestine Elster to be the director of the Publications Unit (until 1991), which published material on many topics besides Southeast Europe. As a UCLA product, she was reliably connected with all the students who participated in Gimbutas' various projects and could be counted on to relay information. Elster played a pivotal role in disseminating information about projects in Southeast Europe (the sites of Achilleion, Anza, Selevac, and Sitagroi), doggedly seeing reports through to publication, like the second volume of the Sitagroi report, which was "in press" for over 20 years (Renfrew and Elster 2003).

15.6 Development of Post-excavation Networks

After the American excavation teams returned home, they continued to stay in contact with one another, albeit intermittently, especially those on opposite coasts of the United States (still before the availability of email). They met at professional gatherings such as conferences of the American Anthropological Association and the Society for American Archaeology. There was some resistance among the Americans working in the New World, who said that if we were working in Europe, we should join a European archaeological society. They failed to grasp that even though our source material was European our research methods and paradigms were indisputably American.

Besides meetings of the professional societies, reinforcement of the network came from informal conferences, called Hleb i Vino, held several times (1976, 1979, and later) in Philadelphia, Berkeley, and Los Angeles. Petar Glumac (see Fig. 15.2) was instrumental in organizing the Hleb i Vino meetings in Philadelphia, where Bernard Wailes, a Europeanist, was on the faculty (Crabtree and Bogucki 2017). The participants at various meetings included Robert Ehrich, Peter Bogucki (the Neolithic in Poland), Peter S. Wells (the Iron Age in Slovenia), Ruth Tringham, Barbara Voytek, Timothy Kaiser, Alice M. Choyke, Ernestine Elster, Arthur Bankoff, Haskel Greenfield, and the author, along with others and visiting Europeans like John Chapman. Some meetings were held on the West coast, at UCLA (organized by Mark Stefanovich), and at the University of California Berkeley. These were happy meetings with intense discussions followed by more discussions fueled by wine. This was a venue for presenting first drafts of papers that could later be polished up for larger meetings or publication (e.g., Kaiser and Voytek 1983).

The impact of the Blocked Funds investment in archaeology is evident in the program from the 1984 Union Internationale des Sciences Préhistoriques et Protohistoriques (UISPP: International Union of Prehistoric and Protohistoric Sciences) conference in Belgrade. Renfrew (1984), Evans (1984), and the author (Rasson 1984a), among others, presented papers. I was honored that Benac came to listen to my presentation. At this conference I re-connected with Juraj Pavúk, who introduced me to archaeologist Frano Prendi (1928–2011) (1966), see also Korkuti et al. (2012) from Albania, which led to further contacts with him and Albanian archaeologist Zhaneta Andrea. Together they had excavated the astonishing water-logged Neolithic (and later) site of Maliq in Albania.

After completing degree work, European students seemed to do better than Americans at finding individual professional niches. Even while still at the university, carefully selected students were given opportunities to advance in their careers; thus, many of the participants in the Blocked Funds' excavations went on to be prominent in archaeology. Yugoslav students who participated in the early combined excavations, like Čaće, Ivanovski, and Stefanovich, became professional archaeologists. Other Europeans who participated with the American team (Jean-Paul Demoule, Serge Cleuziou, Stephen Shennan, Robert W. Chapman) went on to academic careers in archaeology. American academic departments offered few Old World positions and there was more competition for career advancement. Many students associated with the Blocked Funds program went into non-academic roles. Charles A. Schwartz, for example, became a high school teacher, Joan Carpenter (Grasty) and Anne Bohlin-Sterud worked in museums, Robert K. Evans worked for the Smithsonian Institution, Barbara Voytek went into academic administration, and I spent some years in cultural resource management besides university teaching in the United States and Hungary.

15.7 Conclusions

In the 1960s, the excavation of four large Neolithic sites in Yugoslavia came about through a convergence of interest in the Neolithic of Southeast Europe, sites available for excavation, improvements in Carbon-14 dating, a favorable political climate, and ample funding. Practical barriers to be circumvented by the international team were language difficulties, geographical distance (and the cost of air fares), and certain restrictions on fieldwork (no site survey was permitted; topographic maps were not available). Exploring Neolithic society through archaeology was a shared goal. The most important difference that had to be addressed was how to apply and reconcile different methodological and theoretical paradigms among the archaeologists of the participating groups, which materially affected excavation and curation practices. There was no immediate shift in paradigms, rather a meeting of paradigms and mutual compromise.

After the Smithsonian Blocked Funds Program ended there was a solid basis for further collaborative projects. Both Westerners and Yugoslavs shared experiences and ideas in ways that went beyond simple academic exchanges. They learned to work together and discovered what constituted a useful approach to the Neolithic period as well as how to secure sponsorship and funding. In this respect the Blocked Funds Program more than accomplished its cultural and academic goals, creating an informal intercontinental network of archaeologists who have maintained professional and personal contacts decades after the completion of the original projects, including sharing the pain of the break-up of Yugoslavia in 1992 and the subsequent wars.[8]

Acknowledgements Thank you for the advice and inspiration: László Bartosiewicz, Claudia Chang, John Chapman, Barbara Demory, Ernestine Elster, Bob Evans, Joan Grasty, Gene Sterud, and Tim Kaiser. Thanks also to John Chapman and Predrag Novaković for useful reviews, and Gilberto Cardenas for preparing the photographs.

References

Anthony, D. W., (2010). *The lost world of Old Europe; The Danube Valley 5000-3600*. New York: The Institute for the Study of the Ancient World and Princeton University Press.

Anthony, D. W., Bogucki, P., Comşa, E., Gimbutas, M., Jovanović, B., Mallory, J. R., & Milasauskas, S. (1986). The "Kurgan Culture," Indo-European origins, and the domestication of the horse: A reconsideration [and comments and replies]. *Current Anthropology, 27*(4), 291–313. https://doi.org/10.1086/203441

[8]In 1996 and 1997, after the war in Bosnia, I served as a (civilian) translator for the Implementation Force (IFOR) and the Stabilization Force (SFOR), the United States' Army's peace-keeping missions in Bosnia and Herzegovina.

Aranđelović, S. T. (1967). *Štambena svojina u Jugoslaviji*. Beograd: Institut društvenih nauka, Centar za istraživanje društvenih odnosa.

Arcanum Maps. (n.d.). *Habsburg Empire (1869–1887)—Third Military Survey (1:25000)*. https://maps.arcanum.com/en/map/thirdsurvey25000/?layers=129&bbox=2037736.511523675%2C5966207.469494695%2C2243963.6138370805%2C6041574.379383879

Arnold, E. R., & Greenfield, H. J. (2006). *The origins of transhumant pastoralism in temperate Southeastern Europe: A zooarchaeological perspective from the Central Balkans* (BAR International Series, 1538). Oxford: Tempus Reparatum.

Bankoff, H. A., & Winter, F. A. (1990). The Later Aeneolithic in Southeastern Europe. *American Journal of Archaeology, 94*(2), 175–191. https://doi.org/10.2307/505948

Bartel, B., Kondić, V., & Werner, M. R. (1979). Excavations at Kraku'lu Yordan, northeast Serbia: Preliminary report, 1973–76 seasons. *Journal of Field Archaeology, 6*(2), 127–149. https://doi.org/10.1179/009346979791489393

Belošević, J. (n.d.). *Mistehalj. Projekat Arheoloskog Iskopavanja 'Sireg Basena Trebišnjica'*. [Unpublished field report of the Stanford University–Zemaljski Museum excavations in the Trebišnjica Basin, 1967]. Documentation Library, Zemaljski Museum, Sarajevo, Croatia.

Benac, A. (1973a). Obre II – A Neolithic settlement of the Butmir group at Gornje Polje. *Wissenschaftliche Mitteilungen des Bosnisch-Herzegowinischen Landesmuseums, 3*(A), 1–191.

Benac, A. (1973b). Obre I – A Neolithic settlement of the Starčevo-Impresso and Kakanj cultures at Raskršće. *Wissenschaftliche Mitteilungen des Bosnisch-Herzegowinischen Landesmuseums, 3*(A), 327–430.

Binford, L. R. (1962). Archaeology as anthropology. *American Antiquity, 28*(2): 217–225. https://doi.org/10.2307/278380

Binford, S. R., & Binford, L. R. (Eds.). (1968). *New perspectives in archaeology*. Chicago: Aldine Publishing Company.

Borić, D. (2014). Lepenski Vir: Geography and culture. In C. Smith (Ed.), *Encyclopedia of global archaeology*. New York, NY: Springer. https://doi.org/10.1007/978-1-4419-0465-2_1877

Bökönyi, S. (1974). The vertebrate fauna [from Obre]. *Wissenschaftliche Mitteilungen des Bosnisch-Herzegowinischen Landesmuseums, 4*(A), 55–154.

Bökönyi, S. (1976). The vertebrate fauna of Anza. In M. Gimbutas (Ed.), *Neolithic Macedonia: As reflected by excavations at Anza, southeast Yugoslavia* (Monumenta Archaeologica, 1) (pp. 315–363). Los Angeles: Cotsen Institute of Archaeology, University of California.

Bökönyi, S. (1986). Faunal remains. In C. Renfrew, M. Gimbutas, & E. S. Elster (Eds.), *Excavations at Sitagroi: A prehistoric village in northeast Greece* (Vol. 1) (Monumenta archaeologica, 13) (pp. 63–132). Los Angeles: Cotsen Institute of Archaeology, University of California.

Bökönyi, S. (1988). The Neolithic fauna of Divostin. In A. McPherron, & D. Srejović (Eds.), *Divostin and the Neolithic of central Serbia* (pp. 419–446). Pittsburgh, PA: University of Pittsburgh and Kragujevac: National Museum.

Bökönyi, S. (1989). Animal remains. In M. Gimbutas, S. Winn, & D. Shimabuku (Eds.), *Achilleion: A Neolithic settlement in Thessaly, Greece, 6400–5600 BC.* (pp. 315–332). Los Angeles: Cotsen Institute of Archaeology, University of California.

Braidwood, R. J. (1960). The agricultural revolution. *Scientific American, 203*(3), 131–148.

Bugaj, U., Naumov, G., Trzeciecki, M., Chwiej, M., Przybyła, M., Szczepanik, P., & Podsiadło, M. (2014). Archaeological, geophysical, and topographic survey of the Neolithic tells in Pelagonia. *Sprawozdania Archeologiczne, 66*, 221–239.

Chapman, J. (1981). *The Vinča Culture of South-East Europe. Studies in chronology, economy and society* (BAR International Series, 117). Oxford: Tempus Reparatum.

Chapman, J. (1998). The impact of modern invasions and migrations on archaeological explanation. A biographical sketch of Marija Gimbutas. In M. Díaz-Andreu, & M. L. S. Sørensen (Eds.), *Excavating women: A history of women in European archaeology* (pp. 295–314). London and New York: Routledge.

Chapman, J. (2020). *Forging identities in the prehistory of Old Europe. Dividuals, individuals and communities, 7000–3000 BC*. Leiden: Sidestone Press.

Chapman, J. (2021). *A life in Balkan archaeology*. Oxford and Philadelphia: Oxbow Books.

Chapman, J., Shiel, R., & Batović, Š. (1996). *The changing face of Dalmatia: Archaeological and ecological studies in a Mediterranean landscape* (Reports of the Research Committee of the Society of Antiquaries of London, 54). London: Leicester University Press.

Clarke, D. L. (1968). Analytical archaeology. London: Methuen.

Clason, A. T. (1980). Padina and Starčevo: Game, fish and cattle. *Palaeohistoria*, *22*, 142–173.

Conkey, M., & Tringham, R. (1996). Cultivating thinking/challenging authority: Some experiments in feminist pedagogy in archaeology. In R. P. Wright (Ed.), *Gender and archaeology* (pp. 224–250). Philadelphia: University of Pennsylvania Press.

Crabtree, P. J., & Bogucki, P. (Eds.). (2017). *European archaeology as anthropology: Essays in memory of Bernard Wailes*. Philadelphia: University of Pennsylvania Museum of Archaeology and Anthropology.

Ehrich, R. W. (1933). Excavations at Starčevo, Yugoslavia, seasons of 1931 and 1932. *American School of Prehistoric Research Bulletin*, *10*, 33–54.

Ehrich, R. W. (Ed.). (1954). *Relative chronologies in Old World archaeology*. Chicago: University of Chicago Press.

Ehrich, R. W. (1977). Starčevo revisited. In V. Markotic (Ed.), *Ancient Europe and the Mediterranean: Studies presented in honour of Hugh Hencken* (pp. 59–67). Warminster: Aris and Phillips.

Elster, E. S. (1976). The chipped stone industry. In M. Gimbutas (Ed.), *Neolithic Macedonia: As reflected by excavations at Anza, southeast Yugoslavia* (Monumenta Archaeologica, 1) (pp. 257–278). Los Angeles: Cotsen Institute of Archaeology, University of California.

Elster, E. S. (2007). Marija Gimbutas: Setting the agenda. In S. Hamilton, R. D. Whitehouse, & K. I. Wright (Eds.), *Archaeology and women: Ancient and modern issues* (pp. 83–120). Walnut Creek, CA: Left Coast Press.

Evans, R. K. (1984). *Craft specialization in the Chalcolithic of the Balkan Peninsula* [Conference presentation]. AISEE/UISPP (Association for the Study of Southeast Europe), Belgrade, Yugoslavia.

Evans, R. K., & Rasson, J. A. (1984). Part I: Ex Balcanis Lux? Recent developments in Neolithic and Chalcolithic research in Southeast Europe. *American Antiquity*, *49*(4), 713–741. https://doi.org/10.2307/279738

Faier, L., & Rofel, L. (2014). Ethnographies of encounter. *Annual Review of Anthropology*, *43*, 363–377. https://doi.org/10.1146/annurev-anthro-102313-030210

Ferguson, C. W., Gimbutas, M., & Seuss, H. E. (1976). Historical dates for Neolithic sites of Southeast Europe. *Science*, *191*(4232), 1170–1172.

Fewkes, V. J. (1935). On the interpretation and dating of the site of "Belo Brdo" at Vinča in Yugoslavia. *Proceedings of the American Philosophical Society*, *75*(7), 651–672.

Fewkes, V. J., Goldman, H., & Ehrich, R. W. (1933). Excavations at Starčevo, Yugoslavia, seasons 1931 and 1932. *Bulletin of the American School of Prehistoric Research*, *9*, 33–53.

Forenbaher, S., Gaffney, V., Hayes, J. W., Kaiser, T., Kirigin, B., Leach, P., & Vujnović, N. (1994). Hvar-Vis-Palagruža 1992–1993: A preliminary report of the Adriatic Island Project (contact, commerce and colonization 6000 BC–600 AD). *Vjesnik za arheologiju i historiju dalmatinsku*, *86*, 13–52.

Forenbaher, S., Kaiser, T., & Miracle, P. T. (2013). Dating the East Adriatic Neolithic. *European Journal of Archaeology*, *16*(4), 589–609. https://doi.org/10.1179/1461957113Y.0000000038

Garašanin, D. (1954). *Starčevačka kultura*. Ljubljana: Univerza v Ljubljani, Arheološki Seminar.

Garašanin, M. (1951). *Hronologija Vinčanske Grupe*. Ljubljana: Univerza v Ljubljani, Arheološki Seminar.

Garašanin, M. (1961). The Neolithic in Anatolia and the Balkans. *Antiquity*, *35*(140), 276–280. https://doi.org/10.1017/S0003598X00036334

Garašanin, M., Sanev, V., Simoska, D., Kitanoski, B., Findrik, K., Tasev, R. (1971). *Predistoriski kulturi o Makedonija*. Štip: Naroden Muzej.

Gardner, E. (1976). The technology of the ceramics. In M. Gimbutas (Ed.), *Neolithic Macedonia: As reflected by excavations at Anza, southeast Yugoslavia* (Monumenta Archaeologica, 1) (pp. 159–187). Los Angeles: Cotsen Institute of Archaeology, University of California.

Gimbutas, M. (1956). *The prehistory of Eastern Europe. Part I. Mesolithic, Neolithic and Copper Age cultures in Russia and the Baltic area* (American School of Prehistoric Research, Harvard University Bulletin, 20). Cambridge, MA: Peabody Museum.

Gimbutas, M. (1965). *Bronze Age cultures in Central and Eastern Europe*. Paris, The Hague and London: Mouton & Co.

Gimbutas, M. (1974). Introduction—Obre and its place in Old Europe. *Wissenschaftliche Mitteilungen des Bosnisch-Herzegowinischen Landesmuseums*, *4*(A), 5–13.

Gimbutas, M. (Ed.). (1976). *Neolithic Macedonia: As reflected by excavations at Anza, southeast Yugoslavia* (Monumenta Archaeologica, 1). Los Angeles: Cotsen Institute of Archaeology, University of California.

Gimbutas, M. (1986a). Mythical imagery of Sitagroi society. In C. Renfrew, M. Gimbutas, & E. S. Elster (Eds.), *Excavations at Sitagroi: A prehistoric village in northeast Greece* (Vol. 1) (Monumenta archaeologica, 13) (pp. 225–301). Los Angeles: Cotsen Institute of Archaeology, University of California.

Gimbutas, M. (1986b). *The language of the Goddess: Unearthing the hidden symbols of Western civilization*. San Francisco: Harper & Row.

Gimbutas, M. (1991). *The civilization of the Goddess: The world of Old Europe*. New York: Harper Collins.

Gimbutas, M., Winn, S., & Shimabuku, D. (Eds.). (1989). *Achilleion. A Neolithic settlement in Thessaly, Greece, 6400–5600 BC*. Los Angeles: Cotsen Institute of Archaeology, University of California.

Hammel, E. A. (1969). Economic change, social mobility, and kinship in Serbia. *Southwestern Journal of Anthropology*, *25*(2), 188–197.

Harris, E. C. (1979). *Principles of archaeological stratigraphy*. London: Academic.

Harry, K. G., Allison, J. R., & Sakai, S. (2019). New findings from the far western Puebloan region: Papers in honor of Margaret Lyneis. *KIVA*, *85*(4), 281–288. https://doi.org/10.1080/00231940.2019.1692170

Hillman, G. C. (1981). Reconstructing crop husbandry practices from charred remains of crops. In R. Mercer (Ed.), *Farming practice in British prehistory* (pp. 123–162). Edinburgh: Edinburgh University Press.

Jacobson, R. (2018, March). Mother tongue. *Scientific American*, *318*(3), 12–14. https://doi.org/10.1038/scientificamerican0318-12

Kaiser, T., & Voytek, B. (1983). Sedentism and economic change in the Balkan Neolithic. *Journal of Anthropological Archaeology*, *2*(4), 323–353. https://doi.org/10.1016/0278-4165(83)90013-2

Korkuti, M., Demiraj, S., & Lafe, E. (2012). In memoriam of: Frano Prendi (1928–2011). *Studia Albanica*, *1*, 161–167.

Kuhn, T. (1962). *The structure of scientific revolutions*. Chicago and London: The University of Chicago Press.

Legge, A. J. (1990). Animals, economy, and environment. In R. Tringham, & D. Krstić (Eds.), *Selevac: A Neolithic village in Yugoslavia*. Los Angeles: Cotsen Institute of Archaeology, University of California.

Lorber, Č., & Novaković, P. (2020). Internationalisation as a long-term strategic project of the post-war renewal of the Yugoslav archaeology (1950–1971). *Issues in Ethnology and Anthropology, 15*(3), 689–715. https://doi.org/10.21301/eap.v15i3.3

Lyneis, M. (1988). Antler and bone artifacts from Divostin. In A. McPherron, & D. Srejović (Eds.), *Divostin and the Neolithic of central Serbia* (pp. 301–324). Pittsburgh, PA: University of Pittsburgh and *Kragujevac:* National Museum.

Makkay, J. (1985). Diffusionism, antidiffusionism, and chronology: Some general remarks. *Acta Archaeologica Academiae Scientiarum Hungaricae, 37*(1–2), 1–12.

Mallory, J. P. (1989). *In search of the Indo-Europeans: Language, archaeology and myth*. London: Thames & Hudson.

McPherron, A., & Gunn, J. (1988). Quantitative analysis of excavated materials at Divostin. In A. McPherron, & D. Srejović (Eds.), *Divostin and the Neolithic of central Serbia* (pp. 359–378). Pittsburgh, PA: University of Pittsburgh and Kragujevac: National Museum.

McPherron, A., & Srejović, D. (Eds.). (1988). *Divostin and the Neolithic of central Serbia*. Pittsburgh, PA: University of Pittsburgh and Kragujevac: National Museum.

Mellaart, J. (1960). Anatolia and the Balkans. *Antiquity, 34*(136), 270–278. https://doi.org/10.1017/S0003598X00035560

Milisauskas, S. (1986). *Early Neolithic settlement and society at Olszanica*. Ann Arbor: University of Michigan.

Milisauskas, S., Thurston, T. L., & Whitlow, R. (2010). American contributions to European archaeology. *Sprawozdania Archeologiczne, 62*, 35–63.

Milojčić, V. (1949). South-eastern elements in the prehistoric civilization of Serbia. *Annual of the British School at Athens, 44*, 258–306. https://doi.org/10.1017/S0068245400017226

Milosavljević, M. (2020). How archaeological communities think: Re-thinking Ludwig Fleck's concept of the thought-collective according to the case of Serbian archaeology. In J. Roberts, K. Sheppard, U. R. Hanson, & J. R. Trigg *(Eds.), Communities and knowledge production in archaeology* (pp. 14–33). Manchester: Manchester University Press.

Mužijević, R., & Ralph, E. (1988). Geomagnetic surveys at Divostin. In A. McPherron, & D. Srejović (Eds.), Divostin and the Neolithic of central Serbia (pp. 389–418). Pittsburgh, PA: University of Pittsburgh and Kragujevac: National Museum.

Nandris, J. G. (1968). Lepenski Vir. *Science Journal, 1*, 64–70.

Neustupný, E. (2016). Czech archaeology under communism. In L. R. Lozny (Ed.), *Archaeology of the communist era: A political history of archaeology of the 20th century* (pp. 151–166). Cham: Springer. https://doi.org/10.1007/978-3-319-45108-4_6

Novaković, P. (2011). Archaeology in the new countries of Southeastern Europe: A historical perspective. In L. R. Lozny (Ed.), *Comparative archaeologies: A sociological view of the science of the past* (pp. 339–461). New York: Springer. https://link.springer.com/chapter/10.1007/978-1-4419-8225-4_12

Novaković, P. (2015). Binford in the Balkans. In K. Kristiansen, L. Šmejda, & J. Turek (Eds.), *Paradigm found: Archaeological theory—Present, past and future. Essays in honour of Evžen Neustupný* (pp. 124–136). Oxford and Philadelphia: Oxbow Books.

Novaković, P. (2021). *The history of archaeology in the Western Balkans*. Ljubljana: Ljubljana University Press. https://doi.org/10.4312/9789610605393

Novaković, P. (2023). Cohabitation, hybridization and eclecticism of archaeological paradigms: Modern Slovene archaeology and the legacy of the 1980s. *CAS Sofia Working Paper Series, 14*: 72–91.

Pavúk, J., & Bakamska, A. (2018). Zur Typologie und Chronologie der neolithischen anthropomorphen Plastik auf dem Balkan. In S. A. Luca (Ed.), *The image of divinity in the Neolithic and Eneolithic. Ways of Communication* (pp. 47–66). International Symposium, Sibiu, 26–28 October 2017. Suceava: Editura Karl A. Romstorfer.

Piggott, S. (1965). *Ancient Europe from the beginnings of agriculture to classical antiquity*. Edinburgh: Edinburgh University Press.

Popović, V., & Ochenschlager, E. L. (Eds). (1971–1980). *Sirmium: Archaeological investigations in Syrmian Pannonia* (Vols. 1–3). Belgrade: The Archaeological Institute.

Prendi, F. (1966). La civilisation préhistorique de Maliq. *Studia Albanica, 1*, 255–280.

Prinz, B. (1987). *Mesolithic adaptations on the Lower Danube: Vlasac and the Iron Gates gorge* (BAR International Series, 330). Oxford: Tempus Reparatum.

Radovanović, I. (1996). *The Iron Gates Mesolithic* (International Monographs in Prehistory. Archaeological Series, 11). Ann Arbor, MI: International Monographs in Prehistory.

Radovanović, I., & Voytek, B. (1997). Hunters, fishers or farmers: Sedentism, subsistence and social complexity in the Djerdap Mesolithic. *Analecta Praehistorica Leidensia, 29*, 19–31.

Rasson, J. A. (1984a). *Relations between Bosnia and Dalmatia in the Neolithic* [Conference presentation]. AISEE/UISPP (Association for the Study of Southeast Europe), Belgrade, Yugoslavia.

Rasson, J. A. (1984b). Interaction spheres as adaptive mechanisms: Bosnian-Dalmatian relations in the Neolithic [Unpublished doctoral dissertation]. University of California Los Angeles.

Rasson, J. A., Stevanović, M., & Ilić, V. (1999). Living spaces in transition: From rural to urban family life in Serbia. In D. Birdwell-Pheasant, & D. Lawrence-Zúñiga (Eds.), *House life: Space, place and family in Europe* (pp. 177–204). Oxford: Routledge.

Rasson, J. A. (forthcoming). Archaeology in Yugoslavia during the Cold War, 1945–1989. In L. Coltofean, S. Grunwald, F. Link, & K. Roesler (Eds.), *Archaeology during the Global Cold War: Scientific practices and political ideology*. Routledge.

Renfrew, C., Dixon, J., & Cann, J. (1966). Obsidian and early cultural contact in the Near East. *Proceedings of the Prehistoric Society, 32*, 30–72. https://doi.org/10.1017/S0079497X0001433X

Renfrew, C. (1973). *Before civilization: The radiocarbon revolution and prehistoric Europe*. New York: Alfred A. Knopf.

Renfrew, J. M. (1973). *Palaeoethnobotany: The prehistoric food plants of the Near East and Europe*. London: Methuen.

Renfrew, C. (1984). *Sitagroi and the independent invention of metallurgy in Europe* [Conference presentation]. AISEE/UISPP (Association for the Study of Southeast Europe), Belgrade, Yugoslavia.

Renfrew, C. (1987). *Archaeology and language: The puzzle of the Indo-European origins*. London: Jonathan Cape.

Renfrew, C., & Elster, E. S. (Eds). (2003). *Prehistoric Sitagroi: Excavations in Northeast Greece, 1968–1970* (Vol. 2: *The final report*). Los Angeles: Cotsen Institute of Archaeology, University of California.

Renfrew, C., Gimbutas, M., & Elster, E. S. (Eds). (1986). *Excavations at Sitagroi: A prehistoric village in Northeast Greece* (Vol. 1) (Monumenta archaeologica, 13). Los Angeles: Cotsen Institute of Archaeology, University of California.

Rodden, R., Pyke, G., Yiouni, P., & Wardle, K. A. (1996). *Nea Nikomedeia: The excavation of an Early Neolithic village in Northern Greece 1961–1964* (Vol. 1: *The excavation and the ceramic assemblage*) (The British School at Athens Supplementary Volumes, 25). London: The British School at Athens.

Schwartz, C. A. (1976). The vertebrate fauna from Rug Bair. In M. Gimbutas (Ed.), *Neolithic Macedonia as reflected by excavation at Anza, Southeast Yugoslavia* (pp. 364–374). Los Angeles: Cotsen Institute of Archaeology, University of California.

Srejović, D. (1969). *Lepenski vir: Nova praistorijska kultura u Podunavlju*. Beograd: Srpska književna zadruga.

Srejović, D. (1972). *Europe's first monumental sculpture: New discoveries at Lepenski Vir. London:* Thames & Hudson.

Sterud, E. L. (1978). Changing aims of Americanist archaeology: A citations analysis of American Antiquity. 1946–1975. *American Antiquity, 43*(2), 294–302. https://doi.org/10.2307/279255

Sterud, E. L., & Sterud, A.-K. (1974). A quantitative analysis of the material remains. *Wissenschaftliche Mitteilungen des Bosnisch-Herzegowinischen Landesmuseums, 4*(A), 155–279.

Struever, S. (1968). Flotation techniques for the recovery of small-scale archaeological remains. *American Antiquity, 33*(3), 353–362. https://doi.org/10.2307/278703

Struever, S., & Vickery, K. D. (1973). The beginnings of cultivation in the Midwest-riverine area of the United States. *American Anthropologist, 75*(5), 1197–1220. https://doi.org/10.1525/aa.1973.75.5.02a00010

Swartz, B. K., Jr. (1996). The McKern "taxonomic" system and archaeological culture classification in the Midwestern United States: A history and evaluation. *Bulletin of the History of Archaeology, 6*(1), 3–9. https://doi.org/10.5334/bha.06102

Tringham, R. (1971). *Hunters, fishers and farmers of Eastern Europe, 6000–3000 B.C.* London: Hutchinson.

Tringham, R. (1991). On theory: Men and women in prehistoric architecture. *Traditional Dwellings and Settlements Review, 3*(1), 9–28.

Tringham, R., Brukner, B., Kaiser, T., Borojević, K., Bukvić, L., Šteli, P., Russell, N., Stevanović, M., & Voytek, B. (1992). Excavations at Opovo, 1985–1987: Socioeconomic change in the Balkan Neolithic. *Journal of Field Archaeology*, 19(3), 351–386. https://doi.org/10.1179/009346992791548860

Tringham, R., Cooper, G., Odell, G., Voytek, B., & Whitman, A. (1974). Experimentation in the formation of edge damage: A new approach to lithic analysis. *Journal of Field Archaeology, 1*(1–2), 171–196. https://doi.org/10.1179/jfa.1974.1.1-2.171

Tringham, R., & Krstić, D. (Eds.). (1990). *Selevac: A Neolithic village in Yugoslavia.* Los Angeles: Cotsen Institute of Archaeology, University of California.

Tringham, R., & Stevanović, M. (2012). *Last house on the hill: BACH Area Reports from Çatalhöyük, Turkey* (Çatalhöyük Research Project Series, Vol. 11). Los Angeles: Cotsen Institute of Archaeology, University of California.

Vasić, M. (1932–1936). *Преисторијска Винча* (Vols. 1–4). Belgrade: Yugoslav State Printing House.

Whallon, R. (1989). *The Paleolithic site of Badanj: Recent excavations and results of analysis.* Glasnik Zemaljskog *Muzeja Bosne i Hercegovine u Sarajevu, Part A, 44*, 7–20.

White, L. A. (1949). *The science of culture: A study of man and civilization.* New York: Grove Press, Inc.

Willey, G., & Phillips, P. (1958). *Method and theory in American archaeology.* Chicago: University of Chicago Press.

Wiseman, J., & Mano-Zissi, Dj. (Eds.). (1973–1981). *Studies in the antiquities of Stobi* (Vols. 1–3). Princeton: Princeton University Press.

Fifty Years of Networking in Archaeozoology

16

László Bartosiewicz

Abstract

In 1989, two developments of global impact coincided in Europe: the 'Iron Curtain' fell between East and West, and Tim Berners-Lee invented the World Wide Web. In the meantime, the International Council for Archaeozoology (ICAZ) was coming of age, having been founded 18 years earlier at a conference in 1971 in Budapest, Hungary. Archaeozoology is the study of ancient animal-human relationships, relying primarily on the evidence of animal remains recovered during archaeological excavations. Archaeozoologists, holding a special session on domestication, arrived at the 1971 conference from 18 countries. They included the founders of what was to become ICAZ, the first and largest representative body in archaeozoology. By 1971, Hungary was equally accessible to 'socialist' and 'capitalist' scholars in the wake of the Cold War. Some participants already had personal contacts in the opposite half of Europe. Some German colleagues had known each other from before the 1961 erection of the Berlin Wall and archaeological projects with United States participation in non-aligned former Yugoslavia also served as points of personal contact. Personal networks and the keen awareness of the need to crosscut political boundaries have become fundamental to ICAZ. In 1990, the organization had its first meeting outside Europe, further broadening networking opportunities. This chapter is a review of the role personal contacts have played in diachronic diversification within ICAZ during the past 50 years in light of its conferences and recent membership data.

Keywords

Zooarchaeology · International cooperation · Personal networks · Cold War Europe · Global organization

16.1 Introduction

Archaeozoology is the study of excavated animal remains. It differs from paleontology in its focus on ancient animals of cultural relevance (Olsen and Olsen 1981). The analysis of zoological finds from archaeological sites began in the 1850s, marking the first scientific revolution in archaeology that developed in tandem with zoology and geology at the time (Kristiansen 2014). Charles Darwin (1809–1882) praised the pioneering work of zoologist and anatomist Ludwig Rütimeyer (1825–1895), who analyzed animal remains from Swiss lakeshore settlements (Rütimeyer 1861) as "the most important contribution on domestic races ever published" (Burkhardt et al. 1994: 365). The large-scale, regular study of archaeozoological remains only began in the mid-twentieth century, when a new generation of archaeozoologists—mostly paleontologists, veterinarians, and zoologists by training—emerged in post-Second World War Europe. Since they published in flagship international journals in zoology and animal science at the time (e.g., *Säugetierkundliche Mitteilungen, Zeitschrift für Tierzüchtung und Züchtungsbiologie*), much of their work fell outside the purview of the archaeological community. The International Council for Archaeozoology (ICAZ) was founded in 1971 and over the years it has become the largest international body representing archaeozoologists. How has this organization grown in half a century? How did it affect scholarly communities in the participating countries? What was its impact on the development of archaeozoology as a discipline? Interaction between individuals in this narrowly specialized field is a tangible illustration of grassroots efforts in

L. Bartosiewicz (✉)
Department of Archaeology and Classical Studies, Stockholm University, Stockholm, Sweden
e-mail: laszlo.bartosiewicz@ofl.su.se

L. Coltofean et al. (eds.), *Connecting People and Ideas*, Themes in Contemporary Archaeology,
https://doi.org/10.1007/978-3-031-81006-0_16

connecting people and ideas in the ever-changing global political scenario.

Introducing this chapter, it is worth emphasizing that the topic as presented here is focused on networking. The history of ICAZ could have been shown from many different angles under other circumstances. Mapping networks, however, can elucidate the structure of relationships among social entities. In the absence of precise parameters, it is impossible to describe the development of ICAZ using advanced methods of sociometry. How a network is wired is key to understanding its operation. It determines cohesion, knowledge transfer, the diffusion of innovations, news, and rumors. The principles and basic terms of network science, therefore, are worth considering at least metaphorically, even if our historical data are not fit for use in exact modelling.

Distinct actors such as individual scholars in this network are considered *nodes* connected pairwise by *links*. The simplest model of random connections (Erdős and Rényi 1959) may be seen as a static infrastructure such as railroad lines, offering equal access to connected cities. Heavily linked nodes are comparable to *hubs* in more dynamic, active aerial transport that tend to amalgamate additional links, while nodes possessing only a few links are less likely to accumulate new connections. Research has shown that newly entering nodes are more easily attached to already well-linked nodes, that is, established hubs (Barabási and Albert 1999). Although in hindsight it is impossible to reconstruct in detail what the interpersonal structure precisely looked like within ICAZ, reviewing history and drafting quantitative trends in academic content can reveal underlying mechanisms.

Two synonyms for our discipline need to be distinguished as their subtle discrepancy reflects evolutionary trends in the field (Bartosiewicz 2001). Traditional analyses of animal bones consist of morphological identification and documenting human impact on animals. This inductive line of inquiry is archaeozoology. With the promotion of 'processual' or New Archaeology in the mid-1960s in the Anglophone world, the application of scientific methods in archaeology became central to interdisciplinary discourse. Patterns recognized in animal bone assemblages began playing an increasingly important role in testing *bona fide* archaeological hypotheses. This deductive school of thought is zooarchaeology. Zooarchaeology may be seen as applied archaeozoology with a newly found archaeological emphasis (Bobrowsky 1982), although the relationship between these terms is more complex given the ontological difference.

16.2 The Early Years

As mentioned, ICAZ is the first and largest representative body in archaeozoology. Given the small number of experts in this field, personal networking, importantly in the form of physical meeting, played a key role in within-group communication in its early years despite difficulties posed by post-war political animosities in Europe. The six to ten founders represented largely equal nodes forming a small system. Since intensities of their bilateral links are difficult to reconstruct one may presume that they acted as an egalitarian group conforming to the traditional random network model, by which chances of their linking had equal probabilities (Erdős and Rényi 1959).

The roots of the organization can be traced back to the Third International Congress of the Association internationale des musées d'agriculture (AIMA: International Association of Agricultural Museums) held in Budapest, Hungary, between April 18 and 23, 1971. Archaeozoologists took part in Session IV, entitled "Domestication and the History of Domestic Animals." The event was part of celebrating the 75th anniversary of the Hungarian Agricultural Museum and the session on archaeozoology, coincidentally, mustered 75 participants. The proceedings (Matolcsi 1973), organized in six sections, included 40 contributions (24 written in German, 13 in English, and three in French). In keeping with the title of Session IV, most papers focused on questions of domestication from a historical perspective. Many of the international attendees had an opportunity to meet face-to-face for the first time. Even if the conference was not organized with the explicit intention of fostering a new organization, those few who established practically lifelong personal contacts at the event became the founders of ICAZ.

The next meeting of archaeozoologists took place in Groningen, the Netherlands, in 1974. Sixty-three specialists attended while 16 others were unable to participate, some because Western Europe was off-limits to their home countries. The proceedings (Clason 1975), also organized into six sections, contained 48 papers (29 in English, 16 in German, and three in French). In comparison with the Budapest meeting, many of these chapters reflected on newly recognized methodological issues. Importantly, the organizer of the conference, archaeozoologist Anneke T. Clason (1932–2008), took a leading role in forging regular ties among participants, becoming a typical hub personality in the emerging network (Barabási and Albert 1999), in addition to being a hub in the academic world for her Dutch students, especially in the 1980s and early 1990s (Brinkhuizen et al. 2007–2008).

At the Groningen meeting, the idea of a formal organization for archaeozoology was conceived whose structure and format were dictated by what was needed to ensure communication between archaeozoologists on either side of the 'Iron Curtain', a metaphor coined by British statesman Winston Churchill (1874–1965) in 1946 to refer to the political division of post-Second World War Europe. Divergent histories of archaeology and their variegated academic contexts often posed challenges when archaeozoologists from different countries first met, "the various regional traditions of archaeological faunal analysis sometimes seemed like ships passing in the night" (Gifford-Gonzalez 2018: 20, 30).

However, networking aided by common areas of concern and the invariably congenial spirit of ICAZ meetings began to bear fruit over the years.

The general intensification of networking at the time is marked by another symposium that took place outside Europe. In 1975, zooarchaeologists Richard H. Meadow and Melinda A. Zeder organized a session at the meeting of the Society for American Archaeology (SAA) in Dallas (Texas), devoted to archaeozoology in Southwest Asia. This was the first symposium in the western hemisphere that brought together North American and European archaeozoologists. Given the prominent importance of the Near and Middle East in Old World archaeology, the organizers could secure funds for inviting experts from Europe, several of whom had attended the meetings in Budapest and Groningen. The 78 shorter and longer papers in the proceedings of this meeting (Meadow and Zeder 1978) were presented in five sections. Although reflecting the original topic all contributors discussed finds from the Middle East, there was an emphasis on methods that would benefit archaeozoology regardless of geographical location. Overlaps between the participants of the Budapest, Groningen, and Dallas meetings illustrate the expansion of close personal ties consolidating the network that came to form the core of ICAZ.

To ensure official status in this formative period, ICAZ needed affiliation with an organization acknowledged in both Eastern and Western Europe to be able to offer access for as many archaeozoologists as possible. Clason approached the organizers of the 9th Congress of the Union internationale des sciences préhistoriques et protohistoriques (UISPP: International Union for Prehistoric and Protohistoric Sciences, a UNESCO sponsored body) held in Nice, France, in 1976. The UISPP approved the affiliation of ICAZ, thus marking the formal launch of the organization. This is the manifestation of a known phenomenon in network science: the self-accelerating process of preferential attachment to existing hubs while the underlying, small network is still scale free (Barabási and Albert 1999).

When ICAZ became thus consolidated at the UISPP conference in Nice, its first statutes were announced and a self-elected International Committee was set up to further the aims of promoting archaeozoology, standardizing its methods, and fostering international cooperation between its practitioners. Participants agreed to form an Executive Committee, which included (in alphabetical order) Sándor Bökönyi (Hungary), Anneke T. Clason (the Netherlands), Juliet Clutton-Brock (United Kingdom), Marian Kubasiewicz (Poland), Richard Meadow (United States), Nanna Noe-Nygaard (Denmark), Elisabeth Schmid (Switzerland), and Manfred Teichert (German Democratic Republic). In addition to the geopolitical balance, the equal gender distribution of this small group is remarkable. Subsequently, Hans-Peter Uerpmann (Federal Republic of Germany) joined this committee. Along with the growth of the organization, an International Committee was broadened, formed of individuals who represented participating countries. A conscious decision was made that quadrennial international conference venues should alternate between Western and Eastern Europe. Thus, after Groningen, the 1978 conference was organized in Szczecin, Poland by Marian Kubasiewicz. Juliet Clutton-Brock and Caroline Grigson co-hosted the 1982 International Council of ICAZ conference in London, United Kingdom, the first ICAZ event I had an opportunity to attend. Clutton-Brock's and Grigson's increasing engagement with ICAZ ran in parallel with the aforementioned anglophone event at SAA organized by Meadow and Zeder in 1975 (Gifford-Gonzalez 2018: 25).

While ICAZ has no official language, the 2:1 proportion between papers published in German and English in Budapest was reversed at the Groningen meeting. This reflects the general trend of English having become the *lingua franca* of international communication. The ambition to create a German-language forum for the field resulted in establishing the Gesellschaft für Archäozoologie und Prähistorische Anthropologie (GAPA: Society for Archaeozoology and Prehistoric Anthropology) in Konstanz, Germany, in 1994. As shown by its name, however, its scope also includes the study of prehistoric human remains, thereby differing from ICAZ.

16.3 Historical Setting

Mathematical models of network behavior can predict performance based on network size, complexity, and environment (Krackhardt and Carley 1998). This section is an attempt to zoom in on factors behind individual careers against the backdrop of history, relying in part on my personal familiarity with the early years of ICAZ. In the absence of exact data, we can evaluate the environment only in descriptive, historiographic terms. Having reviewed the roles of intersecting networks represented by major professional bodies (AIMA, SAA, UISPP) in the emergence of ICAZ, the importance of individuals becoming hubs tapping into the resources of these networks cannot be overestimated. From this point of view, it is of special interest that the group of experts representing archaeozoology in the 1971 Budapest meeting arrived from 18 countries (Matolcsi 1972). They came from both sides of the Iron Curtain, as by the early 1970s Hungary became equally accessible to scholars from 'socialist' and 'capitalist' countries (Bartosiewicz et al. 2011). This 'East-West balance' at the meeting offered a promising perspective.

In order to shed light on the significance of the 1971 symposium in a historical context, details of research policy and their relevance to the Hungarian organizers are worth considering. In the aftermath of the 1956 anti-Soviet uprising in Hungary and concomitant retaliation, a pragmatic form of socialism evolved; a gradual opening of the country was promoted by the government. Throughout the 1960s, however, two antagonistic interests clashed in the political leadership: the need to open up scholarly communication (especially in lucrative applied sciences) and to maintain political control of foreign relations by limiting private initiatives in academic contacts (Somlai 2009).

As an overture, in 1960—in addition to countries within the Soviet Bloc—Hungary signed research agreements with Iraq and [former] Yugoslavia (Somlai 2009). Shortly thereafter, the Smithsonian Institution inaugurated its Foreign Currency Program to support American archaeological excavations in countries where the United States held large amounts of foreign currency derived from the sale of agricultural commodities. American institutions could therefore cover field-research costs of archeological work in such countries, including Yugoslavia (Smithsonian 1965; also see Rasson, Chap. 15, this volume), but not Hungary. This was possible because the leadership of Yugoslavia defied Soviet hegemony in 1948 and began the country's own independent socialist program, eventually acknowledged in both Cold War political blocks.

16.3.1 Complementary Biographies

The networking aspect of American excavations in Yugoslavia stems from the fact that with no local animal bone specialists available there at the time, Hungarian archaeozoologist Sándor Bökönyi (1926–1994) was invited to participate in United States projects in 1962 (Bartosiewicz 2017), thereby linking Central European morphology-oriented archaeozoology with the Anglo-American anthropology-inspired archaeology. Bökönyi, a veterinarian by training, began working as a full-time archaeozoologist at the Magyar Nemzeti Múzeum (Hungarian National Museum) in 1950. Having a good command of both German and English, he was able to bridge the linguistic gap between the two different schools of thought. Another European archaeozoologist invited to work at United States excavations was Clason, who studied animal remains from pre-Second World War excavations at Starčevo as well as from the 1969 and 1970 field seasons directed by archaeologist Robert W. Ehrich (1908–1992) (Clason 1980).

Following the suppressed 1956 anti-Soviet uprising, Hungary needed improving relations with the world. As a token of consolidation, by the mid-1960s the country gradually opened up toward the 'West' on institutional and even private levels. Aside from the political gesture, keeping abreast with international research and development was recognized as an economic imperative. The United States-based Ford Foundation established direct contacts with the Hungarian authorities in 1963. As a non-governmental organization, however, it tried to avoid the contractual framework of state-level cooperation, maintaining the right to select invitees to the projects that it funded. When this scholarship scheme began in Hungary in 1964, 25 researchers were allowed to spend a year at various United States universities (Somlai 2009). They included Bökönyi, who was hosted by the University of California in Los Angeles in 1966 (Bartosiewicz and Choyke 2002), where he primarily cooperated with Lithuanian born American archaeologist Marija Gimbutas (1921–1994), who worked in Yugoslavia (also see Rasson, Chap. 15, this volume). Having no say in selecting grant holders, rejecting invitations was the only way Hungarian authorities could regulate study trips. By 1967, only four grant holders could leave for the United States (Somlai 2009). In the meantime, Hungarian outreach toward the Middle East increased. This had direct consequences on Bökönyi's research as it allowed him to spend six months in Iraq in 1969 (Bartosiewicz 2010), after the ruling Arab Socialist Baath Party established political ties with the Hungarian government.

Bökönyi's perseverance and good luck in broadening his international network would not have been enough to organize the 1971 conference in Hungary. His peer, János Matolcsi (1923–1983), archaeozoologist at the Magyar Mezőgazdasági Múzeum (Hungarian Agricultural Museum), procured the venue and institution-level logistics. During the Second World War, as a young printshop worker, Matolcsi joined the illegal Communist Party. After 1945, as a career politician, he rose to the rank of minister of agriculture, a post held until 1957. Abandoning politics, he was appointed director-in-chief of the Hungarian Agricultural Museum and served in that position until 1968. During this time, in addition to his management responsibilities, he conducted university studies in agriculture, earned a doctorate in 1966, and subsequently became a full-time archaeozoologist (Bartosiewicz 2015). While Bökönyi was interested in prehistory in Southeast Europe and the Middle East, Matolcsi specialized in protohistoric and historical archaeozoology, focusing on the East European steppe belt and relying on his proficiency in Russian and academic contacts in the Soviet Union for undertaking research. He invested his managerial talent and political network into hosting the Third International Congress of Agricultural Museums in Budapest. The backgrounds and careers of these two archaeozoologists could not have been more different. In fact, their complementary nature benefited ICAZ at the time of its conception (Fig. 16.1). The professional link between Bökönyi and Matolcsi joined two completely different networks (one academic/international, the other political/domestic), indispensable in the physical realization of the first major meeting of archaeozoologists.

Fig. 16.1 Anneke T. Clason (1932–2008), Sándor Bökönyi (1926–1994) and János Matolcsi (1923–1983). (Photograph sources, from left to right: reproduced with permission from Dick C. Brinkhuizen et al. 2007–2008; Judit Lévay; István Takács)

16.3.2 International Participation

By around 1968, Hungary implemented carefully engineered reforms with elements of market economy facilitated by relative political tolerance, a process similar to what went out of [Soviet] control in former Czechoslovakia the same year. It is in this context that Budapest became a meeting ground, equally accessible to 'socialist' archaeozoologists from the closely watched Soviet Bloc (e.g., from Romania, Bulgaria, the German Democratic Republic) and their 'capitalist' counterparts. Some attending the meeting from the Soviet Bloc already had personal links in the opposite half of Europe. In addition, archaeozoologists in the German Democratic Republic (GDR, 'East') and the Federal Republic of Germany (FRG, 'West') were connected by more than language; some had known each other even before the Berlin Wall was built (1961). Clason, who spearheaded the ICAZ initiative, reputedly had Hungarian relatives in the Netherlands. Bökönyi was one of the few East Europeans who had an opportunity to bond with 'Western' colleagues prior to the conference. These disjunct personal clusters, motivated by individual aspirations, became integral components in the emerging network (Barabási and Albert 1999). The group of such core personalities during the subsequent Groningen, Dallas, and Nice meetings formed the International Council of ICAZ.

Clason took on the role of secretary, editing and issuing the annual *Newsletter*. In addition to reporting on the state of the organization, the *Newsletter* directed attention to projects, conferences, and new publications. Issues of standardization were discussed and in 1980, a vocabulary of almost 200 German technical terms and their English equivalents was listed. Most importantly, the *Newsletter* contained annually updated address lists of archaeozoologists. Before the age of digital communication, all information was gathered and circulated by regular mail, in part sponsored by the University of Groningen. Another regular service to the community of researchers was the archaeozoological bibliography collected, edited, printed, and distributed by archaeozoologist Hanns-Hermann Müller working at the Academy of Sciences of the GDR.

It was decided that ICAZ International Conferences should be held every four years. In accordance with the principles of cultivating networks across political boundaries, the venues alternating between Eastern and Western Europe. Increasingly linked, people active in these events developed from simple nodes to hubs within the network, a significant improvement as hubs in a general sense influence the behavior of all systems (Barabási and Albert 1999: 510) which, in this particular case, are represented by the emerging organization of archaeozoologists. Biennial International Committee meetings became occasions when, among others, the state of organization of the upcoming International Conference could be discussed face to face. On the long run, the venues of these smaller business meetings did not alternate between Eastern and Western Europe as strictly as the major International Conferences. However, some participants from Eastern Europe could regularly attend thanks to the gradually improving political situation, even if travel regulations strongly varied between countries.

The 1971 symposium in Hungary was followed by the previously mentioned conferences in the Netherlands (1974), Poland (1978), and the United Kingdom (1982). This alternating pattern between East and West was broken only by the meeting in France in 1986, in want of an appropriate conference bid from an East European country. The next (1990) meeting was planned for the United States.

It would be misleading to suggest that ICAZ has represented all archaeozoologists. Some, like Matolcsi, showed little interest in long-term collaboration. Others may have been skeptical about the future of archaeozoology. As German archaeozoologist Joachim Boessneck noted with palpable resignation, “Our morphological field of work is not in the course of time” (Clason 1975: 7). Some were undeniably disheartened by the mere presence of their domestic competitors in an era when ‘schools’ and their high-status leaders wielded considerable influence in national academic communities. ICAZ, however, indubitably grew into the largest body within the field at a time when personal networking played a key role in maintaining scholarly contacts in the face of politically perpetuated international enmity.

16.4 Sea Change

In October 1988, an ICAZ International Committee meeting was again held in Budapest. It may look like trivia that the worldwide concert tour titled “Human Rights Now!” celebrating the 40th anniversary of the Universal Declaration of Human Rights arrived in town at the same time. A year later, however, the Iron Curtain separating East and West became irrelevant. Another less dramatic looking, but equally important development in 1989 was that computer scientist Tim Berners-Lee invented the World Wide Web. Both events have had an understandable long-term global impact, including on networking in archaeozoology.

In 1990, ICAZ held its first meeting outside Europe, in Washington DC, United States, organized by zooarchaeologist Melinda A. Zeder and Bruce D. Smith of the Smithsonian Institution. While the 1975 meeting in Dallas was a key event in the history if ICAZ it was by no means officially linked to the incipient organization—except for personal networking. In addition to expanding the horizon of the organization by involving many North Americans, generous funding made participation from post-Soviet countries possible at the 1990 ICAZ, lending the conference a distinct air of euphoria. Although animal bones from excavations in the area of the former Soviet Union have been studied since the nineteenth century and in the 1960s important methodological advances were made in its various republics, access to ICAZ became largely available to scholars in some newly formed independent countries after 1991, boosting local research communities (e.g., Lõugas and Rannamäe 2020; Piličiauskienė 2020). While the academic and ethical standards of ICAZ maintained legitimacy, the 1989 political transformation made the sharp East-West distinction politically obsolete. Affiliation with UISPP, which had not really developed beyond a mere formality, lost significance.

As ICAZ came of age, replacing its self-appointed original bodies seemed desirable. With due respect to founders, a consensus emerged that a more formal representational structure with elected officials would benefit the ever-increasing membership. This step would create official positions, stable hubs, without which overgrown, scale-free networks become fragile (Barabási and Albert 1999). During the conference in Washington DC, a decade of concerted efforts began to re-organize ICAZ into a true membership-based professional organization. The statutes were revised and modified. The International Committee accepted these changes to be introduced by 2002.

Organizations are structured along three dimensions: individuals, tasks, and resources (Krackhardt and Carley 1998). By the third millennium, the membership was given, the tasks had been clearly articulated. But ICAZ had no financial resources. A dues-paying general membership was created and the International Committee became an elected representative body chosen from and by registered ICAZ members. In addition to the central position held by the secretary since the beginning, the likewise unpaid posts of president, vice-president, and treasurer were created to be filled by officials elected every four years (International Council for Archaeozoology 2021). Zeder, who as undergraduate student co-organized the 1975 SAA session and later the 1990 International Conference, was elected the first president of ICAZ at its International Conference in Victoria (Canada, 1998).

In 2014, ICAZ gained the status of non-profit organization recognized by the United States Internal Revenue Tax Code. Membership fees are set as low as possible to encourage participation. Importantly, reduced fees are available to students, retirees, unpaid professionals, and researchers from economically disadvantaged countries. Most of the income is invested into sponsoring participation by members of these groups in international conferences. Even if individual expenses usually cannot be fully covered, partial funding tends to be of great help in attracting complementary travel grants and other forms of support. Prioritizing broad and diverse participation in its spending, ICAZ has maintained its commitment to cultivating personal networks and cross-cutting political boundaries, values fundamental to the organization. In addition to voting rights, benefits of membership include access to the *Newsletter* and the database of publications by fellow members as well as reduced annual subscription fees and prices for selected ICAZ related publications (International Council for Archaeozoology 2021). Paralleling the twenty-first century developments in both ICAZ and electronic communication, the *Newsletter* (previously circulated by Clason) and the archaeozoological bibliography (originally distributed by Müller) became on-line services in the form of the biannual digital *Newsletter* and searchable data bases of membership and archaeozoological publications.

Independently of ICAZ, the ZOOARCH listserv—a digital mailing list of names and email addresses—was created in 2000 by two devoted ICAZ members, Umberto Albarella and Jacqui Mulville. This network of archaeozoologists stretches beyond ICAZ. By 2021, its membership counted 1400 individuals world-wide. For over two decades ZOOARCH has remained true to the original mission of its founders, maintaining "a *free* zone (in all senses), where members could help each other on equal terms, and therefore regardless of their status, profession, background, geographic base and language" (Umberto Albarella, personal communication, March 10, 2021). ZOOARCH has no official language (although through its contributions it operates overwhelmingly in English) and contents are not moderated; the administrators interfere with the communication flow as little as possible. Even without a written code of conduct, participants follow principles of mutual support rather than competition, that cultivates fundamental values shared with ICAZ.

16.5 ICAZ in Numbers

The growth of ICAZ resulted in qualitative changes that make diachronic analyses difficult as the nature of the data differs between the early years and the recent past. During the twentieth century, when membership was less tightly defined, conference proceedings offered the most coherent body of information. It became impossible, however, to keep track of the diverse conference publications in the twenty-first century although the increasing membership became easier to follow. Publications and conference participation seem to be positively correlated, reflecting major trends in membership, but precluding the possibility of integrated analyses.

16.5.1 Conference Attendance and Publications

Attendance and published proceedings of the International Conferences can be more easily traced during the early decades, when ICAZ membership was small and informal and the significance of personal networking was evident, as highlighted in the previous sections. The international diversity of conference programs is shown by plotting the number of countries represented against the number of publications/presentations in Fig. 16.2. A slight bias in the graph is due to the fact that presentations before 1990 were counted in published conference proceedings. These works appeared before the availability of electronic publication and can be considered a reliable proxy for the international composition of participants. Subsequently, as single volume (or series) publications became irregular due to the increase of membership, numbers were reconstructed from conference programs, admittedly containing more contributions than would usually be published. All the data, however, show a linear increase in diversity as a function of the number of archaeozoologists involved, supported by a high coefficient of correlation. Small early meetings form a tight cluster in the diagram, while three International Conferences organized in Europe attracted participants from an unexpectedly great number of countries.

There has been a concern that meetings held 'far-away' from a European perspective would be poorly attended. Post-1990 International Conferences hosted outside this continent (in addition to Washington DC; Victoria, Canada, 1998; Mexico City, Mexico, 2006; San Rafael, Argentina, 2014; and Cairns, Australia, 2023), however, fall between the 'European' extremes, showing the considerable attraction of their venues. Local scholars tend to be naturally overrepresented at all international conferences. A study of conference proceedings between 1971 and 2006 showed that the propor-

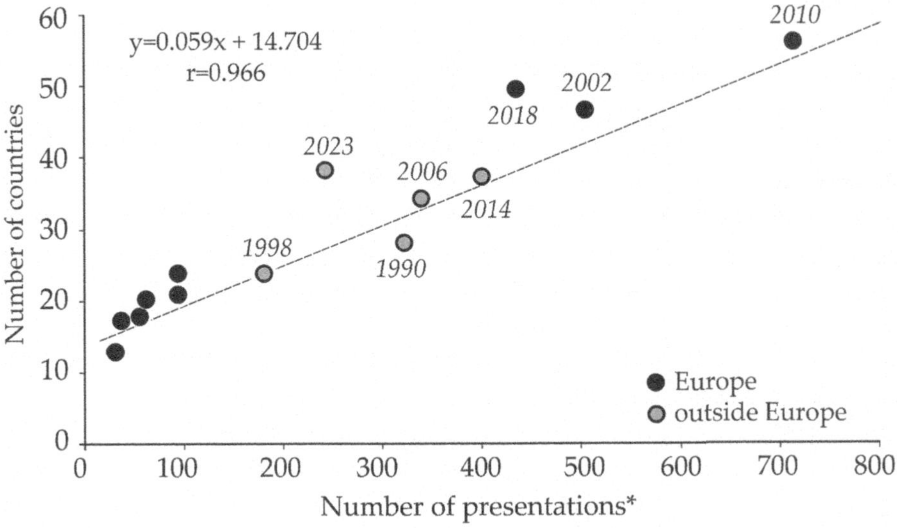

Fig. 16.2 Contributions to ICAZ conference proceedings (number of papers) and conference presentations (between 2002 and 2023) by the number of countries illustrating diversity as the direct function of sample size. (Graph: László Bartosiewicz)

tion of first authors from host countries varied from 5% to 24% (Bartosiewicz 2020). The main purpose of holding these meetings in diverse locations, however, has exactly been to involve local colleagues, thus expanding the ICAZ network. Hosting International Conferences for the steadily increasing membership has grown into a financial and logistical challenge. The main benefit organizing communities can hope for is well-deserved publicity and an increase in personal networking opportunities.

In the beginning, publishing proceedings as single volumes was not particularly difficult. The Budapest volume was edited by Matolcsi (1973), papers delivered at subsequent meetings in Groningen (the Netherlands, 1974) and Szczecin (Poland, 1978) by Clason (1975) and Marian Kubasiewicz (1979), respectively. Proceedings of the 1982 London conference, however, filled four volumes (Clutton-Brock and Grigson 1983a, b, 1984a, b), as did the papers of the meeting in Bordeaux (France, 1986; see Ducos 1987a, b, 1988, 1989). During the next International Conference in Konstanz (Germany, 1994), publication responsibilities were taken over by the journal *Anthropozoologica*, resulting in a volume exceeding 800 pages (Kokabi and Wahl 1997). The numbers of contributions as well as participants, kept increasing. The 2002 International Conference in Durham, United Kingdom, yielded no fewer than 14 thematic volumes, including most of the sessions held during the meeting. This series by Oxbow Books consisted of the titles summarized in Table 16.1.

The increase in the number and complexity of proceedings suggests the enrichment of archaeozoological studies over the first three decades. The publication of comprehensive proceedings became technically and financially near-impossible following the Durham conference. Significant individual volumes have been edited by session organizers, but they tend to be published in a variety of *fora*, targeting specialist readership. Recently they have included special issues of peer-reviewed journals and electronic media, an efficient way of promoting the results of the mass of work carried out on increasingly specialized topics.

16.5.2 Thematic Diversification and Networking

Changes in publication strategy reflect thematic diversification, best expressed by the increase in the number of ICAZ Working Groups (International Council for Archaeozoology 2021). Clusters connected to hubs in a network may be partitioned into groups representing new communities exemplified by these working groups. One of the goals of ICAZ has been to serve as a home organization for such specialized groups. This is reminiscent of thematic working groups in UISPP, called commissions. In ICAZ, however, the emergence of Working Groups was a rather spontaneous, growth-induced development. Initial links with UISPP formed in 1976 remained purely formal, thus direct inspiration seems unlikely. Specialist communities began continuously emerging within ICAZ because of quantitative growth. These working groups are autonomous, researching special themes of shared archaeozoological interest. The underlying thematic variability was evident as early as the 1986 International Conference in Bordeaux; one quarter of the contributions to those proceedings represent topics that later became 'working group staples.' In addition to the long-standing Fish Remains Working Group, several others were gradually established in the years to come (Fig. 16.3). In general networking models, nodes within such communities will be strongly connected to each other (in this case by shared thematic interest), but weakly linked to nodes outside the community (Barabási and Albert 1999). ICAZ working groups,

Table 16.1 The summary of proceedings of the 2002 International Conference in Durham, United Kingdom

2003–2004	2005	2006
Colonisation, migration, and marginal areas: A zooarchaeological approach (Mondini et al. 2003)	Archaeomalacology: Molluscs in former environments of human behaviour (Bar-Yosef 2005)	Beyond 'affluent-foragers': Rethinking hunter-gatherer complexity (Grier et al. 2006)
The exploitation and cultural importance of sea mammals (Monks 2004)	Diet and health in past animal populations (Davies et al. 2005)	Integrating zooarchaeology (Maltby 2006)
Behaviour behind bones: The zooarchaeology of ritual, religion, status and identity (Jones O'Day et al. 2004)	The future from the past: Archaeozoology in wildlife conservation and heritage management (Lauwerier and Plug 2005)	Equids in time and space (Mashkour 2006)
	The zooarchaeology of fats, oils, milk and dairying (Mulville and Outram 2005)	Dogs and people in social, working, economic or symbolic interaction (Snyder and Moore 2006)
	Biosphere to lithosphere: New studies in vertebrate taphonomy (O'Connor 2005)	
	Recent advances in ageing and sexing animal bones (Ruscillo 2005)	
	The first steps of animal domestication (Vigne et al. 2005)	

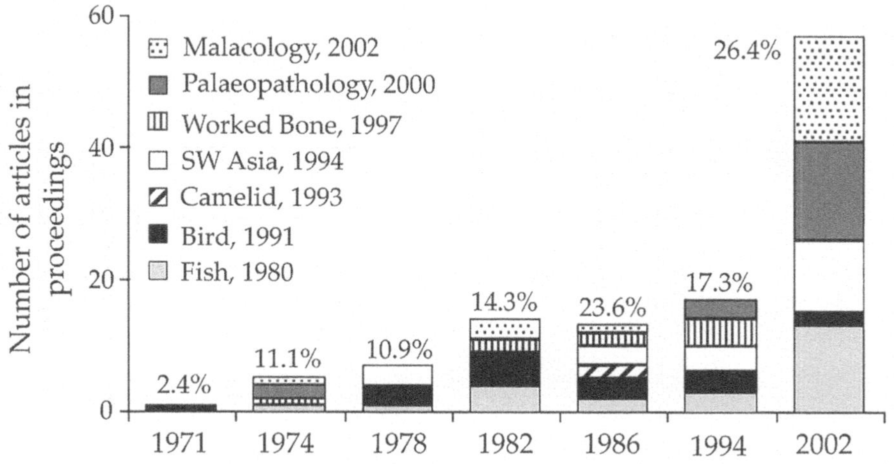

Fig. 16.3 ICAZ working group topics as the percent of all articles in proceedings and the year of establishment of working groups. (Graph: László Bartosiewicz)

however, considerably overlap due to multiple interests among their members, who create personal links between working groups, thereby becoming hubs from the overall perspective. Importantly, working group participants on the 'fringes' of these networks need not be members of ICAZ, but they are encouraged to join the organization, thereby increasing general connectedness.

Figure 16.3 summarizes major working group topics as the percent of all articles in the comprehensive proceedings of the International Conferences held between 1971 and 2002. The legend in this figure also shows the date of establishment for related working groups that, in the long run, have taken over publication responsibilities. Although in the absence of comparable proceedings it is difficult to systematically present working group activities during the last two decades, it should suffice to say that the dynamic development of some of these groups is also remarkable: in addition to the seven listed in Fig. 16.3, their numbers have increased by ten since 2002. Some are linked by geographical (e.g., Baltic Region, Neotropical Zooarchaeology; North Atlantic Biocultural Organization) or chronological (e.g., Roman Period; Modern Era) interests. Others were formed on the basis of state-of-the-art methodology (e.g., Genetics, Proteomics and Morphometrics; Stable Isotopes; Taphonomy) or specialization in particular taxa (e.g., Marine Mammals; Camelids; Microvertebrates) (International Council for Archaeozoology 2021). Working groups tend to include smaller numbers of 30 to 60 specialists who meet according to their own, typically biennial or triennial schedules. Their conference venues tend to be unusually variable, reflecting topical diversity. Many working groups have newsletters or host their own websites.

16.5.3 ICAZ Membership by Region

Global expansion and formalized membership brought a measurable dynamic to ICAZ. Depending on the venue of quadrennial International Conferences, membership oscillates, with a relative majority of participants from within the host country's region. Prior to the so-far-largest 2010 International Conference, which gathered 714 registered participants from 56 of the 61 countries represented in ICAZ (Bartosiewicz 2020), membership peaked at almost 1000 (in recent years membership has numbered 400–600 individuals). This large attendance was possible because the meeting took place in Paris, an easily accessible location for most Europeans and an attractive travel destination for overseas participants (Vigne and Lefèvre 2010). Importantly, the organizers also procured travel grants for many archaeozoologists in need. We may thus presume that the associated 2009 membership list is highly representative of international composition.

In Fig. 16.4 the number of ICAZ registered archaeo zoologists was plotted against the 2010 population of each country contributing to the organization (United Nations Population Division 2010). There is an understandable positive correlation between population size and the number of archaeozoologists employed, although variability between countries is broad, resulting in only a moderate correlation. National archaeologies depend on several factors ranging from the size of the country's territory to investment in heritage management. The analysis of animal remains not only depends on these primary characteristics but also varies with research tradition. It is evident that not all archaeozoologists in large countries such as China, India or Russia have had full awareness of or access to ICAZ membership. At the other extreme, some small countries occupy areas with concentrated archaeological heritages and have funding designated to sustain the analysis of animal remains. What is clear from the top graph in Fig. 16.4 is that the number of archaeozoologists does not follow population size closely in these countries and the overall tendency is degressive: the larger the population, the smaller the relative increase in the number of ICAZ registered archaeozoologists. There is life outside the ICAZ, however. In addition to national organizations

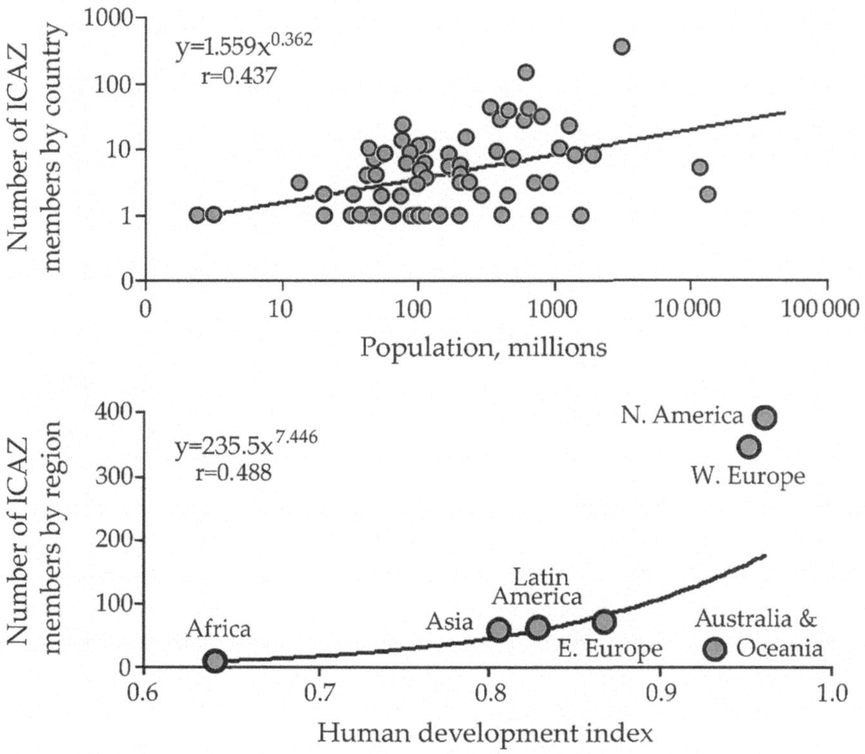

Fig. 16.4 Quantitative characteristics of ICAZ membership in 2009–2010. (Graph: László Bartosiewicz)

in larger countries, unknown numbers of experts may operate in isolation outside our view.

When ICAZ was founded, only the largest countries in Europe employed more than one or two archaeozoologists at a time, but in a generation our community has increased, and supply began surpassing demand even in relatively affluent countries. In a 1993 interview, Clason expressed frustration over this discrepancy in the Netherlands:

> The time that you got a permanent job and could work on what you were interested in is over. For research, universities currently rely on work by unemployed youngsters up to the age of fifty… and pensioners. This keeps new students from engaging in our field. (Bekker 1993: 27)

The arrival of market economy in Eastern Europe also meant that research funding, a token of high prestige for the preceding socialist regime, began to strongly diminish (Bartosiewicz et al. 2011).

In order to put the 2009 membership picture into perspective, I also considered the Human Development Index (HDI) published in the United Nations Development Program's home page (Human Development Report 2010). The lower graph in Fig. 16.4 shows the membership grouped by quasi-continental regions defined in a previous analysis (Bartosiewicz 2020): in spite of the 1989–1990 political changes as well as the concomitant NATO and EU membership of several countries of the former Soviet Bloc, these countries have remained classified as Eastern Europe in this figure. Only the former GDR is counted with Western Europe as part of reunited Germany. Moreover, on a geographical basis, the category of Eastern Europe also includes researchers from Greece, Turkey, and Cyprus who have joined ICAZ relatively recently. The 2010 HDIs were averaged for each of these regions based on the number of ICAZ members from the included countries. HDI is a comprehensive measure of life expectancy, literacy, education, and standards of living with values defined in an interval between 0 and 1. The graph shows that, understandably, more developed regions of Western Europe and North America contribute the most members to the organization (Australia and Oceania, having small populations but a high HDI value, are an outlier in the graph). South America has a long-standing tradition in zooarchaeology and, thanks to the vigorous activity of researchers in several countries, the continent has begun to play an increasingly important role in ICAZ (Mengoni Goñalons et al. 2010).

Unfortunately, the dominance of members from economically developed continents does not mean that individual archaeozoologists would necessarily be particularly affluent themselves. Archaeology in general seems a luxury in regions that face problems of public health and illiteracy. While ICAZ has no power to bridge this gap, a considerable portion of the membership fees is dedicated to helping colleagues in disadvantaged positions participate, regardless of their country of origin. It is of vital importance to maintain or even increase diversity within the organization and help training a cadre of local specialists in all countries.

At the beginning, ICAZ showed a remarkable gender balance. The 2020 list of members indicates greater proportions of women than men involved in archaeozoology in all regions (Table 16.2). Analyzing the complexities of demographic patterns (age structure, forms of employment) behind gender distribution is beyond the scope of this brief review.

In 2020, on average, two-thirds of ICAZ members were women. Large communities in North America and Western Europe are pivotal in influencing this ratio. It is difficult to generalize, however, concerning the position of women in zooarchaeology since, as noted in Latin America (Mondini 2021), female employment in archaeology in general and zooarchaeology in particular have very uneven distributions among the different countries in each region. In addition to the differential availability of training and employment, public engagement by women and the socio- economic status of research positions vary strongly within regions. In certain societies, archaeozoology would classify as a high-status but poorly paid profession, more easily available for talented women than jobs where the proverbial 'glass ceiling' is lower.

The overall 65% female participation falls in line with results in the detailed survey of exemplary precision appraising diversity and demographics of the trade by zooarchaeologist Suzanne E. Pilaar Birch (2015). One may hope that high direct female representation in the organization may lower the risk of the infamous 'Matilda effect', the attribution of achievements by women scholars to their male counterparts.

Table 16.2 Regional summary by gender based on ICAZ membership in 2020

Region	Female	Male	Total	Female %
Eastern Europe	23	7	30	*77*
North America	148	64	212	*70*
Australia and Oceania	19	8	27	*70*
Africa	6	4	10	*60*
Western Europe	73	50	123	*59*
Asia	21	19	40	*53*
Latin America	17	15	32	*53*
Total	**307**	**167**	**474**	***65***

16.6 Conclusions

The past half a century has seen a marked increase in the number of organized archaeozoologists around the world. During the first two decades of ICAZ, personal networking played a key role in forging a community within a group of mostly Old World archaeozoologists due to the relatively small number of nodes and limited outreach. Growth in ICAZ representation by geopolitical region between the 1971 Budapest symposium and the 2020 membership data are summarized in Fig. 16.5. As an artifact of ever-expanding membership, the geographical scope of the organization has widened: all inhabited continents have members who contribute to ICAZ. Along with the eventual improvement of the political atmosphere after 1989, the number of archaeozoologists grew. A remarkable shift is the steady increase of membership from Western Europe and especially North America, accompanied by decreasing participation from Eastern Europe. There is a slow but consistent increase in membership from other regions.

The following four main features characterize the 50 years development of ICAZ: (1) its membership grew by a factor of 6–7 in 50 years and international diversity increased in proportion; (2) developed countries tend to be better represented, although realistic financial comparisons will always be difficult to make at the level of individuals; (3) organizing host countries have consistently gained publicity from International Conferences; (4) the need for specialized working groups has been visible since the beginning and is manifest in their continuous proliferation within ICAZ.

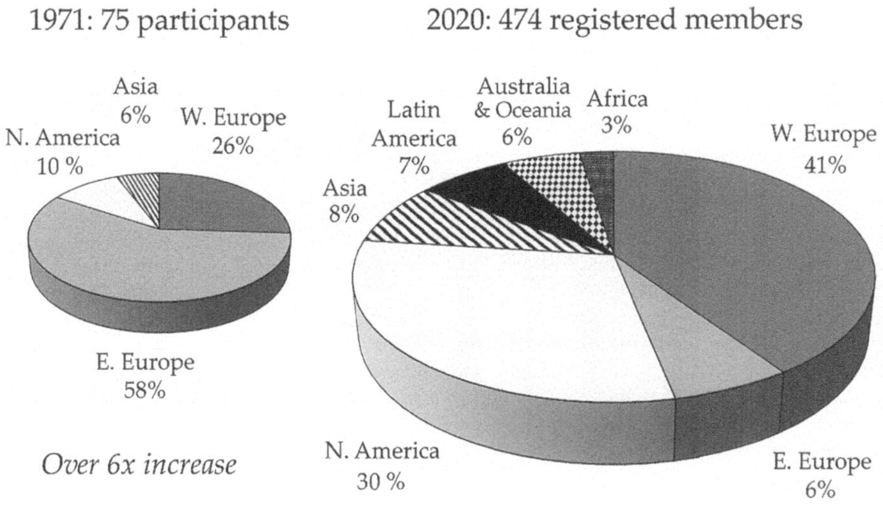

Fig. 16.5 Changes in ICAZ representation by geopolitical region. (Graph: László Bartosiewicz)

The indicators mentioned above impact the methods of our discipline. The membership needs to consider whether our ways of documenting data and analysis are compatible enough on a global level to permit inter-regional and gross diachronic syntheses (Amorosi et al. 1996). While ICAZ can only make recommendations concerning principles and general research methods in the form of its Professional Protocols for Archaeozoology (International Council for Archaeozoology 2021), conscious efforts need to be made by every member to harmonize existing principles and methods in order to harness the powers of this quiet explosion in diversity.

In the long run, archaeozoological networks have emerged across multiple domains (national, thematic, methodological) reflecting close interrelations in the general model (Krackhardt and Carley 1998). Diversification in the field also fostered the organization of smaller specialist communities within which personal networks maintained their initial significance. Thanks to geographical diversification, ICAZ has become an umbrella for the diverse approaches and thematic interests of the global community of archaeozoologists. There is a trend of most personal networking being concentrated in focused working groups specialized in particular time periods, geographical regions or animal taxa. These new networks, however, are interconnected by general activities, such as International Conferences, and linked through individuals—hubs in the overall network—who are active in various working groups as well as the main organization.

Acknowledgements Grateful thanks are due to Sarah Whitcher Kansa, Diane Gifford-Gonzalez, Richard H. Meadow, and Umberto Albarella for kind personal communications regarding membership data, ICAZ history and the ZOOARCH listserv respectively. The English text was edited by Judith A. Rasson.

References

Amorosi, T., Woollett, J., Perdikaris, S., & McGovern, T. (1996). Regional zooarchaeology and global change: Problems and potentials. *World Archaeology, 28*(1), 126–157.

Bar-Yosef, D. (Ed.). (2005). *Archaeomalacology: Molluscs in former environments of human behaviour. Proceedings of the 9th ICAZ Conference, Durham 2002*. Oxford and Philadelphia: Oxbow Books.

Barabási, A.-L., & Albert, R. (1999). Emergence of scaling in random networks. *Science, 286*(5439), 509–512. https://www.science.org/doi/10.1126/science.286.5439.509

Bartosiewicz, L. (2001). Archaeozoology or zooarchaeology? A problem from the last century. *Archaeologia Polona, 39*, 75–86.

Bartosiewicz, L. (2010). Sándor Bökönyi and archaeozoology. In L. Bartosiewicz (Ed.), *Archaeozoology at Pantanello and five other sites* (pp. xviii–xxii) (The Chora of Metaponto, Vol. 2). Austin, TX: The University of Texas Press.

Bartosiewicz, L. (2015). János Matolcsi and the development of archaeozoology (Matolcsi János és a régészeti állattan fejlődése). In A. Kőrösi, & Á. Szotyori-Nagy (Eds.), *Szürkék, rackák, mangalicák* (pp. 29–36). Budapest: Magyar Mezőgazdasági Múzeum és Könyvtár.

Bartosiewicz, L. (2017). Zooarchaeology in the Carpathian Basin and adjacent areas. In U. Albarella, M. Rizzetto, H. Russ, K. Vickers, & S. Viner-Daniels (Eds.), *The Oxford handbook of zooarchaeology* (pp. 99–112). Oxford and New York: Oxford University Press.

Bartosiewicz, L. (2020). International Council for Archaeozoology (ICAZ). In C. Smith (Ed.), *Encyclopedia of global archaeology* (2nd ed.) (pp. 3962–3964). New York, NY: Springer Nature. https://doi.org/10.1007/978-1-4419-0465-2_2112

Bartosiewicz, L., & Choyke, A. M. (2002). Archaeozoology in Hungary. *Archaeofauna, 11*, 117–129.

Bartosiewicz, L., Mérai, D., & Csippán, P. (2011). Dig up-dig in: Practice and theory in Hungarian archaeology. In L. R. Lozny (Ed.), *Comparative archaeologies: A sociological view of the science of the past* (pp. 273–337). New York, NY: Springer.

Bekker, D. (1993). Vaste banen zijn zeer schaars Anneke Clason, een wetenschapper in moeilijke tijden. *Zoogdier, 93*(4), 27–29.

Bobrowsky, P. T. (1982). Olsen and Olsen's identity crisis in faunal studies. *American Antiquity, 47*(1), 180–183.

Brinkhuizen, D. C., Buitenhuis, H., van Es, L. J. M., Kersten, A. M. P., Knol, E., Lauwerier, R. C. G. M., Prummel, W., & Zeiler, J. T. (2007–2008). In memoriam Anneke Clason. *Palaeohistoria, 49*(50), 1–2.

Burkhardt, F., Browne, L., Porter, D. M., & Richmond, M. (Eds.). (1994). *The correspondence of Charles Darwin* (Vol. 9: 1861). Cambridge: Cambridge University Press.

Clason, A. T. (Ed.). (1975). *Archaeozoological studies*. Amsterdam and New York: North Holland and American Elsevier.

Clason, A. T. (1980). Padina and Starčevo: Game, fish and cattle. *Palaeohistoria, 22*, 142–145.

Clutton-Brock, J., & Grigson, C. (Eds.). (1983a). *Animals in archaeology: 1. Hunters and their prey* (British Archaeological Reports, International Series, 163). Oxford: British Archaeological Reports.

Clutton-Brock, J., & Grigson, C. (Eds.). (1983b). *Animals in archaeology: 2. Shell middens, fishes and birds* (British Archaeological Reports, International Series, 183). Oxford: British Archaeological Reports.

Clutton-Brock, J., & Grigson, C. (Eds.). (1984a). *Animals in archaeology: 3. Early herders and their flocks* (British Archaeological Reports, International Series, 202). Oxford: British Archaeological Reports.

Clutton-Brock, J., & Grigson, C. (Eds.). (1984b). *Animals in archaeology: 4. Husbandry in Europe* (British Archaeological Reports, International Series, 227). Oxford: British Archaeological Reports.

Davies, J., Fabiš, M., Mainland, I., Richards, M., & Thomas, R. (Eds.). (2005). *Diet and health in past animal populations. Proceedings of the 9th ICAZ Conference, Durham 2002*. Oxford and Philadelphia: Oxbow Books.

Ducos, P. (Ed.). (1987a). Actes du 5e Congrés International d'Archéozoologie de Bordeaux. *ArchaeoZoologia, 1*(1).

Ducos, P. (Ed.). (1987b). Actes du 5e Congrés International d'Archéozoologie de Bordeaux. *ArchaeoZoologia, 1*(2).

Ducos, P. (Ed.). (1988). Actes du 5e Congrés International d'Archéozoologie de Bordeaux. *ArchaeoZoologia, 2*(1–2).

Ducos, P. (Ed.). (1989). Actes du 5e Congrés International d'Archéozoologie de Bordeaux. *ArchaeoZoologia, 3*(1–2).

Erdős, P., & Rényi, A. (1959). On random graphs, I. *Publicationes Mathematicae, 6*, 290–297.

Gifford-Gonzalez, D. (2018). *An introduction to zooarchaeology*. New York: Springer International Publishing.

Grier, C., Kim, J., & Uchiyama, J. (Eds). (2006). *Beyond 'affluent-foragers': Rethinking hunter- gatherer complexity. Proceedings of the 9th ICAZ Conference Durham 2002*. Oxford and Philadelphia: Oxbow Books.

Human Development Report. (2010). https://hdr.undp.org/en/content/human-development-index-hdi

International Council for Archaeozoology. (2021). *About ICAZ*. http://www.alexandriaarchive.org/icaz/about

Jones O'Day, S., Van Neer, W., & Ervynck, A. (Eds.). (2004). *Behaviour behind bones: The zooarchaeology of ritual, religion, status and identity. Proceedings of the 9th ICAZ Conference, Durham 2002*. Oxford and Philadelphia: Oxbow Books.

Kokabi, M., & Wahl, J. (Eds.). (1997). Proceedings of the 7th conference [Special issue]. *Anthropozoologica, 25–26*.

Krackhardt, D., & Carley, K. M. (1998). A PCANS model of structure in organization. In *Proceedings of the 1998 International Symposium on Command and Control Research and Technology* (pp. 113–119). Conference held in June, Monterey, CA. Evidence Based Research, Vienna, VA. https://www.heinz.cmu.edu/faculty-research/profiles/krackhardt-davidm/_files/1998-pcans-model-structure-in-organizations.pdf

Kristiansen, K. (2014). Towards a new paradigm? The Third Science Revolution and its possible consequences in archaeology. *Current Swedish Archaeology, 22*, 11–34. https://doi.org/10.37718/CSA.2014.01

Kubasiewicz, M. (Ed.). (1979). *Archaeozoology: Volume I. Proceedings of the IIIrd International Archaeozoological Conference held 23th–26th April 1978 at the Agricultural Academy Szczecin–Poland*. Szczecin: Agricultural Academy.

Lauwerier, R. C. G. M., & Plug, I. (Eds.). (2005). *The future from the past: Archaeozoology in wildlife conservation and heritage management. Proceedings of the 9th ICAZ Conference, Durham 2002*. Oxford and Philadelphia: Oxbow Books.

Lõugas, L., & Rannamäe, E. (2020). Investigating Animal Remains in Estonia. *Archeologia Lituana, 21*, 132–141.

Maltby, M. (Ed.). (2006). *Integrating zooarchaeology. Proceedings of the 9th ICAZ Conference, Durham 2002*. Oxford and Philadelphia: Oxbow Books.

Mashkour, M. (Ed.). (2006). *Equids in time and space. Proceedings of the 9th ICAZ Conference, Durham 2002*. Oxford and Philadelphia: Oxbow Books.

Matolcsi, J. (1972). A zoológiai háziállatkutatás állása a budapesti nemzetközi szimpózion tükrében (Der Stand der zoologischer Haustierforschung im Spiegel des budapester Symposions). *Állattani Közlemények, LIX*(1–4), 95–105.

Matolcsi, J. (Ed.). (1973). *Domestikationsforschung und Geschichte der Haustiere*. Budapest: Akadémiai Kiadó.

Meadow, R. H., & Zeder, M. A. (Eds.). (1978). Approaches to faunal analysis in the Middle East. *Peabody Museum Bulletins, 2*, pp. 206.

Mengoni Goñalons, G., Arroyo-Cabrales, J., Polaco, Ó. J., & Aguilar F. J. (Eds.). (2010). *Estado Actual de la Arqueozoología Latinoamericana. Ciudad de México:* Instituto Nacional de Antropología e Historia.

Mondini, M. (2021). Breves apuntes sobre mujeres, zooarqueología y neotrópicos. *Neotropical Zooarchaeology Working Group Boletín Informativo, 12*(Mayo), 4–6.

Mondini, M., Muñoz, S., & Wickler, S. (Eds.). (2003). *Colonisation, migration, and marginal areas: A zooarchaeological approach. Proceedings of the 9th ICAZ Conference, Durham 2002*. Oxford and Philadelphia: Oxbow Books.

Monks, G. G. (Ed.). (2004). *The exploitation and cultural importance of sea mammals. Proceedings of the 9th ICAZ Conference, Durham 2002*. Oxford and Philadelphia: Oxbow Books.

Mulville, J., & Outram, A. K. (Eds). (2005). *The zooarchaeology of fats, oils, milk and dairying. Proceedings of the 9th ICAZ Conference, Durham 2002*. Oxford and Philadelphia: Oxbow Books.

O'Connor, T. (Ed.). (2005). *Biosphere to lithosphere: New studies in vertebrate taphonomy. Proceedings of the 9th ICAZ Conference, Durham 2002*. Oxford and Philadelphia: Oxbow Books.

Olsen, S. L., & Olsen, J. W. (1981). A comment on the nomenclature of faunal studies. *American Antiquity, 46*(1), 192–194.

Pilaar Birch, S. E. (2015). Diversity and demographics of zooarchaeologists: Results from a digital survey. *Ethnobiology Letters, 6*(2), 276–284. https://doi.org/10.14237/ebl.6.2.2015.469

Piličiauskienė, G. (2020). Zooarchaeology in Lithuania. *Archeologia Lituana, 21*, 155–168.

Ruscillo, D. (Ed.). (2005). *Recent advances in ageing and sexing animal bones. Proceedings of the 9th ICAZ Conference, Durham 2002*. Oxford and Philadelphia: Oxbow Books.

Rütimeyer, L. (1861). Die Fauna der Pfahlbauten der Schweiz. Untersuchungen über die Geschichte der wilden und der Haus-Säugetiere von Mittel-Europa. *Allgemeine Schweizerische Gesellschaft für die Gesammten Naturwissenschaften/Société Helvétique des Sciences Naturelles, 19*.

Smithsonian. (1965). Fellowships and grants: Smithsonian Institution Foreign Currency Program in archeology and related disciplines. *Fellow Newsletter, 8*, 4–5.

Snyder, L. M., & Moore, E. A. (Eds). (2006). *Dogs and people in social, working, economic or symbolic interaction. Proceedings of the 9th ICAZ Conference, Durham 2002*. Oxford and Philadelphia: Oxbow Books.

Somlai, K. (2009). Ösztöndíjjal Nyugatra a hatvanas években. Az Országos Ösztöndíj Tanács felállítása. In J. Tischler (Ed.), *Kádárizmus: mélyfúrások* [Special issue]. *1956-ös Intézet és Oral History Archívum Évkönyv, XVI*, 273–314.

United Nations Population Division. (2010). Department of Economic and Social Affairs. https://population.un.org/wpp/

Vigne, J. D., Peters, J., & Helmer, D. (Eds). (2005). *The first steps of animal domestication. Proceedings of the 9th ICAZ Conference, Durham 2002*. Oxford and Philadelphia: Oxbow Books.

Vigne, J. D., & Lefèvre, Ch. (2010). ICAZ 2010 International Conference hugely successful. *ICAZ Newsletter, 11*(2), 1–3.

Index

L. Coltofean et al. (eds.), *Connecting People and Ideas*, Themes in Contemporary Archaeology,
https://doi.org/10.1007/978-3-031-81006-0

The manufacturer's authorised representative in the EU is Springer Nature Customer Service Centre GmbH, Europaplatz 3, 69115 Heidelberg, Germany. If you have any concerns regarding our products, please contact ProductSafety@springernature.com

Printed and bound by CPI Group (UK) Ltd, Croydon, CR0 4YY
06/10/2026
02228554-0001